Evolution of
domesticated animals

Evolution of domesticated animals

Edited by
Ian L. Mason

Animal breeding consultant
formerly FAO, Rome

Longman
London and New York

Longman Group Limited
Longman House, Burnt Mill, Harlow
Essex CM20 2JE, England
Associated companies throughout the world

*Published in the United States of America
by Longman Inc., New York*

© Longman Group Limited 1984

First published 1984

British Library Cataloguing in Publication Data
Evolution of domesticated animals.
   1. Domestic animals – History
   I. Mason, I.L.
   636′.009   SF41

   ISBN 0-582-46046-8

Library of Congress Cataloging in Publication Data
Main entry under title:
Evolution of domesticated animals.
   Bibliography: p.
   Includes index.
   1. Domestic animals – History. 2. Domestic
animals – Origin. 3. Domestic animals – Evolution.
I. Mason, I. L. (Ian Lauder), 1914–
SF41.E93 1983   636.08′2′09   82-22933
ISBN 0-582-46046-8

Set in 10/11pt Linotron 202 Times Roman
Printed in Hong Kong by
Warren Printing Co Ltd

Contents

Contents

Preface

In 1976 Longmans published a reference book entitled *Evolution of Crop Plants*. The book was conceived and edited by Dr N. W. Simmonds. Each species of crop plant was treated in turn in a brief but systematic manner. This book was so successful that the publishers decided to produce a companion volume on domestic animals and asked me to find contributors and edit the book.

Our aim is a single-volume account of the origin and history of domestic animals, species by species, which will not only be a ready source of exact information but will also provide an introduction to the literature. The only recent books in English with a comparable scope are F. E. Zeuner's *A History of Domesticated Animals* (Hutchinson 1963) and H. Epstein's *The Origin of the Domestic Animals of Africa* (Africana Publishing Corporation: New York 1971). In some ways the present book is an up-dating of these classics but it is more concise in order to include more species. It covers the present as well as the past. 'Evolution' in the title refers only to changes since domestication but we have tried to show that this evolution is a continuing process and that it will continue into the future.

In order to make the book authoritative, each chapter has been written by an acknowledged expert but, since the division was by species, the authors were expected to discuss the taxonomy and distribution of the wild ancestors as well as the archaeology, history, genetics and breeding of the domestic descendants. To make the task easier some animals were divided between two authors with different specializations.

I must make it clear that I make no distinction between 'domestic' and 'domesticated'; I consider the two terms to be effectively synonymous although they may have different overtones for different people. The second was chosen for the title in order to emphasize that animals have been consciously domesticated by man; they are not merely 'found around the house'.

One of the most difficult decisions to make was which animals to include as domesticated. In its most developed form the domestic animal exhibits four principal characteristics:

1. Its breeding is under human control.
2. It provides a product or service useful to man.
3. It is tame.
4. It has been selected away from the wild type.

However, not all the animals included here qualify under all four criteria. In fact it would be true to say that I chose the animals because they are commonly considered to be domestic or they are being studied as potential domesticants. Breeding under control is probably the most critical criterion but even this may apply to only part of the population. Thus a few elephants are bred in captivity – the majority are still caught and trained; some domestic cats are bred under control – the majority make their own arrangements. Some animals are included because they have been domestic in the past but are now no longer bred in captivity, e.g. onager (distant past) and muskrat (recent past). In other cases animals are reared, but not bred, in captivity in order to supply a product (e.g. civet) or a service (e.g. fishing by cormorants). The justification for including these animals is that rearing in captivity is a logical first step in the process of domestication. The same would apply to turtles reared for meat and snakes kept for venom production (which are mentioned in the book) and to hawks trained for hunting (which are not). Game birds are a difficult case since they are bred in captivity in order to be released into the wild and the aim is to avoid both tameness and any other deviation from the wild type. The pheasant has been included because of its ancient history and there is a short note on the partridge. On the other hand more recent game birds, such as the bob-white quail, have been omitted.

I have not included animals kept solely for education and entertainment as in zoos and circuses and I have omitted aquarium fish and most pets and cage birds. Some few of these are included because of the interesting variants they have produced (e.g. canary, budgerigar, hamster, goldfish). Laboratory animals are also omitted unless they have some other use (e.g. guinea-pigs for meat) or have been selected far from the wild type (e.g. rats and mice).

The ultimate distinguishing feature of a domestic animal is the presence of a range of genotypes produced by selection. The first genetic changes may be by natural selection, e.g. adaptation to confinement or to poor feed supply; increase in tameness and ease of breeding in captivity. Artificial selection demands control of breeding; that is why this criterion is of the first importance. This human selection has evolved more productive or more specialized breeds on the one hand, and ornamental or 'fancy' types on the other. Nevertheless some animals which have been domesticated for a long time still remain very similar to the wild type (e.g. Bali cattle, Swamp buffaloes).

If I was doubtful whether to include an animal or not I tended to include it in order to clarify the exact status of the animal for the benefit of equally doubtful readers.

Having decided which animals to include the next question was how to arrange them. Neither a strictly zoological nor a strictly agricultural classification was considered appropriate. The primary arrangement is by zoological phyla. The lower vertebrates and invertebrates have not been divided further than into classes but the mammals and birds are subdivided into families and the mammalian families are grouped into orders. The order of species within families and of families within higher groups has been made partly according to agricultural importance and partly according to relationship. This may sound like a terrible mixture but I hope the reader will in fact find it a commonsense arrangement.

The nomenclature of the wild mammals follows that of the British Museum (Natural History) as recorded in G. B. Corbet and J. E. Hill's *A World List of Mammalian Species* (British Museum (Natural History) 1980) and their taxonomy follows G. B. Corbet's *The Mammals of the Palaearctic Region; a Taxonomic Review* (British Museum (National History) 1978). The Latin nomenclature of domestic animals is fully discussed in the Appendix by G. B. Corbet and Juliet Clutton-Brock and, as far as possible, their recommended system is followed in the text. No authorities for Latin names are given in the text but they can be found in the Appendix.

For each chapter an attempt was made to impose a standard format with five main headings and a list of references. While this worked fairly well for the main farm livestock it had to be considerably modified for the non-agricultural animals and even more for the recent and potential domesticants. Authors were asked to restrict their list of references and concentrate on key and source references, especially books and reviews with good bibliographies. There seemed no point in repeating the extensive lists of papers in Zeuner's and in Epstein's books. The illustrations are confined to maps and diagrams – no photographs of animals have been included. Readers requiring these should refer to the remarkable series in Zeuner and in Epstein and in the other sources quoted in the chapter references.

While the responsibility for the content rests wholly with the authors and the editor, I would like to acknowledge the assistance received from others who reviewed some of the chapters, as follows: Dr Caroline Grigson and Dr C. J. M. Hinks (cattle, parts), Dr H. P. Uerpmann (goat, part), Dr A. G. Searle (cat and rat), Dr D. S. Falconer (mouse), Dr. G. A. Clayton (fowl, turkey and goose), Dr. A. Wheeler and Mr W. L. Wilson (goldfish, parts), Mrs Elizabeth Mason (*passim*). I would also like to thank Dr N. W. Simmonds for his help and encouragement in the initiation of the project and Mrs Angela Waters, Librarian of the Animal Breeding Library, Edinburgh, for the assiduity with which she tracked down the most obscure references.

I. L. Mason
Edinburgh
April 1982

N.B. In the reference lists and elsewhere the abbreviation 'FAO' stands for Food and Agriculture Organization of the United Nations and 'IUCN' for International Union for Conservation of Nature and Natural Resources

List of authors

Name and address   **Chapter(s)**

Dr S. S. Ajayi   40
Department of Wildlife and Range Management
University of Ibadan
Ibadan
Nigeria

Dr J. D. Balarin   62
Tilapia Culture Section
Baobab Farm Ltd
Box 90202
Mombasa
Kenya

Professor D. K. Belyaev   23
Institute of Cytology and Genetics
Prospekt nauki 10
630090 Novosibirsk 90
U.S.S.R.

Mr A. L. Berhanu   29

Professor R. J. Berry   38
Department of Zoology
University College, London
Gower street
London
WC1E 6BT, U.K.

Name and address   **Chapter(s)**

Dr M. Bichard   43
Pig Improvement Company Ltd
Fyfield Wick
Abingdon, Oxon OX13 5NA, U.K.

Dr T. H. Blank   17
Cornerstones
Damerham
Fordingbridge
Hants, SP6 3HB, U.K.

Professor S. Bökönyi   18
Archaeological Institute of the
Hungarian Academy of Sciences
Uri U.49, Budapest I
Hungary 1250

Professor J. Bonnemaire   5
Ecole Nationale Supérieure des Sciences
    Agronomiques Appliquées
26 Bld. Docteur-Petitjean
2110 Dijon
France

Dr J. A. Chapman   41
Department of Fisheries and Wildlife
Utah State University
UMC 52
Logan
Utah 84322
U.S.A.

Dr G. A. Clayton   48, 49
Hay Bay Farm Ltd.,
R.R.2, Napanee
Ontario K7R 3K7
Canada

Dr J. Clutton-Brock   22 and Appendix
Department of Zoology
British Museum (Natural History)
Cromwell Road
London SW7 5BD
U.K.

Dr W. Ross Cockrill   8
591 Vale do Lobo
Box 1059
Almansil
Loulé 8100
Algarve
Portugal

Dr G. B. Corbet   Appendix
Department of Zoology
British Museum (Natural History)
Cromwell Road
London SW7 5BD
U.K.

Name and address	Chapter(s)	Name and address	Chapter(s)

Dr Eva Crane 65
International Bee Research Association
Hill House
Gerrards Cross
Buckinghamshire
SL9 ONR
U.K.

Professor R. D. Crawford 42,47,50
Department of Animal and Poultry Science
University of Saskatchewan
Saskatoon
Saskatchewan S7N 0W0
Canada

Dr D. D. Culley 58
School of Forestry and Wildlife Management
Louisiana State University
249 Ag Center
Baton Rouge
Louisiana 70803
U.S.A.

Dr A. de Vos 57
P.O. Box 34
Whitford
New Zealand

Dr I. Douglas-Hamilton 21
IUCN African Elephant Specialist Group
P.O.B. 54667
Nairobi
Kenya

Mr M. Ellis 53
c/o Mr K. Paddock
29 Treforest
Wadebridge
Cornwall PL27 7EN, U.K.

Mr L. J. Elmslie 69
Piazza O. Tommasini 16/25
00162 Rome
Italy

Professor H. Epstein 2, 17, 19
6 Marcus Street
Jerusalem
Israel
92233

Dr E. O. Faturoti 40
Department of Wildlife and Range Management
University of Ibadan
Ibadan
Nigeria

Dr G. A. Feldhamer 41
Appalachian Environment Laboratory
Gunter Hall
Frostburg State College Campus
Frostburg
Maryland 21532
U.S.A.

Dr C.R. Field 12
MAB-FRG Integrated Project in Arid Lands
UNESCO Regional Office for Science and
Technology for Africa
P.O. Box 30592
Nairobi
Kenya

Dr T. J. Fletcher 16
Reedie Hill Farm
Auchtermuchty
Fife KY14 7HS, U.K.

Dr E. González Jiménez 33
Callejón José Gregorio Hernández 22
El Limón
Edo
Aragua
Venezuela

Dr L. M. Gosling 31
MAFF Coypu Research Laboratory
Jupiter Road
Norwich
Norfolk NR6 6SP, U.K.

Major I. Grahame 44
Daw's Hall
Lamarsh
Bures
Suffolk CO8 5EX, U.K.

Dr J. Grau 34
Instituto de Ecologia
Agustinas 641 1°A
Santiago
Chile

Dr Anne Gunn 11
NWT Wildlife Service
Department of Renewable Resources
Yellowknife
Northwest Territories X1A 2L9
Canada

Dr R. O. Hawes 52
Department of Animal and Veterinary Sciences
University of Maine
Hitchner Hall
Orono, Maine 04469
U.S.A.

Name and address	Chapter(s)	Name and address	Chapter(s)
Sir Christopher Lever, Bt Rye Mead House Winkfield Windsor Forest Berkshire, U.K.	54	*Dr M. Plouzeau,* Station de Recherches Avicoles INRA B.P. No. 1 37 Nouzilly France	46
Mrs Elizabeth Mason 8 Ramsay Garden Edinburgh EH1 2NA, U.K.	28	*Dr C. E. Purdom* Ministry of Agriculture, Fisheries and Food Fisheries Laboratory Lowestoft Suffolk NR33 0HT, U.K.	64
Mr I. L. Mason 8 Ramsay Garden Edinburgh EH1 2NA, U.K.	2,10,13		
Dr P. Mongin Station de Recherches Avicoles INRA B.P. No.1 37 Nouzilly France	46	*Professor C. A. Reed* University of Illinois at Chicago Department of Anthropology Box 4348, Chicago Illinois 60680 U.S.A.	1
Dr B. Müller-Haye Research Development Centre Food and Agriculture Organization of the United Nations Via delle Terme di Caracalla 00100 Rome Italy	32	*Mr R. Robinson* St Stephens Road Nursery Ealing London W13 8HB, U.K.	25, 30, 35, 39
Dr C. Novoa Instituto de Zootecnia Apartado 78 Lima Barranco Peru	14	*Dr D. H. L. Rollinson* Animal Production and Health Division Food and Agriculture Organization of the United Nations Via delle Terme di Caracalla 00100 Rome Italy	3
Dr M. A. Ogilvie The Wildfowl Trust Slimbridge Gloucestershire GL2 7BT, U.K.	51	*Dr M. L. Ryder* ARC Animal Breeding Research Organisation Field Laboratory Roslin Midlothian EH25 9PS, U.K.	9
Dr R. C. D. Olivier EcoSystems Ltd P.O. Box 21791 Nairobi Kenya	20		
Mr C. Owen Keythorpe Lodge Keythorpe Leicestershire LE7 9WB,U.K.	26	*Mr S. Drummond Sedgwick* Fisherman's Hill Comrie Perthshire PH6 2LX, U.K.	61
Dr H. F. Peters 17 Sioux Crescent Nepean Ontario K2H 7E4 Canada	6	*Dr R. N. Shackelford* Box 103 Verona Wisconsin 53593 U.S.A.	27
		Professor W. R. Siegfried The Percy FitzPatrick Institute of African Ornithology University of Cape Town Rondebosch 7700 South Africa	55

Professor F. J. Simoons 4
Department of Geography
University of California
Davis
California 95616
U.S.A.

Dr J. R. Skinner 31
MAFF Coypu Research Laboratory
Jupiter Road
Norwich, Norfolk NR6 6SP, U.K.

Dr S. Skjenneberg 15
Postboks 378
9401 Harstad
Norway

Dr Y. Tazima 66
National Institute of Genetics
Yata 1111
Mishima
Sizuoka-ken
411 Japan

Dr O. O. Tewe 40
Department of Animal Science
University of Ibadan
Ibadan
Nigeria

Mr J. D. Turton 36
Commonwealth Bureau of Animal Breeding
and Genetics
Kings Buildings
West Mains Road
Edinburgh EH9 3JX, U.K.

Dr N. H. Valtonen 24
College of Veterinary Medicine
Department of Physiology
Box 6,
00551 Helsinki 55
Finland

Dr N. Wakasugi 45
Laboratory of Animal Genetics
School of Agriculture
Nagoya University
Furo-cho
Chikusa-ku
Nagoya
Japan 464

Dr Jane C. Wheeler 14
The Florida State Museum
Museum Road
University of Florida
Gainesville 32611
U.S.A.

Dr J. F. Wickins 67
MAFF Fisheries Experiment Station
Benarth Road
Conwy, Gwynedd LL32 8UB, U.K.

Dr Gale R. Willner 37
College of Natural Resources
Utah State University
Logan
Utah 84322
U.S.A.

Dr G. W. Wohlfarth 59
Fish and Aquaculture Research Station
Dor
D. N. Hof-Hacarmel
Israel

Dr Z. Woliński 7
ul. Pawia 28 m, 38
01-058 Warszawa
Poland

Dr R. Yamada 63
Department of Genetics
Biomedical Sciences Building
1960 East-West Road
Honolulu
Hawaii 96822
U.S.A.

Sir Maurice Yonge 68
13 Cumin Place
Edinburgh
EH9 2JX, U.K.

Professor Zhang Zhong-ge 56, 60
Department of Animal Science
Beijing Agricultural University
Beijing
China

1

The ~~beginnings~~ of animal domestication

C. A. Reed

University of Illinois at Chicago,
U.S.A.

Human evolution and secondary energy traps

The oldest populations that we assign to the human (hominid) family lived some 14 million years ago, differentiated little but sufficiently from contemporaneous small apes to be recognized as having the potential to evolve into modern humans (Simons 1977). The first evidence of a domestic animal, a dog, is dated between 14 000 and 12 000 years ago (Turnbull and Reed 1974), and the earliest known domestic food animals were sheep somewhat less than 11 000 years ago. We have little, if any, evidence that the cultivation of plants began earlier than 9000 years ago. Thus hominids (humans and their humanlike ancestors), survived for 99.93 per cent of their known history without domestic animals or cultivated crops. This 14 million years of hominid history has been pre-eminently a period of invention and use of secondary energy traps (see Reed 1977) that served, slowly at first but with a quickening pace as time passed, to divert increasing amounts of energy through the hominid population thus increasing its numbers and biomass. Such secondary energy traps had to do with things (the early use of sticks and stones for prying, hammering, and throwing; improvements in twine-making, woodworking, and shaping stones; invention and improvement of projectiles; shaping raw metals; and building windmills and irrigation works), or with chemical processes (the use, and then the making and use, of fire for defence, warmth, and cooking; tanning skins; heating clay to produce pottery; smelting metals; making alloys; synthesizing new molecules to produce fabrics or to kill insects), or with the increasing use of social complexity (such interpersonal complexities as the carrying of food from place of kill or gathering to others at a homesite for sharing with them; maintenance of a family group as a cooperative entity; recognition of kinds and grades of kinship and degrees of responsibility; organization of supra-family groups into tribes and nations). Speech and writing, the latter a very recent offspring of speech, by which the wisdom of experience can be transmitted over many generations, are also such energy savers. Energy saved is energy gained; one need not make the same wasteful mistakes as one's ancestors did.

Energy traps, symbiosis, and domestication

Symbiosis is a well-known biological situation in which at least two different kinds of organisms interact with benefit to both. Domestication is an example of such symbiosis, and thus each of the symbionts is a secondary energy trap. Humans are not the only domesticators; several kinds of ants keep other insects, all suckers of plant juices, as domestic stock from which the ants receive sweet and nutritious droplets. The ants protect (drive away predators), move (put out to pasture), build shelters ('cow' barns) for, and in general care for, their 'cows' with remarkable success. Other kinds of ants are extremely successful horticulturalists, who gather a variety of organic foodstuffs for their underground crops of fungi. Elsewhere (Reed 1977) I have summarized the considerable literature on these patterns of life of ants as farmers and 'cow' keepers; the important thing to understand is that humans are not unique as domesticators.

Humans and their different kinds of domestic animals and plants are excellent examples of symbiosis and thus of mutual secondary energy

1

traps. The human protects, feeds, and cares for the domesticate in numerous ways and is thus a secondary energy trap for the non-human partner; the rewards to the human may be in terms of meat, skins with pelages, leather, milk, fibre, draft power, fly whisks, glue, fertilizer, prestige, and/or companionship.

What is a domestic animal?

In the category of 'domestic animals' we include those whose breeding is or can be controlled by humans, but we exclude most animals in zoos and circuses and many animals (various rodents and primates) in experimental research centres because they have not truly been brought 'into the house'. We must exclude by definition, animals caught wild and tamed, baboons used as goatherds, or macaques used as pickers of coconuts (reviewed in Reed 1977).

In contrast, we should include reindeer because, although they live free for the most part, so that the people of northern Eurasia who herd them must follow the herds to keep alive, the people do intervene by selective killing, by sometimes taking milk, by using them as sledge animals (although snowmobiles are now preferred), and by castrating the older males to diminish excessive fighting.

Although taming was or is necessarily a pathway toward domestication, a tamed animal is not a domestic one; it is an animal that has been taken from its wild environment and has learned by experience that humans are a source of feed and shelter, rather than a source of harm.

Taming merges indistinguishably into domestication. If the originally wild, caught animals were bred in captivity and were selected for particular qualities, such as docility, size, colour, horn size, quality of pelage, and meat production, they are considered to be domestic animals. Such selective changes, made by artificial selection, have a genetic base, even if the breeders in millennia past knew nothing of formal genetics. Thus early in the history of domestication sheep's legs became shorter, goats' horns became twisted, and dogs' brains got smaller, and probably the adrenal glands of all domestic mammals became relatively smaller than in their wild ancestors.

A domestic animal or one descended from a domestic population cannot, by definition, revert to being a truly wild animal. Domesticated animals that return to nature to survive and breed are termed 'feral'. The distinction is a nice one, and intellectually useful, but not necessarily satisfying to a person who has had lambs killed by 'wild dogs'. The wild ancestor of the dog was the wolf, but the dog has been changed sufficiently in characters of bone, brain, and teeth that when it returns to nature it remains a dog – albeit a feral dog – for all of its wild behaviour.

When, why, and how did people first domesticate animals?

Much has been written on the subject of why and how humans domesticated animals, but all is supposition. The 'natural' processes of domestication by primitive people have never been observed by a modern student; all such domestication occurred long before any records were kept. Recent successes at taming or domesticating a variety of animals – including Norway rats, eland, musk oxen, and wolves – have been accomplished in the experimental spirit of modern science and with a purposeful expenditure of time and capital. Additionally, the domesticators could foresee with some clarity the pattern of their experiments and the expected benefits of successful domestication.

Primitive humans had no such advantages – certainly not during the earlier part of the period of animal domestication. Probably taming, and then domestication, occurred without people having been aware of what was happening. Certainly, gatherers and hunters – the people who first domesticated animals – could not have foreseen any uses for those animals other than those they knew already: for meat, bones, and skins. Only later, after long experience and the intensification of a more sedentary life-style, and after the accumulation of random mutations in the domesticates, would secondary uses of animals – such as for milk, wool, motive power, war, sport, and prestige – be realized.

When one considers that hominids had been hunters and consumers of animals for millions of years, the behavioural change required for them to become keepers and conservers of animals was a major cultural revolution. In primitive cultures, the hunters are the heroes and the successful hunters gain extra prestige. People recount the

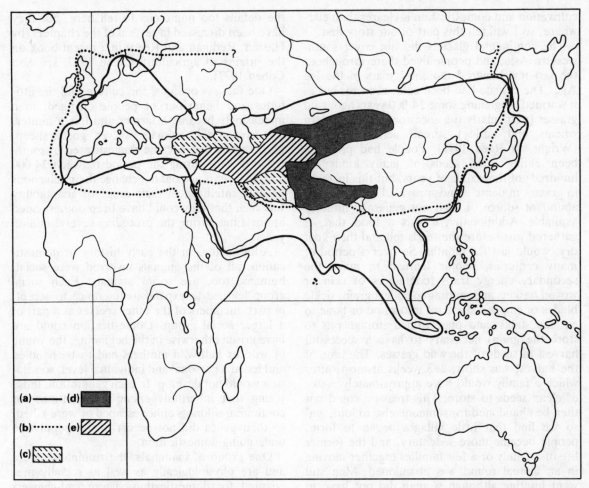

Fig. 1.1 Distribution of wild ancestors of major domestic mammals of the Near East: (**a**) pig, (**b**) cattle, (**c**) goat, (**d**) sheep, (**e**) region of overlap of ranges of sheep, goats, pigs and cattle. (From Isaac 1970, redrawn by permission of Prentice-Hall, Inc.)

hunt, poems are composed, songs are sung; a great hunter probably became a chief, acquired more wives, sired more children, and not only was remembered as a hero long after his death but may also, in time, have become a minor deity. The human male with his ego, did not plan to forsake such rights, privileges, and triumphs to become a herder and cultivator. Some major cultural change must have occurred that the lifestyle of millions of years would be abandoned.

In the first place, such cultural change did not occur until after the evolution of anatomically modern (post-Neandertal) humans and even then not for almost 30 000 years, so the mere emergence of people like ourselves did not automatically result in domestication. The second main factor may have been the world-wide change in environment that accompanied and followed the end of the last glacial period. The earth became warmer, and the continental ice sheets began to melt back, extremely slowly at first, some 14 000 years ago. Soon after this time we find the first evidence of dogs, in southwestern Asia. Three thousand years later, the ice sheets were in full retreat all over the world, so remarkably that geologists proclaim this time of *c.*11 000 years B.P. (before present) as the end of the Pleistocene and the beginning of the Recent.

In southwestern Asia, the record of our story of changing human culture and the beginning of cultivation and domestication is clearer than else-where, so I will tell this part of our story first.

The continental glaciers did not cover south-western Asia, and people lived there throughout the last few hundred thousand years of the Ice Age. The climate had been cold, even so, but as it warmed, beginning some 14 000 years ago, wild grasses (particularly the ancestors of our modern wheats and barley) spread across the hills (Wright, in Reed 1977). People had probably been eating grass seeds of many kinds for hundreds of thousands of years, but this increase in grasses in dense stands provided them with an abundant source of food not generally hitherto available. Additionally, it was a food that, if gathered just before the seeds fell and then kept dry, would last for months. So over a period of many centuries, people learned to make the secondary energy traps (containers of skin or woven baskets and matting to collect grain, sickle blades of chipped stone set into wood or bone to cut the stalks, and plaster-lined storage pits to store the grain) necessary to have a successful harvest of seeds of the wild grasses. The time of the harvest was short, 2–3 weeks at most, after which a family would have approximately 300kg of clean seeds to store. This treasure could not then be abandoned for someone else to loot, and so we find that little villages began to form, people became more sedentary, and the former life of a family or a few families together moving in an annual round was abandoned. Men still went hunting although women did not have to gather so intensively, but stored grain and the daily grinding of some of it became a centre of life that very gradually changed the cultural pattern.

Additionally, the very factors involved with leading a more settled life, without the daily need for women to go gathering, caused a higher birth rate and lower rate of infant mortality. The larger population led to more and larger settlements, diminishing the supply of larger game due to overhunting. Of necessity, people had to depend more and more on the annual supply of stored grain.

These were the circumstances under which we find, by piecing together the archaeological and palaeoenvironmental data surviving over the several thousands of years, that the first cultivation of plants and domestication of animals began in southwestern Asia. The story is a complex one,

the details too numerous to tell here, but they have been discussed in several of the chapters (by Hassan, Redman and Reed) in a recent book on the origins of agriculture (Reed 1977, see also Cohen 1977).

One thing is obvious: the culture and the attitudes and behaviour of people changed; man increasingly became a farmer and less a hunter. The several wild animals (wolves, goats, sheep, pigs, cattle) that became domesticated in southwestern Asia during the period between 14 000 and 8000 years ago did not change their behaviour or their intelligence; their range of adaptability was such that they could have been domesticated at any time during the preceding several million years.

Additionally, in the early history of domestication, all of the animals involved were social; humans, too, are social animals. Each social group learned to expand its tolerance to accept, in part, members of the other species as a part of a larger social group. Domestication could not have arisen otherwise in the beginning; the young of wild or half-wild mothers had to be handled and fed at a personal and individual level, so wildness would not develop. In each generation, those young that inherited genetic combinations for continuing wildness either escaped or were killed, so their genes did not persist in the population undergoing domestication.

One group of mammals, the ruminants, were and are physiologically as well as socially pre-adapted for domestication. These cud-chewers have a four-chambered stomach that allows them to utilize cellulose indirectly as a source of energy. Although a ruminant cannot survive on a diet of wood chips, the bacteria in the rumen and reticulum do break down some of the cellulose in the ruminant's diet into simpler compounds, which the animal then digests. It also derives nourishment from the excess population of bacteria, which pass on into the abomasum and intestine and are there digested. Additionally, ruminants recycle their urea via the circulatory system from the blood back to the rumen, thus conserving nitrogen and allowing them to survive on a diet low in protein. In these ways, all the ruminants are well designed, by a long evolutionary pattern of selective adaptation, for survival on diets on which many other animals would starve (Reed 1969). However, food not utilizable by humans, such as native grasses and straw, may have been the diet provided to early domesticated sheep, goats, and cattle by their human captors. Those individuals that could survive and reproduce on poor diets were those that became domesticated. One family of ruminants, the Bovidae, has furnished us more important domestic animals (sheep, goats, cattle, bantengs, yaks, water buffaloes) than has any other; other ruminant domesticates are camels, llamas, alpacas, and reindeer.

Why did humans domesticate animals? I doubt if at first people knew what they were doing; they had known animals as living creatures to be hunted and killed, or, as with larger predators, to be avoided. At the same time, as is true of more recent but still non-literate peoples, they may have sometimes taken the living young of a killed animal back to camp with them. In particular a little girl might well adopt a young lamb, kid or pig. If so, we have no evidence that during pre-agricultural millennia any such 'pets' survived and reproduced in camp, to become ancestors of domesticated populations. At least we don't have any such evidence until dogs appear, between 14 000 and 12 000 years ago; such primitive pet keeping of wolf pups may have been the beginning of the domestication of wolves into dogs.

Primarily the change from aggressive hunting to protective keeping was a change in attitude, particularly by human males, a change from hunting and killing animals to keeping and protecting them; this change in attitude left no direct trace in garbage heaps, nor in collapsed remnants of village huts, for the archaeologist to discover. In the archaeological record, we can detect only the indirect results of changes in human behaviour following modifications in human thinking.

One such change in attitude is the probability that cattle, like mithan, were domesticated as a result of the process of acquiring and keeping cattle for sacrificial purposes (see Isaac 1970).

What might have been, and what might yet be

The world moves, unfortunately but seemingly irrevocably, towards ever more intense human occupation of all available surface, toward more steel placed vertically, more concrete laid horizontally. Room for wild and domestic animals is decreasing, the way our culture is going – but do we want that kind of cultural poverty? Before our animals are gone, let us think for a moment about the delightful diversity that could have been ours, had we and our ancestors arranged matters differently.

The culture of the ancient Egyptians is interesting in many respects; particularly fascinating to anyone interested in animals are the numerous experiments in taming and domestication that they accomplished (Boessneck 1953, Zeuner 1963). Fortunately for us, they left pictures of what they did. They kept and tamed, even if they did not domesticate, a wide variety of bovids (gazelles, ibex, oryx, addax and Barbary sheep).

The early Egyptians also seem to have tamed hyaenas, which are shown in several paintings that span a period of 250 years. One such painting illustrates hand-feeding, with the hyaenas lying on their backs and men stuffing food into their mouths. Those hyaenas were tame! Hyaenas may have been maligned in our culture; if caught young, fondled happily, fed well on fresh food, and given an occasional bath they might prove to be the best of man's friends.

Probably a wide variety of other carnivores, in addition to cheetahs and ocelots, could be tamed and perhaps domesticated to join the cats, dogs, and ferrets that we already have. All otters, and the sea otter in particular, seem psychologically preadapted for domestication. Seals and walruses also seem eminently domesticable, and perhaps sea lions too, if we could learn how to handle the male during the breeding season.

I suspect that all members of the family Bovidae, which has already contributed all the cattle, sheep and goats to our list, could be domesticated, if we but made the effort; I do not exclude the bison and the African buffalo. The effort is being made with the musk ox and the eland, in large part by interested individuals at personal expense.

The Sirenia (dugongs and manatees) are almost preconditioned to taming, and would not have to be 'domesticated' so much as merely not frightened or disturbed. One could now, I think, domesticate the hippopotamus, probably easier than people in the past domesticated camels or cattle. Why not warthogs? Why not American antelope?

Of the whales, dolphins, and porpoises, probably all – in spite of size of some and the carnivorous behaviour of others – are potential domesticates. Perhaps – some people think it's true – these animals have languages, and perhaps we can learn to talk with them. Certainly they are doing more with their large and complex brains than feeding themselves and squealing noise at each other! We need absolute protection for them first, taming second, and some degree of domestication (perhaps only companionship) later. Of course, that's a big order, mainly because the whales and their allies are big animals.

Clearly, not all adventure and new research lie in outer space or inner atoms; biological adventures and new fruitful discoveries remain here on earth for those who will but seek them.

References

Boessneck, J. (1953) Die Haustiere in Altägypten. *Veröffentlichungen der Zoologischen Staats-sammlung* (Munich), **3**: 1–50

Cohen, M. N. (1977) *Population Pressure and the Origins of Agriculture*. Yale University Press: New Haven, Connecticut

Isaac, E. (1970) *Geography of Domestication*. Prentice-Hall: Englewood Cliffs, New Jersey

Reed, C. A. (1969) The pattern of animal domestication in the prehistoric Near East. In: *Domestication and Exploitation of Plants and Animals*, ed. P. J. Ucko and G. W. Dimbleby, pp. 361–80. Duckworth: London

Reed, C. A. (ed.) (1977) *Origins of Agriculture*. Mouton: The Hague

Simons, E. L. (1977) 'Ramapithecus.' *Scientific American*, **235** (5): 28–35

Turnbull, P. F. and **Reed, C. A.** (1974) The fauna from the terminal Pleistocene of Palegawra cave, a Zarzian occupation site in northeastern Iraq. *Fieldiana Anthropology*, **63**(3); 81–146

Zeuner, F. E. (1963) *A History of Domesticated Animals*. Hutchinson: London

2
Cattle

H. Epstein

The Hebrew University of Jerusalem, Israel

and

I. L. Mason,

formerly, Animal Breeding Officer, FAO, Rome, Italy

Introduction

Nomenclature

The term 'cattle' is derived from the Middle English and Old Northern French *catel*, late Latin *captale*, Latin *capitale*, meaning 'capital' in the sense of chief property (chattel). The term 'common cattle' is used by archaeologists to distinguish them from other bovines.

The wild cattle of Europe were called *urus* in Latin, *ur*, *auer* or *aurochs* in German, and *thur* in Polish, terms related to the Old Slavonic *turu*, old Nordic *thyorr*, Celtic *tarvos*, Latin *taurus* and Greek *tauros* for the domesticated bull.

The scientific name of the aurochs is *Bos primigenius*. The Linnaean binomial, *Bos taurus*, seems inappropriate to a domesticated species, especially as this term, in the meaning of humpless cattle, has been contrasted with *Bos indicus* for humped cattle. These Latin specific epithets are misleading because the domesticated humped and humpless types interbreed readily, are completely fertile and have the same chromosome number $2n=60$.

General biology

Cattle belong to the order Artiodactyla or even-toed hoofed mammals (ungulates), the family Bovidae, and the genus *Bos*. They are subgenerically separated from *Poephagus* (yak), *Bibos* (gaur and gayal, banteng and Balinese, and kouprey), and *Bison*. With all these they will interbreed to produce hybrids which, in general, are fertile in the female and sterile in the male sex.

Cattle are grazers, grazing for about 8 hours a day and resting for the remaining hours. Instead of upper incisor teeth they have a thick layer of the hard palate, the dental pad. They are ruminants with the typical four-chambered stomach.

The horns of cattle are composed of a bony core attached to the frontal bones, and a hard sheath of horny material.

Cattle breed throughout the year. The average length of oestrus is 17.8 hr in cows and 15.3 hr in heifers; the oestrous cycle averages 21.3 days in cows and 20.2 days in heifers. Heifers usually begin to mate at 18 months of age, but in well fed animals of the improved breeds the age at first calving can be very low. Gestation lasts for about 9 months. Cows remain fertile for approximately 12 years, while the life-span is approximately 22 years.

Numbers and distribution

In 1981 the number of cattle, not including buffaloes, in the world was 1210 million, representing an increase of 101 million over the number recorded in 1969–71.

India with 182 million cattle has the largest number in Asia, followed by China with 53 million. In Africa the largest number, i.e. 26 million, is found in Ethiopia. In Europe, France has 24 million head, followed by West Germany with 15 million. The United States has the largest number of cattle in North America, namely 114 million, and Brazil with 93 million head the largest number in South America, Argentine with 54 million coming next. In Oceania Australia has 25 million and New Zealand only one-third of this number (FAO 1982).

In all continents and in every one of the above-mentioned countries the number of cattle has increased since 1969–71, with the relatively largest advance in Australia.

Wild ancestor

All domesticated cattle (*sensu stricto*) of the world are of a single species, derived from the wild aurochs, *Bos primigenius*, which has been extinct for more than 300 years: the last cow died in a forest in Poland in 1627.

Formerly the range of the aurochs extended from the west coast of the Pacific through Asia and Europe to the eastern coastlands of the Atlantic ocean (excluding Ireland), and from the northern tundras southwards into India and North Africa. The original centre of evolution of the animal was in Asia from whence it spread into Europe and Africa during the Pleistocene.

A species ranging over so vast an area and long period of time would be expected to show considerable variability. Yet the aurochs is subdivided into only three local races; the Asiatic *B. primigenius namadicus*, the European *B. p. primigenius* and the North African *B. p. opisthonomus*. These races or subspecies have been distinguished by differences in body size and horn shape, but actually their division denotes little more than geographical range for within each race variability in body size and horn direction is not less than between the three geographical races.

Grigson (1980 and personal communication) refers *B. namadicus* of the Indian subcontinent to a separate species distinct from *B. primigenius* of Europe and the rest of Asia. This separation is based on the supracristal portion of the occiput overhanging or not overhanging the infracristal portion. For *B. namadicus* skulls the former position is claimed, and for *B. primigenius* skulls the latter. However, this classification ignores the fact that the different position of the supracristal portion is merely a result of the static effect of horn shape, weight and direction, i.e. the moment of rotation of the horns, and not a racial characteristic (Epstein 1958). Indeed, quite a number of European and African aurochs skulls show the same position of the supracristal lap as has been observed in crania of the Indian aurochs.

The range of *B. p. namadicus* included practically the whole realm of Neolithic civilization in Asia. In Palestine and Lebanon the animal was hunted by Tiglath Pileser I (about 1100 B.C.). At the time of the Hebrew prophets the aurochs still occurred in the mountain ranges of the Hauran.

In the temple palace of a Hittite settlement on the upper Khabur river basalt and limestone orthostats have been found, some of which depict the hunting of the aurochs. The last report of a wild bull hunt in Assyria dates from the time of Ashurnasirpal (884–860 B.C.). The aurochs reliefs on buildings from later periods seem to be imitations of earlier works.

Fossil remains of the Asiatic aurochs have been found throughout Asia, from Lebanon to China. Skulls from India and China show horn cores 90–100 cm long. The horn sheaths once attached to such cores must have been of huge dimensions.

The African aurochs occurred in Egypt and the Atlas countries. It has been variously described as *B. opisthonomus*, *B. primigenius mauritanicus* and *B. primigenius hahni*. It is pictured in Egypt at the time of the dynastic conquest and in the period of the first king of the 1st dynasty. On an ebony tablet from the second king there is a scene depicting the netting of wild cattle. Wild oxen painted in the tomb of King Sahurā (about 2700 B.C.) are reddish brown with white top- and underlines, similar to the majority of wild cattle depicted in rock tombs at Beni Hasan (*c.* 2000 B.C.). Some of these, however, are completely red, as are the bulls portrayed in hunting scenes in the tomb of Amenemhat and the tombs and mastabas at Saqqara. In wall paintings from the end of the 3rd and the 2nd millennium the hunted oxen are frequently shown with lyre-shaped horns. In Egypt the aurochs seems to have become extinct during the 14th century B.C. Its latest occurrence in the Nile valley is recorded on a scarab of Amenhotep III (1417–1379 B.C.)

In the Atlas countries the aurochs may have survived longer. Its presence in Libya during the 5th century B.C. is attested by Herodotus, who wrote that it was similar to domesticated cattle but black, with thicker hide and horns directed forwards and downwards. The frequent forward direction of the large horns of the African aurochs is confirmed by skulls, early rock engravings and representations of the animal on slate palettes. At the time of the Roman conquest the aurochs was already extinct in the Maghreb owing to the encroachment of the desert on the fertile savannahs, the drying up of the springs and river beds, and the intrusion of domesticated cattle into its grazing grounds.

The European Pleistocene aurochs was a large animal, the bull standing 180 cm at the shoulder; its Holocene descendant was somewhat smaller. In Palaeolithic wall paintings in French and Spanish caves aurochs cows are frequently distinguished from bulls by smaller size, lighter build, a relatively longer and slenderer head and smaller horns. Skeletal remains also indicate that the cow was smaller and more graceful than the bull. The horns were of variable shape. Usually they first curved outwards and forwards, then upwards and inwards with the largest span just below the tips. The horn core of the largest known *B. p. primigenius* skull, from Monte Mario, Rome, has a basal circumference of 50.2 cm and a diameter of 17 cm. The weight of the head with the horns in the aurochs has been estimated at 48 kg. In order to carry this enormous weight, the cervical muscles had to be very strongly developed and the neural spines to be erect, a position responsible for the great height of the withers. In the course of domestication the weight of the horns and of the cranial bones supporting these gradually decreased, the neural spines shortened and sloped backwards, and the withers became lower in consequence. The coat of adult aurochs bulls was black with a whitish stripe along the spine, white curly hair between the horns, and white around the muzzle. Cows and calves were red or reddish brown.

In central Europe *B. p. primigenius* skulls have been found which are about one-third below the normal size but do not differ from ordinary skulls in other respects. It is possible that this dwarfing was due to malnutrition. Such skulls have repeatedly formed the basis for the erroneous claim that a separate species of small shorthorned aurochs occurred side by side with the aurochs proper, and that this alleged species was ancestral to a certain type of domesticated cattle.

Domestication and early history

Early domesticated cattle are classed into two major types, namely humpless and humped. The humpless cattle, again, are separated into longhorned and shorthorned types.

Longhorn cattle

All early wall paintings and rock drawings of domesticated cattle show the characteristic longhorn type which is similar to the aurochs and, clearly descended directly from it. The domestication centre of the Asiatic aurochs may well have been in the southwestern part of its range, for an analysis of the fauna of widely separated sites indicates that the shift from a reliance of human society on wild animals took place in southwest Asia earlier than in other parts of Asia or in Europe (Isaac 1962). In a temple at Çatal Hüyük in southern Anatolia, dated to the 6th millennium B.C., is a wall painting of a bull-jumping scene in a cult action or game in which one person held the bull by the tongue while another one jumped over its back. Remains of cattle from Çatal Hüyük, dated to 5800 B.C., indicate that they were domesticated at least 500 years earlier (Perkins 1969). Brentjes (1967) suggested that domesticated cattle occurred in Anatolia as early as 7000 B.C. The presence of cattle in the Natufian of El Khiam and Mallahah near the Mediterranean coast at about 8000 B.C. is doubtful, but they may already have occurred in a Neolithic context at Belt Cave on the Caspian foreshore and at Sialk on the Iranian plateau, at about 6000 B.C., and at Amouq, east of the Gulf of Iskenderun close to the Syrian–Turkish frontier, during the first half of the 6th millennium B.C. At Fikirtepe their remains date from c.5500 B.C.

Domesticated cattle have been reported from Tepe Sabz at Deh Luran in Iranian Khuzestan, c.5500 B.C., and from Qalat Jarmo on the west flank of the Zagros mountains probably in the late 6th millennium. Horn cores of bulls from prepottery and pottery Neolithic strata of Jericho are indistinguishable from aurochs horns, but a crude clay figurine from the pottery Neolithic phase of the 5th millennium B.C. may represent a domesticated bovid (Zeuner 1963). Certain evidence of the occurrence of domesticated cattle is provided by bovine bones and teeth, considerably smaller than those of the wild aurochs, at Banahilk in northern Iraq, dated to the 5th millennium B.C. (Braidwood and Howe 1960). Bones and teeth of similar cattle have been recovered from the same levels at Shanidar Cave, northwest of Banahilk (Perkins 1960). This evidence is strengthened by the reproduction of a cow's head with forward curving horns, excavated at Arpachiya at the headwaters of the Tigris. The presence of domesticated cattle is also established for Warka in Mesopotamia, early 4th millennium B.C.

In the Nile valley cattle-breeding societies appear at least 1000 years later than in similar cultures in southwest Asia. The earliest Nile valley records of domesticated longhorn cattle are traceable to the Neolithic encampments in Lower Egypt and the Badarian farming villages in Upper Egypt, dated to the second half of the 5th millennium B.C. During the early Amratian age of Egypt, about 3700 B.C., bovine figures occur on white-line pottery and painted vases. Cattle with long lyre- or sickle-shaped horns were common in ancient Egypt. They were large-framed animals, their shoulder height reaching approximately 145 cm and the length of body 170 cm. The horns of a typical Apis bull found at Giza are 62 cm long with a basal circumference of 24 cm. From these longhorn cattle the Hamitic pastoralists of North Africa developed a giant-horned type. They also used to deform the horns of their cattle. Most of the colour patterns and markings known in recent domesticated cattle were already found in this ancient breed.

Rock engravings of longhorn cattle also exist in Ethiopia as well as in Nigeria. Mount Elgon on the Kenya–Uganda border seems to be the southernmost point at which their ancient presence is recorded. They probably reached Ethiopia from Arabia via the Horn of Africa, for cattle with very long lyre-shaped horns are portrayed in rock engravings of central Arabia dated to the 3rd–2nd millennia B.C.

From the eastern shores of the Mediterranean cattle with long lyre-shaped or crescentic horns extended into Mesopotamia where they are depicted in numerous works of art, e.g. a relief carved on a marble bowl from the Sumerian period, c. 2700 B.C., two bulls on the mosaic standard from Ur, a bull's head on the gold and mosaic harp from the great death pit at Ur, and the copper head of a longhorn cow from Tello.

Longhorn cattle are represented in works of art from the Copper–Bronze Age kurgans in the northern Caucasus, and crania of cattle from kurgans in the steppes of southern Russia, dated to the beginning of the Christian era, show similar proportions to aurochs skulls, save for the horns which are a little shorter, thinner and more lateral in direction than those of the wild animal. Domesticated longhorn cattle are present among the faunal remains excavated at Anau in southern Turkestan, early 3rd millennium B.C. Bovine fossils from the Quetta valley of Pakistan have been ascribed to humpless longhorn cattle. The latter are also represented on a large number of

9

seals and by a few statuettes at the ancient Indus valley sites of Mohenjo Daro and Harappa in the 3rd millennium B.C.

There are ancient traces of longhorn cattle all along both shores of the Mediterranean – from Palestine through Egypt to northwest Africa, and from Asia Minor through the Balkans to the Apennine and Iberian peninsulas, extending to the east, north and south as far as the Mediterranean race of man or his culture spread. Throughout this large area wall engravings and statuettes of longhorn cattle strikingly resembling each other have been found in large numbers in stone, bronze, silver and gold. There is little difference in conformation between the recent Highland cattle of Scotland and the sculptures of longhorn cattle made by the ancient Etruscans, between the modern longhorn cattle of Portugal and the breed of the ancient inhabitants of the Nile valley, between the long-horned N'Dama cattle of West Africa and the clay, bronze and silver figurines of longhorn cattle from Mycenae, between the Grey Steppe cattle of the Balkans and the statuetes of transdanubian longhorn cattle of the Hallstatt period.

Cattle may have been domesticated in the southern Balkans independently of southwest Asia (Reed 1977). Considering Banahilk in northern Iraq as the earliest site at which domesticated cattle occurred in southwest Asia, Bökönyi (1974) has suggested that the oldest centre of domestication of the aurochs was not in this part of Asia but in southeast Europe. For the oldest finds of domesticated cattle, radiocarbon-dated to about 6200 B.C., were discovered in pre-pottery Neolithic settlements in Thessaly, in the southern part of the Balkan peninsula, and in Greek Macedonia. From this centre, he holds, domesticated cattle spread into the northern part of the Balkans and subsequently into the Carpathian basin before the end of the 6th millennium B.C. It was only later that they reached southwest Asia, central and eastern Europe and Russia.

However, it is doubtful if Banahilk may rightly be considered as the earliest southwest Asian site at which domesticated cattle were present. The inhabitants of earlier sites, such as Çatal Hüyük, Amouq, Sialk, Belt Cave, El Khiam and Mallaha, may already have been in possession of domesticated cattle, and if in some of these their presence is still uncertain, this may be due to a failing of the archaeological record. Brentjes' (1967)

claim that domesticated cattle occurred in Anatolia as early as 7000 B.C. may still come to be substantiated. Since the cattle at Nea Nicomedeia in the Macedonian plain were much smaller than the wild ox, while those of Çatal Hüyük were almost as large as their wild ancestor, Perkins (1969) holds that domesticated cattle were not introduced into Anatolia from the Balkans but were developed indigenously.

Unless two separate original domestication centres be accepted for the earliest cattle of the southern Balkans and southwest Asia, a very unlikely proposition, the question as to the time and locality of their first occurrence still remains in doubt. Should southwest Asia prove to have been the first centre of domestication of cattle, the Asiatic aurochs must be considered as the wild stock from which domesticated cattle have been evolved. If the southern Balkans are the original centre, the European aurochs would represent the wild source of the longhorn.

With the spread of longhorn cattle from Asia to Africa, *B. p. opisthonomus* cows and calves were doubtless incorporated in various regions and at various times with herds of domesticated longhorn cattle extending into their range. The same must frequently have occurred in Europe with local *B. p. primigenius* cows and calves and throughout Asia with those of the Asiatic aurochs.

At the present time the previously ubiquitous longhorn type has been largely replaced by short-horned or humped cattle throughout its earlier range.

Shorthorn cattle

The humpless shorthorn type of cattle is distinguished from the longhorn type by the size of the horns and conformation of the cranium. The term 'shorthorn' is a literal translation of the Greek *brachyceros*, first applied to this cranial type by Owen in a museum catalogue of 1843; it has no connection with the British Shorthorn breed of cattle. Later, Owen changed the name and called the shorthorn *B. longifrons*.

The horns of typical shorthorn cattle are short, markedly bent, and set close to the head. Owing to the shortness of the facial part, the forehead is longer in relation to the whole skull (hence *longifrons*), and at the same time broader than in the longhorn type. There are two principal theories on the origin of shorthorn cattle. One, repeatedly offered, claims that the shorthorn is descended from a separate species or subspecies of dwarf aurochs. The other considers the shorthorn cattle to be descended from domesticated longhorn cattle.

The theory of shorthorn descent from a dwarf aurochs race is unacceptable. *Bos mastodontis* or *brachyceroides* from Asti in Italy has been shown to be not a wild aurochs but a domesticated animal. Similarly, cranial bones from early recent deposits in Poland, after which *B. brachyceros europaeus* was called, do not belong to a diminutive aurochs race but either to domesticated brachyceros cattle (possibly influenced by longhorn) or to an aurochs cow with the female's relatively weak horns causing a cranial conformation approaching that of the domesticated shorthorn type.

Excavations at Shah Tepé in the Turkmen steppe of northern Iran have yielded bones of an adult shorthorn bull which has been claimed to be wild and been called *B. brachyceros arnei*. But this claim has proved to be fallacious. From a comparison with aurochs bulls and cows as well as with prehistoric and early historical domesticated bulls, cows and oxen, Nobis (1954) arrived at the conclusion that the Shah Tepé skull belonged to a domesticated shorthorn castrate.

The theory of the descent of the shorthorn from domesticated longhorn cattle is based on the effects of small body and horn sizes on cranial conformation. Shorthorn cattle are generally smaller than longhorn cattle. In small cattle, dwarfed or immature, the brain case is larger in relation to the facial part of the skull than in large cattle.

The majority of features and measurements of the skull, whether absolute or relative, by which the shorthorn type of cattle differs from the longhorn type can thus be explained by the small body and horn size of the shorthorn, the natural outcome of the domestication process. Other differences between the shorthorn and longhorn cranial types are due to the adjustment of various cranial bones to the direct effects of body and horn size on the skull and to allometric cranial growth promoted by selection (Epstein 1971).

It is probable that changes in body size, length, thickness and shape of horns and allometric growth towards the shorthorn type of skull occurred more than once in the history of domesticated cattle. While the origin of the shorthorn cattle of Africa and of several shorthorn breeds of Europe from southwest Asia is attested by archaeological and cultural-historical evidence, some of the numerous breeds of cattle in Europe and northern Asia may owe their shorthorn-type skulls not to blood relationship, but to an independent development from domesticated longhorn cattle.

The earliest records of shorthorn cattle are encountered in Mesopotamia, suggesting that Asia Minor received the shorthorn from the southeast. To this day the Kurdi cattle of Kurdistan, including northern Iraq, are of shorthorn type. In Mesopotamia shorthorn cattle began to replace the original longhorn breed approximately 3000 years before the Christian era. Shorthorn cows with their calves at foot are depicted in a milking scene on an early dynastic mosaic frieze in a temple at al-'Ubaid, Ur, *c.* 2900 B.C. The evolution of the shorthorn cattle falls into the period of the urban revolution in Sumer and Elam. The demand of urban society for a regular supply of milk and dairy products called for the development of a type of cattle physiologically adapted for milk production. To this day the outstanding dairy breeds of the world are either entirely or predominantly of shorthorn type. A connection between the increased reliance on milk production and the small size of shorthorn cattle has been suggested by Howard (1962) who has pointed out that in Iron Age sites in Europe there is a great preponderance of the remains of cows of the *brachyceros* cranial type, suggesting a concentration on milk production which would entail the keeping of the animals in byres during the winter months. Shortage of feed and poor living conditions would account at least in part for the small size of these cattle.

Cattle of *brachyceros* type were introduced into Europe during the early Neolithic period. Some arrived as early as 3000 B.C. The type occurs in several Neolithic stations in Switzerland. In some ancient Swiss lake dwellings the *brachyceros* type is found in association with the remains of *B. p. primigenius* and of domesticated cattle larger than the shorthorns. In Britain shorthorn cattle first appeared during the Bronze Age. They became common during the Iron Age of Europe.

From the Neolithic to the Copper and Bronze Ages the cattle of central and eastern Europe decreased in size, and a great number of dwarfed cattle reached central Europe. During the Iron Age the cattle of central and northern Europe grew still smaller, some of them reaching hardly 100 cm at the shoulder. This tendency was interrupted by the importation of Roman cattle into the provinces of the Roman empire but probably not into Britain, but the bulk of European cattle during this period continued to be of small shorthorn type. With the breaking-up of the Roman empire most of the large cattle disappeared again from the provinces. During the period of the migrations they were replaced by smaller stock, such as the Celtic shorthorn and the shorthorn cattle of the Avars. In the Middle Ages the shorthorn cattle of Europe, from the Urals to Britain, degenerated still further (Bökönyi 1974). Yet it was from similar shorthorn stocks that the modern European dairy and beef breeds, e.g. Friesian, Jersey, Shorthorn, Swiss Brown, were later evolved when the Industrial Revolution and growing urban populations necessitated their development.

In North Africa shorthorn cattle began to replace the original longhorn type at about the same time as in southern and central Europe. They entered Egypt by way of the Isthmus of Suez towards the middle of the 3rd millennium B.C. For a considerable time longhorn cattle prevailed over the shorthorn type, but gradually the latter grew in number as every successful military operation of the pharaohs against the rulers of Syria and Palestine brought new shorthorn herds into the Nile country. Remnants of dispersed Asiatic tribes entered the Nile valley as settlers, their shorthorn cattle swelling the ranks of the breed in Egypt. During the Hyksos period (c 1700–1580 B.C.) shorthorn cattle became the predominant breed.

To this day shorthorn cattle are the common type in Egypt. From there they spread into the Nuba mountain area of the Sudan where they have now been replaced by zebu stock, and into Ethiopia where a small number of humpless shorthorn cows are kept by Sheko tribesmen for milk production in a mountainous, rugged and almost completely isolated territory west of Shewa Ghimirra, adjacent to the Sudan border (Alberro and Haile-Mariam 1982). (For cattle migration routes in Africa, see Fig. 2.1).

West of Egypt shorthorn breeds are found from Libya to Morocco; they are used for milk, meat and draught. From Morocco down the west coast

to latitude 14° the continuity of their range is interrupted by the intrusion of humped cattle, but prior to this they had succeeded in establishing a hold in West Africa where they extend from Guinea-Bissau to Nigeria and northern Cameroon. Here they have become fully adapted to the unfavourable climatic conditions, but in several regions they have become dwarfed, standing less than 100 cm at the shoulder and weighing 70–135 kg. Their economic importance to the peasants is negligible; living in the neighbourhood of the villages in a half-wild state, their only usefulness arises from their skins, in which corpses are buried.

Asia Minor was the starting point of the extension of shorthorn cattle not only to Africa but also to the Balkan peninsula, the coastlands of the Adriatic, Switzerland, France, the Channel Islands and the British Isles, where typical shorthorn breeds still occur, or have only recently become extinct, some of them unimproved, like the Buša, Hérens and Tux-Zillertal, and others improved, like the Angeln and Jersey.

Towards the north shorthorn cattle have spread far into the present area of the Soviet Union, and east through Kurdistan and Iran to Tibet and Mongolia.

Humped cattle

Humped cattle are classed into thoracic-humped and cervico-thoracic-humped types according to whether the hump is situated on the shoulder above the 1st to 9th thoracic vertebrae, or on the posterior part of the neck, i.e. above the 6th/7th cervical to 4th/5th thoracic vertebrae. The majority of the cattle with thoracic humps are referred to as 'zebu', although not all are pure zebu; some, such as the longhorned Fulani cattle of West Africa, include a humpless longhorn strain. For cattle with cervico-thoracic humps, Mason has proposed the general term 'zeboid'. These include the cattle with cervico-thoracic or cervical humps depicted on ancient monuments in Mesopotamia, India and Egypt, for which in archaeological literature the term zebu is in common use, several breeds in central and east China and other border areas of the zebu territory in Asia, the Sanga cattle of eastern and southern Africa, and the recent crossbreds of zebu and humpless cattle.

Zebu cattle generally have a narrow body and

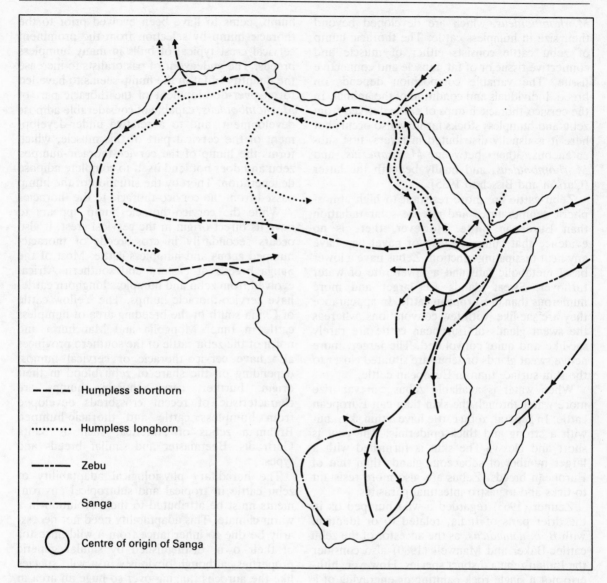

Fig. 2.1 Probable migration routes of cattle in Africa

slender skeleton, and their rump slopes roof-like laterally and caudally. Their crania are as variable in conformation as those of humpless cattle. The skull of typical Indian thoracic-humped zebus is long and narrow, with a convex profile of the head, even surface and flat orbits. In bulls the convex profile and receding forehead are more pronounced than cows. In zebu breeds with horns projecting forwards these characteristics are absent. In general, the direction of the zebu

horns is lateral or upright with a backward tendency. In profile the horns are usually set well behind the forehead.

The majority of humped cattle are distinguished from humpless cattle by the bifid spinous processes of their thoracic vertebrae, but uncleft spinous processes have been observed in some pure zebu cattle.

In contrast with camel humps which are almost entirely composed of fat, the hump in cattle consists of muscle and connective tissue in which variable quantities of fat are embedded. The hump is formed by *Musculus trapezius* and

M. rhomboideus, which are developed beyond their size in humpless cattle. The thoracic hump of zebu cattle consists either of muscle and connective tissue or of fat, muscle and connective tissue. The variable composition depends on breed, individuals and condition of the animal. In the cervico-thoracic humps of cattle derived from zebu and humpless stocks fat may also occur, but here it is usually distributed in layers, first subcutaneous, then between *M. trapezius* and *M. rhomboideus*, and finally beneath the latter (Curson and Bisschop 1935).

Zebu cattle are more resistant to high atmospheric temperatures and intense solar radiation than European cattle. However, there is no evidence that hump, dewlap or navel flap have any heat-dissipating function. Zebus have a lower basal metabolic rate and a lower rate of water turnover. Sweat glands are larger and more numerous than in European cattle. In appearance they are sac-like, with few convolutions, whereas the sweat glands of European cattle are rarely sac-like and quite convoluted. The larger, more active sweat glands of zebus are situated closer to the skin surface than in European cattle.

When water is available zebus can vaporize more water through the skin than can European cattle. In general, zebu cattle have a thin skin, but with a strong and thick epidermis. The coat is short and glossy. The skin is furnished with a larger number of sebaceous glands than that of European breeds. Zebus are also more resistant to ticks and to gastro-intestinal parasites.

Zeuner (1963) regarded a wild humped ox in the drier parts of India, related to or identical with *B. p. namadicus*, as the ancestor of the zebu cattle. Baker and Manwell (1980) also consider the Indian zebu a distinct species. However, hitherto not a single rock painting or engraving of a wild thoracic-humped or cervico-thoracic-humped zebu, which could serve as evidence for this theory, has been found in India. It must therefore be assumed that the hump of the zebu, like the fat tail and fat rump in sheep, was developed by artificial selection. It is improbable that the thoracic and the cervico-thoracic humps were evolved independently, as both zebu types have many important characteristics in common which distinguish them from humpless cattle; they have a similar geographic origin and appear successively in the same areas – Mesopotamia, India, Africa. The cervico-thoracic, mainly muscular,

hump seems to have been evolved prior to the thoracic hump by selection from the prominent cervical crest typical of bulls in many humpless breeds. The endeavour of pastoralists to increase the quantity of fat in the hump seems to have led to an over-development of the thoracic part of *M. rhomboideus*, capable of considerable adipose development, and to a relative under-development of the cervical part of this muscle, which forms the hump of the cervico-thoracic-humped zebu and does not lend itself to complete adipose degeneration. Thereby the situation of the hump passed from the cervico-thoracic to the thoracic.

While the cervico-thoracic hump appears to have its direct origin in the cervical crest, it also occurs secondarily in crossbreeds of thoracic-humped zebus and humpless cattle. Most of the Sanga breeds of eastern and southern Africa, evolved from zebu and humpless longhorn cattle, have cervico-thoracic humps. The Yellow cattle of China south of the breeding area of humpless cattle in Inner Mongolia and Manchuria and north of the zebu cattle of the southern provinces also have cervico-thoracic or cervical humps, depending on the share of zebu blood in their origin. Further, cervico-thoracic humps are characteristic of recent crossbreds developed from humpless cattle and thoracic-humped Brahman zebus of America, such as Santa Gertrudis, Beefmaster and similar breeds and types.

The hereditary physiological adaptability of zebu cattle in tropical and subtropical environments must be attributed to their evolution in a warm climate. This adaptability need not necessarily be due to inheritance from a wild ancestor of their own, distinguished by similar genetic properties, although obviously in a wild species like the aurochs ranging over so huge an area in Asia, North Africa and Europe, the southern representatives would be genetically better adapted to a warm climate, and the northern to a cold one.

Wherever cervico-thoracic-humped cattle make their appearance in prehistoric and protohistorical times they precede thoracic-humped zebus chronologically. From the geographical position of the early centres of their occurrence it may be inferred that cervico-thoracic-humped cattle were evolved in the eastern steppes of the Great Salt desert of Iran. They first appear in the chalcolithic site of Rhana Ghundai in northern Baluchistan, towards the end of the 4th millennium B.C. From here they spread east into the valleys of the Indus and Ravi where they are represented at Mohenjo Daro and Harappa on numerous seals from the period between 2500 and 1500 B.C. They also extended into southern Baluchistan where they are represented on pottery of the Quetta-Pishin valley between 2500 and 1500 B.C. On the Makran coast of Baluchistan cervico-thoracic-humped cattle are copiously represented on painted pottery and by clay figurines from the Kulli culture of the Kolwa region and, north and east of Kolwa, in the contemporaneous Amri-Nal culture of Sind and the head of the Nal valley.

From the Makran coast they doubtless reached the north coast of the Persian Gulf, for the same type of humped cattle is also encountered at Tell Agrab, Ur, Susa and Arar (Larsa) in southern Mesopotamia and the adjacent region of Iran, towards the end of the 4th and during the first half of the 3rd millennia B.C. However, compared with humpless cattle, humped cattle do not seem to have been numerous in this area.

Cervico-thoracic-humped cattle were probably introduced into Egypt from the region of the Somali coast which they had reached by sea from the coastlands of the Persian Gulf. The earliest paintings of such cattle date from the middle of the 2nd millennium B.C. They are found on tomb walls and other monuments. Save for the single exception of an Egyptian bronze weight, thoracic-humped cattle do not appear in Africa prior to the first centuries A.D.

In the Quetta–Pishin valley at the head of the Bolan pass in northern Baluchistan zebu cattle with thoracic humps and large horns occurred, along with cervico-thoracic-humped zebus, during the period 2500–1500 B.C. In the Deccan of south-central India thoracic-humped cattle appear to have been introduced by Neolithic settlers from Baluchistan by way of Gujarat and Maharashtra (Allchin 1963).

In southern Babylonia the original humpless as well as the cervico-thoracic-humped cattle were completely replaced by thoracic-humped zebus during the middle of the 2nd millennium B.C. A few thoracic-humped cattle seem to have reached Sumer still earlier, as indicated by the occasional occurrence of this type in ancient seals and sculptures.

Thoracic-humped zebu cattle reached Africa at a relatively late period. In Somalia and Ethiopia they occur along with the camel and horse not before the 4th century A.D. In larger numbers they were imported into eastern Africa after the Arab invasion of A.D. 669.

From the long interval between the arrival of thoracic-humped cattle in southern Babylonia and eastern Africa it may be inferred that, as in the case of the cervico-thoracic-humped type, the centre of evolution of the thoracic-humped zebu lies not west but east of Mesopotamia.

Typical Sanga cattle have long horns and small cervico-thoracic humps. They were evolved either in Ethiopia or in central east Africa (or both) in the 1st millennium B.C. by crossing the original longhorn humpless cattle of the Hamitic pastoralists with cervico-thoracic-humped cattle coming from India via southern Arabia. The evolution of Sanga cattle may have continued in Ethiopia and the Horn of Africa when the thoracic-humped zebus were imported in the 1st millennium A.D. These Sanga cattle were then carried by the Bantu people in their migrations to southern Africa. These migrations appear to have been comparatively late. Dates given by Epstein (1971) are: 700 A.D. Bantu crossed the Zambesi; 900 A.D. remains of Sanga cattle at Zimbabwe; 1400 A.D. Hottentots crossed the Orange river; 1652 A.D. Dutch bought first cattle from the Hottentots. (These dates may have to be revised if Thorp, 1979, is correct in his claim that cattle were present in Zimbabwe in the 6th century A.D.) From these Hottentot cattle were developed the Africander breed, first as a draught animal and later for beef. At one time the Sanga was ubiquitous from Ethiopia through the lake district of East Africa south to Namibia and South Africa. In most areas it was subsequently replaced by thoracic-humped zebus owing to their greater resistance to rinderpest and their higher milk yield. North of the Zambesi the Sanga now remains only in isolated pockets (see Fig. 2.2).

Thoracic-humped zebus have now displaced the Sanga from most of northeast and east Africa as far south as the Zambesi. Further north it was the shorthorn humpless cattle which they displaced and the zebus of northern Sudan show traces of this cross. In Egypt the effect of zebu introduction was only slightly to modify the pre-existing humpless shorthorn.

From the Sudan the zebus moved into West Africa in a broad belt between the desert to the north and the tsetse-fly area to the south.

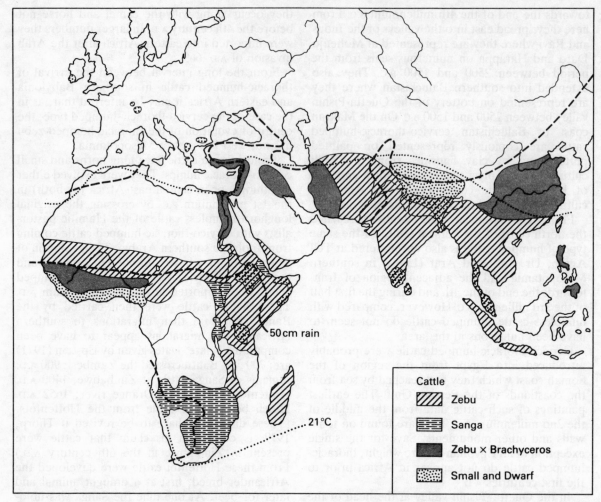

Fig. 2.2 Distribution of cattle types in the Old World. The blank areas are either occupied by humpless cattle (Europe, northern Asia, North Africa, South Africa) or are devoid of cattle through presence of tsetse flies (Central Africa) or of deserts (Sahara, Kalahari, Arabia). The temperature and rainfall figures are annual means

Crossing with the humpless cattle produced various local populations of intermediates such as the Djakoré of Senegal, Bambara of Malawi, and Borgu of Nigeria and Benin. Furthermore the Fulani breeds appear to have derived their long horns from the Hamitic Longhorn.

In Asia zebus remained in southern Iran and Arabia and in Iran and Afghanistan. From there they spread to the northeast into Azerbaijan and

to the northwest into Russian Central Asia. From the Indo-Pakistan subcontinent they have now spread into all the countries of southeast Asia and into southern China. On the frontiers of their distribution, where they encountered humpless cattle, they formed intermediate breeds with small humps cervical or cervico-thoracic in position. Thus were formed the Yellow cattle of central China as well as the intermediate types of Iran and Iraq (see Fig. 2.2).

Recent history and present status

Development of breeds in Europe

Until recently local breeds of cattle were typical of each country and often of every corner of each country in Europe. Their differing characteristics

16

can be accounted for by alternating migrations, and periods of stability during which selection for specific colour types and productivities was possible, together with some degree of natural selection imposed by the environment.

Cattle were the first farm animals used for work and have always been the most important draught animals the world over. It was not until the ox was replaced by the horse as a draught animal, when improved agricultural practices made possible better feeding of cattle, and when enclosure of common land opened the way to controlled breeding, that serious selection for the production of milk and meat was initiated among European breeds. The work of Bakewell (in the 18th century) in selecting the English Longhorn breed of cattle is well known. Unfortunately this took place when the important products of the bovine were work and tallow (for making candles) so that the Longhorn was selected for size and fatness.

Bakewell's methods were no more than common sense but they appeared revolutionary to his contemporaries. He collected foundation material from all over the country, recorded the performance of his breeding animals and their progeny and retained for further use those animals coming nearest to his requirements and leaving the best offspring. This concentration on a few good animals inevitably led to some inbreeding and so a particular type, colour and conformation tended to become fixed. The result was a breed as we understand it today.

Although the Longhorn was the dominant breed in Britain in the 18th century, it was soon displaced by the Shorthorn, a dual-purpose milk/beef animal. In the 17th and 18th centuries red-and-white cattle were imported from The Netherlands to Yorkshire and Lincolnshire and formed the basis of the Durham or Shorthorn breed. This breed spread rapidly and virtually eliminated the Longhorn which today remains only as a rare breed of beef type. The Shorthorn was selected in two directions – for beef or for milk – and in 1905 this division was recognized by the formation of the Dairy Shorthorn Society. Nevertheless both varieties continue to be registered in *Coates' Herdbook* which, first published in 1822, is the oldest cattle herdbook. So successful was the Shorthorn that by 1908 a census showed that 64 per cent of cattle in Britain belonged to this breed and as recently as 1936 50 per cent of the cattle were Shorthorns.

At the end of the 19th and beginning of the 20th centuries, further Dutch imports of their now black-and-white breed formed the basis of the British Friesian. Although it is called a dairy breed, the Friesian produces more beef as well as more milk than the Shorthorn; it rapidly expanded at the expense of the Shorthorn and now accounts for over 60 per cent of British cattle, while the Shorthorn is reduced to less than 1 per cent.

Other beef breeds such as the Hereford and Aberdeen-Angus were developed and improved in the wake of the Shorthorn.

Changing circumstances are making the small fat beef breeds obsolescent, especially the Beef Shorthorn. Firstly the demand is now for lean meat rather than fat meat and secondly, beef cattle are now finished on a higher nutritional plane and on this feeding system early-maturing animals get fat at too low a weight. Present demand is for animals which grow rapidly and efficiently and produce the maximum amount of lean meat. Thus, selection aims for the British beef breeds are changing. However, the same result can be achieved more quickly by importing the large beef breeds from the European continent. The Charolais and Limousin are now the most popular beef breeds in England after the Hereford, and several other Continental breeds are represented in Britain.

Many local breeds still remain in Britain but, compared with the breeds mentioned above, their numbers are small and many are reduced to the status of rare breeds. Three of importance should be mentioned. The Jersey and Guernsey, originating in the Channel Islands, were selected for the high butterfat content of their milk. The Ayrshire is intermediate between Friesian and Jersey in size, milk yield and butterfat content; it has lyre horns.

In the Baltic countries the *brachyceros* cattle formed a population of red dairy cattle differentiated by selection in the 19th century to produce the Angler (in east Schleswig), Danish Red (on the Danish islands), Estonian Red and Latvian Brown. This group of breeds has been responsible for producing new breeds in Eastern Europe such as the Bulgarian Red, the Red Steppe (Ukraine) and the Romanian Red. In Sweden the local breeds, crossed with British Shorthorn and Ayrshire at the end of the 19th century, produced the Swedish Red-and-White which is now the dominant Swedish breed with a milk yield approaching that of the Friesian. This breed was subsequently imported into Norway to cross with local breeds and with imported Ayrshires to form the Norwegian Red, which has now absorbed all the other local breeds and is effectively the only breed in the country. In Finland also, the Ayrshire was imported and has had a lasting influence – the improved Finnish Ayrshire has almost displaced the original native Finnish cattle and is now the dominant breed.

In central Europe the local breeds at the beginning of the 19th century were mostly red and of the shorthorned type. They were grouped as the Central European Red and there were local breeds of this type in White Russia, Lithuania, Poland, Czechoslovakia and central Germany. Further east and south the original *brachyceros* type had been replaced (from the 5th century A.D. onwards) by longhorned grey-white cattle from the steppes of southern Russia. These came via, if not from, Podolia and are called the Grey Steppe or Podolian group. They are larger and therefore better draught animals than the *brachyceros* which they completely displaced in Hungary and in the lowlands of Albania, Bulgaria, Greece, Romania, northwest Turkey and Yugoslavia. They also penetrated into eastern and southern Italy and remained as the dominant breed in the Ukraine.

These various shorthorn and longhorn breeds of central and eastern Europe have, in turn, been almost completely replaced (except in the Ukraine and Italy) by the Simmental and Brown breeds from Switzerland.

Early in the 19th century the cattle of the Simme valley in the Bernese Oberland, known as Berner cattle, were little more than an improved landrace of various colours – red-and-white, black-and-white, or red. Improvement measures were started, using pure breeding, and the modern *Simmentaler Fleckvieh* was developed as a triple-purpose breed with a yellow-red and white coat and a white face. It soon spread to cover the whole of the western half of Switzerland. Exports to Germany and France (Jura – to form the Montbéliard breed) started in the 18th century and during the 19th and early 20th centuries the improved breed was exported not only to southern Germany and eastern France but also to Austria, Czechoslovakia, Hungary, Italy (Friuli), Romania, Russia and Yugoslavia. In these coun-

tries they formed, by crossbreeding, the various national Simmentals, Fleckvieh or local Red Pied (highland) breeds. More recently the Simmental has been exported overseas (especially to the Americas and to Africa) where it is used as a single-purpose beef breed.

At about the same time the grey-brown breed of the Schwyz canton was similarly improved to form the *Schweizerisches Braunvieh* (Swiss Brown), which is now the only breed in eastern Switzerland and has spread to Austria, southern Germany, Hungary, Italy, Romania, western Turkey, Russia and Yugoslavia (Slovenia), all of which now have their national Brown breeds. The Brown cattle, like the Simmental, were triple-purpose.

The Yellow cattle (*Gelbvieh*) of central and southern Germany also arose by crossing with the Simmental. They are now almost extinct except for the Yellow Franconian (German Yellow).

In northern Germany, Denmark, the Low Countries and France there were lowland breeds in contrast to the highland breeds so far discussed. The most influential of these was the Friesian, which developed in Friesland in The Netherlands. Originally, the majority of cattle in The Netherlands were red-and-white like the ancestors of the British Shorthorn breed. When most of the Dutch cattle were destroyed by plague at the end of the 18th century, black pied cattle were imported from Jutland. Whatever their ultimate origins the cattle of Friesland were soon recognized for their milking ability and their influence spread to all the other black pied populations of the Atlantic seaboard. The Friesian or Black Pied Lowland is now the dominant breed in northern Germany and France, Poland, Italy and the British Isles and there are large populations in Belgium, Spain, Portugal, Sweden and Denmark. It is now penetrating the stronghold of the highland breeds in central and eastern Europe. Its success outside Europe is equally noteworthy.

The original lowland cattle population in northwest Europe also included a Red Pied Lowland type which was represented by separate breeds in The Netherlands, Germany, Belgium and France. These breeds are now very similar having all been strongly influenced by the Dutch Meuse-Rhine-Yssel.

Turning now to the Mediterranean countries, the stream of shorthorned cattle from the Middle

East spread into the Balkans and Italy. They are small, red-brown or black, usually triple-purpose, hardy animals. They remain as the dominant type in Turkey although much crossed with Swiss Brown in the west. Elsewhere the purebred populations are much smaller and have been pressed into the mountains or isolated on islands. This applies to the Greek Shorthorn, the Buša of the highlands of Yugoslavia and Albania, the native Sardinian cattle and a few relict breeds in northern and central Italy. The dominant breeds in Italy are now the Friesian and the Swiss Brown. The Swiss Brown (*Bruna Alpina*) first displaced the local breeds and in its turn is being displaced by the Friesian.

In central Italy the Podolian (Grey Steppe) breeds, which were single-purpose draught breeds, are turning their size to advantage and are now developing as single-purpose beef breeds. These are the Romagnola, Marchigiana, and Maremmana. A fourth is the Chianina which has a mixed longhorn–shorthorn ancestry. There is now a great deal of competition among the terminal crossing beef breeds and in the long run the Maremmana may come out best since it has retained some hardiness and can thrive outside on poor land while the others are adapted to intensive indoor systems.

The Longhorn of England (now nearly extinct) and the Highland cattle of Scotland may owe their long horns to the original cattle of *primigenius* type introduced into Britain in the Neolithic. Longhorned cattle are also found in the other most westerly part of Europe, namely the Iberian peninsula, and these probably have a similar origin; most of them are primarily draught animals.

There are also longhorn breeds in central and southern France – namely the Salers, Gascon and Aubrac. Of these the Gascon is grey-white in colour and has red calves which seems to relate it to the Grey Steppe type.

The British Shorthorn, as the first 'improved' dual-purpose breed, was much sought after as an improver in the 19th century but it has not left much mark on the European breeds. The most successful breed with Shorthorn blood is undoubtedly the Normandy breed of northern France, which is on a par with the Friesian in terms of numbers. Its dished face suggests a relationship with the neighbouring Jersey breed.

In central France there are several breeds which, although not numerous, are important for their world-wide influence. They retained their draught function much later than in Britain, so that when the demand for large beef breeds developed, they were ready to supply it and they were followed by the Simmental and the white Italian breeds. The Charolais is the best known and is now widely exported. It was first imported into the U.S.A. which it entered by the back door (Mexico) in 1950. It came to Britain in 1961 (after prolonged effort to prevent it) and is now known all over the world. The other French breeds in the same category are the Limousin, the Maine-Anjou and the Blonde d'Aquitaine.

It is in the countries of the European Economic Community that the evolution of the cattle population has been most closely studied. Cunningham (1976, 1979) made a survey which showed that of the 30 million cows in the E.E.C. 80 per cent are dairy (or dual-purpose) and the rest are beef cows. Forty separately identifiable populations each of more than 100 000 cows account for 95 per cent of the total. In the thirty dairy subpopulations more than half the cows belong to the eight national Friesian populations. The non-Friesian breeds tend to be red or red-and-white in colour, to be more heavily muscled but lower in milk production than the Friesians, and to occupy inland rather than maritime areas. Numerically most important among the non-Friesian breeds are the Simmental derivatives, Normandy, Red Pied Lowland, and Swiss Brown derivatives. These account for over 90 per cent of the dairy cows, the rest being Danish Red, Danish Jersey, German Yellow, Irish Shorthorn and British Ayrshire, Guernsey and Jersey. The major current change is the continued grading to Friesian of these minor breeds (except the German Yellow). At the same time two of the European types are being crossed with improved American derivatives: Cunningham (1979) calculated that in less than a decade the European Friesian population will be genetically more than 50 per cent Holstein-Friesian. At the same time the European Brown cattle are being extensively crossed with American Brown Swiss. Other changes are the use of Montbéliard bulls on the other two Simmental derivatives in France (Abondance and Pie Rouge de l'Est) and of Normandy bulls on the Maine-Anjou.

The ten identifiable beef cow populations account for over 90 per cent of all beef cows in

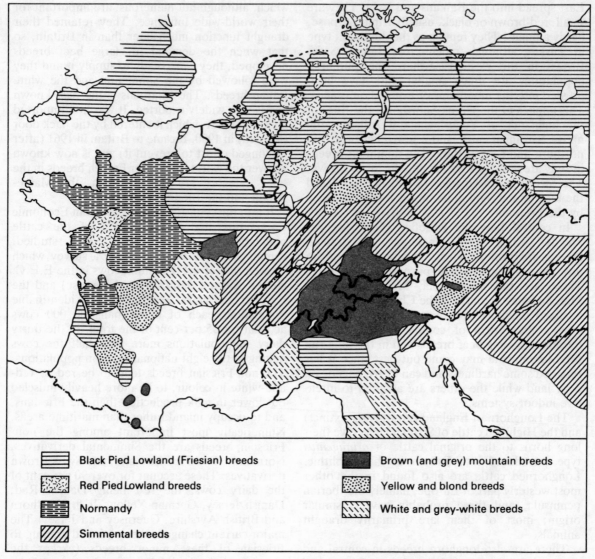

Fig. 2.3 Distribution of chief cattle breeds in western Europe. (From Mason 1971)

Legend:
- Black Pied Lowland (Friesian) breeds
- Red Pied Lowland breeds
- Normandy
- Simmental breeds
- Brown (and grey) mountain breeds
- Yellow and blond breeds
- White and grey-white breeds

the Community. The populations in Britain and Ireland are crossbred, often an Aberdeen-Angus or Hereford crossed onto a Friesian or Shorthorn cow. In France beef cows are Charolais, Limousin or Blonde d'Aquitaine and they are bred to bulls of their own breed. In Italy they are Chianina or Marchigiana (with smaller populations of Romagnola and Maremmana), also bred pure (see Fig. 2.3).

Cattle populations in other continents

Asia All the indigenous cattle of the Indo-Pakistan subcontinent are zebus bred primarily for draught but also for milk. They are of different types in different parts of the country – the grey-white shorthorned type of the northwest and centre, the longhorned type in the west and north, the diminutive hill type in the mountains (and in Sri Lanka), and the fast-trotting Mysore type in the south of India. While the majority of cattle are *desi*, or unimproved, there are tracts which have developed their own special breeds within each group.

Outside these groups are the specialized dairy breeds of Pakistan – the Sahiwal and Red Sindhi – and the Gir of Kathiawar. Although counted as dairy breeds these have a very low milk yield compared with the improved dairy breeds of Europe. The latter have been imported for crossbreeding and there are now many research and development programmes for producing new breeds or promoting crossing schemes by which the tropical adaptation of the zebu can be combined with the high yield of the European breed. This is usually a Friesian or Jersey but the Brown Swiss and Danish Red have also been used.

Outside India and Pakistan there has been little breed development among zebu cattle. In southwest Asia there are few cattle in the arid areas. Where feeding and disease control are adequate they are being upgraded and replaced by European breeds (chiefly Friesian).

In southeast Asia there has been no breed formation. Cattle are used primarily for work but in many places the water buffalo is more important. Any improvement that has been made has been by the use of Indian zebus – Ongole for work and Sindhi or Sahiwal for milk – and now by European dairy breeds, as in India.

North of the Himalayas the indigenous cattle are humpless and are collectively known as the Turano-Mongolian group; they have horns of intermediate length. Those of Russian Central Asia and Siberia have been extensively crossed with Friesian and in some places with Hereford. This process is only just starting in China where there are many local types of draught cattle. In the north they are of the Mongolian type, in the south zebus and in the centre of an intermediate type known as Chinese Yellow. The humpless cattle of Tibet are dwarfed.

Africa The pattern of cattle types in Africa has not yet been significantly disturbed from that resulting from the migrations shown in Fig. 2.1 and illustrated in the distributions of Fig. 2.2. It can be recapitulated as follows:

Humpless	1. N'Dama (longhorned) in Guinea, Guinea Bissau and the contiguous area of neighbouring countries (including all of Gambia).

2. Kuri (longhorned) around Lake Chad.
3. West African Shorthorn along the coast from Liberia to Cameroon including some inland areas.
4. Shorthorned cattle in the Maghreb, Libya and Egypt, the last showing traces of zebu blood.

Humped	1. Zebus in eastern Africa from Sudan to Malawi and Madagascar.

2. Zebus in West Africa between the humpless zone and the Sahara.
3. Intermediates in a zone between the West African Zebu and the humpless cattle.
4. Sangas (zebu and humpless longhorn origin) in parts of eastern Africa (Nilotic in southern Sudan and Ankole on Uganda–Zaire–Tanzania borders) and in southern Africa south and east of Malawi.

In West Africa considerable breed change is taking place. The small N'Dama and West African Shorthorn have long been under pressure from the larger zebus to the north. They have been protected only by their natural resistance to trypanosomiasis (carried by the tsetse fly) which is endemic in West Africa south of 14°N (approximately) and in all of Central Africa. This pressure has been increased by the Sahel drought, which forced the zebus south, and by the recently acquired ability to protect them from trypanosomiasis by drugs. However, recent studies (FAO/ILCA/UNEP 1980) show that the small cattle, because of their trypanotolerance, are just as productive as the larger zebus. Measures are now being taken to quantify their attributes and prevent their loss by attrition.

Their merits have long been appreciated in Central Africa to which they are being imported to provide a cattle population in areas where tsetse flies had hitherto precluded any. There are now large populations in Zaire (the first import dating to 1904) and considerable ones in Gabon, Congo and the Central African Republic.

The greatest influence of the European breeds has naturally been in the temperate zones of northern and southern Africa. In Algeria and even more so in Tunisia there has been widespread crossing of the Brown Atlas breed with European dual-purpose breeds, especially the Friesian and the Brown Swiss. In South Africa, European dairy breeds predominate and the beef cattle include British breeds and Simmental as well as the local Africander (an improved Sanga). In the highlands of Kenya, British dairy breeds now flourish. Elsewhere the use of European breeds has been only on a small or experimental scale.

North America Most of the British breeds were taken to America during the 19th century. As in Britain a sharp distinction was drawn between beef and dairy breeds and this distinction became even sharper than in Britain. The beef industry was soon dominated by the Beef Shorthorn, Aberdeen-Angus and Hereford. The major dairy breeds are the Friesian, Ayrshire, Swiss Brown, Jersey and Guernsey. Among these, the Friesian soon became the most numerous and now represents 90 per cent of the dairy cow population. The official name became Holstein-Friesian under the mistaken impression that some of the initial stock came from Holstein and this is a common abbreviation of the breed name. In fact, the change of name has proved useful. Intense selection has produced a breed with a higher milk yield than the various European strains and with a correspondingly lower aptitude for beef production. The Swiss Brown also was changed from a triple-purpose to a single-purpose milk breed designated American Brown Swiss to distinguish it from its parent breed.

Beef breeders in the United States continued the same lines of selection as the British beef breeders. Their cattle became smaller and more compact. Selection aims have now changed to a larger animal and at the same time the large beef breeds are being imported from Europe. Among dairy breeds the influence is reversed – the Holstein-Friesian and American Brown Swiss are being used on a large scale to increase milk yield in their European relatives.

In the southern states of the U.S.A. British beef breeds have not flourished and zebu breeds were imported from India via Brazil in the second half of the 19th century. They now form the

Brahman breed (possibly with some European blood) which is an outstanding tropical beef breed. It has been used in crosses with British breeds to form several new zeboid breeds of which the best known is the Santa Gertrudis developed from a Brahman × Beef Shorthorn cross in Texas during the period 1910–40.

South and Central America Until recent deliberate importations most cattle movements followed in the wake of movement of peoples. Thus when the Spanish and Portuguese colonized South and Central America and the southern part of North America they populated them with the breeds from the Iberian peninsula. They were forced to develop by natural selection some adaptation to the local environment. The resultant cattle are called criollo (or crioulo in Portuguese). Many of these live in a tropical environment and although they have been there only a few hundred years compared with the millennia which the humped cattle have been in the tropical areas of Asia they have developed some tropical adaptations similar to those of the zebu.

In the 19th century the cattle composition started to change by the importation of cattle considered more suitable for the environment or for the production of milk or beef. In the tropical area the criollo was gradually replaced by zebu cattle from India and now some of the best beef breeds in Brazil are direct descendants of Indian zebus. These have been imported into the other tropical countries. Further south and particularly in the pampas of Argentina and Uruguay, the similarities to the ranges of North America led to the importation of Aberdeen-Angus, Hereford and Shorthorn. It was soon shown that these breeds were well adapted and they became as well known in South America as they were in the north. Now the large European breeds are being introduced.

Australia The first import of beef breeds (in the early 19th century) was naturally of British breeds, and specifically the Beef Shorthorn. It did not flourish in the tropical areas of northern Queensland and the Hereford was brought in. Finally, after more than a century of struggling with British breeds, the solution of introducing a tropically adapted breed was made. By chance a few zebu bulls (from a circus) left crossbred progeny and their superiority led to the development of a new breed – the Droughtmaster – based on an early zebu × Shorthorn cross. Deliberately designed breeding programmes involved the import of American Brahman, Santa Gertrudis and Africander. The Africander has already fathered another local breed – the Belmont Red. Now it is estimated that the majority of cattle in northern Queensland carry zebu blood.

Dairy breeds were imported from England but it is now recognized that in the tropical areas some zebu blood would be desirable. Accordingly new tropical dairy breeds are being developed by crossing, Jersey with Sahiwal and Sindhi to form the Australian Milking Zebu, and Friesian with Sahiwal to form the Australian Friesian-Sahiwal.

Current improvement methods

After the period of the early improvers in Britain came the stage of consolidation and conservation. During this period, which is only now ending, the emphasis was on the results of the early improvers rather than on their methods. Too often prime consideration was given to the incidental results rather than the main results, i.e. to breed type rather than to performance. Breed standards were set up and breed societies and herdbooks were established to protect the interests of breeders and preserve the purity of their breed. The emphasis on standards was dangerous because there is little correlation between appearance and performance. At the same time it was difficult to select effectively, especially for milk production, in the small herds which were the rule.

Two developments completely changed the situation. Starting in the 1940s, artificial insemination became the standard method of breeding dairy cattle. In the 1950s the development of freezing semen, thus enabling it to be stored indefinitely, added immense flexibility to the technique. At the same time the theories of population genetics were applied to animal breeding and made it possible to determine the most efficient breeding system. All current national cattle selection schemes are based on these two foundations, plus a nation-wide recording scheme. National schemes differ slightly but all have in common the principle of comparing daughters of the bull under test with the daughters of other bulls milking in the same herd at the same time. This is to overcome the

effect of environmental (including managemental) differences between herds which may have more effect on milk yield than genetic differences between bulls. If bulls were used in one herd only these two effects would be confounded; by the use of artificial insemination each bull can be used in many herds and by calculating the result as some sort of 'contemporary comparison' the environmental effect can be minimized. The best bulls must, of course, be used for breeding future bulls for testing.

By such integrated breeding schemes milk yield is being continuously increased in the countries of the developed world at a rate of about 1 per cent per annum. The selection is made within the existing breeds. Only the more numerous breeds within a country can efficiently operate a selection scheme. Once a breed falls to such low numbers that it cannot test a group of several bulls each year it loses out in the productivity race and its numbers fall even lower. This accounts for the number of local dairy breeds now in the rare breed class. The successful breeds are eagerly sought after and in fact, on a world view, since the days of breed formation the major source of genetic improvement of cattle has been by grading up or displacing existing stock by improved breeds.

Improvement of beef cattle is easier than that of dairy cattle since the important characters – size and growth rate – can be measured in both sexes prior to breeding age. Thus only a performance test rather than a progeny test is necessary. (A progeny test may be performed if carcass characters are important; but tests for estimating carcass composition in the live animal are being developed). However, selection of beef breeds along these lines has led to a major problem – the large calf which is associated with the large body size causes difficulties in calving which may mean loss or damage to cow or calf or both. Attempts are being made to overcome this difficulty by selecting those breeds and those bulls within breeds which produce the minimum dystocia rate in relation to their size.

This problem may be approached in another way. What is wanted is an efficient animal – one which produces a maximum amount of lean meat from a given amount of food consumed. But efficiency of food conversion is difficult to measure. Every item of food eaten must be measured and its composition must be exactly known.

The most efficient animal is not necessarily the largest and most rapidly growing one. Whilst there is a correlation between absolute growth rate and efficiency, that between relative growth rate and efficiency may be higher. If selection could increase efficiency by basing it on growth rate divided by average weight (or some corresponding index) this would obviate the necessity for measuring food intake.

In the case of dual-purpose breeds the normal procedure is to performance test bulls in order to estimate their beef-producing ability and to progeny test for milk only those which rank highest in the performance test.

Breed conservation

The evolution of the bovine species is now taking place more rapidly than at any time in the past. The rapid improvement in the productivity of a few breeds means that many others have become extinct or are threatened with extinction. Lauvergne (1975) estimated that of 148 indigenous breeds in Europe and the Mediterranean countries only 33 are holding their own. Cunningham (1979) showed that of the 200 or so breeds present in the E.E.C. countries at the beginning of the century only 40 populations of significant size remain and, of these, 8 are Friesians.

These changes make good economic sense and will therefore not easily be halted. Nevertheless there has been a reaction to the disappearance of breeds and in several countries (notably in the United Kingdom, France and The Netherlands) private organizations have been established which are trying to conserve the threatened local breeds. In other countries (e.g. Hungary, Bulgaria, the Soviet Union) governments have taken action. The justifications for breed conservation are several. Rare breeds may contain unique genes or gene combinations which should be studied before they are lost. In particular, they may have qualities which will be needed in the future, when breeding aims change as a response to changing environments or market demands.

Many breeds have historical and cultural associations with particular countries or regions which justify their conservation in the same way that ancient monuments are conserved.

The above arguments apply primarily to Europe and North America. In other continents, action must be taken before breeds reach a threatened status. Given an artificial insemination system, replacement crossing with a European breed is all too easy but it is a course of action which ought to be resisted until the relative values (in total dairy merit, production of milk and beef per cow per year or some similar measure) of the local breed, the crossbred and the pure imported breed have been evaluated.

Blood groups

The study of blood groups and other biochemical polymorphisms of blood and milk has shed some light on breed relationships (see Braend 1972). In general the results are in accordance with known history but some discrepancies can be explained only by resort to hypotheses involving selection, inbreeding or genetic drift. The most recent attempt to use biochemical polymorphisms to throw light on breed relationships and breed history has been made by Baker and Manwell (1980). They studied figures from the literature on gene frequencies for ten protein polymorphisms. The data were obtained from 196 breeds which were classified into 10 major groups as follows:

N	North European – Scandinavian and British breeds (excluding Channel Island and British Friesian).
L	Pied Lowland – Red Pied Lowland, Black Pied Lowland (including all Friesian derivatives and Belgian Blue-White).
R	European Red – Baltic Red, Central and East European Red and Flemish.
C	Channel Island – Jersey, Guernsey, Normandy and Breton Black Pied.
U	Upland *brachyceros*–Pinzgau, Simmental and derivatives, Yellow, Brown and Grey Mountain, Italian and Balkan *brachyceros*, North African, Egyptian, Near Eastern and Turkish breeds; several minor breeds in central Europe.
M	*Primigenius–brachyceros* mixed – Charolais, Limousin, Blonde d'Aquitaine, Modicana, Piedmont, Brown and Simmental derivatives in eastern Europe; and several minor breeds.
P	*Primigenius* – Portuguese, Criollo, Podolian and Grey Steppe.
Z	Indian Zebus.

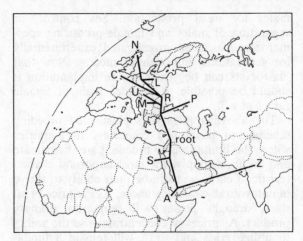

Fig. 2.4 Superposition of the phylogenetic tree for ten major cattle breed groups onto a map of Europe, western Asia and northern Africa. For key to letters, see text. (From Manwell and Baker 1980)

S   African Sangas.
A   African zebus.

For all markers except carbonic anhydrase II there were significant differences among breed groups. In other words the data on these protein polymorphisms confirmed the groupings made on the basis of morphological similarity and breed history.

Manwell and Baker (1980) go on to construct a phylogenetic tree for the ten breed groups. They then superimpose this tree on a map of Europe and the Middle East as shown in Fig. 2.4. The North European group was arbitrarily located in Scotland, the Pied Lowland group in north Germany and the Channel Island group in the Channel Islands. This brings the location of the European Red to southeast Poland which appears to be a reasonable compromise between the Baltic, central and eastern Europe. The upland *brachyceros* can be centred appropriately in the Alps and the arc for the primigenius breed group can reach out to near Podolia in the Soviet Union or swing over to Italy for its Podolian cattle. The root of the tree is located in the Fertile Crescent. The Sanga and African zebus can just be fitted into Africa; although far from their present centres of distribution, their locations on the map are near to their centres of origin in northeast Africa. The Indian zebus can comfortably be located in India and the authors consider

that the great difference from European breeds in the gene frequencies for these ten polymorphisms justifies their classification as a distinct species, *B. indicus*. We do not think that this is sufficient proof. Manwell and Baker do not discuss the taxonomic position of the African zebu which on their chart is almost halfway to the European breed but in appearance can be almost identical with an Indian zebu.

Feral populations

Compared with goats on very many islands, sheep in New Zealand or camels and buffaloes in Australia, there are few populations of feral cattle. They can live perfectly well on their own but they are probably too valuable for their owners to allow them to stray or to leave them unmolested if they do. A few feral herds have been described (see Mason 1979) in Colombia, in the western Pyrenees (Bétiso in France and Betizuak in Spain), on Japanese and New Zealand offshore islands, in the Seychelles, and in Spain. There is also the Chillingham herd of White Park cattle in England.

These few examples show that cattle, after thousands of years of domestication, can still fend for themselves and that several different varieties in different climatic and topographical situations have succeeded in doing so, even in small herds where the rate of inbreeding must be high. There has been no tendency to revert to a common type and small body size seems to be the only characteristic common to the populations.

Future prospects

There can be no doubt that cattle will continue to be the major farm animals for the production of milk and meat in all parts of the world and, in developing countries, also of work. Only in the Middle East is mutton preferred to beef and only in south and southeast Asia have cattle a serious competitor, the water buffalo, as a source of milk and work.

Milk yield will continue to increase in the major dairy breeds which are subject to sophisticated progeny testing schemes based on artificial insemination and milk recording. But the time may come when additive genetic variation in milk yield (or in total dairy merit) within a single breed is exhausted and then there will be a need for a systematic crossbreeding system between breeds in order to exploit the non-additive variation. Whilst some degree of heterosis is exhibited when dairy breeds are crossed, the Friesian has a yield so much higher than that of other breeds that its crossbreds, while yielding more than the mean of the two parental breeds concerned, do not yield more than the pure Friesian. Thus there will soon be a need for a breed with a yield comparable with that of the Friesian but unrelated to it. The most likely candidates appear to be the red and red-and-white breeds of Scandinavia (excluding the Norwegian Red which has Friesian blood).

Among specialized temperate beef breeds there is plenty of scope remaining for improvement by selection within breeds and also for an increase in the diversity between breeds – especially between hardy adapted breeds and specialized breeds for use in a terminal crossing system. The current need for the latter is to produce animals which are economically efficient and are rapidly growing without a high rate of calving difficulties.

If an early indication of performance could be obtained, especially for dairy bulls, the process of selection would be immensely speeded up. Blood groups seem to have no future for this purpose. Other biochemical tests have been explored, so far without success. Nevertheless there is a possibility that future research may reveal some enzyme or other metabolite whose concentration in blood or other tissue could be used to predict future milk yield or growth rate.

Other technological shortcuts are also being explored. Embryo transplantation following superovulation has so far been exploited chiefly to produce several offspring of particularly outstanding cows, or to introduce new breeds into a country without the expense of importing adult animals. Used in these ways it is unlikely to have much influence on the evolution of the cattle species. If, however, the fertilized egg could be divided into many genetically identical embryos before transplanting it into the brood cows, i.e. if cloning were possible as in plants, then embryo transplantation might approach artifical insemination in its effect on cattle breeding. At the moment it is not possible to divide the egg at later than the 4-cell stage so the maximum is four identical embryos. Embryo transfer may also be useful in order to produce offspring of predominantly one sex – females for milk production or males for meat production. Sex control, or separation of male- and female-producing spermatozoa, has been investigated experimentally for generations, but without success. Now that blastocysts can be sexed before implantation it should be possible to produce male or female calves at will.

The above applies to sophisticated breeding schemes in developed countries. In countries where the limitations to breeding are the climate and the food supply, the most important characteristic of cattle will still be their ability to thrive in unfavourable circumstances. As the price of oil rises, draught power will remain a primary product. As prices of concentrates rise the ability to utilize grass and straw will remain a fundamental requisite.

What is needed are breeding programmes to improve the productivity of adapted breeds without decreasing their adaptation. Such programmes are starting but are difficult to apply in the absence of artificial insemination, recording schemes and advisory services. With settled agriculture something can be achieved but with nomadic animal husbandry this is much more difficult. The present tendency in the tropics is to import temperate improved breeds and to neglect the local ones. This policy has worked well in many subtropical and dry tropical regions and the Friesian is now spreading over the world, but it will only succeed as long as management – and particularly feed supply and disease control – improves to match the improved breed. In harsher environments new breeds based on a crossbred foundation or local breeds improved by sophisticated techniques are likely to be the most successful.

References

Alberro, M. and **Halle-Mariam, S.** (1982) The indigenous cattle of Ethiopia, Part I. *World Animal Review*, No. 42: 27–34

Allchin, R. (1963) Cattle and economy in neolithic south India. In: *Man and Cattle*. Royal Anthropological Institute, Occasional Paper No. 18

Baker, C. M. A. and **Manwell, C.** (1980) Chemical classification of cattle. 1. Breed groups. *Animal Blood Groups and Biochemical Genetics,* **11**: 127–50

Bökönyi, S. (1974) *History of Domestic Mammals in Central and Eastern Europe*. Akadémiai Kiadó: Budapest

Braend, M. (1972) Studies on the relationships between cattle breeds in Africa, Asia and Europe: evidence obtained by studies of blood groups and protein polymorphisms. *World Review of Animal Production*, **8**: 9–14

Braidwood, R. J. and **Howe, B.** (1960) *Prehistoric Investigations in Iraqi Kurdistan*. Studies of Ancient Oriental Civilisations, vol. 31. University of Chicago Press

Brentjes, B. (1967) Die Tierwelt von Chatal Hüyük. Zur Geschichte der Haustierwerdung aus dem 7. und 6. Jahrtausend v.u.Z. *Säugetierkundliche Mitteilungen*, **15** (4); 317–32

Cunningham, E. P. (1976) The structure of the cattle populations in the E.E.C. In: *E.E.C. Seminar on Optimisation of Cattle Breeding Schemes*, Dublin, Ireland, Nov. 1975, pp. 13–27

Cunningham, E. P. (1979) Cattle populations in relation to their ecological environment. In: *The Future of Beef Production in the European Community*, eds J. Bowman and P. Susmel, pp. 153–69. Martinus Nijhoff: The Hague

Curson, H. H. and **Bisschop, J. H. R.** (1935). Some comments on the hump of African cattle. *Onderstepoort Journal of Veterinary Science and Animal Industry*, **5** (2): 613–739

Epstein, H. (1958). Die Unbrauchbarkeit einiger anatomischer Merkmale für die Rassengeschichte europäischer Langhornrinder. *Zeitschrift für Tierzüchtung und Züchtungsbiologie*. **71** (1): 59–68

Epstein, H. (1971). *The Origin of the Domestic Animals of Africa*, Vol. 1. Africana Publishing Corporation: New York. Edition Leipzig: Leipzig

FAO (1982) *FAO Production Yearbook 1981*, Vol. 35. FAO: Rome.

FAO/ILCA/UNEP (1980) *Trypanotolerant Livestock in West and Central Africa* (2 vols). FAO Animal Production and Health Paper No. 20. FAO: Rome

Grigson, C. (1980) The craniology and relationships of four species of *Bos*. 5. *Bos indicus* L. *Journal of Archaeological Science*, **7**: 3–32

Howard, M. M. (1962) The early domestication of cattle and the determination of their remains. *Zeitschrift für Tierzüchtung und Züchtungsbiologie*, **76** (2–3): 252–65.

Isaac, E. (1962) On the domestication of cattle. *Science*, **137** (3525): 195–204

Lauvergne, J. J. (1975) *Pilot Study on Conservation of Animal Genetic Resources*. FAO: Rome

Manwell, C. and **Baker, C. M. A.** (1980) Chemical classification of cattle, 2. Phylogenetic ties and specific status of the zebu. *Animal Blood Groups and Biochemical Genetics*, **11**: 151–62

Mason, I. L. (1971) Comparative beef performance of the large cattle breeds of Western Europe. *Animal Breeding Abstracts*, **39**: 1–29

Mason, I. L. (1975) Factors influencing the world distribution of beef cattle. In: *Beef Cattle Production in Developing Countries*, ed. A. J. Smith. Proceedings of a conference held in Edinburgh, 1–6 Sept 1974. Centre for Tropical Veterinary Medicine, University of Edinburgh.

Mason, I. L. (1979) *Inventory of Special Herds and Flocks of Breeds of Farm Animals*. (Draft). FAO: Rome

Mason, I. L. and **Maule, J. P.** (1960) *The Indigenous Livestock of Eastern and Southern Africa*. Commonwealth Agricultural Bureaux: Farnham Royal, Bucks, England

Nobis, G. (1954) Zur Kenntnis der ur- und frühgeschichtlichen Rinder Nord- und Mitteldeutschlands. *Zeitschrift für Tierzüchtung und Züchtungsbiologie*, **63** (2): 155–194

Perkins, D. (1960) The faunal remains of Shanidar cave and Zawi Chemi Shanidar: 1960 season. *Sumer*, **16**: 77–8

Perkins, D. (1969) Fauna of Çatal Hüyük. Evidence of early cattle domestication in Anatolia. *Science*, **164** (3875): 177–9

Reed, C. A. (ed.) (1977) *Origins of Agriculture*. Mouton Publishers: The Hague and Paris.

Rouse, J. E. (1970). *World Cattle*. University of Oklahoma Press: Norman

Thorp, C. (1979) Cattle from the early Iron Age of Zimbabwe–Rhodesia. *South African Journal of Science*, **75** (10): 461.

Zeuner, F. E. (1963) *A History of Domesticated Animals*. Hutchinson: London.

3
Bali Cattle

D. H. L. Rollinson

*formerly FAO Livestock Planning Adviser,
Indonesia*

Introduction

Nomenclature

The Bali is a domesticated banteng, *Bos javanicus*, (synonyms: *B. banteng, B. sondaicus*). Although very similar to cattle in general characteristics the banteng was often classified with the gaur and the kouprey in the subgenus *Bibos*, characterized by long dorsal spinal processes of the thoracic vertebrae 3–11. Groves (1981) uses the name *Bos javanicus domesticus* for the Bali ox. Recent Australian literature tends to refer to these cattle as banteng cattle.

General biology

Bali cattle are remarkably uniform in type. The hair colour is very distinctive, usually reddish brown but with a clearly defined bilaterally symmetrical white area on the hindquarters which extends along the ventral surface, white 'socks' extending from the hooves to just above the carpus and tarsus, and white hair around the muzzle, in the ears and on the tail. There is also a well-defined narrow band of black hair running along the backbone from behind the shoulders to the tail, and the hair on the switch of the tail is black.

In entire bulls, the red hair colour begins to darken at 12–18 months of age from front to rear until at maturity it is almost black. The white hair on the body remains unchanged and the black band along the back may still be discernible. This colour change is due to the influence of male sex hormones. One of the more interesting characteristics of the species which was noticed by Raffles (1816) is the change which takes place in the appearance of the male banteng after castration, the colour in a few months becoming red.

A proportion of calves (8%) are born with black rather than red hair although the white rump and legs remain the same. These are known in Balinese as *bulu indjin*. It is reported that *indjin* bulls can be distinguished by grey hair in the ears and the pigmentation of the skin of the lower jaw. In these animals both males and females have black pigmented skin which is inherited as a single autosomal dominant gene (Darmadja and Oka 1974). Pigmented males do not change colour on castration. A few mature females possess scattered patches of white interspersed with their red hair and are known as *bulu tultul*. They are considered off-type and are not usually used for breeding.

Whatever its colour, the hair is fine, short and glossy and the skin is partially pigmented. The hairs lie parallel to the skin and the coat thus offers resistance to solar radiation while allowing maximum air movement (Moran 1973). The skin of Bali cattle was studied by Jenkinson (personal communication) during a world survey of the sweat glands of cattle breeds: from these comparisons, the volume of the sweat glands appeared to be small ($1.92 \pm 0.91 \mu^3 \times 10^6$) – second only to the Hallikar cattle of India. The length of the sweat gland was also short (518μ). The cutaneous evaporation rate in Bali cattle is higher than in Brahman crossbred cattle suggesting differences in methods of heat loss (Moran 1973). Bali cattle were also found to utilize water most efficiently when compared to water buffalo, Shorthorn and crossbred cattle (Moran *et al*. 1979) and to lose less body weight (4.9%) than cattle (6.1%) when dehydrated (Siebert and MacFarlane 1975).

Bali cattle are smaller than European cattle and also smaller than the larger types of zebu. Calves weigh 15–17 kg at birth (range 12.6–18.0 kg) and during the first 6 months grow more slowly than other breeds, averaging 250–350 g per day in the first year. The growth rate of females is lower than that of males from 8 months of age onwards. In Australia rates of 0.22–0.27 kg per day were only about one-third those of Brahman–Shorthorn crossbred calves (Kirby 1979). Although growth is slow, they achieve satisfactory mature weight: females weighing 248–315 kg and males 412.5 ± 48.8 (Payne and Rollinson 1973, Copland, 1974, Subandriyo *et al.* 1979)

Adaptation to poor-quality feed, response to supplementary feeding and conformation make Bali cattle useful meat producers. Carcasses produce better hindquarters than forequarters and the disposition of fat is largely mesenteric (60–70%), the remainder being subcutaneous. The carcass fat is yellow in colour and there is virtually no marbling of the meat and little or no intramuscular fat. Carcass composition data show a surprisingly high killing-out percentage for average cattle from peasant holdings of about 55–57 per cent. The meat : bone ratio is in the range 85 : 15 to 87 : 13. The meat is dark red in colour and appears to be more elastic than ordinary beef.

Bali cattle are not usually milked and their milk production is low. Payne and Rollinson quoted an average production of 3 litres per day but Kirby (1979) measured milk production by a tritiated water technique at 1.87 litres per day. Milk consumption of Bali calves was only 40 per cent of that of Brahman-cross calves, relative to body weight, although efficiency of conversion of milk to body solids or live weight was similar.

Reports of age at sexual maturity vary from 13 to 18 months for female Bali cattle and to 21 months for males. Bulls are first used for service at about 3 years of age and there is usually one bull available for 30 females. Under village conditions (Payne and Rollinson 1973, Copland almost entirely by smallholders each owning one and rarely more than two animals, and breeding bulls are used on a communal basis.

Bali cattle have a reputation for high fertility and average calving rates of 80–90 per cent are achieved both under traditional and experimental conditions (Payne and Rollinson 1973, Copland 1974, Kirby 1979). The gestation length is about 286 days (Davendra *et al.* 1973).

The oestrous cycle is apparently of the same duration as in other cattle but oestrus is said to last for up to 48 hours, which may explain the high fertility of the breed. Age at first calving is about 30–36 months in Bali although environment greatly affects this. In Australia, Bali heifers do not normally conceive under 36 months of age, while in Sabah the average age of first calving was 32.4 ± 4.8 months (Copland 1974). Similarly, there is a wide variation in the reported calving interval from 328 ± 155 days to 582 days. The average Bali cow in Bali produces six calves in her lifetime. Multiple births are rare (Payne and Rollinson 1973).

Wild banteng are reported to mate in September and October, calves being born in May, June or July. In Australia, Bali cattle mate in September to December and calve from June to September (Kirby 1979).

The number of chromosomes of Bali cattle ($2n = 60$) is the same as that of other cattle types (Kirby 1979)

Bali cattle respond well to the close handling and attention that is practised by smallholders with one to three animals and they can readily be trained. Under extensive herd management they tend to be timid by nature, nervous and easily upset and this temperament causes losses from injury, especially among young stock when mustered or disturbed. Animals which have become accustomed to handling will revert to the shy nervous temperament in less than 3 weeks. The bond between dam and calf is weak and predisposes to cross suckling of calves and failure to protect them from predators (Kirby 1979).

Geographical distribution and economic importance

Bali cattle occur principally in the Indonesian archipelago where in 1975 the population numbered some 1.4 million or 14.5 per cent of the total cattle population. Bali cattle now make up virtually the entire cattle population of the islands of Bali and Timor and are also found in large numbers in south and southeast Sulawesi (Celebes) and the island of Lombok and in smaller numbers in east Java and in Sumatra (South Sumatra and Lampung Provinces), and in east Kalimantan (Borneo) (see Fig. 3.1).

Small herds are also present in Malaysia, in the Philippines and at the Coastal Plains Research Station near Darwin in northern Australia. A feral population of Bali cattle of perhaps 1000–2000 still exists in the Cobourg Peninsula of northern Australia.

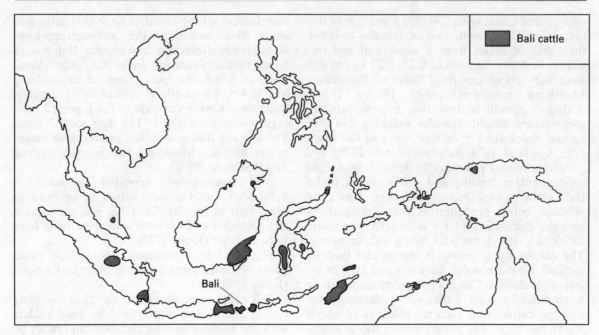

Fig. 3.1 Current distribution of Bali cattle

Bali cattle have considerable economic importance in the islands in which they predominate, providing a high proportion of all animal traction for food crop (rice) production. Their moderate size and great agility make them particularly suitable for work in small fields of irregular shape, and they are considered to be easy to train for work. Training commences when they are about 2 years of age, and they are normally worked in pairs for about 5 hours per day. One pair can carry out all the cultivations on 3 ha of land.

An export trade in live Bali cattle from Indonesia for slaughter in Hong Kong started to develop in the late 1940s and reached a peak in 1973 when 63 000 head were shipped. Only castrated males exceeding 350 kg were exported and supplies were obtained mainly from the islands of Bali, Lombok and Timor. This trade, in addition to earning foreign exchange, provided smallholders with an incentive for livestock production. Exports have, however, declined considerably in recent years (17 400 head in 1977) due to the combined effects of increasing meat prices in Indonesia and reduced prices of Australian cattle in Hong Kong.

Wild ancestor

The wild ancestor of the Bali cow is the banteng, *Bos javanicus*, of which three subspecies are found in southeast Asia: *B. j. birmanicus* on the mainland, *B. j. lowi* in Borneo and *B. j. javanicus* on Java. A fourth subspecies *B. j. porteri* from central Thailand is not considered a valid subspecies (Lekagul and McNeely 1977).

The main difference between the banteng and Bali cattle is one of size. Banteng are heavier (600–800 kg) than Bali cattle (250–450 kg); the skull of the Bali is relatively broader and the horns are much shorter and thicker, turning out at the tips. The banteng skull measures 500–550 mm in length compared to the Bali at 434–514 mm.

Reciprocal crosses between banteng and common cattle are possible. Female hybrids are fertile but male hybrids are usually sterile, although fertile males occasionally occur (Gray 1972). The Y chromosome is apparently identical with that of European cattle rather than with that of the zebu and is submetacentric. The sterility of F_1 male hybrids is caused by structural differences in the Y chromosome, resulting in disturbances of meiosis (Steklenev and Nechiporenko 1979) sperm development ceasing at the

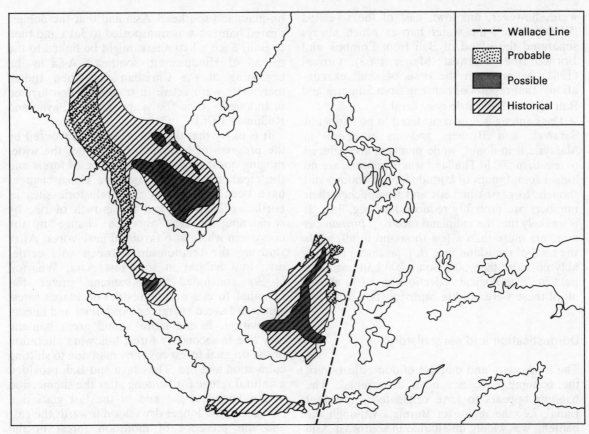

Legend:
- - - Wallace Line
⋯⋯ Probable
■■ Possible
▨▨ Historical

Fig. 3.2 Historical and current distribution of the wild banteng

secondary spermatocyte stage (Kirby 1979). Infertility also exists in one-quarter and three-quarter Bali bulls. No problems were found in mating Bali bulls to Brahman/Shorthorn cows, but attempts to mate Brahman/Shorthorn bulls to Bali cows at first proved unsuccessful, the breeds remaining voluntarily segregated in the paddocks (Kirby 1979). The Madura cattle are thought to be derived from crosses between the Bali and/or the banteng and either Indian or Javanese cattle (Payne and Rollinson 1976). Crossbreeding between domestic cattle and wild banteng was certainly an accepted method of husbandry in the early 19th century for Raffles (1816) noted that nondescript domestic cows in Java were sometimes driven into the forest in the hope of their mating with wild banteng in order to improve the breed.

There is no record of banteng west of the chain of hills which runs southwards from Nagaland through Manipur to the Arakan Yoma. Banteng formerly ranged from Burma, eastwards through Thailand, central and southern Indochina, northern Malaysia, Borneo and Java (see Fig. 3.2). The limit of spread in a southeasterly direction was determined by the rise and fall in sea level during the Würm glacial times (70 000–40 000 years ago) of the Pleistocene epoch. During the glacial periods Malaysia and the islands of the Sunda Shelf were connected by dry land and could have permitted migration of the banteng from Malaysia to Sumatra, Borneo, Java and Bali. Various authors mention the prehistoric existence of banteng on the islands of Sumatra and Bali (Meijer 1962, Halder 1976), although it is conceivable that the large connecting areas of tropical rainforests impeded extensive and long-lasting settlement in the Malay peninsula and Sumatra (Halder 1976). Banteng were, however, unknown east of the so-called 'Wallace line', the water barrier which always separated the island of Bali from Lombok and Borneo from Sulawesi (Meijer 1962). Groves (1981) believes, on the basis of skull examinations, that reports of banteng from Sumatra and Bali refer to Bali cattle gone feral.

The banteng is now considered to be extinct in Sarawak and Brunei, and its presence in Malaysia is in doubt, while numbers are reduced to less than 500 in Thailand where banteng are no longer found south of latitude 8°N. Although still thought to persist in Laos and Kampuchea their numbers are probably reduced (see Fig. 3.2). It is unlikely that the mainland race *B. j. birmanicus* numbers more than a few thousand in all, while the total of the island race *B. j. javanicus* is probably no more than 500. Some 200 banteng were held in 38 zoological collections in 1976; almost all of these were bred in captivity (IUCN 1978).

Domestication and early history

The time, place and purpose of domestication of the banteng have not been determined. The banteng appears to have originated in 'further India', i.e. the region of Burma. Although the banteng was widely distributed in southeast Asia it is only on the island of Bali that domesticated Bali cattle became established as the predominant breed.

Domesticated cattle were apparently present at Non Nok Tha, a prehistoric site in northeast Thailand, which was occupied at least by 3500 B.C. and possibly earlier, and some of the bones could have belonged to banteng; the evidence is by no means conclusive since the bones were very similar to those of the zebu (Higham and Leach 1971). There can, however, be little doubt that domesticated cattle were present at several sites in northeast Thailand earlier than the domestication of the water bufflo in that region around 1600 B.C. It is also thought that domesticated cattle were not then used for traction (Higham 1979).

Since convincing evidence of the presence of wild banteng on the island of Bali in prehistoric times is lacking, it seems likely that domestication of the banteng occurred in Java and the domesticated variety was later taken to Bali. Nevertheless it is also possible that domestication occurred

on mainland southeast Asia and that the domesticated banteng was transported to Java and then to Bali. Such a hypothesis might be linked to the spread of Hinduism in southeast Asia in the beginning of the Christian era when Indian merchants were active in trade and first arrived in Indonesia some 500 years ago (see Payne and Rollinson 1976).

It is likely that domestication was favoured by the progress of human societies from the wide-ranging quest for food to clearance of forest and the creation of open grassland. These changes have been detected in the prehistoric sites of northeast Thailand, and the growth of rice by swiddening brought about a change in the ecosystem which also favoured herbivores. After studying the relationship between wild cattle, man and habitat in southeast Asia, Wharton (1969) concluded that banteng prefer flat savannah forests within the drier deciduous forest zones and avoid evergreen rainforests and human population. In the more humid areas banteng may live in secondary forest following alteration of the original forest cover by man due to shifting cultivation and fire. Thus Java and Bali provided a natural refuge for banteng after the submersion of Sundaland at the end of the last glaciation because of the longer dry season towards the east and the presence of monsoon forest in the lowlands. These conditions were not found in Borneo, Sumatra or Malaysia.

Bali cattle are now used for traction, riding and meat but if Higham is correct that the early domestic cattle in northeast Thailand were not used for traction, then riding and pack carriage may have been a significant reason for domestication, meat being a useful by-product.

Recent studies on the haemoglobin types of Bali cattle indicate that there is a high frequency of Hb B, a fairly high frequency of Hb C and a complete absence of Hb A (Bachman *et al*. 1978, Namikawa and Widodo 1976).

Recent history and present status

The uniformity of Bali cattle suggests that selection for type has been consciously applied for some hundreds of years and they appear to have been developed through a long domestication period, since prehistoric times. One important factor in recent years has been an absolute ban on introduction of exotic cattle into the island of Bali and the prohibition of crossbreeding. When this regulation was introduced is uncertain, but organized selection of Bali cattle was implemented about 40 years ago. Inspections in each district were followed by inspection of the best bulls in the area in the hope that only well-selected animals would be entered in the register of prize bulls. This basic system is still continued. Another factor which has had some influence over the last 20–30 years has been the imposition of live-weight limits on cattle for export and slaughter. Entire males are not permitted to be exported since organized selection was started.

The nucleus of a pure breed thus established has been used to introduce Bali cattle to other parts of Indonesia and considerable spread has taken place in the last 150 years. Cattle from Bali were introduced into Sulawesi in 1890 and further importations were made between 1920 and 1927. They were introduced into central and west Java from about 1907 and to Timor between 1912 and 1920. Bali cattle were imported from Indonesia to Singapore in the late 1930s for slaughter and they first entered Malaysia as animals escaping slaughter. Subsequently eight heifers and two bulls were imported into West Malaysia in 1951 (Davendra *et al.* 1973). Offspring from this nucleus were supplied to several states and also to Sabah, Sarawak and Brunei. A single group of twenty Bali cattle was imported from Badung in Bali to the Port Essington settlement on the Cobourg peninsula of northern Australia in January 1849 (Calaby 1975) but within the year the settlement was abandoned and the cattle were allowed to fend for themselves. They became feral and were not rediscovered until 1960. In the intervening period of 111 years this herd grew considerably but is unlikely to have exceeded 3000 head. The feral herd in the Cobourg peninsula is now estimated to number approximately 1000 head (Kirby 1979).

There is little information about other feral herds or individuals but some feral Bali cattle are said to exist in the Sanghir Islands, northeast of Sulawesi and some are also present in wildlife reserves in Java (Groves, personal communication). The compatibility of feral Bali cattle with their environment has been noted by several authors (see Kirby 1979).

Future prospects

Bali cattle possess a number of significant characteristics, notably high fertility, good dressing percentage, minimal fat deposition, efficient water utilization, ability to thrive in hot humid environments and to maintain body condition on poor-quality pastures. While still very important as draught animals for which they are well suited in very small fields, and where their lighter weight causes less risk of damage to bunded slopes, there is strong interest in Bali cattle for meat production in places where climate and feed supply are not suitable for exotic breeds. There is considerable interest in crossbreeding of Bali cattle to take advantage of some of their excellent qualities. It is to be expected that these characteristics will be studied and exploited for many years to come.

References

Bachman, A. W., Campbell, R. S. F. and **Yellowlees, D.** (1978). Haemoglobins in cattle and buffalo. *Australian Journal of Experimental Biology and Medical Science*, **56** (5): 623–9

Calaby, J. H. (1975) Introduction of Bali cattle to northern Australia. *Australian Veterinary Journal*, **51**: 108

Copland, R. S. (1974) Observations on banteng cattle in Sabah. *Tropical Animal Health and Production*, **6** (2): 89–94

Darmadja, D. and **Oka, L.** (1974) Melanism and its inheritance in Bali cattle. *1st World Congress on Genetics applied to Livestock Production*, Vol. 3, Contributed Papers, 115–19

Davendra, C. T., Choo, L. K. and **Pathmasingam, M.** (1973) The productivity of Bali cattle in Malaysia. *The Malaysian Agricultural Journal*, **49** (2): 182

Gray, A. P. (1972) *Mammalian Hybrids. A check-list with bibliography.* Commonwealth Agricultural Bureaux: Farnham Royal, Bucks, England

Groves, C. P. (1981) On the agriotypes of domestic cattle and pigs in the Indo-Pacific Region. *Proceedings of the Indo-Pacific Prehistory Conference*, (1979), Puna, India

Halder, U. (1976) Ökologie und Verhalten des Banteng (*Bos javanicus*) in Java. *Zeitschrift für Säugetierkunde*, **10**: 9–124

Higham, C. (1979) The economic basis of prehistoric Thailand. *American Scientist*, **67**: 670–9

Higham, C. F. W. and **Leach, B. F.** (1971) An early centre of bovine husbandry in southeast Asia. *Science*, **172**: 54–6

IUCN (1978) *Red Data Book*, Vol. 1: *Mammalia*. International Union for Conservation of Nature and Natural Resources: Gland, Switzerland

Kirby, G. W. M. (1979) Bali cattle in Australia. *World Animal Review*, No. 31: 24–9

Lekagul, B. and **McNeely, J. A.** (1977) *Mammals of Thailand*. Association for the Conservation of Wildlife: Bangkok

Meijer, W. C. P. (1962) *Das Balirind*. Die neue Brehm-Bücherei No. 303, Ziemsen Verlag: Wittenberg

Moran, J. B. (1973) Heat tolerance of Brahman cross, buffalo, banteng and Shorthorn steers during exposure to sun and as a result of exercise. *Australian Journal of Agricultural Research*, **24**: 775–82

Moran, J. B., Norton, E. W. and **Nolan, J. V.** (1979) The intake, digestibility and utilisation of a low-quality roughage by Brahman cross, buffalo, banteng and Shorthorn steers. *Australian Journal of Agricultural Research*, **24**: 775–82

Namiakawa, T. and **Widodo, W.** (1976) Electrophoretic variations of haemoglobin and serum albumin in the Indonesian cattle including Bali cattle (*Bos banteng*). *Japanese Journal of Zootechnical Science*, **49** (11): 817–27

Payne, W. J. A. and **Rollinson, D. H. L.** (1973) Bali cattle. *World Animal Review*, No. 7: 13–21

Payne, W. J. A. and **Rollinson, D. H. L.** (1976) Madura cattle. *Zeitschrift für Tierzüchtung und Züchtungsbiologie*, **93** (2): 89–100

Raffles, T. S. (1816) *The History of Java*. Historical Reprints, Oxford University Press: Kuala Lumpur, 1965, 2 vols

Siebert, B. D. and **MacFarlane, W. V.** (1975) Dehydration in desert cattle and camels. *Physiological Zoology*, **48** (1): 36–48

Steklenev, E. P. and **Nechiporenko, V. Kh.** (1979) [Chromosome complements of hybrids of the banteng (*Bos (Bibos) javanicus*) with domestic cattle (*Bos (Bos) taurus typicus*).] *Tsitologiya i Genetika*, **13** (1): 31–3. (In Russian: summary in English)

Subandriyo, P. S., Zulbardi, M. and **Roesyat, A.** (1979) Performance of Bali cattle as work animals and milk and beef producers. *Indonesian Agricultural Research and Development Journal*, **1** (1/2): 9–10

Wharton, C. H. (1969) Man, fire and wild cattle in southeast Asia. *Proceedings, Annual Tall Timbers Fire Ecology Conference*, No. 8, Mar. 14–15, 1968, Tallahassee, Florida, pp. 107–167

4
Gayal or mithan

F. J. Simoons

Department of Geography, University of California, Davis, U.S.A.

Introduction

Nomenclature

The terms 'gayal' (Bengali and Hindi) and 'mithun' (Assamese) are both used to refer to a domesticated bovine of the hill regions of eastern India, Bangladesh, and adjacent Burma. In English, the spelling 'mithan' is more common. Considerable confusion has resulted from the use of two distinct names for the same animal, and some writers have concluded that two animals are involved and not one. The consensus of informed scholars today, however, is that there is a single domesticated species, by whatever local names it may be called.

The mithan is placed by taxonomists in the genus *Bos*. Many place it in the subgenus *Bibos*, along with one, and sometimes two or three, wild southeast Asian bovines: the gaur (*Bibos gaurus*), banteng (*Bibos javanicus*) and kouprey (*Bibos sauveli*), as well as the domesticated form of banteng, Bali cattle. Though the designation *Bos frontalis* is commonly used for the mithan, the animal is also sometimes called *Bibos frontalis, Bos gaurus frontalis*, or, by those who view *Bibos* as a subgenus of *Bos*, *Bos (Bibos) frontalis* (Bohlken 1961).

General biology

The mithan is strongly built. Bulls average about 150 cm at the shoulder and cows several centimetres less. The average adult weighs 400–500 kg (Tamhan *et al.* 1977). Young animals are brown but darken with age. Adult cows are typically dark brown in colour and adult bulls black. Both young and adults have white stockings on all four legs. Piebald or white mithan also occur. The mithan has a small dewlap, and a pronounced dorsal ridge on the crest of its shoulders. Its horns are of considerable girth. They extend mainly outward and are straight or gently curving. Horns of adult males average about 35 cm in length.

In most of mithan country there is no deliberate selection within the mithan species (Simoons and Simoons 1968), and no mention of breeds of mithan. Deliberate breeding of mithan and common cattle does occur in Bhutan, and the mention of distinctive types or breeds of mithan in Bhutan and Arunachal Pradesh (Tamhan *et al.* 1977) suggests that selective breeding within the mithan species may also occur in these territories.

The mithan prefers cool, forested places, and is typically found at elevations from 600 to 2700 m. It is both browser and grazer but given an opportunity it seems to prefer to browse. It has a strong craving for salt, and has an oily sweat that seems to serve as an insect repellent.

Males are capable of reproducing at 1 year of age, but full maturity of behaviour comes later, after 27 months of age. In the female, first oestrus can occur at an age of roughly 15 months (Scheurmann 1975) but first calving is recorded as 42–48 months by Tamhan *et al.* (1977). The oestrous cycle averages about 28 days, with the duration of oestrus from 1 to 4 days. Oestrous phenomena generally resemble those of European cattle. The gestation period is 293–303 days, with oestrus returning 21–53 days after birth (Scheurmann 1975). The newborn calf weighs 19.5–33 kg.

The mithan has 2n = 58 chromosomes (Fischer 1969).

Distribution and numbers

The mithan has a quite limited geographic distribution: the north-south hills and low mountains that separate India and Bangladesh from Burma, and the east-west highlands that lie below the high Himalayas of easternmost India (see Fig. 4.1). It also occurs in small numbers in the hills of northern Burma. At lower elevations the principal bovines are common cattle and water buffalo; at higher elevations in the Himalayas, yak and yak/cattle hybrids.

In view of the isolation of mithan country and the incompleteness of census data, one cannot be certain of its total numbers. Using available data for India and Burma (J. L. Anderson, personal communication to I. L. Mason; Swaminathan 1981) and making allowances for other areas, one might estimate that the total mithan population is at least 150 000. More than half the total numbers are found in the Indian Union territory of Arunachal Pradesh (the former North-East Frontier Agency).

Wild ancestor

There have been three hypotheses concerning the origin of the mithan. One is that it is a domesticated gaur. The second is that it is a hybrid resulting from the crossing of bull gaur and zebu cow. The third is that it is descended from a wild Indian bovine that is now extinct. The gaur, whose numbers and territory have been greatly reduced by hunting in recent centuries, formerly occurred in much of India and across mainland southeast Asia as far as Indo-China and Malaya. Thus the mithan's distribution is well within that of the gaur, though not of the other southeast Asian bibovines. The mithan, moreover, resembles the gaur in many ways, though the latter is a larger, more massive animal (average bulls: 170–175 cm at the shoulder). Where gaur have been present in the nearby forest, male gaur have bred freely with female mithan, and there seems to be no sterility barrier between the two.

Common cattle (zebus) have been rare in the heart of mithan country until recently, though they sometimes breed with mithan. On the other hand, there is little in the appearance of the pure-bred mithan that resembles the common cattle of the area. There is also a sterility barrier between common cattle and mithan; female hybrids are fertile, but male hybrids are usually sterile (Gray 1972).

Recent studies have shown the mithan to differ from zebu cattle in haemoglobin genotype (Lalthantluanga *et al.* 1975a, b; Lalthantluanga

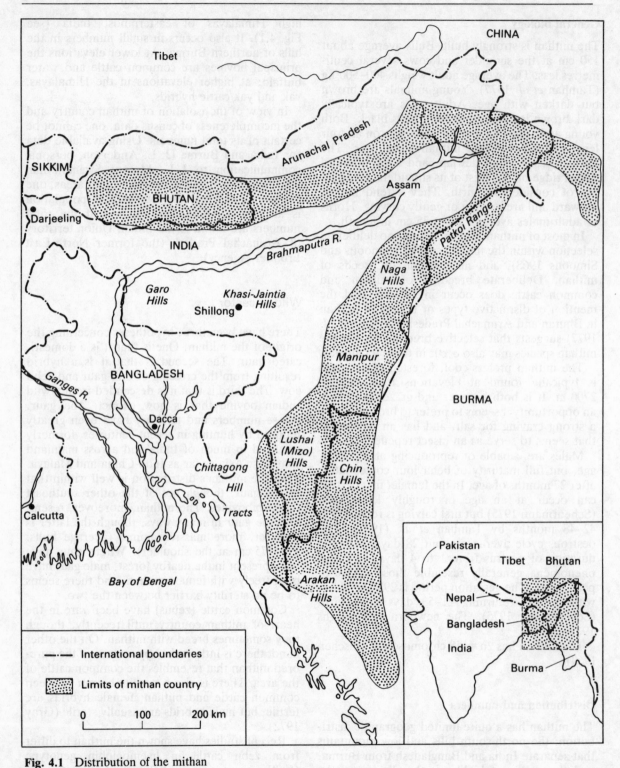

Fig. 4.1 Distribution of the mithan

and Barnabas 1974), and to be different from European cattle but identical with gaur in karyotype (Fischer 1969). Considering also the native traditions of origin and the absence of any evidence that a separate wild form of mithan ever occurred, one may reasonably conclude that the mithan is a domesticated gaur and is descended entirely or mainly from one race of gaur, *Bos gaurus gaurus*, of India and Burma.

Domestication and early history

There is no clear historical record of the mithan before the arrival of Europeans. In addition, the early archaeology of mithan country remains *terra incognita*. In attempting to reconstruct the history of domestication, which presumably occurred within the animal's present-day range, one is left mainly with native peoples' traditions, and observations of present-day practices from which certain deductions may be made.

Native traditions are that the mithan is an early domesticate, that it derives from a related wild animal, that it was attracted to man by its craving for salt, and that human motives were to obtain an animal for sacrifice and to provide flesh. One suspects that the gaur, a forest animal, first came into contact with humans around their agricultural clearings, the often-isolated cultivated plots cut from a heavy forest cover. Then, as may have been the case with the reindeer, the mithan was drawn into a more permanent association by humans who recognized its craving and regularly provided it with salt.

In this regard, one notes that most modern mithan keepers keep mithan under systems of free ranging, with animals permitted to go untended about the forest. With some peoples, especially in the south, mithan go back to the village each evening and are fed salt and confined for the night. With other groups, they remain on their own in the forest for lengthy periods – weeks or months. To retain links with his mithan under such conditions, a man from time to time enters the forest, calls his animals, and feeds them salt.

There is no reason to doubt that such free ranging has characterized mithan keeping since antiquity or that then, as is generally the case today in mithan country proper, mithan owners made no deliberate efforts at breeding their animals. Thus the evolution of the species would have come about largely from unplanned selective pressures within the system of free ranging, and from human choice of animals for slaughter. Under free ranging in former times there was ample opportunity for regular interbreeding with wild gaur. Such interbreeding would tend to perpetuate the ancestral size, power, and wariness. At the same time, free ranging would permit those mithan who were so inclined – perhaps those animals best able to fend for themselves against predators – to return to the wild. This would have a contrary effect, leading those mithan remaining in a domesticated state to be smaller and more amenable to human control. Within that domesticated group, moreover, the animals likely to be sacrificed first would be the larger and more difficult ones. In any case, the mithan today is notably smaller and less powerful than its wild ancestor; it differs in horn shape and size, and in certain other anatomical features; and it has evolved into the gentlest of all bovines. Interbreeding with common cattle, now commonplace, particularly in Bhutan and in the mithan's southern range, further encourages the mithan in evolutionary divergence from its wild ancestor.

Another present-day observation that may be relevant in the matter of domestication is the tendency for mithan to be owned by village chiefs and other important, well-to-do men. This is partly because the animals often damage cultivated fields, and the owner must be able to pay for such damage. If such a concentration of ownership has prevailed since antiquity, men of wealth and influence may have played a special role in the domestication process. They would not only have been better able to bear the costs of ownership, but they would have had a special need for large, impressive sacrificial animals, such as the mithan. For ordinary men, pigs, dogs, and chickens would have sufficed.

Recent history and present status

Most peoples of mithan country are tribal, and through history have largely remained distinct from the neighbouring lowland populations, whether the Buddhists of Burma, Hindus of Assam, or Moslems of Bangladesh. Most mithan people are swidden cultivators who use hand implements for cultivation. Such groups had no need for plough animals, and the mithan is not as a rule used for this purpose in the heart of its distribution. Most groups of the area, moreover, are of non-dairying tradition (Simoons 1970), and do not milk their mithan. The animal's primary traditional use is for sacrifice, at which time its flesh is eaten.

The role of the mithan in sacrifice differs in detail from group to group, but in general this animal as well as others is offered to a variety of unseen supernatural beings, great and small, good and evil, to protect man, to assure as far as possible an existence free of catastrophe, to eliminate illness, and to provide for his continuing fertility and that of his animals and crops. The occasions requiring animal sacrifice are numerous and varied (Simoons and Simoons 1968).

Ownership of mithan confers prestige, and many wealthy men are quick to inform the visitor about the numbers of mithan they own. In addition, the sacrifice of mithan on important occasions, social and religious, is usually commemorated by various forms of public display. The skull and horns of sacrificed mithan are commonly displayed on the verandah wall or elsewhere in or near the house. Some groups allow the sacrificial posts to which the animals were tied to remain standing; since one post is erected for each animal slaughtered, the status of the house owner is clearly revealed. Some groups also permit the sacrificer certain rights of house decoration, such as horns mounted on the roof. They may also allow a man rights of personal display, as the right of the sacrificer and members of his family to wear cloths of particular patterns. Against this background, it is understandable that mithan skulls and horns are prominent among the art motifs of mithan-keeping peoples, and that they symbolize wealth.

Besides the great value of the mithan in sacrifice and the prestige structure, it is utilized by many of the hill peoples for payment of social and legal obligations. Throughout the area, fines for violation of societal norms are commonly assessed in mithan, and ransom and tribute were assessed in mithan too.

The mithan also figures in systems of exchange, usually as the highest value in the traditional tables of barter. Also, for some groups, it is a notable item of trade. Trade is particularly important between Arunachal Pradesh and Bhutan; certain peoples of Arunachal Pradesh who are rich in mithan, trade their animals westward to the Bhutanese; the Bhutanese have numerous common cattle, but need mithan bulls for crossing, to obtain sturdy plough oxen.

In the southern range of the mithan, from the Naga hills southward, the most interesting institutions involving mithan are the Feasts of Merit. These are a series of graded feasts carried out over a period of several years. Performance of the feasts, the most important of which require mithan sacrifice, permits a man to advance in social and economic position, and it raises his status in the afterworld. Equally important are the rights of display (see above). Among various Naga peoples the performance of the feasts is important to village well-being, too, for it is believed that the fertility of the feast-giver is transmitted to the village crops, cattle, and women. Involved here is that megalithic complex which is believed formerly to have extended from the Naga hills into southeast Asia. Even though to some Western observers the mithan may seem to serve few purposes they deem 'useful', it is very useful in an economic sense within the context of the cultures where it is found.

Future prospects

In many places, contact with the outside world has led to changes in traditional belief systems in which mithan played so prominent a role. In part, this derived from the influence of Christian missionaries, and in part from education and broader national contacts, especially in the post-colonial period. In any case, the need for keeping mithan for prestige and sacrificial purposes is being undermined.

In a hopeful development, the government of India has recently set up a national centre for mithan study in Arunachal Pradesh (Swaminathan 1981). This could lead to the development of superior types of mithan that not only better compete against other bovines in the Indo-Burman hill country, but that may come to be introduced to similar ecological niches in other parts of the world.

References

Bohlken, H. (1961) Der Kouprey, *Bos (Bibos) sauveli*

Urbain 1937. *Zeitschrift für Säugetierkunde*, **26**: 193–254

Fischer. H. (1969) Die Chromosomensätze des Bali-Rindes (*Bibos banteng*) und des Gayal (*Bibos frontalis*). *Zeitschrift für Tierzüchtung und Züchtungsbiologie*, **86**: 52–7

Gray, A. P. (1972) *Mammalian Hybrids. A check-list with bibliography*. Commonwealth Agricultural Bureaux: Farnham Royal, Bucks, England

Lalthantluanga, R. and Barnabas, J. (1974) Hemoglobin alpha chain allelic variants in gayal (*Bos gaurus frontalis*). *Folia Biochimica et Biologica Graeca*, **11**: 65–9.

Lalthantluanga, R., Gulati, J. M. and Barnabas, J. (1975a) Hemoglobin genetics in bovines and equines. *Indian Journal of Biochemistry and Biophysics*, **12**: 51–7

Lalthantluanga, R., John, M. E., Barnabas, S. and Barnabas, J. (1975b) Molecular forms, comparative structure and evolutionary trends of hemoglobins in the descent of *Bovinae* and *Caprinae*. *Indian Journal of Biochemistry and Biophysics*, **12**: 331–9

Scheurmann, E. (1975) Beobachtungen zur Fortpflanzung des Gayal, *Bibos frontalis* Lambert, 1837. *Zeitschrift für Säugetierkunde*, **40**: 113–27

Simoons, F. J. (1970) The traditional limits of milking and milk use in southern Asia. *Anthropos*, **65**: 547–93

Simoons, F. J. and Simoons, E. S. (1968) *A Ceremonial Ox of India: the mithan in nature, culture, and history*. University of Wisconsin Press: Madison

Swaminathan, M. S. (1981) Keynote address. *Animal Genetic Resources Conservation and Management*. FAO Animal Production and Health Paper No. 24: 6–14. FAO: Rome

Tamhan, S. S., Prasad, M .C., Ghosh, S. S. and Roy, D. J. (1977) Mithun: the animal of the N. E. hills. *Indian Farming*, **27** (6): 27–29

5
Yak

J. Bonnemaire

*Ecole Nationale Supérieure des
Sciences Agronomiques Appliquées,
Dijon, France*

Introduction

Nomenclature

The yak is a type of long-haired and bushy-tailed cattle which lives in the high mountains of central Asia and especially on the Tibetan plateau. It belongs to the subfamily Bovinae of the family Bovidae and in the past it was often regarded as a separate genus or subgenus, *Poephagus*, close to the genera *Bison* and *Bos*. The name *poëphagus* was first used by Aelianus in the 3rd century A.D. and means 'grass eater'. Linnaeus, who knew only the domestic yak, named it *Bos grunniens* on account of its grunt. Przewalski, from his travels in Tibet, described the wild yak that emits the characteristic grunt only during the mating season and added the term *mutus* because of its usual muteness.

It was these two names that were finally kept – *Bos mutus* for the wild yak and *Bos grunniens* for the domestic yak (Corbet 1978). In fact, these two animals are interfertile and we can consider that they belong to the same species.

The adoption of the word 'yak' (from Tibetan *g-yag* = domestic yak) in naming this animal in western languages owes its origin to Turner's travel to Bhutan and Tibet in 1783 (Boulnois 1976). But it is only one of the names given to the yak in different central Asian and East European languages. Thus, in Tibetan, the wild yak is called *bron* (male) or *bron-bri* (female) whereas the domestic female is called *bri-mo*. Bazin (1970) notes that before the 11th century, Turkic people from the mountainous areas neighbouring Tibet called the wild yak *qotuz* or *qotäz* (the bulky, the massive), a word which subsequently spread from what is today Chinese Turkestan to Anatolia; then, having developed its breeding in the north of central Asia, breeders called the yak *sarliq* (the butter producer). These terms, as well as breeding techniques, were then taken over by the Mongolians who were the Turks' eastern neighbours in the 13th century; lastly, more recently, the Kirgiz changed the name of this animal into *topos* or *topoz* (the all round). As long ago as the 18th century, the Chinese used the term *mao niu* ('hairy cattle') to designate the yak.

General biology

The yak shows a general morphology relatively close to that of other Bovinae, which caused naturalists to hesitate for a long time about its right position in zoological classification. For instance, in 1760 Gmelin called it 'Vacca grunniens, villosa, cauda equina', a description repeated by Buffon in 1767 who put it together with the bison and called it 'cow of Tartary', while Pallas, in 1777, described it as a 'horse-tailed buffalo'.

In fact, the yak at first sight looks like a common hardy breed of cattle of rather small size. But it is distinguishable from them by specific characters, especially the emission of a distinctive grunt and some morphological features. Its body is fairly long, rather short-legged, compact and with particularly developed forequarters (the yak has fourteen or fifteen pairs of ribs – like the bison – instead of the thirteen of domestic cattle). It has a very characteristic bushy tail with very long profuse soft silky hair, something like the tail of a horse. There is a lateral fringe of fairly long hair (20–60 cm) on the flank and the lower part of the body. The withers are high and prominent on account of elongated neural spines. The short slender neck is without dewlap. The head is relatively heavy and bulky,

with characteristic curved horns. The udder is small and is covered with hair.

The yak is unfavourably affected by high air temperatures but it shows an outstanding adaptation to living at a high altitude in the open air, all round the year under a very cold and rather dry climate, and with times of very poor feed. The following characteristics are related to this adaptation: plentiful coat; ability to move easily across all types of country; ability to graze very close to the ground; thick skin; large thoracic cavity; respiratory exchange potential higher than in cattle (enlarged alveolar area; larger, more numerous erythrocytes with higher haemoglobin content); ability to scoop out the snow for feeding during the winter (Denisov 1958). It has been noted in the Kirgiz mountains that the percentage live-weight loss during the winter was less in yaks than in local cattle and sheep on the same pastures.

Sexual maturity of yaks comes rather late (Denisov 1958). Under traditional management, domestic females reach puberty between 16 and 40 months of age and males at about 2 years of age; with improved feeding, sexual maturity is reached at 14–18 months. The yak bull can mate all round the year but the cow has a seasonal anoestrus (in winter) which is more or less strongly marked according to feeding and maintenance conditions. The oestrous cycle length seems to be nearly the same as for cattle, but heat tends to be shorter and less marked. Gestation length (258 days on average, with a range 224–284 days) is significantly shorter than in cattle. Calving takes place during spring and summer. Under the usual husbandry in Tibet and the Himalayas, adult yak cows generally give birth only once every two years. The live weight of newborn calves is in the range 8–25 kg (12–18 kg as a general rule) (Ivanova 1956, Denisov 1958). Twins are very rare.

The yak has the same chromosome number ($2n = 60$) as other *Bos* and *Bison* species, and karyotypes are very similar.

Yak can be crossed with several other species of *Bos* and *Bison*, giving single hybrids as well as triple hybrids (Gray 1972).

1. Yak x domestic cattle (*Bos 'taurus'*). This is the most frequent cross, certainly because of the presence of domestic cattle all round the distribution area of the yak. Reciprocal crosses

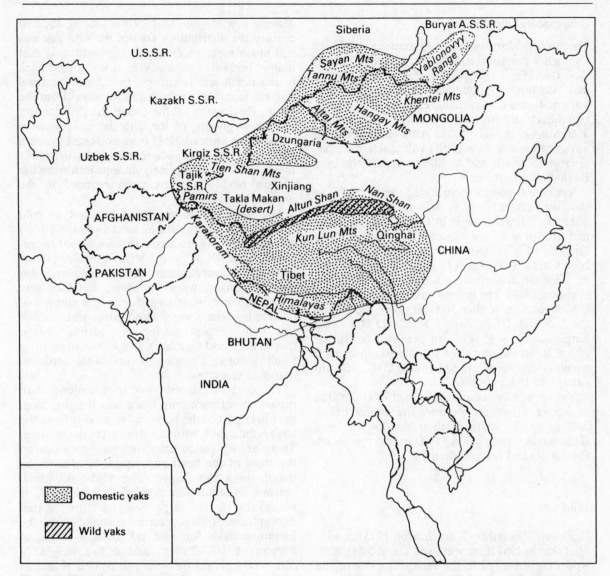

Fig. 5.1 Distribution of the yak

are possible. In these hybrids, males are sterile and females fertile. In appearance they are intermediate between the parents; they show heterosis in weight and size, and in milk production.

2. Yak x zebu cattle (*Bos 'indicus'*). The characteristics of this cross are the same as indicated above. These hybrids are widespread on the southern slopes of the Himalayas.

3. Yak x American bison (*Bison bison*). Reciprocal crosses are possible. Male hybrids are sterile and female hybrids are of low fertility. The conformation and hair of the yak seem to be dominant.

4. Yak x gayal (*Bos frontalis*). F₁ hybrids are intermediate between the parents. Male hybrids are sterile (with normal sexual behaviour) and female hybrids show normal ovogenesis.

Concerning triple hybrids, three kinds have been obtained with the domestic yak: male American bison x female yak-gayal; male European bison x female yak-gayal; and male banteng x female zebu-yak.

Geographical distribution

The area of distribution of domestic yak covers the higher parts of the mountain system of central Asia (see Fig. 5.1), especially the Tibetan plateau and surrounding mountains (Himalayas, Karakoram, Kunlun, Nan Shan, etc.) and the mountain mass which stretches from the Pamirs and the Tien Shan in the west to the Altai mountains, the Sayan mountains (where the yak meets the reindeer and camel), and to Mongolia and the lake Baykal area in the east.

Yaks are generally included with cattle in statistical tables, so it is difficult to know their numbers. However, it is in China that the large majority of yaks are concentrated, especially in Qinghai (4.6 million in 1979, according to the New China News Agency) and in Tibet, but also in Sichuan, Xinjiang, Kansu and Yunnan (Epstein 1969). The second yak-breeding country is Mongolia (more than 500 000 yaks); then come the U.S.S.R. (from Tajik S.S.R. – 15 000 – and Kirgiz S.S.R. – 75 000 – in the west to Buryat A.S.S.R. in the east) and Nepal; countries with smaller populations are Bhutan (52 000), Afghanistan, and India (18 000).

Yaks generally live at elevations of 3000–5000 m or above. They are occasionally found below 2500 m in the south of their distribution area (Himalayas) and below 1500 m in the north (Mongolia and Buryat Mongolia).

Wild yak

Bannikov (quoted by Wunschmann 1975) considered that in 1964 there were still about 3000–8000 wild yaks scattered in the remotest parts of the highlands of northern Tibet (see Fig 5.1). Other authors are more pessimistic about the reduction of this small stock under the pressure of hunters on the one hand and limitation of their territory on the other. IUCN (1974) classified wild yak as endangered, but efficient protection and control are obviously very difficult to implement in these regions.

During the last Ice Age, the wild yak existed in northern Siberia, and during the Neolithic it was hunted in the Pamirs from which it has now disappeared. Indeed in the Afghan Pamirs there is a rock engraving – probably from the Neolithic age – that shows horsemen armed with bows chasing a wild yak (Dor 1976). During the last century the distribution area of the wild yak was still much more extensive to the south and east than at present. Progressively, it was forced back to the north and northwest by Tibetan pastoral nomads from the south and Mongolians from the north. Particularly, at the end of the 19th century and the beginning of the 20th the main caravan route which crossed Tibet from northeast to south cut across the area where wild yaks originally lived. Wild yak completely disappeared from the eastern part and were much reduced in the western part.

The wild yak is very well adapted to cold (tolerating temperatures as low as −40 to −50 °C) and to a roaming life across extended poor rangelands situated at very high altitude (above 4500 m). It very much likes water-places for bathing during warm weather. Females and young animals live in large herds (the number of head can reach several hundred) whereas adult yak bulls, except during the mating season (September and October), live either alone or in small groups. Crossing occurs with herds of domestic yak cows.

The coat of the wild yak is of uniform dark brown colour with long black hair hanging from the lower part of the body and especially from the bushy tail, and with a silver-grey dorsal line. There are also such hairs with pale grey tips on the front of the head and especially around the muzzle in a thin fringe. The wild yak, which reaches its adult size towards 8 years old, is remarkable for its large size and strong sexual dimorphism. Indeed, records available in the literature show for wild yak bulls a height at withers of 1.70–2.08 m and a live weight of 550–1000 kg, and for wild yak cows a height at withers of about 1.45–1.55 m and a live weight of 300–350 kg (i.e. two or three times less than the bull's weight). The horns of the bull can reach a length of 0.8–1 m with a base circumference of 40–50 cm; they are sometimes used as milking pails by Tibetans.

Domestication and history

According to Zeuner (1963) the domestication of the yak appears to be fairly old and to have taken place when agriculture first appeared (Neolithic age), at the same time as the domestication of cattle. Epstein (1974) considers that it was effected in Tibet probably by the introduction into domestic cattle herds of wild yak calves whose dams had been killed by hunters, and then by a progressive increase of the number of yaks in these herds because of their better adaptation to the hard Tibetan climate than that of cattle. Palmieri (1976) considers that, besides this capture of young animals, taming by giving salt (that is confirmed by Tibetan tales) also played a part in the domestication of the yak; he thinks that the dependence created by the giving of salt could have taken place mostly in the southern valleys of the Himalayas (where there is no natural salt such as exists on the Tibetan plateau).

The domestic yak (Bonnemaire 1976) is slightly smaller than the wild one and the differences between the sexes are less obvious. Anyway, the size and weight of domestic yaks are very variable according to regions and food variations during the year. Full-grown males generally weigh 300–550 kg and height at withers is about 1.25 m (range: 1.15–1.35); the females weigh 180–350 kg and measure about 1.10 m (range: 1.00–1.25) at the withers.

Domestication certainly led to diversification of the colours of the coat, the breeders' choice being partly based on aesthetics. Moreover, yak hair and especially the tail, which has been an article of trade since early times was even more sought after when it was white (and hence easier to dye), and Pallas noticed that the Tibetan and Mongolian stockbreeders used to choose breeding animals chiefly according to this character. Also, this diversification of colours (very variable according to regions: white, black, brown, red, grey, tawny – self or parti-colour) might have been increased in yak by crossing with domestic cattle (Phillips *et al.* 1946). In some regions, polled yaks have been selected: in Mongolia for instance, they constitute the majority of the stock; on the other hand, they are less sought after by the Tibetans.

One of the oldest human uses for this animal (whether domestic or wild) concerned the tail (Boulnois 1976). It sold at a high price and it formerly led to a flourishing trade in countries far from the areas where the animal lived. In some languages (Sanskrit, Old Turkish, some north Indian modern languages) the same word is used for the tail and the whole animal. The tail has been used by the Romans as a fly-whisk or a duster and by the old Turks as an emblem of power or nobility (fixed on a staff); it has also been used as an aigrette in the religious symbolism of Indian and Buddhist ceremonies and as an emblem of royalty; traditionally, the wealthy Chinese used to dye it red and use it to decorate their caps.

The yak has always played an important part in Tibetan mythology: we can find it associated with mountains, with men, gods and devils, with legends on the origins of kings, with religious festivals. Besides, some of its products (butter and other dairy products, meat, heart, horns) are used in Tibetan medicine. Yak butter is used in some Buddhist ceremonies (e.g. butter lamps).

Present status

The domestication of the yak allowed it to be exploited for many purposes in the material life of central Asian pastoral societies. Yaks can easily graze where cattle and horses can walk only with the greatest difficulty. With a small cost for its maintenance, the domestic yak produces nearly as much milk as the local cattle: an average milk production of 600 kg per lactation (with a range from 120 to more than 1000 kg) with half generally sucked by the calf. The milk of the yak is richer than that of cattle: 6.5 to 7 per cent of fat (golden coloured and with large fat globules), 5.3 per cent of protein, 4.6 per cent of lactose and 17.4 per cent of total dry matter. This milk is drunk but is also used to make butter, many kinds of cheese and other milk products often with sophisticated techniques that make long preservation possible (Bonnemaire and Teissier 1976). Yak's milk and milk products have a particularly important place in Tibetan life.

Because of religious taboos about slaughter, yak meat is of minor importance throughout the Tibetan cultural area, but it is of great value in other countries such as Mongolia and the U.S.S.R. In this last country, the cost of meat production from yaks appears to be two to five times lower than that from cattle in the same mountain regions. If properly fed with milk by their dams, growing yaks weigh about 200–260 kg when they are 18 months old (varying according to sex and food); if the dam has been milked twice a day during lactation, the corresponding weight is not more than 150–170 kg (Denisov 1958). Yak meat is dark-coloured and rather coarse-fibred. At some periods in the year, the Tibetans sometimes bleed yaks, which is a convenient way of harvesting protein in a breeding system where the annual growth is very low. It should be observed that live-weight gains in the growing yak occur only when fresh grass is available. During the winter, feeding conditions are very difficult and lead to live-weight losses and often death of young calves.

The hair, plucked or shorn in the spring (400–1400 g per animal), is used to make ropes, tents or felt. The leather has many uses: for thongs, shoes, bottles and even boats (in Tibet). The dry dung is very valuable as fuel in highlands where no shrubs or trees can grow. The yak is also used for draught and riding but mostly as a pack animal, due to its uncommon agility through difficult mountainous ranges and also its ability to carry burdens – usually of 50–80 kg but sometimes more than 250 kg.

The domestic yak is also famous because of its widespread use in crossing with cattle (mostly on the periphery of its distribution area): in the F_1 hybrids (which can live at lower altitudes than the yak itself – 1500–3000 m), males are sterile and females fertile. Given the backcrossing possibilities with the F_1 females, herds are often very complex in their genetic composition (Bonnemaire and Teissier 1976). The F_1 hybrids, in appearance intermediate between their parents, are most interesting because they show heterosis in weight and size (males are particularly large and are very efficient draught and pack animals) and in milk production and fertility in the females. Indeed F_1 yak-cattle hybrid cows can often calve almost every year where yak cows calve only every 2 years on average and they give more milk which is of intermediate composition between yak and cattle milk. These characteristics make them very desirable animals in moderately high zones. Hybrids obtained by crossing a common cattle bull with a yak cow are more vigorous and sought-after than the reciprocal inverse hybrids. It is sometimes difficult to obtain the reciprocal hybrids because the yak bull does not willingly mate with a domestic cow, and also there are problems of viability at birth and retention of the placenta probably due to the difference in pregnancy length between the two species.

As the distribution area of the wild yak became restricted the domestic animal showed some extension. For instance, when the Kirgiz settled in the Pamirs in the 19th century, they introduced the domestic yak, the wild yak having vanished several centuries earlier (Dor 1976). The domestic yak has been introduced even into regions far away from its original area, with greater or less success (Bonnemaire 1976). Thus, in the U.S.S.R., yak have recently been successfully introduced from the Pamirs into the Caucasus in order to exploit pastures at elevations of 2500–3700 m (Mochalovskii and Abdusalamov 1973). Trials of acclimatization with a few animals have also been carried out but not followed up in other mountainous countries, e.g. France (1854) and Scotland. In North America, experiments have shown the yak's ability to adapt to local conditions and have confirmed heterosis in hybrids. In Canada, yaks were introduced between 1909 and 1928 as part of an experiment primarily concerned with crossing American bison and cattle (Cattalo project) in an attempt to improve their hardiness (see Ch. 6). In Alaska, an experiment of yak breeding (and crossing with Galloway cattle) was made from 1919 to 1932.

Future prospects

Important scientific work on domestic yak has been done principally in the Soviet Union and in Mongolia. Yak breeding (in the open air all the year round) is still being developed and improved in the mountainous regions of these countries where this animal's economic efficiency has been clearly shown (in purebreeding or in crossing with cattle). Selection seems to be especially orientated towards size and live-weight gain (or absence of loss) in different seasons and also towards production of hybrids with various improved cattle breeds. In these countries the main emphasis is on meat production in difficult ecological conditions. However, in Mongolia, yak milk production is still important and is being improved. China also promotes yak breeding for meat and milk production; in this country crossing with Hereford bulls has been attempted in Qinghai since 1975. However, in many regions where the yak was traditionally bred, it is still serving many purposes, giving milk, work, fuel, fibres, leather, and meat, in surroundings where most of the other domestic species appear to be quite unfit or less fit than the yak. Any future improvement work must take this situation into account.

References

Bazin, L. (1970) Les noms du 'yak' chez les peuples Turcs et Mongols. *Beiträge zur alten Geschichte und deren Nachleben*, Festschrift für Franz Altheim (6 Oct. 1968), **2**: 213–21

Bonnemaire, J. (1976) Le yak domestique et son hybridation. *Ethnozootechnie*, **15**: 46–77

Bonnemaire, J. and **Teissier, J. H.** (1976) Quelques aspects de l'élevage en haute altitude dans l'Himalaya Central: yaks, bovins, hybrides et métis dans la vallée du Langtang (Népal). *Ethnozootechnie*, **15**: 91–118

Boulnois, L. (1976) Le yak et les voyageurs et naturalistes occidentaux. *Ethnozootechnie*, **15**: 7–22.

Corbet, G. B. (1978) *The Mammals of the Palaearctic Region: a taxonomic review*. British Museum (Natural History); London. Cornell University Press: Ithaca, New York

Denisov, V. F. (1958) [*Domestic Yaks and their Hybrids*]. Seljhozgiz: Moscow. (In Russian)

Dor, R. (1976) Note sur le yak au Pamir. *Ethnozootechnie*, **15**: 126–132

Epstein, H. (1969) *Domestic Animals of China*. Commonwealth Agricultural Bureaux: Farnham Royal, Bucks, England.

Epstein, H. (1974) Yak and chauri. *World Animal Review*, No. 9: 8–12

Gray, A. P. (1972) *Mammalian Hybrids. A check-list with bibliography*. Commonwealth Agricultural Bureaux: Farnham Royal, Bucks, England

IUCN (1974) *Red Data Book*, Vol. 1, *Mammalia*. International Union for the Conservation of Nature and Natural Resources: Gland, Switzerland

Ivanova, V. V. (1956) [*Hybridization of the yak with domestic cattle and its prospects*]. Dissertation, Timirjazev Agricultural Academy: Moscow. (In Russian)

Mochalovskii, A. and **Abdusalamov, Sh.** (1973) [The Pamir yak in the Caucasus.] *Molochnoe i myasnoe Skotovodstvo* No. 8: 21. (In Russian). Abstract in *Animal Breeding Abstracts*, **42**, No. 163

Palmieri, R. (1976) Domestication and exploitation of livestock in the Nepal Himalaya and Tibet: an ecological, functional, and culture historical study of yak and yak hybrids in society, economy and culture. Ph.D. dissertation, University of California: Davis

Phillips, R. W., Tolstoy, I. A. and **Johnson, R. G.** (1946) Yaks and yak-cattle hybrids in Asia. *Journal of Heredity*, **35**: 162–70, 206–15

Wunschmann, A. (1975) Les bovinés. In: *Le Monde Animal en 13 volumes*, ed. B. Grzimek, **13**: 351–54. Stauffacher: Zurich

Zeuner, F. E. (1963) *A History of Domesticated Animals*. Hutchinson: London

6

American bison, and bison – cattle hybrids

H. F. Peters

*Alberta Agriculture,
Edmonton, Canada*

Bison

The American bison, popularly known as 'buffalo', exists in two freely interbreeding forms, the plains bison (*Bison bison bison*) and the wood bison (*B. b. athabascae*), and their hybrid. Reciprocal crosses of the American bison and the European bison, or wisent (*B. bonasus*), have occurred, and the hybrids of both sexes are fertile and show heterosis in growth (Gray 1972).

The bison is a ruminant. The head and forequarters are massive, the head low-slung and the hump of the shoulders supported by tall dorsal spines on the thoracic vertebrae. Bison have a dense coat of fine hair with a component of long, coarse fibres extending over the head, shoulders and forelegs. Christopherson and Hudson (1978) confirmed the superior cold-tolerance of bison in contrast to that of cattle as determined by insulation properties of the hair coat and lower critical temperatures.

Bison bulls attain adult size at about 6 years of age and the cows at about 4 years. Banfield (1974) gave the following weights and measurements of adult bison: weight, males 460–720 kg and cows 360–460 kg; shoulder height, males 167–182 cm

and cows 152 cm. Larger bison do exist, however, and the mature bison bull may range up to 1270 kg in body weight. The wood bison outweighs the plains form.

Bison cows are usually sexually mature at 2 years and generally calve at 3 years, though a small proportion of yearling bison of both sexes are capable of reproduction. Mature bison have an average conception rate of about 67 percent up to the onset of old age at 12–15 years and in some instances to over 30 years of age. Breeding occurs over several months, with a cycle length of about 3 weeks. The gestation period is 270–280 days, calves weighing about 18 kg at birth. Twins are rare.

Bison has the same chromosome number as *Bos*, $2n = 60$. However, the Y chromosome differs, being a small metacentric in cattle and an acrocentric (Basrur and Moon 1967) or telocentric (Bhambhani and Kuspira 1969) in the bison.

McHugh (1972) estimated the number of bison in North America at about 30 000, though present-day estimates range as high as 80 000 (H. W. Reynolds, personal communication). McHugh (1972) estimated that the original population must have numbered 30 million. Bison served an important role in settlement of the American West through supply of meat and hides, and were ultimately reduced to the verge of extinction.

There are today in private herds over 400 bison in Canada and several thousand in the United States. In addition, there about 7000 bison in Canadian parks and about 3000 in parks in the United States (Meagher, personal communication). There are about 450 plains bison and 120 wood bison in Elk Island National Park, Lamont, Alberta; 700 wood bison in the MacKenzie Basin Sanctuary, near Fort Providence, Northwest Territories; about 6000 'hybrids' (of mixed plains and wood bison ancestry) in Wood Buffalo National Park, and small numbers of plains and wood bison in other institutional herds (Reynolds, personal communication). There are about 2000 bison in Yellowstone Park. These are the interbred descendants of plains bison introduced from Texas and Montana in 1902 and the original wild population of mountain (wood) bison (Meagher 1973).

Hudson (1977) has described the effects of stress on bison in round-ups at Wood Buffalo National Park. These include mortality from gorings and self-mutilation, separation of cows and calves, shock, heart failure, shipping fever, pulmonary emphysema, and other physiological disorders. Research is being done to develop techniques which will minimize the stress and attendant losses from handling for management requirements and in bison ranching. Jennings (1978) regards the bison as a wild animal, citing the requirement for higher and stronger fences, corrals, chutes and gates than those commonly used in cattle ranching.

Bison made slower feedlot gains and had lower carcass fat content than contemporary Hereford cattle (Peters 1958). However, bison are efficient in the use of forage and natural grazing lands.

The high spinal process and heavier forequarters of the bison in contrast to domestic cattle constitute no disadvantage in meat content. Bison differ from cattle in having less abdominal wall muscle and much more weight in the muscles connecting the neck to the thoracic limb (Berg and Butterfield 1976). This represents an advantage in proportion of the muscles identified as the most expensive in the commercial beef carcass.

The specialty market for 'buffalo' meat in Canada and the United States has led to an increasing interest in bison for private ranching as well as for controlled hunting in the parks. Aside from the scientific question of whether *B. b. bison* and *B. b. athabascae* are sufficiently distinct to warrant zoological recognition as separate subspecies, they should be kept as separate breeding populations for maintenance of the species in its present forms as distinct wildlife resources. There is also the prospect for increased efficiency of range utilization through multiple-use management in game ranching.

Bison × cattle crosses

Cattalo

C. J. Jones of Kansas, Charles Goodnight of Texas, and Mossom Boyd of Bobcaygeon, Ontario were three of the early hybridizers of cattle and bison. Boyd contemplated 'taking the fur and hump of the bison and placing them upon the back of the domestic ox'. His herd was dispersed, however, and in 1916 the Canadian Experimental Farms Service selected sixteen

cows and four bulls from his herd to begin the 'cattalo' experiment.

None of the animals purchased from Mossom Boyd left progeny in the experiment, and new crosses of bison with cattle were made. A number of yak-cattle crossbreds were produced, but they demonstrated no apparent advantage and this aspect of the programme was discontinued in 1928.

The cattalo experiment was conducted at Wainwright, Alberta, until 1950 (Deakin *et al.* 1935, 1941, Logan and Sylvestre 1950), and at the Dominion Range Experiment Station, Manyberries, Alberta thereafter (Peters 1964). At Manyberries the cattle × bison hybrid cows and one-quarter bison cattalo surpassed domestic cattle in hair-coat density and ability to forage on upland winter range during periods of low ambient air temperatures and concurrent high wind velocity.

The species cross gave best results when domestic bulls were used on bison cows, the reciprocal cross causing losses of both cows and calves associated with excessive amounts of placental fluids ('hydramnios'). The first-cross hybrids were sterile in the male and generally fertile in the female: 60 percent of 250 matings of hybrid cows with domestic bulls resulted in live calves born at Wainwright (Logan and Sylvestre 1950). The calves from these hybrid cows were small at birth but surpassed contemporary Hereford calves in weaning weight (Peters and Slen 1966). There was some indication of an abnormal sex ratio at birth, with a deficiency of males.

The primary defect in sterile or infertile cattalo bulls was the absence or low concentration of spermatozoa. However, the percentage of yearling cattalo bulls with 'fair' to 'good' semen density improved with selection, reaching 57 percent by 1964 when the project was terminated to provide facilities for research on other problems of beef cattle breeding. At that time the level of bison ancestry in the cattalo growing stock was 14 percent (Peters 1964).

Further research and development work with the hardy, long-lived cattle × bison hybrid cow in order to augment production of ¼–bison market cattle is warranted.

American breed

The American breed, developed by Art Jones of Portales, New Mexico for adaptation to ranges of the American southwest, is reported to be one-half Brahman, one-quarter Charolais, one-eighth bison, one-sixteenth Hereford and one-sixteenth Shorthorn (Fowler 1975). A test of the breed is needed to determine its genetic status and productivity.

Beefalo

World-wide interest was aroused by press reports in 1973 that a rancher, D. C. Basolo of Stockton, California, had produced a herd of 5000 hardy 'cow-buffalo hybrids' (Beefalo) which were cheaper to feed (reaching market weight sooner than cattle), were more resistant to disease, and reproduced readily. The Beefalo was given great publicity. However, it shows little superficial resemblance to bison and researchers found no bison-specific markers in the blood of 'full-blood' Beefalo (Stormont *et al.* 1977). Belgian researchers reported that the Y chromosome of the Basolo hybrid Beefalo was derived from *Bos taurus* (Lenoir and Lichtenberger 1978).

Most of the Beefalo bulls in the Basolo herd were said to be three-eighths bison, three-eighths Charolais and one-quarter Hereford, though other domestic breeds were used in some of the crosses. Beefalo meat was also reported to contain less fat than regular beef. This attribute would be expected in cattle × bison crosses, but a relatively high lean meat content is also found in domestic cattle crosses that include the Charolais breed.

The American Beefalo Association define full-blood American Beefalo as those animals containing three-eighths American bison and five-eighths domestic cattle. They have recently incorporated into their registry breeding stock derived from the herd of Jim Burnett, Luther, Montana, who has reportedly had some success in obtaining bulls with improved spermatogenesis by backcrossing to the bison (Burnett 1975).

A report from Botswana (Makobo et al. 1981) shows that Beefalo × Tswana calves at 18 months of age were not significantly different from purebred Tswanas and were lighter than crosses of Tswanas with Brahman, Simmental or Italian breeds. These results suggest that the Beefalo, whatever its origin, needs to be further tested before the claims made for it can be validated.

References

Banfield, A. W. F. (1974) *The Mammals of Canada*. National Museums of Canada. University of Toronto Press: Toronto and Buffalo

Basrur, P. K. and **Moon, Y. S.** (1967). Chromosomes of cattle, bison, and their hybrid, the cattalo. *American Journal of Veterinary Research*, **28** (126): 1319–25

Berg, R. T. and **Butterfield, R. M.** (1976) *New Concepts of Cattle Growth*. John Wiley: Toronto

Bhambhani, R. and **Kuspira, J.** (1969). The somatic karyotypes of American bison and domestic cattle. *Canadian Journal of Genetics and Cytology*, **11**: 243 –9

Burnett, Jim. (1975). How I got fertile buffalo/Hereford bulls. *Buffalo, 3*: 3–6

Christopherson, R. J. and **Hudson, R. J.** (1978). Northern animal agriculture. *Agriculture and Forestry Bulletin*, University of Alberta: Edmonton, **1** (2): 3–5

Deakin, A., Muir G. W. and **Smith, A. G.** (1935). Hybridization of domestic cattle, bison and yak. Report of the Wainwright experiment. *Publication* No. 479, Canada Department of Agriculture: Ottawa

Deakin, A., Muir, G. W., Smith, A. G. and **MacLellan, A. S.** (1941) Hybridization of domestic cattle and buffalo (*Bison americanus*). Progress report of the Wainwright experiment 1935–41. *Mimeographed Publication*, Canada Department of Agriculture: Ottawa

Fowler, S. H. (1975) The American breed: North America's most cosmopolitan beef breed. *Focus on Beef*, Sept., pp. 29–32

Gray, A. P. (1972) *Mammalian Hybrids. A check-list with bibliography* (2nd edn). Commonwealth Agricultural Bureaux: Farnham Royal, Bucks, England

Hudson, R. J. (1977) Terms of reference: reaction of bison to roundups project. *1977 Annual Report, Bison Research*, Wood Buffalo National Park: Parks Canada

Jennings, D. C. (1978) *Buffalo History and Husbandry*. National Buffalo Association: Custer, South Dakota. Pine Hill Press: Freeman, South Dakota

Lenoir, F. and **Lichtenberger, M. J.** (1978) The Y chromosome of the Basolo hybrid beefalo is a Y of *Bos taurus*. *Veterinary Record*, **102**: 422–23

Logan, V. S. and **Sylvestre, P. E.** (1950) Hybridization of domestic beef cattle and buffalo. A progress statement. *1937–49 Progress Report, Animal Husbandry Division*, Canada Department of Agriculture: Ottawa

Makobo, A. D., Buck, N. G., Light, D. E. and **Lethola, L. L.** (1981) A note on the growth of Beefalo crossbred calves in Botswana. *Animal Production*, **33**: 215–17

McHugh, Tom (1972) *The Time of the Buffalo*. Alfred A. Knopf: New York

Meagher, M. M. (1973) *The Bison of Yellowstone National Park*. National Park · Service Scientific Monograph Series No. 1

Peters, H. F. (1958) A feedlot study of bison, cattalo and Hereford calves. *Canadian Journal of Animal Science*, **38**: 87–90

Peters, H. F. (1964) A review of animal breeding research at Manyberries, past and present. *Field Day Hi-Lites*, Experimental Farm: Manyberries, Alberta

Peters, H. F. and **Slen, S. B.** (1966) Range calf production of cattle × bison, cattalo, and Hereford cows. *Canadian Journal of Animal Science*, **46**: 157–64

Stormont, C. J., Morris, B. and **Suzuki, Y.** (1977) Analysis of bison-derived genetic markers in the blood of Beefalo and other bison–cattle hybrids. *Animal Blood Groups and Biochemical Genetics*, **8** Supplement 1: 33

7
European bison

Z. Woliński

*European Bison Pedigree Book,
Warsaw Zoological Garden,
Poland*

Introduction

The European bison or wisent, *Bison bonasus*, of the family Bovidae and closely related to the American bison, has probably evoked the greatest interest among scientists as a result of the international effects to protect it from extinction. Two subspecies were once distinguished: the lowland or Białowieza wisent – *B.b. bonasus* – and the mountain or Caucasian wisent – *B.b. caucasicus*. This distinction is now solely of historical significance because the mountain wisent in its pure form has been extinct for more than half a century.

The distinguishing characteristics of European bison are: length of body (including head) 250–350 cm, height at shoulder 180–200 cm, weight of adult up to 800–1000 kg; the hindquarters are relatively high. The horns are dark in colour, almost black, of medium length and slender, curving forwards and inwards. The mane of the males is larger than that of the females. The colour of the coat is brown (Lydekker 1912).

When living in natural or semi-natural conditions mating takes place in autumn, mainly from August to October, and births occur after some 9 months' gestation in early summer, mainly from May to July; usually only one, very rarely two, calves are born. The European bison reaches sexual maturity at about 3, rarely 2 years of age. The life expectancy is up to 30–40 years (Asdell 1964). Chromosome number is $2n = 60$.

Wild wisent

It is difficult to establish exactly the former area of distribution of the European bison in its natural state, since it was preceded by an allied, extinct form – *B. priscus*. In historical times the range of the bison extended from western and southern Europe in the west, through central Europe to Poland and Russia, including the Caucasus, in the east (Lydekker 1912).

European bison (like the aurochs) were once favourite objects of hunting due to their value as a source of meat and the possibility they offered to prove courage when confronting such powerful animals with only primitive weapons. Descriptions of hunting the European bison can be found in the old German *Nibelungenlied* and in the sagas and chronicles of other nations.

As a result of hunting and also of the depletion of the virgin forests, which are the wisent's natural habitat, its range declined rapidly. The bison became extinct in France around the 11th century and in Britain around the 12th century. Precise dates are known for Germany (1755) and Hungary (1790) (Raczyński 1978).

In the 19th century the wisent was present in only two regions: in the Białowieza primaeval forest where its number varied between 500 and 1900 head, and – as was discovered only about 1830 – on the northern slopes of the Caucasus. In the latter region the herd amounted to some 800–1000 head, before it was finally killed off before 1927 by the local population in the upheavals following the First World War.

Possibility of domestication

It would be an exaggeration to claim that the European bison was ever domesticated, but it has enjoyed the protection of the Polish monarchs since the 16th century. They saw in it not only an object for hunting but also an animal of great cultural value. The first royal zoological garden to exhibit European bison and aurochs, was created at Ujazdów, a Warsaw suburb, towards the end of the 16th century by King Sigismund III Vasa. There were several more such wisent parks in Poland at this time.

It is only in the case of the Białowieza herd that one can speak of any far-reaching human interference in the life of the European bison. True it was never fully domesticated but, as the result of its only rarely being the object of hunting and by being systematically fed, it gradually lost its fear of man. When forest guards of the Russian tsar in the 19th century appeared with their carts on forest paths with fodder for the wisents, they were followed by these enormous beasts. The same thing happened in August 1915, when, after the retreat of the Russian army, hungry Prussian soldiers entered the forest using similar carts; this time, however, some 500 of the total of about 770 European bison were killed by them (Zabiński 1960). The remainder were killed by marauders of both armies and by the local population between the end of 1918 and April 1919.

Recent history

A certain number of European bison were sent from the Białowieza forest by the Russian tsars in the 19th century as gifts to royalty, wild-animal breeders, zoological gardens and also commercial companies. About fifty-four pure-blood European bison, the offspring of these animals, were alive at end of 1922 according to a head count.

The idea was born in Poland and almost simultaneously also in Germany and in the United States, to unite the efforts of men of science and breeders the world over to save the last European bison. It was a Polish naturalist, Jan Sztolzman, who first presented this idea to the International Congress on Nature Protection meeting in Paris on 2nd June 1923. As a result of this appeal the International Society for the Protection of the European Bison was created during 25–26 August 1923 in Berlin.

In the period preceding 1939 three parts of the European Bison Pedigree Book were published, the first such documentation concerning non-domesticated animals. Following the Second World War the publication of this book was assigned to Poland. Sixteen issues have been prepared in Warsaw and Białowieza which include data from 1937 to 1976. At the same time two authoritative European bison research centres were formed in Poland: the European Bison Research Centre at Warsaw Agricultural University and the Mammals Research Institute of the Polish Academy of Sciences in Białowieza (Krysiak 1960).

In the Białowieza Institute, studies, among others, on hybridization of European bison and domestic cattle have been carried out. The length of gestation was markedly influenced by the paternal species. The F_1 hybrids were intermediate in appearance but resembled the domestic cattle breeds used for crossing (Polish Red or Black Pied Lowland) more closely than wisent. The male hybrids (of all generations) were sterile; on the other hand 100 per cent of the F_1 females were fertile when mated with domestic bulls. All hybrid calves had a high rate of growth and development, both before and after birth, especially during the first year of life, a clear sign of heterosis. Their body weight (up to 300 kg at the age of 6 months) was considerably higher than that of the parental cattle breed. The hybrids delivered to State livestock breeding centres are used for production of slaughter animals which have a large mass of meat and relatively thick skin (Krasińska and Pucek 1967).

Present status and future prospects

The data in the hands of the editors of the European Bison Pedigree Book (referring to 1 January 1977) indicate that wisents were present in 211 breeding centres scattered around the world (193 in Europe including the Soviet Union, 12 in North America, 3 in South America and 3 in Asia). About 1075 wisents were living in 190 so-called closed breeding centres – zoos, enclosures in reserves and forests, and 'wildparks'; about 790 wisents were semi-wild in 21 centres in Poland and in the Soviet Union. Each year about 190 wisent calves are born. The world total of European bison reached 2000 head in 1978. Species whose number is above 2000 are no longer classified by IUCN as threatened with extinction (Pilarski *et al.* 1981).

There is, however, little possibility of expanding the number of wild-living herds of European bison (with the exception perhaps in the extensive forest areas in the Soviet Union). For this reason recent years have seen an overproduction of wisents in many breeding centres, e.g. in West Germany, Great Britain and other countries. The surplus animals are in many cases killed and either delivered to animal disposal units or, preferably, are used as food for human beings or zoo animals.

It seems quite likely that the European bison, while continuing to be a wild species, will exist in the near future in two different population groups. The first group, comprising those held in conditions approaching the wild state as closely as possible (and registered in the Pedigree Book) should be maintained at above the level of 2000 head. As is the case in the wild-living herd in Białowieza forest, human interference in this group should be as small as possible.

The second group will comprise all breeding surpluses, i.e. European bison used for scientific research, for crossbreeding with domestic cattle and other species of Bovidae, for hunting in special centres and even as a source of meat. It must be stressed that the two latter possibilities have never seriously been taken into account by world science.

The fact remains, however, that threat of extinction of the wisent, which was the principal problem after the First World War, has currently turned into a seeming surplus of this interesting species.

References

Asdell, S. A. (1964) *Patterns of Mammalian Reproduction* (2nd edn). Cornell University Press: Ithaca, New York.

Borodin, A. M. *et al.* (1978) [*Red Data Book of the U.S.S.R.*] Central Research Laboratory on Nature Conservation: Moscow. (In Russian)

IUCN (1966) *Red Data Book*, Vol. 1: *Mammalia*. International Union for Conservation of Nature and Natural Resources: Gland, Switzerland.

Krasińska, M. and **Pucek, Z.** (1967) The state of studies on hybridisation of European bison and domestic cattle. *Acta Theriologica, Białowieza*, **12** (27): 385–89

Krysiak, K. (1960) *The European Bison* (Bison bonasus). State Council for Conservation of Nature in Poland: Warsaw

Lydekker, R. (1912) *The Ox and its Kindred*. Methuen: London

Mohr, E. (1952) *Der Wisent*. Akademische Verlagsgesellschaft Geest und Portig K. -G.: Leipzig

Pilarski, W. *et al*. (1981) *European Bison Pedigree Book 1975–1976*. National Council for Nature Conservation: Warsaw

Raczyński, J. (1978) *Zubr*. PWRiL Publishers: Warsaw

Zabiński, J. (1960) *The European Bison* (Bison bonasus). State Council for Conservation of Nature in Poland: Warsaw

8

Water buffalo

W. Ross Cockrill

International Consultant,
Animal Production and Health,
Portugal

Introduction

Nomenclature

Buffaloes form two groups: the Asian (genus *Bubalus*) and the African (genus *Syncerus*). Along with cattle (*Bos*) and bison (*Bison*) they belong to the tribe Bovini of the family Bovidae. Matings between cattle, the Asian buffalo and the African buffalo are invariably sterile.

Bubalus bubalis, the Asian water buffalo, achieved domestication at least 4 000 years ago. All domestic buffaloes in the world today are descended from the Asian and not the African form. *Syncerus caffer*, the wild African buffalo, can be tamed and has bred in captivity, but it has never been domesticated.

White (1974) observes that in literature, art and history there has been much confusion in the use of the terms 'buffalo' and *bubalus*. In Latin, *bubalus* originally referred to an antelope; it was later applied to the wild bison or the wild aurochs and this use continued through the Middle Ages in Europe. The North American bison has been indelibly misnamed the 'buffalo'.

In China buffaloes are *shui niu* (i.e. water

cattle) while common cattle are *huang niu* (i.e. yellow cattle). The water buffalo is known as the *kerbau* in Malaysia, *karbouw* in Indonesia, *carabao* in the Philippines and *camus* (pronounced jamoos) in Turkey (as well as *manda*).

MacGregor (1941) classified water buffaloes into two groups: the Swamp buffaloes of southeast Asia and the River buffaloes of the Indian subcontinent. The Swamp buffalo closely resembles the wild arni of India (*Bubalus arnee*) in general conformation and habit. Archaeological, anatomical and historical evidence supports the contention that all breeds of domestic buffaloes are descended from the arni.

General biology

Buffaloes are normally black or grey, but they are occasionally brown, white or pied (see pp. 57–8). Hair whorls are seen in all buffalo breeds and in all conditions of skin colour; they are particularly distinctive in the Swamp buffalo. Figurines and depictions of the buffalo by artists and sculptors from the earliest times emphasize the rugate horn striations and the hair whorls. Whorls are legally recognized as distinguishing marks in Malaysia, Kampuchea and the Philippines; they are believed to be uniquely individual in number, location, size and shape and they are an important and convenient factor in registration.

Naturally hornless or poll buffaloes are very rare. Occasional hornless individuals have been recorded in Indonesia (Sumbawa and Sulawesi), Peninsular Malaysia, Papua New Guinea, Turkey and Iraq. There is some evidence that hornlessness may be associated with infertility which would account for its rarity. In both Sumbawa and Papua New Guinea hornless males lacked testicles.

The water buffalo is a wallowing animal. The River breeds of the Indian subcontinent favour the clear water of streams and pools but the Swamp types of China and southeast Asia like to wallow in deep slime in a mudhole which they will prepare by digging over the area with their horns.

Buffaloes are not heat tolerant and they can suffer acute distress if left for even a few hours in direct sunlight. Sweat glands are much less numerous than in cattle and the heat-regulating mechanism appears to be less efficient. In hot climates there must be provision for rest in the shade or in the wallow, or a routine of sluicing and spraying.

Buffaloes are not normally vocal and do not low or bellow. The voice is a querulous grunt. The domestic buffalo has a remarkably long productive life. It may produce ten or more calves in a lifetime, go on yielding a significant amount of milk to 18–20 years or, as a working animal, remain capable of a good day's labour in the yoke to the age of 30 years or even more.

The buffalo is an efficient utilizer of poor-quality feed, perhaps the most efficient of all domestic animals with the possible exception of the camel.

The buffalo female attains sexual maturity at a rather more advanced age than cattle. The age at first oestrus varies greatly with breed, geographical location and nutritional status, from 15–18 months in the Egyptian buffalo to 30–32 months in the Turkish and 36 months in the small Swamp buffaloes of China. The age range at first calving is 20–52 months, again varying with breed and management. Duration of oestrus is 15–48 hours depending on breed, season of the year and age; and the period between oestruses is generally 21 days. Undetected silent heat and failure to inseminate during the hot months of the year result in seasonality of breeding. In the economy of milk production seasonal variation in breeding is of great importance (Pandey and Raizada 1979) but seasonal breeding can be controlled under appropriate management. The gestation period of the water buffalo is generally quoted as 10 months and 10 days. The observed range is 298–340 days and the great majority of pregnancies last 308–320 days. It lasts about 2 weeks longer in Swamp than in River breeds. The incidence of twinning in all breeds of buffaloes is very low (0–0.6%). The calving interval varies between different breeds and under different systems of management. Averages of 430–530 days are reported.

Karyotype

The Swamp buffalo of Thailand, Malaysia and Australia has the diploid chromosome number 48 and the River breeds of Malaysia (Murrah), Turkey and Europe, 50. In the latter case there are 5 pairs of metacentric or submetacentric chromosomes and 25 pairs (including the sex chromosomes) of acrocentric chromosomes. This gives the 60 chromosome arms typical of the Bovidae. The Swamp buffalo lacks one of the smallest chromosome pairs but the first chromosome pair has longer arms and is metacentric instead of submetacentric. Thus the amount of chromosomal material is similar in the two types. In both cases the X chromosome is the longest of the rod-shaped and the Y one of the smallest. The Swamp X River cross has 49 chromosomes, but does not suffer from any loss of fertility. The first pair is disparate in size and one of the small chromosomes is unpaired (Fischer and Ulbrich 1968, Ulbrich and Fischer 1967, Toll and Halnan 1976).

The karyotype of the Sri Lanka buffalo is similar to that of the Murrah in number and morphology of chromosomes. Thus, although the Sri Lanka buffalo has been classified as the Swamp type based on phenotype and habitat, on the basis of similarity of karyotype it might be more appropriate to classify it as the River type (Bongso *et al.* 1977). Alternatively it might be argued that the difference in chromosome number does not separate the River from the Swamp type, but is of geographical origin and separates the buffaloes of southeast Asia from those of India, Sri Lanka and further west.

The Cape variety of the African buffalo (*Syncerus caffer caffer*) has 52 chromosomes; as in *Bubalus* the X is the largest and the Y the smallest of the acrocentric chromosomes. The Congo buffalo (*S. c. nanus*) on the other hand has 54 chromosomes (Heck *et al.* 1968, Ulbrich and Fischer 1967). The karyotypes of *Syncerus* and *Bubalus* are quite different from that of cattle which have 60 chromosomes.

Distribution

There are about 122 million water buffaloes in the world in some 40 countries compared with a total world cattle population of around 1 214 million. Their distribution is shown in Table 8.1. In the majority of buffalo-owning countries, and in all those in which buffaloes make an important contribution to the agricultural economy, numbers are increasing, sometimes, as in Brazil, at a phenomenal rate. In India and Pakistan, which together have approximately 77 million domestic water buffaloes, they now supply almost 70 per cent of all the milk used for human consumption. China, which has the second largest establishment with over 18 million, utilizes its buffaloes almost exclusively as work animals. Until very recently China has not attempted to exploit the outstanding potential of the buffalo for milk and meat production. In most countries

Table 8.1 Numbers of domestic buffalo by region and by country

Region	Country	Number (in thousands)
Indian subcontinent (chiefly River type)	India	61 500
	Pakistan	11 794
	Nepal	4 267
	Bangladesh	1 600
	Sri Lanka	843
	Bhutan	27
Far East (chiefly Swamp type)	China	18 854
	Thailand	6 299
	Philippines	2 850
	Indonesia	2 506
	Vietnam	2 378
	Burma	1 950
	Lao	879
	Kampuchea	404
	Malaysia	293
	Brunei	14
	Singapore	3
Near and Middle East (River type)	Egypt	2 347
	Turkey	1 031
	Iraq	228
	Iran	220
	Afghanistan	35*
	Syria	2
Soviet Union	Azerbaijan, Georgia, Armenia and North Caucasus	340
Europe (River type)	Romania	227
	Italy	160*
	Yugoslavia	67
	Bulgaria	48
	Greece	2
	Albania	2
South America	Brazil	550
	Venezuela	8*
	Trinidad	8
Oceania	Australia (Swamp, feral)	250*
	Papua New Guinea	1.5*

* These figures are author's estimates; the rest are from FAO (1982)

meat is a by-product of work and milk production, but the potential is such that meat production is now emerging as perhaps the most important aspect of buffalo productivity for the future as the buffalo comes to be recognized as a truly multi-purpose domestic animal.

Wild species

There are three wild species of *Bubalus*. They are the tamarao (*B. mindorensis*) of Mindoro in the Philippines, the small anoa (*B. depressicornis*) of Sulawesi, Indonesia (both of which are in danger of extinction due to poaching and hunting with modern weapons and the widespread destruction of their habitat), and the Indian wild buffalo (*B. arnee*) known as the arni. The arni is considered to be the progenitor of all the breeds of domestic buffaloes. It identifies with the Swamp breed in all characters save tameness and size. As late as the beginning of the 19th century it was widespread throughout the Indian subcontinent east of latitude 80 °E and north of the Godavari river, as well as in Sri Lanka and the southeast Asia mainland. The arni now exists in its true wild form only in a few parks and reserves in Assam, Nepal and Burma. In these areas domestic female buffaloes are frequently served by arni males, while tamed arnis may occasionally be introduced into domestic buffalo herds. The crossbred offspring are highly prized as milking animals.

From time to time statements have been made concerning the existence of wild buffaloes in various islands of south and east Asia (Sri Lanka, Java, Sarawak, Sumatra). Such statements can be discounted. These buffaloes are feral animals which originated with domestic animals 'gone bush'.

In former times the range of the arni comprised a large part of India and extended east into mainland southeast Asia and south China and west into Mesopotamia. The wild buffalo is frequently featured in Mesopotamian mythology showing its subjugation to the god-hero Gilgamesh. The fullest Akkadian text dates from the 7th century B.C. but the story and the buffaloes date from long before this. Great numbers were killed by the Assyrian King Ashur-nasir-pal (1049–1031 B.C.) in the hunting grounds near the Euphrates.

That the wild arni existed in the Middle East in the 7th century A.D. is shown by the evidence of a chased-silver plate of the period in the Bibliothèque Nationale, Paris. On it is depicted Khosrau II, King of Persia (A.D. 590–628) hunting buffaloes on horseback. The wild buffaloes are clearly delineated. The site of the hunt may have been Iran or Afghanistan.

Evidence is accumulating of the presence of buffaloes in China some 7000 years ago, when they were hunted. These were presumably wild arni. From the Hemudu site at Yu-yao in Zhejiang province, skulls of water buffaloes were unearthed. The Lingzhou site of the late Neolithic age dating from about 5000 B.C. yielded a considerable number of bones of the animal (Ran Xiancui, personal communication).

Domestication and early history

Time and place of first domestication

It is usually assumed that water buffaloes were first domesticated as work animals in Mesopotamia during the Akkadian dynasty and/or in the Indus valley civilization of Harappa and Mohenjo Daro some time prior to 2500 B.C. As Zeuner (1963) indicated, archaeological evidence for their history as domesticated beasts is curiously scanty and is mainly restricted to seals and bone findings.

The Indus valley civilization of Mohenjo Daro produced a great many representations of water buffaloes, all apparently of males. They include a small copper sculpture and many steatite seals in the National Museum in New Delhi. A steatite seal in the Lahore Museum shows a male Swamp-type buffalo at a feeding trough which surely indicates domestication.

Buffaloes were first tamed and then domesticated as work animals. The milking of work buffaloes might have been practised to provide offerings to the lunar goddess to ensure crop fertility: the circular formation of the horns rendered the animal sacred by association with the full moon. The buffalo itself may have been a sacrificial animal in the Mohenjo Daro civilization as it is today in parts of southeast Asia. Seals show the animal walking, rearing, feeding, fighting, being watered and fed, but never being milked or working as a plough animal or in harness. A copper model of a two-wheeled cart from Harappa and many small clay models or toys provide evidence of the early use in the Mohenjo

Daro civilization of wheeled vehicles and it is tempting to assume that the working buffalo would be used in cartage, but there is as yet no concrete evidence of this.

Brentjes (1969) states that the oldest known buffalo pictures come from the multicoloured ceramics of the Nal culture (of south Baluchistan). However, the current view of the British Museum is that the Nal culture was contemporary with the Indus valley civilization (Knox, personal communication).

Male buffaloes appear frequently on cylinder seals from the Akkadian period of ancient Mesopotamia. These seals would appear to provide evidence of domestication by 2350–2150 B.C. They depict such scenes as a bearded god-hero holding a rearing buffalo by a foreleg and a horn possibly in some form of contest, a god-hero lifting a buffalo by the hind legs, buffaloes being watered by the side of a stream by two god-heroes and lions attacking buffaloes. Their presence in company with men or gods in human form is, perhaps arbitrarily, taken as evidence of domestication.

According to Boehmer (1974) evidence for the presence of the domestic water buffalo in Mesopotamia is restricted to two periods: the late 3rd millennium B.C. and the Sasanian period (A.D. 224–651) onwards. He suggests that the animal was introduced from the Indian subcontinent on both occasions and that it died out in the interval.

The Mesopotamian and Indus valley cultures were ancient riverine civilizations similar to those of the Yangtze river and the Yellow river in China. The climate of central and north China during the Shang dynasty (c. 1766–1123 B.C.) was probably warmer than it is at the present time and this would explain the presence of buffaloes as far north as Anyang in the province of Henan. The bones of water buffaloes have been discovered there, together with those of cattle, dogs, pigs, sheep and goats all of which were evidently domesticated. Identification of the water buffalo among the Shang remains of around 1400 B.C. was substantiated by finding the typical triangular horn core and the distinctive metacarpus (Creel 1936). White (1974) also notes that the tamed buffalo was known in Shang China in the 2nd millennium B.C.

Its cult role, which reflects its economic importance, can also be traced in China according to Brentjes (1969). The heads of buffaloes are represented on handles of cult vessels and on cultic pillars of the Shang epoch.

There is osteological evidence for domestic buffalo in southeast Asia in the 2nd millennium B.C. For instance, at the Ban Chiang site in northeast Thailand, buffalo bones appear suddenly at about 1600 B.C. at the same time as the inception of wet rice cultivation. This suggests introduction of the tame animal from somewhere else rather than the taming of a local wild animal. Presumably the animals were used for working the fields (Higham and Kijngam 1979).

Indeed Epstein (1969) suggests that the domesticated buffalo was introduced into China prior to the Shang dynasty from southeast Asia. It appears more likely, however, that the habitat of the arni in prehistoric times extended across north India, west to Mesopotamia and east to south and central China and that domestication of the indigenous wild arni took place in all these locations. However, the question is clearly far from settled. If buffaloes were domesticated in Mesopotamia in the 3rd millennium B.C. why were they not seen in the countries to the west (whose environment has proved suitable) until A.D. 723?

The greatest development of breeds has been in India and Pakistan. This is thus a centre of diversity and would therefore suggest an early domestication centre and a centre of intense human selection since domestication. In southeast Asia, on the other hand, the Swamp buffalo still resembles the wild and so a more recent domestication is indicated.

The domestic buffalo is supposed to have been brought to the Philippines by Malay immigrants in the period 300–200 B.C.

It is of interest to note that, in the Bayon and the Baphuon at Angkor Thom, Kampuchea, there are bas-reliefs of the 12th century A.D. in which Swamp-type buffaloes are depicted as cart animals and participating in ceremonial processions.

Westward spread

The buffalo may have reached south Russia by way of Persia (Zeuner 1963). In 538 B.C. the Medes and the Persians conquered the Iranian plateau and Mesopotamia. The geographical position of Iran lying between Iraq (Mesopotamia) and Pakistan (Indus valley) is such that Swamp-type buffaloes must have existed in the country long before the Christian era. In the British Museum there is a bronze sceptre from Luristan in western Iran, dating from 900 to 750 B.C., which carries the heads of three water buffaloes and three buffalo calves.

Domestic buffaloes were unknown in ancient Judaea, Egypt, Greece, or Rome. They were first observed in Palestine in A.D. 723 (by St Willibald) but they may have been known there earlier since they are represented in the ruins of Sebastiye which date from the last decades B.C. They were probably brought from Mesopotamia by the Arabs. From Palestine they entered Egypt, again with the Arabs, although some authorities say not until the 9th century A.D. (Epstein 1971).

Buffaloes came to Europe in advance of the Muslims and by the beginning of the 13th century there were large numbers in Thrace and Macedonia. Either they had been brought by returning Crusaders or they entered from southern Russia. However, it seems probable that the first buffaloes in Europe may have been those of Italy. There are good reasons for discounting the story that they first came to Europe in A.D. 595 as a gift from the Khan of the Avars to the Lombard leader Agilulf (Epstein 1971). The *bubali* referred to in the Latin text by Paulus Diaconus which is the source of this story could well have been bison. On the other hand, White (1974) believed that the writer was well able to distinguish between buffalo and bison and that Indian buffaloes were ensconced in the Danube valley (the home of the Avars) at that time. But White goes on to admit that there was no sure trace of buffaloes in Europe for the next 400 years. It is also possible that buffaloes were first brought by the Arabs to Sicily after their conquest of that island in A.D. 827 and that from there they spread to the mainland. But even this is only a theory and does not explain where the buffaloes came from since there are no records of their presence in Tunisia (the nearest part of Africa and the country from which the Arab invasion was mounted) at this early date. By 1154 buffaloes were numerous and valued in the Campagna di Roma. During the fascist era Italy's buffalo stocks declined, lacking government support, but in the last 35 years numbers have greatly increased and today there are some 160 000, mainly in Campania, with the greatest concentrations in the provinces of Caserta and Salerno.

Buffaloes appeared in Turkey, the traditional bridge between the east and the west, some centuries after the beginning of the Christian era and coincident with the westward spread of Islam. White (1974) comments that the buffalo having been firmly established in Anatolia by the 9th century at the latest, the way was open to Thrace, the Balkans and southern Italy. But it was not until the 14th century that the Turks started their conquest of the Balkan peninsula and by then buffaloes were already present. In their campaigns the Turks used the buffalo as a draught animal for hauling heavy battering rams, equipment and food. The animal became established in Yugoslavia, Albania, Hungary and Romania, and Muslim communities perpetuated the stocks.

Introductions were made into France, Germany, Spain and England in the Middle Ages but buffaloes failed to establish themselves in any of these countries.

Recent history and present status

Types and breeds

Swamp type Horns are massive, heavily striated and grow outward from the head laterally and upwards to form a semi-circle as in the wild arni.Two white or light grey chevrons or gorgets are generally present, one below the line of the jaw and the other on the front of the brisket. Grey stockings are usual. Overall colour is dark slate grey; a red tinge in the long hairs of the coat is common.

Swamp buffaloes are mainly used for work, exhibiting great muscular power and versatility. Small-scale milk production from the working animal is usual: the potential exists for a much greater yield of milk. The potential for meat production is outstanding.

Colour variations of skin and coat are encountered in all buffalo breeds, the frequency being greatest in the Swamp type. The white Swamp buffalo is found in many areas, perhaps most notably in south China, Indonesia (Bali and Timor) and Thailand. Random counting by the author in Hunan province, south China, put the incidence of albinoid Swamp buffaloes at 8 per cent; in the north of Thailand the proportion may be as high as 20 per cent. The skin is pink or of a light yellow tinge and the sparse hair is straw-coloured. These animals are often assumed to be albinos but horns, hoofs, buccal mucosa and irides are pigmented and there is no sensitivity to light. They are properly referred to as albinoid. True albinos are much rarer than in other mammalian species. A short lived albino calf of the Swamp type was born in Sabah (Malaysia) in 1975 (Soo Phin Chia, personal communication). Piebald and skewbald Swamp buffaloes occur in Australia, Brazil and Indonesia (Timor and Sulawesi). In Sulawesi, in particular, the Toradja people breed their buffaloes for a highly desired spotted characteristic. These spotted animals are used for fighting and for sacrifices.

The Swamp type is widespread throughout southeast Asia and southern China. It is also found in Nepal, Sri Lanka and parts of south India. Body and horn conformation are contant but size and markings vary with location. These local varieties should not be accorded breed status.

River type The River buffaloes are usually black, have curled or sickle-shaped horns and are primarily dairy animals. A certain proportion (under 5% in most breeds) are brown. Chevrons are absent except in the Surti breed. Piebald animals are extremely rare but some breeds have white markings on head and feet. In Brazil some breeders of River buffaloes have been selecting for colour types and a white-face strain has been one result. The centres of origin of the chief River breeds and varieties are shown in Fig. 8.1. To the eye the most obvious differences between breeds are in the shape, size and curvature of the horns.

Murrah group. These breeds have been selected for tightly curled horns set close to the head and for high milk yield. The original breed is the Murrah of Haryana state, India, and particularly the area around Delhi. From this has developed the Nili-Ravi of the Sutlej and Ravi river valleys in Pakistan and India. White markings on head, legs and tail switch are common. The Kundi breed of Sind, Pakistan, is also a Murrah derivative differing from the Murrah chiefly in its smaller size.

Other Indian breeds. The recognized breeds of Gujarat are the Surti, Mehsana and Jafarabadi. The Surti is distinguished by two white chevrons under the neck, as in the Swamp type. Horns are sickle-shaped. The Mehsana appears to have arisen from a cross of Surti and Murrah; its

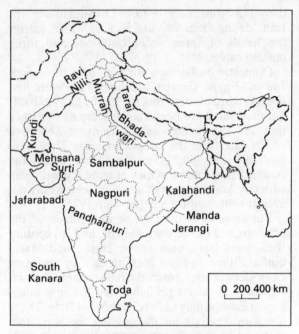

Fig. 8.1 Map of India and Pakistan showing centres of origin of chief buffalo breeds and varieties. (Redrawn from Cockrill 1974)

characteristics are intermediate. The Jafarabadi has broad heavy horns which are downswept and turn upwards at the points. The Bhadawari and the Tarai are improved local types in Uttar Pradesh. Horns are sickle-shaped and upswept. The Bhadawari is usually brown. The Nagpuri of northeast Maharashtra and the Pandharpuri of southeast Maharashtra are characterized by extremely long backward-sweeping horns. The Manda, Jerangi, Kalahandi and Sambalpur of Orissa and its borders are local breeds which are not yet recognized or adequately described. The Toda of the Nilgiri hills (Tamil Nadu) is interesting because of its similarity to the Swamp buffalo and its important use in ritual. Likewise the South Kanara of Karnataka is (or was) of primary interest for buffalo racing.

Desi. The above are the recognized breeds or improved local types of India and Pakistan. In fact, a great many of the buffaloes of the subcontinent are breedless, nondescript, unimproved local types called *desi*. They are the result of haphazard, uncontrolled breeding. Horns may be of any shape. In south India they may have Swamp as well as River characteristics. Their size and productivity are less than those of the improved breeds. They are chiefly work animals used for ploughing, harrowing, road haulage and general draught purposes.

Middle Eastern and Mediterranean. These buffaloes are of River type; they are black (or dark grey) in colour and have small to medium-sized sickle-shaped horns which are directed backwards with the point curving inwards and upwards. They are primarily milk producers or dual-purpose milk/meat animals. The principal populations are in Iran, Iraq, Soviet Union (Transcaucasia), Turkey, Egypt and Europe (Albania, Bulgaria, Greece, Italy, Romania and Yugoslavia). Several of these populations have received infusions of blood from India to improve milk yield, notably Iraq in 1918 (probably Surti and Nili-Ravi) and Bulgaria since 1962 (Murrah, Surti, Jafarabadi). In Italy, on the other hand, there are no records of the introduction of any buffaloes since the 11th century and the Italian herd has thus been a closed herd for a thousand years. The Egyptian buffalo appears to have a degree of resistance to rinderpest to which disease other buffaloes are highly susceptible.

Herdbooks for the water buffalo breeds are, as yet, relatively few. They are maintained for the Murrah in India; for the Mediterranean in Italy; for the Swamp, Mediterranean, Murrah and Jafarabadi in Brazil; and for the Swamp in Australia. For more details on breeds, see Mason (1974a).

Blood groups The inherited blood characters are valuable in studying breed relationships and genetic variation within breeds. Three polymorphic protein systems (haemoglobin, transferrin and albumin) and a number of red cell antigens have been detected in the buffalo (Rendel 1974). It has been noted that Mediterranean buffaloes and Murrahs in Bulgaria possess characteristic specific genetic polymorphism of serum albumins, transferrins and amylases. Makaveyev (1968) observed a close trend in the frequency of the transferrin, albumin and amylase alleles and some of the erythrocytic antigens between the Bulgarian buffaloes and those of the Murrah breed, and concluded that this was immunogenic proof of the common origin and genetic relationship of Bulgarian (i.e. Mediterranean) and Indian Murrah buffaloes.

Recent introductions

Swamp buffaloes were first introduced into Australia's Northern Territory in 1826 when sixteen animals were brought from Timor. In 1838 eighteen buffaloes were obtained from Kisar Island in Indonesia. A further forty-nine animals were received in 1843. When the settlement at Port Essington was abandoned in 1848, the breeding animals were left behind. They spread and multiplied (Tulloch 1974). Finding the environment suitable they became feral and spread to the west of the Cobourg peninsula toward Darwin. The buffaloes of Timor are larger and heavier than those of the Northern Territory at the present time. Animals weighing over 1000 kg are not unusual. The smaller size of the Australian feral Swamp buffalo is attributable to the many years during which the buffaloes were shot in great numbers virtually for the value of the hides alone, and in an attempt to annihilate the feral population which was regarded as a pest. The hunters selected the largest animals and preferred big males. The feral population is probably around 250 000 at the present time. Since 1969 there have been successful ventures at the 're-domestication' of these feral Swamp buffaloes; large numbers are now being ranched in extensive fenced pastures in the Northern Territory. In the past decade there have been exports to Costa Rica, Nigeria, Papua New Guinea and Venezuela.

During the German administration of New Guinea, Swamp buffaloes were introduced into what is now Papua New Guinea from the Philippines (1900–1903). Further introductions were made in 1906 from Indonesia and in 1913 again from the Philippines. The buffaloes were used to haul copra carts in the plantations and to pull cargo barges in waterways. Many became feral between the two world wars. During the Second World War the livestock population was almost eliminated by the Japanese troops and all buffaloes were killed except for a few feral ones. Since then there have been numerous introductions from Australia and New Britain and at the present time there are some 1500 buffaloes in Papua New Guinea. These are being successfully ranched as beef animals while small numbers are being trained in harness and as plough and pack animals.

There have been many introductions of water buffaloes into Africa over the centuries but, apart from Egypt, nowhere in the African continent has the buffalo flourished for long. From 1594 to 1729 there were frequent introductions into Portuguese East African territories from Sri Lanka and Goa. These small buffalo colonies invariably succumbed to disease, were eliminated by war or simply disappeared through mismanagement. In the present century small introductions into Madagascar, Zanzibar and Zaire seem to have been equally unsuccessful but recent importations into Mozambique, Nigeria, Tanzania and Uganda have resulted in thriving herds. There was a small feral herd in the marshes of Ischkeul near Mateur in Tunisia apparently descended from a few domestic animals coming from Naples in the mid-19th century. It was destroyed in 1956.

There have been many importations into the Americas during the past century. Around 1890 Swamp buffaloes reached French Guiana from Indo-China. Some were re-exported to Surinam in 1895. Brazil, with some 500 000 water buffaloes, now has the largest and most rapidly expanding population. Fonseca (1977) states that the first importation may have been in 1890 when buffaloes from French Guiana were landed on the island of Marajó. In 1895 one male and two females of the Mediterranean breed arrived in Brazil from Italy. Other early importations were from Egypt in 1902 and 1907, Trinidad in 1906, Java in 1907, India in 1908 and 1918–22 and Italy around 1920. The principal breeds in Brazil are the Murrah and the Jafarabadi of India, the Mediterranean of Italy and the Swamp buffalo of the Far East. The River breeds are referred to as Preto, meaning black, and the Swamp type as Rosilho, meaning roan (although it is in fact grey). There have been further introductions in the past 35 years from Italy, India and Trinidad. The Preto breeds are seen as excellent dual-purpose milk/meat animals while the Rosilho is a dual-purpose work/meat animal. There is much appreciation of the qualities of the water buffalo in Brazil, and a very active interest in increasing stocks (Ponce de Leon Filho, personal communication).

In addition to Brazil there are buffalo stocks in South America in Bolivia, Colombia, French Guiana, Guyana, Peru, Surinam, Trinidad and Venezuela; in Central American in Costa Rica and Panama (introduced in the 1970s from Trinidad); and in the U.S.A. in Florida and Louisiana (from Guam in 1978). Most of these populations are still small and, with the exceptions of Trinidad and Venezuela, do not exceed a few hundred, but all are thriving and are increasing at rates of up to 10 per cent per annum. The total population in Trinidad at the present time is around 8000. Jafarabadis were brought from India about 1900. Further importations of Murrahs, Jafarabadis, Nagpuris, Nili-Ravis, Surtis and Bhadawaris followed in 1924, 1931, 1938 and 1949. Extensive crossbreeding has taken place and a beef type has evolved called the Buffalypso for which breed status is claimed.

Buffaloes were first introduced to Venezuela in 1922 from Trinidad and were located in the Orinoco delta. They bred well and numbers increased rapidly. However, they were allowed to become feral and they were hunted to extinction for sport, the last one being killed in 1948. In 1960 there was a revival of interest in the possibility of the controlled ranching of buffaloes in the vast swamps of Lake Maracaibo and the Orinoco delta and there were numerous small introductions from Trinidad between 1960 and 1971. Many private cattlemen began to show interest in buffaloes and during the 1970s there were substantial imports of up to 200 a time from Australia, Bulgaria, Italy and Trinidad. These have all adapted well to the environment and the population is increasing. Further importations are planned. The buffaloes are ranched as beef animals in areas which provide an ideal habitat for them but which are totally unsuited to zebu, criollo and European breeds of cattle.

Productivity

Buffaloes are easily trained. Their placid tranquil temperament, affinity for man and intelligence facilitate the process. They are widely used to plough, harrow and grade the fields; in road haulage; as pack and riding animals; to thresh by trampling the rice sheaves; to provide the power for grain, oilseed and sugar-cane mills; to puddle clay and for a variety of other purposes such as raising water from wells. Buffalo teams were used for military purposes to haul heavy ordnance and supplies in the Turkish campaigns. Buffalo cavalry charges made a decisive contribution to the final victory of Siam over Burma in the wars of the 18th century.

Buffaloes may yield less milk than cows maintained under identical conditions but in India, as one example, the average milking buffalo gives much more milk than the average milking cow. The average annual milk production of 491 kg for the buffalo in India is 2.8 times higher than the annual average of 173 kg for the cow (Shah 1967)

In well-managed buffalo dairy farms yields per lactation of 1000 kg are common. In the best enterprises in Italy, with twice-daily machine milking and a lactation period of 280 days, an average daily yield of 8–10 kg is normal. State and military farms in India and Pakistan record similar yields. In Bulgaria buffaloes often attain a yield of 2000–2500 kg per normal lactation period of 300 days, while exceptional yields of 4000 kg have been recorded. In India, there are instances of Murrah buffaloes producing 5000 kg in a single lactation.

The milk is rich in both butterfat and in non-fat solids, but is otherwise closely similar in chemical composition and physical properties to cow's milk. Butterfat content is usually 7–10 per cent but many individual animals show a very high percentage and it is not uncommon to find a 15 per cent butterfat content.

In most of the countries in which buffaloes are of economic importance their value and importance as work animals or as milk producers and their longevity, have distracted attention from their outstanding potential as meat producers.

Future prospects

The River breeds are highly prized milch animals and increasing numbers are being maintained for milk production. With breeding, management and feeding steadily improving, yields are increasing.

As a wallowing animal and a powerful swimmer the buffalo is well adapted to swamp or marsh environments and to areas which are seasonally inundated. There are such areas covering vast expanses of land in many parts of the world, e.g. in China, the Amazon valley, the marshes of southern Iraq, the floodplain of south Kalimantan in Indonesian Borneo, the 'top end' of the Northern Territory of Australia, the Laguna de Petén of Guatemala, the Orinoco delta of Venezuela, the basins of the Mucelo and the Pungué rivers of Mozambique, Bendel state in Nigeria and the Okefenokee of Florida, U.S.A.

Such areas provide only low-grade grazing and cattle do not do well on them. Buffaloes, however, can be ranched successfully and reared for slaughter at 2–3 years where cattle, if they survive, will not reach slaughter maturity till they are 4 or more years old.

The buffalo has a great future as a meat producer. Until very recently almost the only meat reaching the market came from aged animals at the end of a long productive life. It has now been established in Australia, Bulgaria, Italy, Thailand, Trinidad and Yugoslavia that when buffaloes are reared and managed for slaughter at 12–18 months of age and at an optimum live weight of 300–350 kg, depending on breed, a satisfactory yield of high-grade meat can be obtained at much less cost than with cattle. This is mainly due to the outstanding feed conversion efficiency of the buffalo and the fact that there is no need for an expensive finishing procedure. Daily weight gains, carcass yields and muscle:bone ratios generally equal or exceed those of cattle reared under identical conditions. The meat is lean, tender and palatable. It contains less fat than that of cattle. The classic procedure of mating dairy cows with beef bulls and raising the calves for early slaughter can profitably be applied to buffaloes. The production of buffalo meat as veal or beef from buffalo dairy herds by the use of beef-type males, e.g. of the Swamp type, on such milch breeds as the Kundi, the Surti and the Mediterranean, offers opportunity for development. It is suggested that such countries as India and Pakistan should encourage and subsidize the rearing of buffalo calves by small farmers. The young animals would enter feedlots or beef herds destined to help satisfy the insatiable demand for meat. This should be a highly profitable development.

Milking buffaloes in multiple ownership are sometimes maintained in large conurbations which cover many hectares. Examples are the Aarey Milk Colony, Bombay, India, which has a population of 12 000 and the Landhi Colony near Karachi, Pakistan, which has some 40 000. Such large colonies are uneconomic and wasteful of good genetic material. The buffaloes are usually maintained for only one or sometimes two lactations and are then slaughtered. Very few calves are reared, most being allowed to die of starvation. The colony system which is highly profitable is likely to continue in spite of its crude inefficiency and the malpractices which it perpetuates, but there is a trend to put dry females to graze and after calving to bring them back to the colony. Attempts are also being made to rear the calves to around 4 months and market the veal.

The value of the buffalo as a working animal has been emphasized in recent years due to fuel shortages and steeply rising costs. Most of the world's rice is grown by peasant farmers and the buffalo is more efficient than the machine in raising rice which is a staple food for half the world's human population. It has been shown in the Philippines, for instance, that, under similar circumstances, there is no significant difference in the yields obtained by rice growers using buffaloes and those using tractors, but the tractor has a shorter working life, is more expensive to maintain and produces no dung. Only in countries where manpower availability is low and oil fuel is cheap is there an economic argument for relegating the work animal to history. In the last decade, as the advance of agricultural mechanization in most of the developing world has slowed, the number of working buffaloes has increased. This trend seems likely to continue.

Until very recently, artificial insemination and the storage and use of deep-frozen buffalo semen presented many difficulties. Conception rates were generally so low that artificial insemination was not economically viable. Research, conducted mainly in India, Pakistan and Bulgaria, has resulted in improved techniques bringing the procedures to the level of efficiency attained in cattle. Artificial insemination is now a highly important tool in improving the quality and productivity of water buffalo breeds and crossbreeds, and its use is becoming widespread in almost all buffalo-owning countries. When combined with extensive milk recording the use of artificial insemination makes possible the efficient progeny testing of males for milk production and so opens the way to the increase in milk yield which has occurred in the dairy cattle breeds of the western world using these techniques.

The domestic water buffalo population of the world provides a major source of farm power, milk and meat. Farm level output can greatly be increased through improved herding and management and the conservation, preservation and use of selected, superior germ plasm. Rational selective breeding under conditions embodying better nutrition and improved health control can yield spectacular results in no more than two buffalo generations especially where, as in India and Pakistan, the genetic potential is high. The means exist for improving the quality and productivity of water buffaloes. As the technical procedures are extended and are widely applied, genetic improvement and greatly enchanced yields of quality milk and meat and of work output will be achieved in all the many countries which have significant and increasing water buffalo populations (FAO 1978).

References

Boehmer, R. M. von (1974) [The appearance of the water buffalo in Mesopotamian history and its Sumerian characteristics]. *Zeitschrift für Assyriologie*, **64**: 1–19. (In German)

Bongso, T. A., Kumaratileke, W. L. J. S. and **Buvanendran, V.** (1977) The karyotype of the indigenous buffalo (*Bubalus bubalis*) of Sri Lanka. *Ceylon Veterinary Journal*, **25**: 1–4, 9–11

Brentjes, B. (1969) [The water buffalo in the civilizations of the ancient Orient]. *Zeitschrift für Säugetierkunde*, **34**: 3. (In German)

Cockrill, W. Ross (ed.) (1974) *The Husbandry and Health of the Domestic Buffalo*. FAO: Rome

Cockrill, W. Ross (1976) *The Buffaloes of China*. FAO: Rome

Creel, H. G. (1936) *The Birth of China: a survey of the formative period of Chinese civilization*. Cape: London

Epstein, H. (1969) *Domestic Animals of China*. Commonwealth Agricultural Bureaux: Farnham Royal, Bucks, England

Epstein, H. (1971) *The Origin of the Domestic Animals of Africa*. Africana Publishing Corporation: New York

FAO (1978) *Report of the FAO/SIDA/Government of India Seminar on Reproduction and Artificial Insemination of Buffaloes*. FAO: Rome

FAO (1982) *FAO Production Yearbook 1981*, Vol. 35. FAO: Rome

Fischer, H. and **Ulbrich, F.** (1968) Chromosomes of the Murrah buffalo and its crossbreeds with the Asiatic Swamp buffalo (*Bubalus bubalis*). *Zeitschrift für Tierzüchtung und Züchtungsbiologie*, **84**: 110–14

Fonseca, W. (1977) *O Búfalo: sinonimo de carne, leite, manteiga e trabalho*. Ministerio da Agricultura: São Paulo, Brasil

Heck, H., Wurster, D. and **Benirschke, K.** (1968) Chromosome studies of members of the subfamilies Caprinae and Bovinae, family Bovidae; the musk ox, ibex, aoudad, Congo buffalo and gaur. *Zeitschrift für Säugetierkunde*, **33**: 172–9

Higham, C. and Kijngam, A. (1979) Ban Chiang and northeast Thailand; the palaeoenvironment and economy. *Journal of Archaeological Science*, 6: 211–33.

MacGregor, R. (1941) The domestic buffalo. *Veterinary Record*, 53 (31): 443–50

Makaveyev, Ts. (1968) [On the study of biochemical polymorphism and blood groups of the water buffalo (*B. bubalis*)]. *Animal Science. Sofia.* 5 (6): 3–20. (In Bulgarian)

Mason, I. L. (1974a) Species, types and breeds. In: Cockrill (1974) (op. cit.) pp. 3–47

Mason, I. L. (1974b) Genetics. In: Cockrill (1974) (op. cit.) pp. 57–81

Pandey, M. D. and Raizada B. C. (1979) Overcoming summer sterility in buffaloes and cows. In: *Buffalo Reproduction and Artificial Insemination*. FAO Animal Production and Health Paper No. 13. FAO: Rome

Rendel, J. (1974) Blood groups and protein polymorphisms. In: Cockrill (1974) (op. cit.) pp. 82–7

Shah, M. K. (1967) Buffalo, the mainstay of the dairy industry. *Indian Farming*, 17 (8): 39–42

Tol, G. L. and Halnan, C. R. E. (1976) The karyotype of the Australian Swamp buffalo (*Bubalus bubalis*). *Canadian Journal of Genetics and Cytology*, 18: 101–104

Tulloch, D. G. (1974) The feral Swamp buffaloes of Australia's Northern Territory. In: Cockrill (1974) (op. cit.) pp. 493–505

Ulbrich, F. and Fischer, A. (1967) The chromosomes of the Asiatic buffalo (*Bubalus bubalis*) and the African buffalo (*Syncerus caffer*). *Zeitschrift für Tierzüchtung und Züchtungsbiologie*, 83: 219–23

Ulbrich, F. and Fischer, A. (1968) [The chromosomes of the Turkish and southeast European water buffalo (*Bubalus bubalis*)]. *Zeitschrift für Tierzüchtung und Züchtungsbiologie*, 85: 119–22. (In German)

White, L. (1974) Indic elements in the iconography of Petrarch's *Trionfo della Morte*. *Speculum, Journal of Medieval Studies*, 49 (2): 201–21

Zeuner, F. E. (1963) *A History of Domesticated Animals*. Hutchinson: London

9
Sheep

M. L. Ryder

ARC Animal Breeding Research Organisation, Edinburgh, Scotland

Introduction

Nomenclature

The Latin name *Ovis aries* is used for the domestic sheep to distinguish it from the wild type, the nomenclature of which is confused (see p. 65). The Latin specific names given by early workers to historical domestic types are not recognized today since such types represent varieties or breeds and not different species.

General biology

The hoofed herbivores walk on 'tip-toe' as a specialization for speed, and cloven hoofs are well adapted to walking on soft ground as well as to climbing the stony slopes of the recent natural mountain home of the sheep and goats (family Bovidae, subfamily Caprinae). Sheep can be distinguished from cattle by their narrow, hairy and cleft upper lip which allows them to graze closer and more selectively than cattle. They also have only one pair of nipples compared with two pairs in cattle.

The cud-chewing habit is probably the single most important factor contributing towards the evolutionary success of the ruminant group. Domestic sheep must nevertheless spend 9–11 hours of each day grazing, and 8–10 hours, mostly at night, ruminating (Hafez 1968). Cud-chewing is associated with a four-chambered stomach and a specialized digestive system involving the fermentation of cellulose by micro-organisms in the rumen, which allows the animal to derive nutriment from fibrous material.

Wild sheep breed in November and December and have a gestation period of about 5 months. They have adapted to high latitudes and cold climates by either delaying the breeding season (which is controlled by day length) or by extending the gestation length (Geist 1971). There appears to be no consistent association of breeding season with latitude in domestic sheep, and one wonders whether the onset has been hastened by selective breeding. The main season lasts from the beginning of September until the end of November in the northern hemisphere, and ewes quickly adjust to a transfer between hemispheres.

The Merino and most tropical breeds can breed all the year round, and this has been attributed to their evolution in latitudes with little seasonal change. Among temperate breeds the Dorset Horn is notable for its long breeding season, which starts as early as June. The Finnish Landrace is in season from early October until mid-May, which could be regarded as an adaptation to high latitudes.

Rams can produce sperm throughout the year, although there is a tendency towards quiescence during the summer months in breeds with ewes having a restricted season. In some breeds both sexes can mate as young as 6 months. Ewes are polyoestrous during the breeding season, the heat period recurring in cycles at intervals of about 17 days. The cycle length ranges from 16.4 days in the Scottish Blackface, for instance, to 17.5 days in the Merino. The duration of heat is 30–36 hours with little breed variation. The length of *post-partum* anoestrus is related to the degree of mammary stimulation. If a ewe is 'dried off' immediately after birth, the first heat can occur as early as 1 month afterwards, but in a ewe suckling lambs the anoestrous period can last up to 12 weeks. Although there is considerable individual variation in the length of gestation within breeds, ranging from 143 days to 159 days in the Rambouillet Merino, for example, early-maturing breeds tend to have a shorter mean gestation period. The average for the Southdown and Dorset Horn breeds is only 144 days, compared with 151 days in the Merino.

The fecundity of domestic sheep varies considerably with breed (in their typical environment) suggesting some degree of genetic control. The more fertile breeds such as the Finnish Landrace have lambing percentages between 200 and 300 per cent, Longwools at least 175 per cent, Down breeds 150 per cent, and hill breeds such as the Cheviot and Scottish Blackface less than 100 per cent, although this figure can exceed 150 per cent with better husbandry and feeding, indicating a strong nutritional influence.

The diploid chromosome number of domestic sheep is 54. These comprise 26 homologous pairs plus the sex chromosomes. They are divided into four groups decreasing in length: A (3 pairs); B (8 pairs); C (6 pairs) and D (10 pairs – including the sex chromosomes). Earlier reports that the number was 60 arose from the classification of the six large A-chromosomes as 12 small ones. (For the chromosome number of wild sheep, see p. 65).

The wild ancestor

The main home of wild sheep today is the mountain ranges of central Asia, from which they extend westwards into Europe, and eastwards into America (Fig. 9.1). This spread took place in the Pleistocene period, and is indicative of the success of the group, much of which is due to behavioural and reproductive adaptations so that morphological changes in, for instance, horns and coat colour are of social significance rather than adaptations to the environment (Geist 1971).

Within this area there are four main types of sheep: the argali of central Asia, the bighorn of North America, the urial west of the argali and west of this the mouflon which extends into Europe; the European mouflon is restricted to the islands of Corsica and Sardinia. The mouflon sheep on these islands are now considered to be feral domesticates (Poplin 1979) but they appear no different from truly wild sheep and so must have been introduced by man at a very early date. Compared with other species the wild sheep has retained a wide distribution, the only marked

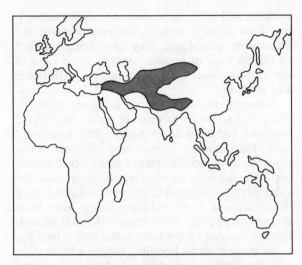

Fig. 9.1 The Old World distribution of wild sheep. The western arm represents the mouflon; south of the Caspian this changes to the urial which continues into the southeastern arm, while the northeastern arm represents the argali. (From Harlan 1976, after Isaac 1970)

decrease having occurred in Europe. The size ranges from 36 kg in the Cyprus mouflon to 200 kg in the Siberian argali. Wild sheep are generally brown in colour but the North American sheep include all-white and all-black varieties.

These popular names are useful since, although the status of genus *Ovis* is well established, and there is little dispute about the numerous races or varieties of wild sheep, what constitutes a species is not clear. According to one view since all wild (and domestic) types apparently interbreed freely, they should be regarded as a single species. Geist (1971), however, points out that interbreeding in zoos is artificial, and that the natural geographical isolation of the different races meant that they never met to evolve reproductive barriers. Much of the difficulty of classification therefore arises because the differences are social rather than anatomical.

Some authors divide the genus into only five species, while others, e.g. Geist (1971) and Corbet *et al.* (1980), recognize six. There is *Ovis orientalis* (mouflon) with two races, *O. o. laristanica*, the Asiatic mouflon, and *O. o. musimon*, the European mouflon. Then there is *O. ammon* (argali) which has nine races. Geist uses *orientalis*

for the urial, which has thirteen races; Corbet uses the more usual name *vignei*. *Ovis canadensis* (bighorn) is the fourth species, but in addition to this (with eight races) both Corbet and Geist recognize *O. nivicola* (the snow sheep) of Siberia with three races, and *O. dalli* (the thinhorn) of Alaska with three races

The last three species were almost certainly never domesticated. Since domestic sheep first appeared in the urial area of southwest Asia, it has long been thought that the urial was the ancestor of domestic sheep, but the mouflon is thought to have contributed towards European breeds, and the argali towards those of Asia. The chromosome numbers will be detailed below during the discussion of domestication.

Domestication and early history

Sheep were domesticated entirely within the prehistoric period by primitive peoples living at the end of the Mesolithic (Middle Stone Age) period. They were almost certainly settled plant cultivators, and not primarily hunters, the sheep being attracted to man by his crops.

The first animals to be domesticated after the dog were the goat and the sheep. Whether the goat or the sheep was domesticated first is not yet clear, because of the fragmentary nature of the skeletal remains, and the difficulty of distinguishing sheep and goat bones. But the goat would have been initially more useful in helping to clear forest, after which sheep would have been economically superior.

It was long thought that the main ancestor of domestic sheep was the urial. Studies of chromosome number were first made with the mouflon, and when the diploid number was found to be 54 (as in domestic sheep) it was thought that all sheep must have the same chromosome number. Improved techniques and access to a greater range of wild sheep showed, however, that the mouflon is the only wild type with $2n = 54$ chromosomes; the urial has 58, and the argali 56. For a time this led to the conclusion that the mouflon was the sole ancestor of domestic sheep, despite the fact that the argali and urial will hybridize with domestic sheep.

Since wild sheep with different chromosome numbers were available for domestication, and since hybrids between different wild types have intermediate numbers, it seems surprising that there appear to be no modern breeds with intermediate numbers. Hybridization between the argali (56) and the mouflon (54) threw light on this puzzle when it was found that the hybrid ewes, which themselves had 55 chromosomes, produced ova with 27. This suggests prezygotic selection towards a lower chromosome number, and shows that the 54 chromosomes of modern domestic sheep need not have come solely from the mouflon, but could have arisen by hybridization between the mouflon and another wild type followed by selection for a reduced chromosome number.

The domestication process was almost certainly so gradual that it probably began long before there is evidence for it, and may have been preceded by a form of 'game management'. The first evidence for the domestication of sheep comes from sheep remains dated *c.* 9000 B.C. at Zawi Chemi Shanidar in Iraq. The evidence for domestication rests on a high proportion of young animals, and a change in the economy. Other sites of early domestication are Jarmo, also in Iraq, dated *c.* 7000 B.C. and Tepe Sarab in Iran, dated *c.* 8000 B.C.

The domestication of herd animals, along with the cultivation of plants, appears to have begun on the grassy, open-forested hills of the 'Fertile Crescent' which extends from Palestine northwards through Lebanon, across southern Turkey, and then southwards through the foothills of the Zagros mountains which straddle the Iraq–Iran border. The three adjacent areas particularly associated with sheep domestication are the Kermanshah plateau in Iran (at Tepe Asiab and Tepe Sarab); in Luristan (Iran) which descends from the Kermanshah plateau to the plain (sites Ali Kosh and Tepe Sabz); and the western foothills of the Zagros mountains in Iraq (Zawi Chemi Shanidar and Jarmo).

The initial domestication was almost certainly not planned. Indeed it is often easier to obtain food by hunting (farming can be harder work) and many peoples did not domesticate any animals in their fauna (e.g. the North American Indians did not domesticate the bighorn). The stimulus is therefore likely to have been ecological rather than economic. The ecological factors involved were probably: 1. the desiccation caused by the retreating ice cap which caused man to share 'oases' with the dwindling numbers of wild animals, and at the same time to 'seek' an alternative food supply; and, 2. the extinction of the Pleistocene megafauna may have caused a shift to the hunting of smaller animals for food and therefore a closer association of man with sheep and goats.

Although the first use of domestic sheep is likely to have been to provide meat, once blood and milk had been developed as other sources of food the animals would have become more valuable alive, so that meat consumption almost certainly declined after domestication. But the value of milk could not have been foreseen before domestication. Nor could the sheep have been domesticated for its wool since in the wild animals this is obscured by the hairy outer coat which has to be made finer by breeding before a fleece suitable for textile use is obtained. At the same time it is wrong to think of different species supplying different needs since most domestic animals have more than one use, and it is common to use all products with no wastage.

Most domestic species form social groups in the wild state, and Zeuner (1963) considered that the initial phases of domestication constituted social overlap between man and the animal. The scavenging action of wild dogs which led to their domestication was symbiosis, but man was a social parasite on reindeer.

The herding sheep and goats went through a taming rather than a parasitic stage, and whereas Zeuner considered that these were domesticated before the cultivation of plants, Reed (1969) regards settled agricultural villages as an essential precursor and there is now appreciable evidence that they in fact came first. Hunting could not have led directly to nomadic pastoralism since the latter usually has some dependence on agricultural communities.

The following stages of domestication are from Zeuner (1963) modified by Ryder (1966):

1. Loose ties of animals with man, but no control of breeding.
2. Confinement and breeding in captivity, with separation from the wild type allowing a distinct domestic race to emerge. This stage was reached during the Neolithic period.
3. Selective breeding by man for certain features with the occasional mating to wild forms. Selection was already practised by the Bronze Age, because by 3000 B.C. in the Middle East distinct 'breeds' can be recognized.

4. The gradual intensification of the development of different breeds with desirable economic characters – wool, milk and meat, coupled with disregard for, and even elimination of, wild and primitive domestic types.

The domestication of sheep (and goats) was an unconscious and gradual strengthening of an association between two social species (man and sheep) pre-adapted by their respective evolutions to be of mutual benefit. Clues on the actual initial process of domestication come from ethnographic and animal behaviour studies. Young animals may have been caught as hunting decoys or as pets, and suckled by women, a practice with modern ethnographic parallels. A man could have returned home with a lamb, and his wife then raised it with the children. The young animal would become attached to the foster mother by the process of 'imprinting'. Since imprinting must take place within a few hours of birth, a human wet-nurse would have been essential. The key to domestication is therefore not man the hunter, but woman in the home.

The genetic consequence of domestication

The changed environment following domestication allowed a greater range of variation (mutations) to survive through reduced natural selection. The new types surviving were selectively bred by man, at first unconsciously, and then fixed by unintentional inbreeding. As well as increased inbreeding, domestication allowed more assortative mating, and later, more outbreeding (crossbreeding).

The migrations of man allowed more outbreeding, with many more different races, than was possible in the wild state. Geneticists recognize that a combination of moderate inbreeding alternating with occasional wide outbreeding, has the effect of producing numerous relatively uniform families. Such a population is more suitable for selective breeding and so the conditions were more favourable for breed formation than for the development of races in the wild form.

Castration of undesired males is essential for effective selective breeding, and there is evidence from intermediate horn sizes that this was carried out in the Neolithic period. The practice may have been preceded by the preferential killing of young males, for which there is evidence on early Neolithic sites in the Middle East. Since undesirable males would have been killed, this could have been the origin of selective breeding.

Reed (1969) pointed out that the automatic elimination of aggressive animals would lead to submissiveness, and that those animals that bred best in captivity would contribute most to the gene pool. Artificial selection by man, in addition to being usually more intense than natural selection, often acts in the opposite direction. The combined effect of all these features was that animals changed more following domestication, and evolved faster than under natural selection.

Most of the major changes in sheep, e.g. horn variation, the lengthening of the tail, and the development of a white, woolly fleece had taken place by the time that illustrations and records first appeared about 3000 B.C. The changes in question (colour and horns) appear to be controlled by relatively few genes.

Many characters of economic importance such as fleece type and milk yield are controlled by numerous genes and so show continuous variation. Selective breeding for such characters was a much more gradual process. The breeding achievements of early man become somewhat less amazing when we realize that he lived very close to his animals and is likely to have observed simple truths such as 'like begets like' long before Mendel or even Bakewell. The unconscious selection involved in cherishing particularly desirable individuals probably led to what Darwin called methodical selection. Ryder (1966) suggested that just as Palaeolithic hunters appear to have had an anatomical instinct, so the Neolithic farmer may have acquired a breeding sense.

The spread of sheep keeping

The prehistoric period Although many sites yield bones which indicate where and when sheep were kept, relatively few have adequate remains to indicate either age and sex ratios, or the importance of sheep relative to other livestock.

At Zawi Chemi Shanidar in Iraq about 9000 B.C. 50 per cent of the bone remains were from sheep (over 40% being immature) and 42 per cent from goats (one-quarter immature). In southwestern Iran between 10 000 and 5500 B.C. in addition to settled villages cultivating cereals, there were seasonal pastoral camps which suggest the existence of transhumance between two grazing areas. This early indication of a suppos-edly specialized form of nomadism suggests that transhumance (at any rate with sheep) is more primitive than true nomadism.

At Tepe Sabz in Iran about 7000 B.C. sheep and goats constituted 67 per cent of the bone remains, and the goat predominated until about 5500 B.C. when both species became more common. During this period only 40 per cent of the goats and sheep reached 3 years of age. There was then a decline in skin-working tools with an increase in spindle whorls used in wool spinning, and after 4000 B.C. sheep predominated. The excess of young males on many early sites may indicate differential mortality rather than killing, but could have created a surplus of ewe's milk which led to the origin of milking.

Sheep had reached Syria by 6000 B.C., and Egypt and other parts of North Africa by 5000 B.C., and this proves diffusion since wild sheep are not native to Africa. The Tassili rock paintings in the Sahara show horned sheep about 4000 B.C. with a smooth coat apparently little different from that of the wild type.

There were domestic sheep in Afghanistan before 7000 B.C. and they had reached Balu-chistan and the Indus valley by 6500 B.C. During the 3rd millenium B.C. in the Quetta valley of Pakistan sheep and goat remains predominated, but in the Neolithic of India which lasted until 1000 B.C. up to 85 per cent of the remains were from cattle. There were sheep on the Yellow river plain of China by 3000 B.C. and diffusion is indi-cated by the similarity of pottery with that from Turkestan.

Turkey had sheep by 6500 B.C. and Crete by 6000 B.C. when sheep/goats formed 75 per cent of the remains. By the end of the early Neolithic, however, cattle had caused the proportion of sheep/goats to be reduced to 61 per cent. Sheep had reached Greece by 6000 B.C. where sheep or goats initially formed up to 80 per cent of the bones. This proportion had fallen to less than 50 per cent by 4500 B.C.

One route of entry of New Stone Age farming into Europe led from Asia Minor up the Danu-bian valley; a second possible route was the Axios valley from Greece. A third main route depended on coastal trade via the Mediterranean, and the Atlantic coasts of Europe, to the western parts of the British Isles. The Danubian mixed farmers gradually spread westwards by clearing the lightly-forested, but highly fertile, loess land of central Europe. Domestic sheep had therefore diffused to Corsica by 5500 B.C. and reached Scan-dinavia, the Atlantic and Britain by 4000 B.C. the last two by way of the western route, the livestock being carried in skin boats (see Fig. 9.2).

During the spread of agriculture through Europe cattle and pigs predominated owing to the forest cover; only later when grazing had become available did sheep and goats again become predominant. Where it was possible to distinguish sheep from goats, sheep were more numerous. Ten British Neolithic sites surveyed by Ryder (1981) had a mean of 44 per cent cattle, 44 per cent pigs, and 12 per cent sheep/goats.

About 4000 B.C. a copper-using culture appeared in Greece in which sheep/goat keeping was important. Since the metal technology probably came from southwest Asia, a new bigger type of sheep could have been introduced into Europe about this time; it could also have been the fleeced type seen in Bronze Age textiles if a fleece had not already evolved in Europe (see p. 70–71). Murray (1970), traced the spread of sheep across Europe with seven copper-using cultures, the last of which flourished in France from 2400 to 2000 B.C. There is a stone relief in Malta of about this date showing a sheep with mouflon-like horns.

The next main influx of a different people occurred about 2500 B.C. when nomadic pastor-alists who could have introduced yet another type of sheep spread across the north European plain. The resulting blend of cultures led to the Bronze Age which in northwest Europe lasted from 2000 to 500 B.C. Because of the nomadism, this period, compared with the preceding Neolithic age, had few habitation sites yielding animal bones. Ryder (1981) was able to quote only only British site; it had 77 per cent sheep/goats , 15 per cent cattle, 4 per cent pig and 4 per cent horse. But for the first time textile remains from human burials indicate what the fleece was like. Further evidence of a fleece comes from rotund sheep figurines. This may have been the time when European transhumance, for which there is evidence in Italy in the Neolithic, became established.

The use of iron began in Anatolia about 1500 B.C. and had reached Britain by 500 B.C. The prehistoric Iron Age saw appreciable change in sheep type (see p. 71), but little change in

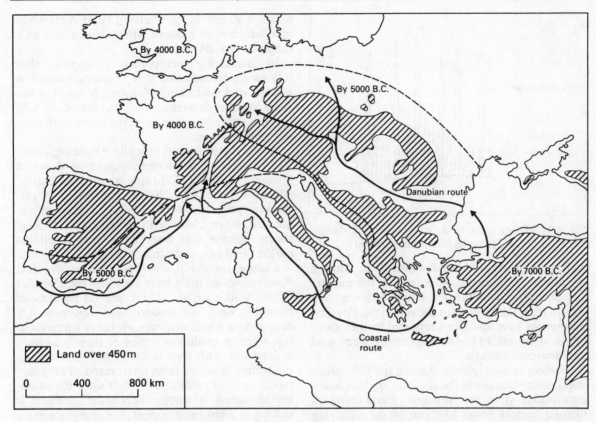

Fig. 9.2 Map showing the spread of domestic sheep
into Europe. (From Ryder 1983)

husbandry although the period in Britain was
characterized by settled mixed farming in which
sheep predominated, and which formed a recognizable precursor of later farming developments.
A survey of seventeen sites by Ryder (1981)
indicated a mean of 27 per cent cattle, 44 per cent
sheep, 14 per cent pigs, and 10 per cent horses.

Changes following domestication. In the search
for morphological changes after domestication,
the big changes in behaviour have been
neglected. Modification of behaviour was
important in the adaptation of wild sheep to a
new environment, and this ability to change
behaviour may have made sheep amenable to
domestication.

Major morphological changes following domestication are as follows: a reduction in the length
of the legs followed more recently by a thickening
of the limb bones; a lengthening of the tail; a
reduction in the size of horns; and the development of breeds that are either hornless in the

ewes, or hornless in both sexes (there are also
breeds in which both sexes are horned). Whether
the lack of horns in some mouflon and urial ewes
is due to a mutation in these wild types, or to
crossbreeding with polled domestic ewes, has not
yet been settled.

The coat of the wild sheep comprises a hairy
outer coat of kemps which obscures a woolly
undercoat (Fig. 9.3), and the whole coat moults
each spring. Since domestication the outer coat
has become finer, and the tendency to moult has
almost disappeared. Most modern domestic
sheep are mainly white, whereas in wild sheep
only the belly is white. The upper parts of wild
sheep have brown agouti kemps as well as all-
black and all-white patches, and the underwool
is grey by dilution. Domestic sheep have a smaller
heart, and a smaller eye socket than wild sheep,
and the brain is about 20 per cent smaller. There
is also a reduction in brain size of about 8 per cent
between primitive domestic and modern breeds.

Since remains from the first domestic sheep are
indistinguishable from those of the wild ancestor
it is necessary to resort to indirect non-morpho-

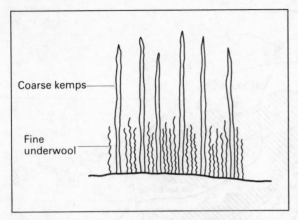

Coarse kemps

Fine underwool

Fig. 9.3 The structure of the coat of the wild sheep. (From Ryder and Stephenson 1968)

logical criteria to establish domestication. Among the possible criteria are differences in the numerical proportions of the sexes and different age groups from those found in the wild type. Physical differences have also been reported in the orientation of crystals of bone substance between wild and domestic animals.

Changes in the skeleton. Among the difficulties of following changes in the skeleton are that many excavations yield few remains. Early workers likened ancient bones to those of the surviving primitive breeds of the country in question, e.g. the Soay of Britain, and this was done from a general impression, or at best on a few measurements. Today it is realized that many measurements are necesssry together with statistical treatment, and that the material is usually inadequate. It is therefore going to be an enormous task to distinguish differences in size due to breed, from within-breed variations due to sex, diet or genetic variation.

There is evidence of hornless ewes in the Neolithic of Iran (7000 B.C.) and on the continent of Europe, but not of hornless rams until after the Middle Ages.

One of the first changes was a reduction in body and horn size. Wild sheep have large forequarters, the difference being exaggerated in rams by a mane and throat fringe, but the full development of the hindquarters is probably a recent development for meat.

It is unlikely that it will ever be possible to follow changes in tail length except in surviving breeds. At birth the Soay (Bronze Age) has a mean tail length of 60 mm, compared with a

mean of 90 mm in the Shetland (Iron Age) while in a selection of modern British breeds it had a range of 120–190 mm.

By measuring metapodials it appears that following domestication sheep became less tall as a result of a shortening of the leg bones, but the characteristic slenderness of the bones of wild sheep is very obvious in domestic sheep until after the Middle Ages.

Bökönyi (1974) used other limb bones to calculate changes in the withers height of sheep in Hungary with the following results: Neolithic, 60 cm; Bronze Age, 70 cm; Iron Age, 60 cm; Roman period, 70 cm; Middle Ages, 56 cm; modern, 80 cm. My own measurements of Soay sheep indicate that Soay ewes have a withers height of 52 cm, and rams 56 cm.

Changes in the fleece. The coat of the first domestic sheep must have been the same as that of the wild ancestor. There are no remains of Neolithic wool but modern 'hair' sheep with a coat of kemp and wool are probably survivals of this stage of evolution; many of them also have a short tail. This type is found in hot areas and the simple coat has been interpreted as an adaptation to a hot environment. But since the stimulus to breed a woolly fleece for clothing is lacking in a hot environment, the sheep are more likely to represent the first animals to have reached these areas.

Man may have begun to select for less hairy fleeces while skins were still used in clothing. The first evidence of change is found in wool from textile remains from the European Bronze Age. This is brown, and there are woolly as well as hairy fleeces, but the latter are much less hairy than the wild type, although the hairy fibres are fine kemps. The similarity of this wool to that of the small brown Soay sheep of St Kilda off northwest Scotland suggests that the Soay is a survival of Bronze Age sheep, and this conclusion is supported by the similarity of bone remains with the skeleton of the Soay. The primitive features shared by the Soay with the wild ancestor are a short tail, a white belly, and an annual moult (which necessitates harvesting the wool by plucking).

The evolutionary changes one can propose from these remains are summarized in Fig. 9.4. The first change was a narrowing of the coarse, outer-coat kemps of the wild sheep (top) to produce the hairy medium fleece of the hairy

Soay. Further narrowing of these fine kemps by selective breeding, changed them into wool fibres of medium diameter, the result being the generalized medium fleece of the woolly Soay (centre).

The same two fleece types predominated in the Iron Age, when the main difference was an increased range of colour, black, white and grey animals being found in addition to the brown of the Soay, and the wild-pattern white belly is rare. Although records and illustrations indicate white sheep in the Middle East about 3000 B.C. during the Bronze Age (see below) no textiles or sheep remain. Europe, however, has sheep with this fleece type and range of colour, e.g. the Northern Short-Tail in Scandinavia, and these appear to be survivals from the Iron Age. They have a short tail and a tendency to moult, and the skeleton is little different from that of the Soay. The type is represented in Britain by the Orkney and Shetland breeds. The stimulus to breed white sheep may have come from the development of dyes, and one way of doing this is to select for greater and greater areas of white wool in piebald sheep. But wild-pattern sheep have been observed to mutate to all-white animals, and these have mutated to self-colour (black or brown) which in turn have mutated to grey or roan (the latter being rare in sheep).

The invention of iron shears in the Middle East about 1000 B.C. may have stimulated selection for fleeces with continuous growth, and this change appears to have been associated with the appearance of heterotype hairs. These are intermediate between kemp and wool, being thick and kemp-like in summer, but thinner and wool-like in winter, when they continue to grow, whereas kemps cease to grow prior to the spring moult. These hairy fibres are a dominant feature of carpet-type fleeces such as the Scottish Blackface, and the finding of some rudimentary hairs in some Scythian wool from central Asia dated 400 B.C. suggests that these originated from the fine kemps of a hairy medium type of fleece, the change being from kemp to hair instead of from kemp to medium fibres (Fig. 9.4).

The three fleece types at the right of Fig. 9.4 all appeared in antiquity, but discussion of their evolution must follow the historical details to be given below.

The sheep of the ancient civilizations About 3000 B.C. while Europe was still in the Stone Age,

Bronze Age civilizations were developing in the Middle East. So from Mesopotamia and Egypt, there is evidence of sheep in artistic representations and records written on clay tablets. The earliest kind of sheep depicted in Sumer had corkscrew horns (the ewes sometime being polled), a tail of medium length, and apparently a hair coat, although some were shown with a fleece. The presence of corkscrew horns has been confirmed by remains and they persist in such breeds as the West African Uda, and the Hortobāgy variety of the Hungarian Racka. Another type of sheep represented was a wooled animal with a short, thin tail and 'normal' coiled horns. The sheep are often shown being milked, and records indicate that milk and wool were more important products than meat. In one count of remains there were 57 per cent pigs, 23 per cent sheep or goats and 19 per cent cattle.

The third main type of sheep had coiled horns, apparently no fleece and a fat tail. This mode of food storage probably evolved in an arid area of the Middle East, and it is of interest that the earliest fat-tailed animals depicted had a short broad tail since it has been suggested that the tail of sheep was bred longer in order to increase its fat-carrying capacity. During Babylonian times from 1800 B.C. and Assyrian times from 1300 B.C. the main sheep depicted had a fleece and a long fat tail.

It is difficult to summarize in this brief account the considerable literary evidence of sheep and particularly wool from 2000 B.C. onwards in Babylonia, where it was of immense importance (being listed along with corn and oil as one of the three main products). There are records of flocks of sheep numbering thousands, as well as an indication that twins were desirable. Meat sheep were distinguished from wool sheep, and the three main grades of wool: first, second (having a yellow discolouration) and mountain, are suggestive of the modern world-wide divisions into Merino, crossbred and carpet. As today there was subdivision into other sorts, and a distinction was made between white and coloured (black and brown) wool. The oldest surviving weight was used for weighing wool, and from this one can deduce that the average fleece weight was 1 kg.

Illustrations and records between 750–500 B.C. give evidence of the marking of animals with the symbol of the owner as well as the use of a belt

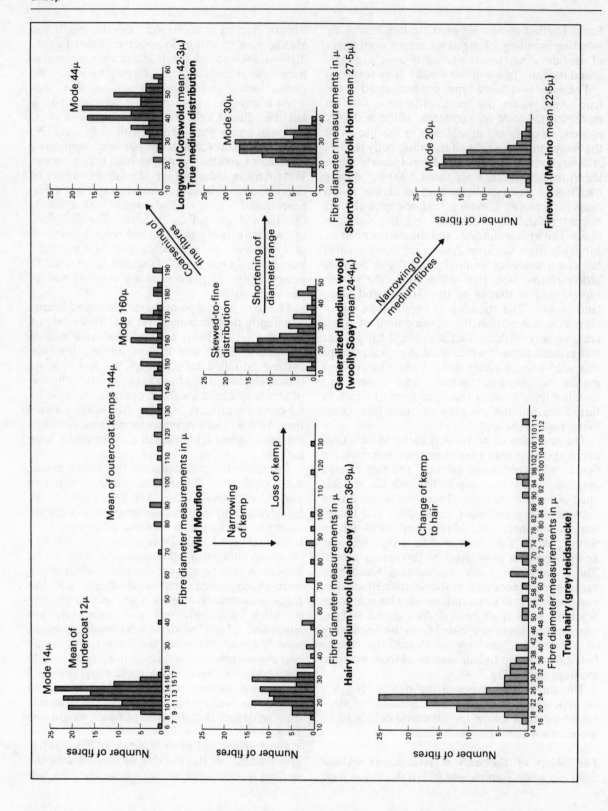

Fig. 9.4 Changes in fibre diameter distribution during fleece evolution. The coat of the wild sheep is shown at the top left. The hairy medium type of fleece represented by the surviving hairy Soay appears to have arisen from the wild type by a narrowing of the outer coat kemps. Further narrowing of these kemps to produce wool fibres of medium diameter could have given the generalized medium type represented by the woolly Soay.

A change of kemp to hair could have produced the true hairy type (bottom left) seen in the modern Scottish Blackface. Other modern types (on the right) could have arisen from the generalized medium, the longwool (true medium) by a coarsening of the fine fibres, the fine wool (Merino) by a narrowing of the medium fibres, and the shortwool by a shortening of the diameter range. (From Ryder 1983)

to support the fat tail. There are references to the shearing as well as the plucking of animals, in addition to terms indicating different ages, conditions, and methods of raising. Variety was indicated by references to sheep from different areas, and there are clear records of white, black, brown, dappled (piebald) and yellow sheep, but whether this means 'tan' colour or a yellow discolouration is not clear.

Sheep and wool were less important in Egypt, and there are fewer surviving records. The first kind of sheep depicted about 2500 B.C. have corkscrew horns and a hair coat, as in Mesopotamia, which suggests diffusion from that country, but the tail is long and thin. These are shown black, white and piebald. One wool sample I measured from this period was white and hairy (carpet) type. After 2000 B.C. the type depicted has normal horns and a fat tail, but still apparently no fleece. Not until after 1000 B.C. are sheep shown with clear wool staples. However, of three white samples of wool dated 14th century B.C. examined two were of hairy medium type and one a generalized medium wool; although white, they were in fact coarser than roughly contemporary samples from the Danish Bronze Age.

The third great civilization of antiquity, that of the Indus valley in Pakistan which flourished about 2500 B.C., had sheep but apart from an illustration of one with corkscrew horns by Epstein (1971) nothing is known of their type or

husbandry. The Rig Veda written about 1000 B.C. refers to the shearing of sheep and the weaving of wool, but cattle later predominated in the Indian subcontinent. The earliest records of sheep in China of about 1300 B.C. indicate that they were kept by possibly nomadic tribes who were not Chinese. In the steppe region of central Asia, Bronze Age transhumance gave way to true nomadism during the 1st millenium B.C. Here the Scythians were the first people to emerge into history (see Scythian wool on p. 71 above), and among fine artistic representations of sheep is one from the Ukarine of the 4th century B.C. showing polled ewes with a medium-length tail and a short curly fleece. This appears to be more like Greek sheep (see p. 74) than the hairy type referred to above.

Persia was greatly influenced by Assyria, and the sheep on the relief at the Achaemenid palace of Persepolis of about 500 B.C. may represent Assyrian tribute. The animals have long fat tails, and the wool staples are shown in exceptional detail. The points have the true posterior-ventral direction each one having a curl at the tip as in the first (or hogg) fleece.

The Bible provides many records of the sheep of the nomadic Israelites before they settled in Palestine about 1400 B.C. The accent is on husbandry rather than sheep type although there are references to white sheep, and it is clear from the story of Jacob's ability to breed spotted lambs that he knew which of his ewes and rams were carrying the recessive gene for the spotting character and would throw such lambs. Although not conclusively mentioned in the Bible, fat-tailed and broad-tailed sheep continue to be illustrated in the area until after the Roman period. The first literary distinction between the two was made by Herodotus in the 5th century B.C.

The people of Phoenicia were the leading seatraders between the 14th and 4th centuries B.C., but there is no conclusive evidence supporting past assertions that they spread a fine-woolled sheep around the Mediterranean. This appears to have been done by the Greeks (see p. 74 below). The Phoenicians may have introduced a hairy thin-tailed sheep to northwest Africa since fat-tailed sheep appear to have spread along the coast to the area around the Phoenician settlement of Carthage (Fig. 9.5). Among sheep illustrations in Tunisia dating from the 4th century B.C. to about A.D. 300, only one has a (long)

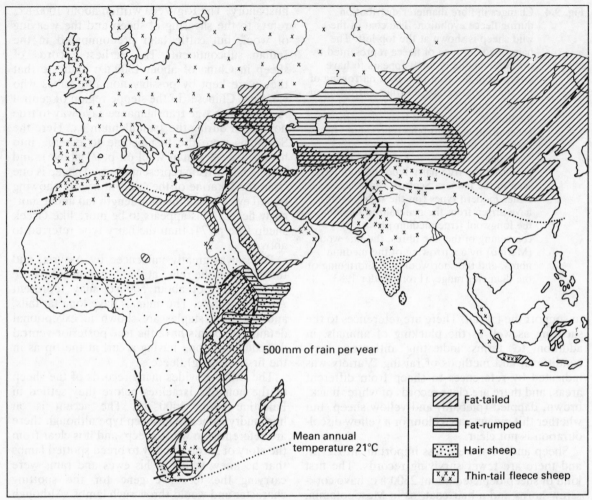

500 mm of rain per year

Mean annual
temperature 21°C

Fat-tailed

Fat-rumped

Hair sheep

Thin-tail fleece sheep

Fig. 9.5 The distribution of the main sheep types in
the Old World (From I. L. Mason)

thin tail, seven have broad tails, and five have
long fat tails.

From about 1500 B.C. the pastoral Hittites of
Anatolia have left records of sheep husbandry
and the value of wool, which was ten times that
of the meat of a sheep. A stone relief of the 14th
century B.C. illustrated by Epstein (1971) shows
horned sheep with long thin tails. Much earlier,
people from this area had settled in Crete and
here, also about 1500 B.C. there flourished the
Minoan civilization, the wealth of which, as in
Babylonia, was based on wool. Clay tablets from
Knossos give censuses of flock size and wool
production by wethers (fleece weight 0.73 kg) but
there is nothing to indicate the type of fleece or

sheep. The Minoans influenced the Mycenaean
civilization of mainland Greece, but it was not
until the classical period about 500 B.C. that sheep
sculptures become sufficiently detailed to indicate
fine-woolled fleeces (see reference to Ukraine on
p. 73). The three main products of Greek agriculture were sheep, olives and wheat, and considerable evidence on sheep husbandry comes
not only from the scientific writings of Aristotle,
but from Greek literature.

Wool remains from a Greek colony in the
Crimea dated the 5th century B.C. confirm the
fineness of the fleece (in an area associated with
fine wool in antiquity through the legend of the
Golden Fleece). The Greeks established similar
colonies throughout the Mediterranean (Fig. 9.6)
and no doubt took sheep with them. Asia Minor
was another Greek area noted for fine wool.

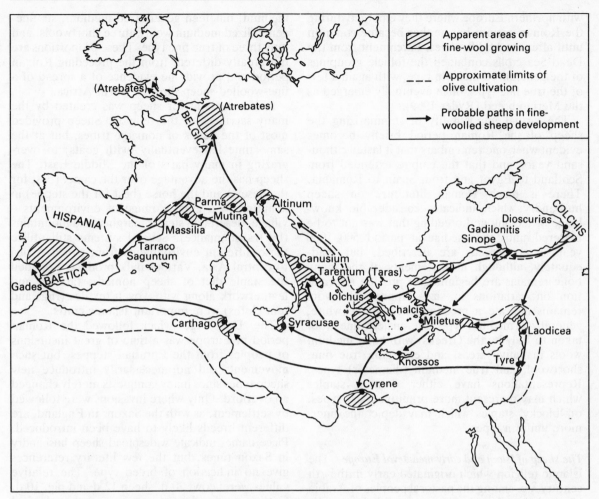

Legend:
- Apparent areas of fine-wool growing
- --- Approximate limits of olive cultivation
- → Probable paths in fine-woolled sheep development

Fig. 9.6 Map showing the ancient areas of fine-wool production deduced from classical literature, and the possible links between them. (From Carter 1969)

This fine wool may have originated much earlier in the Mesopotamia–Palestine region, but the earliest remains from Daliyeh (in the latter area) date no earlier than the 4th century B.C. They did, however, include some true fine wools. Most remains, from Bar-Kokhba and Murraba'at, date from the 1st century B.C.–A.D. (i.e. during the Roman period) and are of immense interest in lacking natural pigment and being fine to the naked eye. Microscopic examination, however, shows them to have a proportion of medium fibres. This skewed diameter distribution makes them identical in structure to the generalized medium wool of the coloured Euro-

pean Soay and Orkney–Shetland types. The generalized medium wool lies in a unique evolutionary position between the hairy medium wool from which it evolved by a narrowing of kemps to produce medium fibres, and three modern fleece types (see above p. 71).

Further narrowing (by breeding) of the medium fibres to produce fine fibres could have produced the symmetrical distribution of the true fine type (Fig. 9.4, p. 72) and support for this change can be gained from the existence of a few true fine wools at that time. If, however, the fine fibres had become coarser and changed into medium fibres then the symmetrical distribution of the true medium wool would have been obtained. A shortening of the diameter range would have produced the distribution of the short-wool. The last two are associated particularly with northern Europe where they emerged during the Roman period, but neither became common until after the Middle Ages. Parchment from the Dead Sea crolls contained the follicle groupings of the generalized medium type, with again a few of the true fine type which eventually emerged as the Merino breed (Ryder 1969).

The virtual impossibility of summarizing the sheep of the Roman period briefly becomes evident when one remembers that it lasted a thousand years, and that the empire extended from Scotland to Egypt and from Spain to Romania. There is much Roman literature on sheep husbandry which indicates considerable knowledge of feeding and breeding that was not to be bettered until the mediaeval period 500–1000 years later. Breeds are described, but inadequately, although measurements of the many bone remains are beginning to provide information on variations in size. Appreciable wool remains indicate a predominance of the white, generalized medium type (the ancient fine wool taken to Italy by the Greeks) with hairy medium wools in native areas, and emerging true fine, shortwool, and true medium (diameter) types. Representations have either pointed staples which may represent more primitive hairy types, or 'blocky' staples which may depict the finer, more uniform types.

The sheep of Islam and early mediaeval Europe The Islamic religion which originated early in the 7th century A.D. among the pastoral peoples of Arabia has had a profound effect on world sheep. The pastoral Arabs expanded and within about 100 years established an empire extending from Spain to China and from Russia to Zanzibar. Arab agricultural works concentrate on plant cultivation, so that one must rely on observations of traditional methods to gain details of sheep husbandry. This means that there are no breed descriptions and no historical confirmation of the suggested contribution of Arabic sheep breeders to the development of the Merino in Spain.

The avoidance of naturalism in Islamic art means few sheep illustrations, although at least one zodiacal sign for Aries depicts a fat-tailed sheep.

Wool remains measured by Ryder (1969) from Egypt, Nubia and Libya and dating from the 6th to the 14th centuries A.D. were mostly white and comprised: four hairy (carpet), seventeen hairy

medium, nineteen generalized medium, six fine-generalized medium wools, three shortwools, and only three of true fine type. These proportions are not greatly different from the preceding Roman period and provide no evidence of a spread of a fine-woolled sheep through North Africa.

A big demand for sheep was created by the many sacrifices of the religion. Sheep provided most of the needs of nomadic tribes, but at the same time led eventually (with goats) to overgrazing in many parts of the Middle East. The sheep has the advantage over the camel (best for the desert) and the horse (best for the steppe) in adapting to a greater range of environments – village, desert and steppe margins, and mountains (by transhumance). Several systems still utilize these different environments from North Africa to central Asia. Various milk products provided the staple diet of sheep nomads; tanning and leatherwork along with woven textiles (rugs and tent cloths) were important Islamic crafts.

The 'Dark Age' which followed the Roman period in Europe was a time of great incursions of peoples from the Eurasian steppes, but such movements did not necessarily introduce new sheep types since many conquests merely changed an overlord. Only where invasions were followed by settlement, as with the Saxons in England, are different breeds likely to have been introduced. Place names indicate widespread sheep husbandry in Saxon times, but the few literary references give no indication of breed type. The relative values were: cow, 20 d, sheep 12 d and pig, 10 d. Cattle predominated among archaeological remains, whereas previously sheep had been more common, but the skeleton was not greatly different from that of earlier sheep. The range of fleece types, too, was similar: ten hairy medium, ten generalized medium, thirteen fine-generalized medium wools, two of true medium type, and eight true fine wools. Norse wools comprised a relatively greater proportion of the coarser type.

Recent history and present status

The Middle East

Sheep form an integral part of the cultural heritage of this region, and their numbers are increasing so that there are nearly as many sheep (177 million) as people (220 million). Most are located in Turkey, Syria, Iraq, Iran and Afghanistan, and 70 per cent are kept under nomadic or transhumant systems, which are the only way of using livestock to exploit areas largely unsuitable for cultivation. The main sheep kept are of fat-tailed type, and milk (plus its products) provides the main food of the nomads as well as 46 per cent of the livestock production. Skins and wool (mostly of carpet type) provide 6 per cent of the income. Few sheep are sold live, mainly for religious sacrifice.

In Turkey 80 per cent of the 32 million sheep belong to fat-tailed breeds. Of these the Karaman of the Anatolian plateau form 62 per cent and the Dağliç further west, 17 per cent. The Karayaka is a thin-tailed but coarse-woolled breed kept on the Black Sea coast (3%) and in the northwest is the thin-tailed and finer-fleeced Kivircik (8%). This appears to have been the ancestor of three Balkan breeds (see p. 79).

The region was under Arab influence throughout the Middle Ages, and latterly was part of the Turkish Ottoman empire which lasted until 1918. The main sheep (also in Cyprus) remains the Middle Eastern fat-tailed type; in Lebanon, Palestine, Jordan and Syria as the brown-faced, polled Awassi breed it forms 80 per cent of the population. In Iraq the Arabi forms 30 per cent, and the Kurdi 10 per cent (both fat-tails) along with the Awassi.

Iran was also part of the Arab empire from the 7th century until Persian rule was re-established in the 15th century. Although on the trade route between east and west, the seas to the north and south, and mountain ranges within the country form barriers which have allowed many breeds to develop. All have coarse (carpet) wool, but there is only one thin-tail (the Zel). The remaining breeds are fat-tail, usually horned in the rams only, and having at least some black on the face and legs. Whereas the modern breeds have a long fat tail, 16th- and 17th- century illustrations show sheep with a short fat tail.

Afghanistan, too, lies on trade routes, and has been occupied by various peoples including the Persians from which it gained independence in 1747. The most primitive breed is the Turki with a brown hair coat and a fat rump, forming 17 per cent of the population. Another fat-rumped breed is the Afghan Arabi (different from the Iraqi Arabi) which forms 12 per cent of the population, and along with all other breeds has carpet wool. The Karakul (producing lamb skins)

(27%) and the Ghiljai (29%) are fat-tails.

The Arabian peninsula is one of the least-disturbed areas and contains sheep with the most primitive 'hair' coat, e.g. in North Yemen. Some are fat-rumped and therefore short-tailed, and could be recent importations from East Africa. Another 'hair' breed in Yemen has a long fat tail. In Oman there are fleeced sheep with thin tails of short to medium length, which may be remnants of the second wave of sheep into the area. This would imply that a fleece developed before a fat tail. The main breeds of Arabia, the Hejazi and the Nejdi, illustrate the distinction noted by Herodotus between a short (broad) fat tail (in the former) and a long fat tail (in the latter). Both have a hairy (carpet-wool) fleece.

The whole of North Africa was under Arab influence, and became part of the Ottoman empire in the 16th century. Fat-tailed breeds with a hairy (carpet wool) fleece extend westwards as far as Tunisia, (Fig. 9.5, p.74), while Algeria and Morocco have breeds with the same hairy fleeces, but thin tails. This distribution suggests that the fat-tail spread along north Africa from the Middle East, and that any sheep transported by sea are likely to have been thin-tails (see Phoenicians, p. 73). There is certainly no evidence that the thin-tail spread along North Africa. The Beni Ahsen breed of Morocco has wool almost as fine as that of the Merino which supports the suggestion of Phoenician involvement, and although this (and other breeds) could have been introduced into Spain by the Arabs, this is unlikely to have been the sole origin of the Merino since there is evidence of fine-wooled sheep in Spain going back to Roman times. At the same time the fine-woolled sheep now in North Africa could have come more recently from Spain.

Asia

Most sheep are concentrated in the central steppe and mountainous regions. Here nomadic pastoralism and transhumance respectively became highly developed. Asiatic wool is predominantly of carpet type, but was used in the development of felt by nomads who are also noted for milk products.

The sheep of Siberia have short fat tails and wool with a range of colour. This is mostly of carpet type, but there are some hairy medium fleeces. In Kazakhstan the sheep mainly have long fat tails, while in Turkmenistan they are mainly of fat-rumped type, except the Karakul (kept to produce lamb-skins) which has a broad tail. This breed is said to have originated from a cross between the fat-rumped and long fat-tailed types. In discussing the origin of the fat-rumped sheep Epstein (1971) pointed out that the Asiatic breeds have horns, few caudal vertebrae, and a black, brown or grey, coarse-woolled fleece. The African type is quite distinct in being mainly polled, having more tail vertebrae, and a white 'hair' coat as well as a black head. That the Asiatic and African types each had a long, fat-tailed sheep on one side of their range, and a thin-tail on the other, led him to suggest that each originated independently as a cross between these other types.

What seems more likely is that the fat-rumped sheep evolved directly from the short-tailed type while the coat was still at the 'hair' stage of evolution. The fleece possessed by the Asiatic fat-rump can be explained by selective breeding in a cold environment (which could have evolved before the fat rump, see p. 77). The fat tail could then have originated either by a cross between the fat-rump and the long, thin-tail, or by selection of the fat rump for a longer and fatter tail.

The most primitive sheep of the Indo-Pakistan subcontinent are the coloured hair breeds of south India which include at least two short-tailed breeds. Overlapping with these and covering the centre and north India are thin-tailed carpet-woolled breeds some of which have coloured fleeces and several of which have short tails. The majority of the sheep of the subcontinent, however, are in the dry hills of the northwest (Pakistan, Kashmir and northwest India) and, besides the thin-tailed breeds, they include fat-tailed breeds which were probably introduced from the Middle East by Muslims. With one exception (the Lati breed in the Salt Range of northwest Punjab) they are not found east of the Indus. In Nepal the breeds appear to have hairy medium rather than true hairy fleeces, and the Kagi is of particular interest in having a short tail, horns in the rams only, and both generalized medium and hairy medium fleeces with a colour range. These are all characteristics of European Iron Age sheep (see p. 71). Another breed, the Bhyanglung (=Tibetan), has a colour range and

four horns, a characteristic which in Europe is associated with the primitive breeds of the northwest.

The sheep of China (including Tibet) are concentrated in the higher regions of the north and west. The earliest illustration shows the 'hair' type. The sheep of Tibet are short and thin-tailed and have both hairy and hairy medium fleeces. The fat-tailed carpet-wool type prevails in Mongolia and in north and northwest China. There is an intermediate type in central China where the Tibetan and Mongolian types meet. One-third of sheep in China are thin-tailed (Tibetan), one-half are fat-tailed and one-tenth are fat-rumped (in Xinjiang).

Despite a humid tropical climate apparently unsuitable for sheep there are 4.6 million in southeast Asia, 90 per cent of which are in Indonesia, and notably in Java. The basic type in the region has a short, thin tail, with a coarse carpet-type fleece. East Java also has a breed with a short, fat tail, which may have been introduced by the Arabs since there is a stone relief showing such a broad tail dated A.D. 800 in Java.

Africa south of the Sahara

There are four main types of sheep. The most primitive is the thin-tailed hair sheep of West and Central Africa which closely resembles in appearance the sheep of ancient Egypt. The tail hangs to the hocks. The hair sheep of northern Sudan is similar but has a long cylindrical tail which may indicate some fat-tail admixture. The fat-rumped Somali sheep of the Horn is a hair sheep and its short tail is a primitive character. The whole of eastern and southern Africa from Ethiopia south is inhabited by fat-tailed hair sheep. They are very variable in tail form and in extent of woolly undercoat. They may be the direct descendants of the hair breeds of Arabia.

European settlement of South Africa began in 1652 with a Dutch supply station at the Cape. Fat-tailed sheep were obtained from the Hottentots, and other sheep were later imported from Holland and Bengal. In 1789 two Spanish Merino rams and four ewes were taken to South Africa. The government established a Merino stud in 1804 and more Merinos were introduced by the British after 1806. By 1846, of 3 million sheep in South Africa, half were Merinos, and today wool exports are second to gold in importance. Many Australian Merinos were imported before 1929.

Somali sheep imported in 1868 were the basis of the Blackhead Persian breed which numbered 2 million in 1951 but has declined since. They have been exported to East Africa and to tropical America. The Dorper breed has been formed by crossing the Blackhead Persian with Dorset Horn.

Europe

The predominant sheep type in the Balkans is the carpet-woolled Zackel type, among which are pockets of the finer 'Ruda' type. The Turkish Kivircik is said to have given rise to three Ruda breeds: the Thraki of Greece, the Karnobat of Bulgaria, and the Tsigai of Romania (see p. 77). Recent fleece investigations indicate a lack of distinction between the Zackel and the Ruda types. First, the Ruda breeds are no finer than generalized medium, and second, not all Zackels have a true hairy fleece, some breeds have hairy medium as well as generalized medium individuals. Indeed, this range within a breed can perhaps be interpreted as 'evolution in progress' from the hairy medium to the true hairy on one hand, and to the generalized medium on the other.

Two Balkan breeds in particular, the hairy Karakachan of Macedonia, which is predominantly grey, and the finer-fleeced Chalkidiki of the peninsula with the same name in Greece, having a range of colours, appear to be relics of the Iron Age type (see p. 71) which probably remained common throughout the Middle Ages. Other types forming the same relics are the so-called 'Grey Alpine' (Ryder 1983) which includes the now extinct Swiss Bündner; the hairy, black-faced Heath breeds of the north European plain, in which grey predominates, and which appear to have affinities on the one hand with the Zackel of the Balkans, and on the other with the Northern Short-tailed type. Further west the sheep of Corsica and Ushant have the same colour range, if not the same colour inheritance, as the Northern Short-tail, and the latter group, which extends from Finland to Iceland (which it reached about A.D. 900) is noted for large litters (in the Finnish Landrace), grey fleeces (in the Gotland) as well as white, fine (generalized

medium) wool (in the Shetland breed). It is noteworthy that the two best-known European sheep with large litters, the Chios of Greece and the Finnish Landrace, were both originally kept as 'house sheep' in which strong selection pressure can be applied.

The systematic use of mediaeval and later paintings as a source of evidence on European sheep type was begun by Ryder (1964), one of the richest sources of material being Flanders. Virtually all illustrations show short-woolled sheep with a white face and no horns; any with horns appear to be rams. There are occasional coloured animals, but the tail is invariably shown long. In the early Middle Ages Flanders was noted for fine wool, but later English wool gained the reputation for being the finest.

Italy today has over thirty breeds of sheep. Those of Sardinia and Sicily are milk sheep; Sicily also has the triple-purpose Barbaresca derived from the fat-tailed Barbary of North Africa but now showing little trace of fat tail. In southern Italy all sheep are milked, being divided into the carpet-woolled Moscia, and fine-woolled Gentile di Puglia, which probably dates back to the ancient fine wool later crossed with Spanish Merinos. The breeds of central Italy are triple-purpose and include the fine-woolled Sopravissana which is kept in a transhumant system. In northern Italy most breeds have medium or coarse wool, and they are divided into meat and milk types. The Alpine group is interesting (though now few in numbers) having mostly lop-eared breeds: they extend from the French Alps into Yugoslavia. Also in this group are the White Swiss Mountain, the Carinthian of Austria, and the Bergschaf of Germany.

Further north the German plateau had a native 'mixed wool', then there was the Heath type (*Heidsnucke*) on the northern plain, which gave way to Marsh sheep in the coastal areas from which the modern East Friesian Milk sheep developed. The most important breed in Germany is the Merino–Landschaf – a fine-wool in the centre and south kept on a migratory system. It originated from the crossing of the local breeds in the south (especially Württemberg) with the Spanish Merino from 1780 on. These breeds had been improved by the Flemish sheep in the 16th century. Merino sheep had reached Prussia in 1748 and these, with another introduction in

1765, were used to improve the local sheep in Silesia and Saxony into Electoral Merinos. From 1775 onwards Merinos were imported into Austria and these developed into a type with more compact body as well as neck and skin folds.

During the 19th century falling wool prices as a result of cheaper Australian production, coupled with an increased demand for meat, led to the development of the German Mutton Merino (*Merino Fleischschaf*) from crosses with the French Précoce Merino (see below).

The south of France has half the sheep population, and half the breeds of the country including most of the native ones. It has all those with carpet wool, all the milk sheep (20% of the population) and most of the fine wools which form 10 per cent of the population.

The variability of French sheep makes it difficult to place them into groups, but one-quarter belong to native mountain breeds. At the western end of the Pyrenees the breeds have a carpet-type fleece, and the Manech of the Basque area, which may be related to the Spanish Churro, is superficially similar to the Scottish Blackface. In the central Pyrenees the breeds have white faces and horns, with medium wool, often reaching 58s quality, which suggests Merino influence. At the eastern end the breeds appear more primitive, with medium wool that has a tendency to be coloured.

From the Massif central to the foothills of the Alps there is the Causses-Lacaune group which has white-faced polled breeds with medium-quality wool, some of which are milked to produce Roquefort cheese. There are several Alpine breeds kept in transhumance, which are unrelated to the Alpine type of other countries. The white-faced, polled Limousin with wool of 50s–56s quality is an important native breed of central France, but the similar Berrichon and Charmoise breeds contain influence from British sheep, notably the Rommey.

There are breeds on the Mediterranean coast with Merino influence, the original fine-woolled stock probably dating back to an ancient Greek introduction. Merino imports from Spain began in the early 18th century and following experiments begun by Daubenton in 1767 a large import was made in 1786. The flock established then at Rambouillet is still in existence. The types developed are the Arles Merino which transhumes into the Alps in summer, and the Précoce. A cross with the English Leicester produced the Ile de France.

Iberia has three main types of sheep: the fine-woolled Merino, a coarse-woolled type (the Churro) which is milked, and a cross between the two (Bordaleiro, Cruzo or Entrefino). Although the Merino was important in the history of Spain, as well as in the world history of sheep husbandry, its origin (including the name) and early history are uncertain. The name first appears in the 15th century, but the type can be traced back to the 13th century. Arab involvement in the development of the Merino is inconclusive, but there is evidence of fine-woolled sheep in Spain in Roman times (see p. 75). The four main Merino types in Spain were the fine-woolled Infantado and Paular, and the superfine Escurial and Negretti.

At the height of the Merino's importance it was a capital offence to export sheep, but the king began their exodus by sending some to Sweden in 1723. They reached Germany in 1748, France in 1767, Austria in 1775, Britain in 1787 and South Africa in 1789.

During the Middle Ages Britain competed with Spain as a supplier of fine wool (see measurements of wool remains, p. 81). Ryder (1964) explained the origin of the present diversity of British breeds in terms of several waves of introduction. The Neolithic sheep probably had a hair coat like the wild type; the Bronze Age sheep was brown with a white belly and had hairy and woolly types as in the modern Soay. Iron Age sheep had the same fleece structure with a range of colours, black, white and grey in addition to brown, but whether these arose by local evolution or introduction is not clear. The second main introduction probably came about Roman times. These sheep were larger, and had a long tail, as well as being white-faced and mainly hornless, at any rate in the ewes. Crosses of such sheep with the indigenous type could have given rise to the white- or tan-faced horned breeds that are still associated with Scotland and the western parts of Britain.

The Roman sheep probably had links with the ancestor of the Merino on the continent, and could in England have given rise, on the one hand to the mediaeval shortwool, now probably represented by the Ryeland breed, and on the other, to the mediaeval longwool, which may have been like the modern Romney. The third main stock was black-faced, horned and hairy. This appears

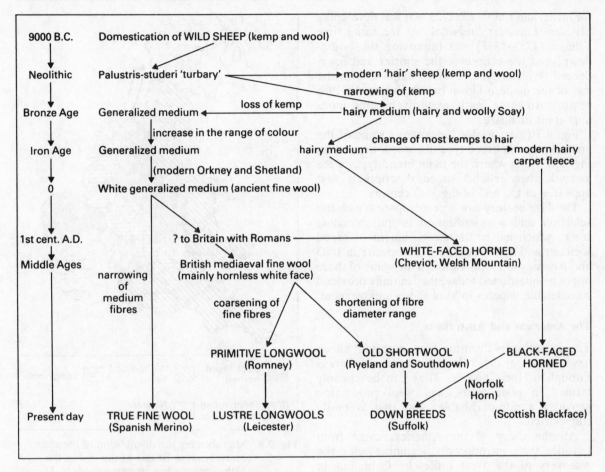

Fig. 9.7 Suggested lines of evolution of fleece and breed types. The dates given are approximate based on the earliest date from the beginning of each period at the centre of domestication in southwest Asia, although, for instance, the Bronze Age fleece evidence comes entirely from northern Europe at a late date. (From Ryder 1983)

to have a link with the similar Heath breeds of the north European plain, and the fact that this type emerged into recent history in the east and north of England, areas occupied by the Danes before the Norman Conquest, suggests that the Danes brought it. But it is also possible that it evolved within Britain from the Iron Age type. On its own it gave rise to the black-faced breeds of the Pennine hills, which led to the Scottish

Blackface, and this type appears to have influenced the Down breeds.

Recent measurement of mediaeval wools from towns ranging from Southampton to Aberdeen indicate that the predominant type was not the true fine wool (7%) nor the shortwool (6%) but the generalized medium wool (36%) The next most common type was the hairy medium wool (27%) and there were 10 per cent true hairy wools. There were also 8 per cent of finer, generalized medium wools, and 6 per cent true medium wools.

Wool fibres remaining in parchments had earlier indicated a coarsening of the wool during the 16th and 17th centuries, and a predominance of medium wools (longwools) during the 18th century. This was when longwools were replacing the shortwools (in order to get a bigger carcass for meat) and Bakewell (1725–95) was developing his New Leicester longwool. At the same time Ellman (1753–1832) was improving the South-down, and the eclipse of the shorter and finer-fleeced Ryeland type became complete with the rise of the modern Down breeds during the 19th century. In these, early maturity became more important than size.

Fig. 9.7 (from Ryder 1969) shows some of the main changes in breed type discussed; the map in Fig. 9.8 shows where the main breed types were located when reliable breed descriptions first appeared at the end of the 18th century.

The 19th century saw a preoccupation with the selection and standardization of purebreeding stock, which led to the establishment of breed societies and flock books, e.g. Shropshire in 1883 and Romney in 1895. Although the value of these might be questioned today, they initially provided considerable impetus to local sheep improvement.

The Americas and Australasia

These continents illustrate how important sheep have been in the colonization of new areas throughout their history. They can be cheaply farmed in poor areas, and wool production provides a non-perishable crop that is easily transported.

All the sheep of the Americas came from outside, their introduction beginning after the discovery of the West Indies by Columbus in 1492. The first mainland area colonized from Spain was Central America. The initial main stock was the hairy (carpet-wool) Churro breed from Iberia, and crosses of this with the Merino gave rise to the criollo (creole) type. The Merino itself did not become important until the 19th century. A third type introduced was the West African 'hair' sheep.

Sheep reached Chile via Peru in 1540. After South American independence about 1850, the Hampshire, Romney, Suffolk and Cheviot were introduced, followed by the Corriedale in 1920 (see p. 83), which predominates in Uruguay. Merinos were introduced into Argentina about 1812 but after 1890 the emphasis changed to meat, and the Lincoln became more important.

Merinos were introduced to Mexico after 1521, but they did not reach Texas and California until about 1700. The eastern part of North America was colonized from northern Europe during the 17th century, and the first sheep were British

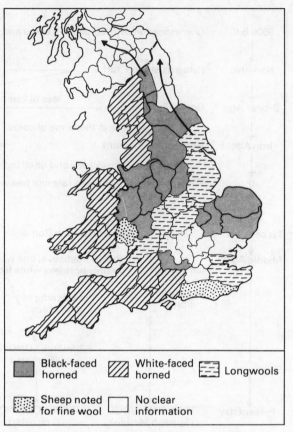

Fig. 9.8 Map showing the distribution of the main breed types in Britain at the end of the 18th century when county records had begun to appear. (From Ryder 1964)

breeds of longwool and Down type. Movement westwards began about 1800, and about that time sheep-ranching with Churros or poor Merinos developed in the southwest.

After 1850 the Spanish area in the west received Merinos from the east, and they were kept as 'range' flocks. Different Merino types developed were the wrinkly Vermont, the fine-woolled Delaine of Texas, and the eventually more popular Rambouillet from France. During the 19th century, the Cotswold, Cheviot, Oxford, Shropshire, Hampshire, Scottish (Highland) Blackface, and Dorset Horn were introduced from Britain.

Settlement of Australia began at Sydney in 1788 and the first sheep were fat-tails from India and South Africa, but included some Spanish Merinos from which the vast wool-growing industry of today developed. By 1840 independent settlements had been established in the coastal towns that later became state capitals, and the main pastoral expansion inland took place between 1840 and 1870. The small and fine-woolled (70s) Saxony Merino was introduced in the 1820s, and this remains on the tablelands of New South Wales and in Tasmania.

The medium-woolled (64s) Peppin Merino was developed in southern New South Wales in the 1870s, using Rambouillet 'blood', and the Peppin was used in turn to develop a large, strong-woolled (60s) strain in South Australia.

The British settlement of New Zealand began in 1840, and in 12 years Merino sheep had spread across the South Island. Pasture improvement caused a change to British breeds, beginning with longwools in the 1850s. The Southdown was introduced in the 1860s, and the emphasis shifted from wool to frozen lamb in the 1880s. Today the 'Canterbury lamb' industry is based on lambs from Romney ewes by Southdown rams. Efforts to combine wool and meat led to the fixing of the Merino–Lincoln crossbred as the Corriedale breed in 1882. A backcross to the Merino was later developed in Australia as the Polwarth breed.

Settlement of the North Island began in 1843, and by 1886 all grazing areas were occupied. There the Romney soon became the most popular breed, and recently 80 per cent of the sheep in the North Island, and 40 per cent in the South Island, were Romneys, with only 2 per cent Merinos, which are confined to the hills. A hairy mutant of the Romney has been developed into the Drysdale breed producing carpet wool.

Feral populations

In St Kilda the island of Soay (which is Norse for 'sheep island') gave its name to the Bronze Age sheep which have probably survived there since prehistoric times. In 1931, after the islanders had left, 100 sheep were transferred from Soay to the main island Hirta, and these have provided a feral study population (of about a thousand animals), as well as a source of animals for more detailed study on the mainland.

The Iron Age sheep with a range of colours is represented by the native Orkney breed which survives in Orkney only on North Ronaldsay. Although not truly feral, since (c. 2000) sheep have owners, they receive little attention living almost entirely on seaweed, being confined to the seashore by a wall encircling the island. Animals have been removed for study, and a reserve population has been established on Linga Holm, another island in Orkney.

Returning to St Kilda, a second feral population remains on the almost inaccessible island of Boreray. These 400 sheep are descendants of Hebridean Blackface sheep kept by the islanders, but which could not be removed when they were evacuated in 1930. They represent the Hebridean Blackface of 1870–1930 and must contain 'blood' from the now extinct Hebridean variety of the Old Scottish Shortwool, which was probably similar to the Orkney (above). The Blackface sheep on Boreray in fact 'fossilize' the 19th-century Scottish Blackface which was the result of a cross between the Old Scottish Shortwool, and the Blackface sheep which came from England in the 18th century.

Another semi-feral population is that of the original Swedish Landrace on Lilla Karlsö off the large island of Gotland in the Baltic. Here about a hundred ewes and six rams, horned in both sexes and showing the range of colours already mentioned, survive as the Goth sheep, the name 'Swedish Landrace' now being given to the white variety, and 'Gotland' to the uniform grey variety bred to produce fur-skins. Norway has a half-wild sheep population in the southwest, which is protected from hunting, the breed of which is not reported.

Australia does not appear to have any feral sheep, but New Zealand has three island population: Arapawa, Campbell and Pitt, plus five mainland flocks: Hawkes Bay, Hokonui, Marlborough, Raglan Harbour and Waianakarua. Most appear to have originated from Merinos in the 19th century, but some populations now contain influence from British breeds. They vary in number from a few hundred to 3000 head. No feral flocks are recorded for the Americas, but there are feral sheep on Hawaii. These have been likened to Shetland sheep, some being coloured, but they no doubt contain influence from the Merinos introduced from Australia before 1929 since one wool sample measured was of Merino quality.

Future prospects

Apart from changes in eating habits, pressure to

reduce the current high consumption of meat in Western Europe is coming from several sources: the high cost, imbalance with developing countries, and a medical belief that it would be healthier to eat less meat. But any overall reduction in meat consumption in Britain could affect the sheep more than other livestock since less sheep meat than any other meat is eaten; it forms only 10 per cent of the total and consumption is declining.

Yet the sheep is an economic animal in terms of energy and labour input. Ruminants eat mainly grass: it forms 60 per cent of the diet of cattle, and an even greater proportion in sheep. So compared with other livestock the sheep would appear to have a greater potential for survival in difficult circumstances, and where arable crops are unlikely to compete successfully. Its role as a universal provider the world over and throughout history has been demonstrated in the present account. Sheep can live in harsh environments unsuitable for other purposes ranging from the subpolar regions to tropical deserts.

Sheep also produce wool as a desirable by-product which meat subsidies in Britain have tended to devalue so that only 10 per cent of the income from the ewe is provided by wool. But the human population has increased faster than the sheep population; so cheapness is not the only reason why synthetic fibres are used as a substitute for wool. The Soviet Union is one of the few areas in which sheep numbers have increased. In the U.K, they tend to remain constant, and in Australia they have actually fallen. This is because it tends to be uneconomic to keep Merino sheep solely for wool production. Although there have been attempts to improve the carcass of fine-woolled sheep, and attempts to find an alternative to the costly process of shearing, diversification in Australia has often meant a move from sheep into beef. Ecologists have recently pointed out more serious limitations to sheep numbers in arid areas of Australia where overgrazing has destroyed the natural plant cover. Unless proper management is started soon the process might well become irreversible.

Sheep numbers have declined enormously in the United States since the Second World War through competition from synthetic fibres and the unpopularity of sheep meat. Yet sheep in the United States can convert low-grade fodder into food and wool with little energy cost, and if necessary the sheep population could be quickly expanded tenfold using only range land unsuitable for cultivation. So long as cheap energy remains available, wool in the United States will remain a luxury product, but if conservation is demanded, with the greater use of renewable resources, then sheep have a promising future.

In Britain the entire sheep industry is subordinated into an integrated system designed for the production of fat lamb, despite a falling demand for sheep meat, and a large import bill for clothing wool. Yet there is no biological reason why Merino influence should not be introduced to make the British wool-clip finer. One threat to the future of sheep is the uniformity brought about by specialization on one aspect such as fat lamb. This means a lack of variation on which future selective breeding for new uses would depend. It is increasingly being realized that populations of uneconomic and therefore declining breeds must be kept for possible future developments that are currently unforeseen.

The sheep appears to be under less of a threat than other livestock, but the chances of survival would be better with a more generalized animal like that in mediaeval England which gave milk and wool, and manure for arable land, only providing meat after a relatively long life.

References

Bökönyi, S. (1974) *History of Domestic Mammals in Central and Eastern Europe*. Akadémiai Kiadó: Budapest

Carter, H. B. (1969) The historical geography of fine-woolled sheep. *Textile Institute and Industry*, 7: 15–18; 45–48

Corbet, G. C. and **Hill, J. E.** (1980) *A World List of Mammalian Species*. British Museum (Natural History): London

Epstein, H. (1969) *The Domestic Animals of China*. Commonwealth Agricultural Bureaux: Farnham Royal, Bucks, England

Epstein, H. (1971) *The Origin of the Domestic Animals of Africa*, Vol. 2. Africana Publishing Corporation: New York

Geist, V. (1971) *Mountain Sheep*. Chicago University Press

Hafez, E. S. E. (1969) *Adaptation of Domestic Animals*. Lea and Febiger: Philadelphia

Harlan, J. L. (1976) Plant and animal distribution in relation to domestication. *Philosophical Transactions of the Royal Society London*, B. **275**: 13–25

Isaac, E. (1970) *The Geography of Domestication*. Prentice-Hall: New Jersey

Mason, I. L. (1967) *The Sheep Breeds of the Mediterranean*. Commonwealth Agricultural Bureaux: Farnham Royal, Bucks, England

Mason, I. L. (1969) *A World Dictionary of Livestock Breeds, Types and Varieties*, 2nd edn. Commonwealth Agricultural Bureaux: Farnham Royal, Bucks, England

Murray, J. (1970) *The First European Agriculture*. Edinburgh University Press

Poplin, F. (1979) Origine du mouflon de Corse dans une nouvelle perspective paléontologique: par marronage. *Annales de Génétique et de Sélection animale*, **11**: 133–43

Reed, C. A. (1969) The pattern of animal domestication in the prehistoric Near East. In: *The Domestication and Exploitation of Plants and Animals*, ed. P. J. Ucko and G. W. Dimbleby, Duckworth: London

Renfrew, A. C. (1972) *The Emergence of Civilization*. Methuen: London

Ryder, M. L. (1964) The history of sheep breeds in Britain. *Agricultural History Review*, **12** (1): 1–12, (2): 65–82

Ryder, M. L. (1966) The exploitation of animals by man. *Advancement of Science*, **23** (103): 9–18

Ryder, M. L. (1969) Changes in the fleece of sheep following domestication (with a note on the coat of cattle). In: *The Domestication and Exploitation of Plants and Animals*, ed. P. J. Ucko and G. W. Dimbleby. Duckworth: London

Ryder, M. L. (1981) Livestock. In: *The Agrarian History of England and Wales*, ed. S. Piggott, Vol. 1, Part I, pp. 301–410. Cambridge University Press: Cambridge.

Ryder, M. L. (1983) *Sheep and Man*. Duckworth: London

Ryder, M. L. and Stephenson, S. K. (1968) *Wool Growth*. Academic Press: London

Trow-Smith, R. (1957) *A History of British Livestock Husbandry to 1700*. Routledge and Kegan Paul: London

Trow-Smith, R. (1959) *A History of British Livestock Husbandry 1700–1900*. Routledge and Kegan Paul: London

Zeuner, F. E. (1963) *A History of Domesticated Animals*. Hutchinson: London

10
Goat

I. L. Mason

*Formerly Animal Breeding Officer,
FAO, Rome, Italy*

Introduction

The domestic goat, *Capra hircus*, derives from the wild goat or bezoar, *C. aegagrus*, but other wild species of *Capra* are also called 'wild goats'. The name is sometimes extended to the related tahr of the genus *Hemitragus* (= half goat). The so-called Rocky Mountain goat (*Oreamnos americanus*), on the other hand, is not a goat but a goat-like antelope (tribe Rupicaprini).

The female is the nanny-goat and the male is the buck or billy, but the terms 'she-goat' and 'he-goat' are commoner in scientific literature.

Taxonomically goats belong to the tribe Caprini of the family Bovidae. The other members of the Caprini are the sheep (*Ovis*), the tahr, and two monospecific genera closely related to *Capra* – *Ammotragus lervia*, the Barbary sheep or aoudad of the Sahara, and *Pseudois nayaur*, the blue sheep of the Himalayas. Neither will hybridize with *Ovis* but fertile hybrids have been obtained from the cross of male aoudad and female domestic goat (Gray 1972).

Morphologically goats differ from sheep in the presence of beard and caudal scent glands in the male. On the other hand sheep have suborbital

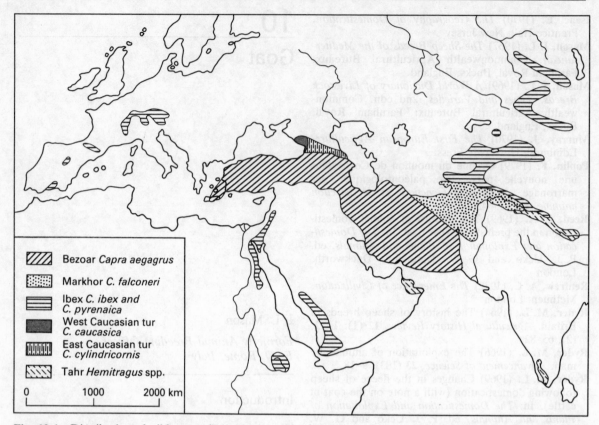

Fig. 10.1   Distribution of wild goats. (From Mason
              1981, redrawn from Harris 1962)

Legend:

- Bezoar *Capra aegagrus*
- Markhor *C. falconeri*
- Ibex *C. ibex* and *C. pyrenaica*
- West Caucasian tur *C. caucasica*
- East Caucasian tur *C. cylindricornis*
- Tahr *Hemitragus* spp.

0   1000   2000 km

tear glands and lachrymal pits in the skull as well
as foot glands, but goats may have these in the
forefeet. In domestic breeds the easiest way to
distinguish the two species is by the carriage of
the tail – held erect in goats and pendent in sheep.

In fighting, sheep butt each other with much
greater force than goats. They have evolved a
new skull shape in which the foramen magnum is
rotated to a point below the point of impact. The
brain is protected from damage by a greater skull
thickness due to the increased extent of the
cornual and frontal sinuses and the septa within.
At the same time the horns of sheep are much
stouter and more closely curled than the slender
vertical horns of goats (Schaffer and Reed 1972).

In very general terms, while the sheep is an
animal of grassy plains the goat is primarily a
browser and prefers a mountainous habitat. It is
also well adapted to semi-desert conditions and
can survive on very sparse fodder.

It produces the same products as the sheep but

fewer breeds are developed for wool production
(in fact only the Angora and the Cashmere and
their derivatives). Its milk yield is, in general,
higher than that of sheep.

The goat is polyoestrous and in the tropics it
can breed all the year round. However, there is
some evidence that tropical breeds can be divided
into seasonal and aseasonal breeds (Devendra
and Burns 1983). In temperate regions the
breeding season is normally in autumn. The
oestrous cycle is 20–21 days in length and oestrus
lasts 2–3 days. In temperate climates the age of
first kidding is normally 18 months but some
breeds if well fed will breed as kids. Litter size
averages 1.5–2 in most breeds but this varies very
much according to breed and feeding. Some small
tropical breeds regularly produce three or even
four kids. Pregnancy duration averages 150–152
days in large European breeds but only 145–146
in small tropical breeds (Devendra and Burns
1983).

The diploid chromosome number of the
domestic goat is 60. The goat is found all over the
world but it is commoner in warmer, drier regions

and in developing rather than in developed countries. This is because of its importance as a subsistence animal for the local population rather than as a producer of meat, milk or wool for export.

Wild species

The tahrs are divided into three species: *Hemitragus jayakari*, the Arabian tahr, an endangered species in the mountains of Oman; *H. jemlahicus*, the Himalayan tahr; and *H. hylocrius* in the Nilgiri hills of southern India. The tahr has short stout horns and no beard. Its diploid chromosome number is 48 (Hsu and Benirschke 1969). It will mate with the genus *Capra* but no offspring result – although abortions have been reported (Gray 1972).

The genus *Capra* is divided into six species (Corbet 1978; Corbet and Hill 1980).

1. *Capra aegagrus* – the bezoar or wild goat and ancestor of the domestic.
2. *Capra ibex* – the ibex, with subspecies in the Alps, central Asia, the Near East and Ethiopia.
3. *Capra caucasica* – the west Caucasian tur. This species has also been called the Kuban or west Caucasian ibex, *C. ibex severtzovi*.
4. *Capra cylindricornis* – the tur of the eastern Caucasus.
5. *Capra pyrenaica* – the Spanish ibex or Spanish wild goat with a national reserve in the Sierra de Gredos.
6. *Capra falconeri* – the markhor of Afghanistan, Pakistan and Tajikistan.

The name *C. prisca* was formerly used for a species which was reputed to be the ancestor of European breeds with twisted horns. It has now been shown that the osteological remains on which it was based are actually the bones of domestic goats which had been wrongly dated (Epstein 1971, Bökönyi 1974).

The ibexes and turs have not been domesticated. (It is true that there are pictures of the Nubian ibex, *C. ibex nubiana*, being *fed* under control in ancient Egypt but there is no proof that it was *bred* in captivity.) The markhor *may* have influenced some of the breeds of central Asia but this is becoming less likely. Harris (1962) suggested that the domestic Nubian goat may have inherited its convex profile from the Abyssinian ibex, *C. ibex walie*. However, Epstein (1971) points out that, when the Nubian goat does have horns, they lack the knots which are characteristic of the ibex. Furthermore, the so-called Nubian goat probably does not in fact originate from Nubia, and certainly not from Ethiopia, and the convex profile is characteristic of goats in the Middle East and India.

The chromosome number of the ibex and the markhor is the same as that of the domestic goat, namely $2n = 60$. The other species have not been examined. The domestic goat gives fertile offspring when crossed with bezoar, markhor, ibex or eastern tur. The same applies to the bezoar × markhor cross. The ibex produces offspring when crossed with the markhor or the eastern tur but their fertility has not been reported. Likewise the western tur will cross with the eastern tur, domestic goat and ibex but the fertility of the crosses has not been reported (Gray 1972).

Bezoar

The name 'bezoar' for the wild goat, *C. aegagrus*, is a corruption of the Persian *pád-zahr* meaning counter poison. It applied to the concretion found in the stomach of the bezoar (and other ruminants) and believed to be an antidote against poison. The Persian for mountain goat is *pázan*, which has the same origin, and gives the alternative names 'pasan' or 'pasang' for the bezoar.

Corbet (1978) accepts five subspecies:

1. *Capra a. aegagrus* – the Persian wild goat, in the mountains of Iran, Turkey and the Caucasus.

The distinguishing feature of this species is the long sabre- or scimitar-shaped horns of the male. These are laterally compressed in cross-section so that the anterior edge forms a sharp keel for some distance above the base. Further up the horn there are irregular knots on the front edge. The horns of the female are shorter but of similar shape.

Epstein (1971) describes the bezoar as follows:

Capra hircus aegagrus is a relatively slender
animal, with a shoulder height of up to 95 cm.
The beard of the male is very long, occupying the
whole width of the chin. The hair on the neck and

shoulder is elongated in winter when a soft under fur is developed in the colder parts of the animal's range. The general colour of the upper parts of the pelage is brownish grey in winter, reddish brown in summer. The under-parts and inner sides of the buttocks and thighs are either white or whitish. The face, a nucho-dorsal stripe, the tail, a collar on the neck expanding into a breast-plate, the throat, chin, beard, front of limbs except the knees, and a flank stripe are all blackish brown, becoming nearly black on the beard, face and some other parts. The knees, the hind and inner surface of the fore-legs and corresponding parts of the hind-legs, including the hocks, are white or whitish. A certain degree of individual variability is displayed in the extent of the black and white markings. Females and young males lack most of the dark markings.

2. *Capra a. blythi* – the Sind wild goat whose range extends from western Sind through Baluchistan to southern Afghanistan. It is smaller than the type species, the ground colour of the coat is paler and the markings are darker. The horns are less sharply curved, often nearly straight, and the knots are few or absent.
3. *Capra a. cretica* – the *agrimi* of Crete. This subspecies is smaller again and averages only 85 cm in height at withers. The *agrimi* was formerly present in the Ida and Lassithi mountains but is now restricted to the Levka Ori (White mountains) in the southwest. Here a national park has been formed to protect it. It has been introduced into three islands off the north coast of Crete (as well as elsewhere in Greece). In 1963 there were about 300 in western Crete, 55 on Theodorou Island, 20 on Agii Pantes Island and 20 on Dia Island.
4. *Capra a. picta* – the wild goat of the island of Antimilos (Erimomilos) and Samothrace. It was called *picta* because of the sharpness of its colour markings. Its numbers were estimated at about 100 in 1963. The goat on Samothrace shows evidence of considerable contamination with the blood of the domestic goat.

The goat on the island of Gioura (*jour-ensis* or *dorcas*) is certainly a feral domestic goat; and some authorities believe that the wild goats on the other Greek islands are also feral and derived from very early domestic stock (Uerpmann, personal communication).
5. *Capra a. turcmenica* – This variety from the

borders of Turkmenistan and Iran is doubt-fully distinct from *C. a. blythi*.

Markhor

This wild goat now has a very scattered distribution in the high mountains of Kashmir, western Pakistan, northeast Afghanistan and southern Tajikistan and Uzbekistan. Although it does not extend into Iran its name derives from the Persian and means 'snake-eater'.

The markhor is much larger than the bezoar and it has a long mane on the neck and chest. It has huge, vertically directed, spiral horns. These are heteronymously twisted, i.e. the right horn is twisted in an anti-clockwise direction and the left horn clockwise. The coat colour is grey in winter and reddish brown in summer. Legs and under-parts are creamy white and there is a dark stripe along the back and the lower part of the front legs.

Because of its discontinuous distribution and the variation in horn form from a very open spiral to a close straight screw, the species has been divided into six or seven subspecies. Some authors (Schaller and Khan 1975, Schaller 1977) would reduce the number to three – the larger flare-horned markhors in the north of Afghanistan and Pakistan, the smaller screw-horned markhors in the south and the Tajik markhor in Afghanistan and the Soviet Union. The position of the so-called *C. f. chialtanensis* is discussed below (p. 90).

The horns are still valued as hunting trophies so that all subspecies are vulnerable and at least two (*jerdoni* and *megaceros*) are classified by the International Union for Conservation of Nature and Natural Resources (IUCN) as endangered.

Domestication and early history

The goat was probably the first ruminant to be domesticated. Bökönyi (1974) suggests that the principal reasons for this were that the wild goat was present in those regions of southwest Asia where agriculture was developing and that the goat is an extremely hardy animal which could withstand the rigours of being 'reduced to the state of domestication' as Zeuner (1963) so aptly puts it. The goat would also be more suitable than sheep in a forest environment. The species concerned was undoubtedly *C. aegagrus*. Its horns have the anterior keel of domestic goats and their scimitar shape is mirrored in that of many modern breeds. Those of the ibexes, the turs and the markhor are quite different. The bezoar sometimes has a slight twist at the end of the horn and it is presumably selection from this beginning that has given rise to the homonymously twisted horns of many modern breeds (i.e. with the right-hand horn twisted clockwise).

The time of domestication was before 7000 B.C. and the place was probably the slopes of the Zagros mountains on the borders of present-day Iran and Iraq. The evidence for time and place is based on the identification of bones from archaeological sites. Much doubt remains because it is difficult to distinguish between wild and domestic forms. It is also difficult to distinguish between bones of sheep and those of goats so that many reports from excavations speak of 'ovocaprids'. Skulls and horns of sheep and goats are comparatively easily separated but the bones of the extremities are very similar. However, there are differences which can be detected if the sample is large enough.

The ibex can be distinguished from the true goat because the horns of the former lack the sharp anterior keel of the latter. The distinction between bezoar and domestic goat is naturally more difficult. It depends on finding remains in which the bones of one particular age predominate (due to selective slaughter of a domestic animal) or in finding the bones in sites which are outside the range of the wild species.

Because of these uncertainties there should be more 'probablys' and 'apparentlys' in what follows. However, while further discoveries may well place the date of first domestication earlier in time, they are unlikely, given the distribution of the wild goat in relation to early civilization, to move the place geographically outside the Fertile Crescent.

Middle East

The earliest osteological remains which can be identified with certainty as belonging to domestic goats were found at Tepe Ali Kosh, Deh Luren, Khuzestan, southwest Iran. They date from before 7000 B.C. Here they are already outside the habitat of the bezoar in the mountains and the identification of the remains as being from domes-

ticated animals is based on the high proportion of bones from yearlings. However, according to Bökönyi (1976), the earliest domestic goats were found at Asiab on the Iranian plateau, about 8000 B.C. Goat remains precede those from sheep. Domestic goats from succeeding centuries have been recorded in other sites in the Near East. They were probably present at Cayönü in southeast Anatolia about 7000 B.C. They were certainly present at Jarmo in Iraqi Kurdistan which site is dated from about 6500 B.C. to the end of the 6th millennium. They were also present in Palestine (Beidha, El Khiam and probably Jericho) in the 7th millennium and, early in the 6th millennium, at Sialk on the Iranian plateau, at Belt Cave on the shore of the Caspian Sea (in Iran), at Hassuna in the upper Tigris valley and at Amouq on the gulf of Iskenderun (both now in Turkey) (Reed 1969, Epstein 1971). Uerpmann (1979) points out that the Upper Pleistocene distribution of *C. aegagrus* extended as far south as Beidha in southern Jordan. Thus goats could have been domesticated in Palestine, or elsewhere in the Fertile Crescent and not only in the Zagros range.

During the Neolithic period domestic goats, with scimitar horns like the wild bezoar, spread over a large part of the Old World. They were represented in vase paintings from ancient Mesopotamia in the 4th millennium.

Goats with twisted horns are also depicted in ancient Mesopotamia from the 4th millennium B.C. The best example is the statuette found at Ur of a goat with vertical corkscrew horns standing on its hind legs to browse on a tree. It has been wrongly called the 'ram in the thicket'. These were presumably derived from the straight-horned goats by selection – as mentioned, there is sometimes a twist at the end of the bezoar's horn. If the horns are not very long they may be sabre-shaped in the female and twisted in the male.

Goats had reached Egypt by the 5th millennium B.C. and the small short-eared sabre-horned generalized type was occasionally illustrated in the tomb paintings of the 4th millennium. However, by the Hyksos period (1730–1570 B.C.) the sabre-horned goats appear to have been completely replaced by goats with spiral or corkscrew horns which had originally entered Egypt from the east about 3500 B.C. The twisted-horn goats of Egypt had horns diverging at a much wider angle and often extending laterally from the skull. Both angle of divergence and degree of twist vary considerably. Goats played an important role in the economy of ancient Egypt. Their flesh was eaten (at least by the poorer classes) and their skins were used to carry water. Probably they were also milked. Selection affected the ears as well as the horns and long drooping ears became the fashion. These ancient Egyptian screw-horn goats had fairly long legs and short hair and their colour was red, black, fawn, white or pied.

From Egypt the goat moved south and west and a dwarf goat is recorded from Shaheinab near Khartoum in the Sudan from 3300 B.C. The screw-horned lop-eared type later spread in the same directions and is illustrated from Nubia and from Libya (*c.* 2650 B.C.).

Europe

Since Europe (except for Crete and possibly other Greek islands) had no wild goats proper – it was the territory of various species and races of ibex – goats must have come in already domesticated from southwest Asia. Goats are common in the remains from the Neolithic Swiss lake-dwellings and here it is usually a scimitar-horned animal – the so-called 'turbary' goat. Mainly scimitar-horned goats were also excavated at the early Neolithic sites of northeast Yugoslavia and Hungary (Bökönyi 1974). From the middle Neolithic goats with twisted horns became more frequent in central and eastern Europe. The so-called 'copper' goat had twisted horns. In the Bronze Age this type became dominant in Austria and Germany but scimitar horns remained in Switzerland, Hungary and Scandinavia (Zeuner 1963).

In ancient Greece goats with scimitar and with twisted horns are both represented on coins and seals. During the Roman empire hornless goats first appeared and the size of the animals generally increased. They decreased again up to the 14th century A.D. after which they again increased in size. Animals with scimitar horns were still found – mostly in the northern part of central Europe – a case of survival in border areas (Bökönyi 1974). Both scimitar and twisted horns occur in modern European breeds but the moderately twisted horns predominate. True corkscrew horns are found, however, in the Agrigento breed of Sicily, and in the Ulokeros breed of Greece.

Origin of screw-horned breeds

Epstein (1971) described the great variation in the horn shape of modern Iraqi goats. Generally the males have laterally directed horns with an open twist while the female horns may be scimitar-shaped or twisted. However, some animals have vertical screw-like horns. In fact, all intermediates are present between the original untwisted scimitar through the scimitar with tips turned inwards, the open homonymous twist and the spiral, to the tight corkscrew. It is thus easy to imagine how the corkscrew horn may have developed from the original bezoar shape by artificial selection.

It is not necessary to implicate the other wild species – the markhor – because of its twisted horns. Such involvement is unlikely because the screw-horn goats of Europe, Africa and the Near East have homonymously twisted horns while those of the markhor are heteronymous, i.e. the right-hand horn is twisted anti-clockwise. However, animals with heteronymous horns are present in many Asiatic domestic breeds in the area between the Caucasus and Mongolia. These include the Kurdi of northern Iraq, the down goats of Russian Central Asia and the Cashmere goat of Mongolia. Animals with heteronymous and with homonymous twists occur in the same herd (Epstein 1969).

Up to 1975 there was general agreement that the gene (or genes) for heteronymous twist was derived from the wild *C. falconeri*. The heteronymous twist of the markhor is dominant to the homonymous twist of the domestic goat. Domestic goats also differ from the markhor in the shape of the cross-section of the horn. In the domestic goat the base of the horn has an anterior keel so that the cross-section is pear-shaped. The markhor horn has keels both fore and aft but the major keel is posterior. In this case it is the domestic character which is dominant. The heteronymous horns of domestic goats have a more vertical direction than those of the markhor and have a much more open screw. The same features together with the anterior keels occur in the horns of the subspecies *C. f. chialtanensis* described from the Chiltan range near Quetta (Pakistan). It was suggested by Roberts (1969) and others that this may in fact be a cross between markhor and domestic goats. Schaller and Khan (1975) and Schaller (1977) go further and state categorically that the so-called Chiltan markhor

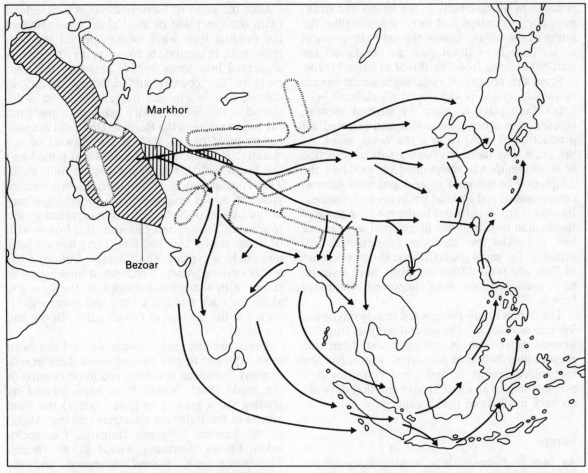

Fig. 10.2 Possible dispersion routes of domestic goats from the domestication area to south and east Asia. (From Devendra and Nozawa 1976)

is in fact a variety of the bezoar, *C. aegagrus*. In addition to the horn features mentioned above they point out that the Chiltan goat lacks a ruff and that in general its coat colour resembles that of the bezoar more than that of the markhor. Furthermore bezoar and Chiltan goat are said to occur in the same herds on several hills south of the Chiltan range. If this is confirmed it shows that heteronymous horns can exist as a variant within the wild species, *C. hircus*, and therefore it is not necessary to search further for the origin of this trait within domestic breeds.

Devendra and Nozawa (1976) have suggested the routes by which domesticated goats reached eastern from western Asia (see Fig. 10.2). One

is the northern route via Afghanistan and Turkestan to Mongolia and northern China; the other is through the Khyber Pass to the Indian subcontinent and hence to southeast Asia. A sea route was also available.

Recent history and present status

The first domestic goats must have had the same morphology as the wild bezoar. However, as we have seen, very early in their history types were developed with lop ears, twisted horns, long hair and colours more varied than the wild animal. These must have arisen as mutations which were artificially selected through the ages by the breeders. There is no evidence of association between environment (especially climate or vegetation) and hair length, ear length or coat colour, either between or within breeds. Unlike sheep, goats show no variability in tail length and thickness nor in dewlaps and fatty deposits (like the hump of the zebu). Almost the only excrescences which have been developed are the tassels (or wattles) hanging from the throat of some breeds.

From this reserve of variation different breeds or populations were developed which have been taken from place to place by migrant owners, crossed with other populations and selected for production of milk, or in a few cases, wool. At the same time there has been a natural selection at work which has maintained or modified the adaptation of the goat to arid and mountainous environments and to local parasites and diseases. Its effects are less obvious to the eye than those of the human modifications in external appearance but, if looked for, they can be detected. For instance, the small goats living in the tsetse zones of East and West Africa are much more resistant to trypanosomiasis than imported European breeds.

The details of this process are largely unknown. We can only deduce the general outlines from the known movement of peoples and from the present distribution of goat types. For only a few recognized breeds in developed countries is it possible to give an exact history and that does not go back much more than a century.

Europe

As indicated above there is little trace of the primitive turbary goat remaining in Europe today. If it has descendants they should be sought in Switzerland and Norway but in those countries it has been improved out of all recognition. Their goats retain the short ears, straight facial profile and sabre horns of the wild goat but in terms of milk production they are among the most advanced breeds.

Nearly all the goats of Scandinavia are in Norway. This is the only country of northern or central Europe which has retained its original type of goat and turned it into a high-yielding milk goat without the introduction of Swiss blood. The Norwegian goat is usually long-haired but there is some variation among the local varieties. Colour is also variable but some varieties have characteristic colours. Some are horned and some are polled. The improvement in milk yield is being continued by a modern programme based on milk recording and progeny testing.

Like the goats of Norway those of Switzerland (with the exception of the Valais) have retained the original type while being selected for high milk yield. In contrast to Norway, they have been separated into seven breeds native to different parts of the country and differing sharply in colour, hair length and horns. The most widespread is the Saanen which takes its name from the Saane valley in the Bernese Oberland but now covers most of the west and northwest of the country. It is white, short-haired and polled and the breed with the highest recorded milk yield. The Toggenburg is a close second in terms of yield. It comes originally from Obertoggenburg in the canton of St Gallen and has spread to the cantons of Schwyz and Lucerne. It is brown with white stripes on face and feet. Long hair on back and thighs is typical. The Chamois Coloured has a very striking pattern – a chestnut brown ground colour with sharp black stripes on the face and along the back and black belly and lower legs. It occurs in the cantons of Graubunden, Berne and Freiburg.

These are the three breeds which have been most exported to give the improved dairy breeds in many European countries and in other parts of the world. Such breeds have been formed by grading (to a greater or lesser extent) the local breeds in the following countries (among others) to the Saanen: Belgium, Bulgaria, Czechoslovakia, France, Germany, Great Britain, Israel, The Netherlands, Poland, Romania, and the Soviet Union. In all these countries new national breeds have been formed whose name includes the descriptive terms 'Saanen', 'White' or 'Improved'. The Toggenburg has not been exported so much as the Saanen but it has given rise to new breeds in Britain, Germany and The Netherlands. The Chamois Coloured has contributed to the *Bunte Deutsche Edelziege*, the French Alpine and the Italian *Camiosciata delle Alpi*. The fourth breed is the Appenzell which is similar to the Saanen but smaller and with longer hair.

These four Swiss breeds also share a characteristic which shows the difficulty that man may run into when he tries to interfere with nature. The breeders of these four breeds have always attempted to produce polled animals. Unfortunately the dominant gene for hornlessness has a recessive pleiotropic effect which causes many females to develop as intersexes and some to be turned into sterile males. If this turns out to be a question of two very closely linked genes rather than a case of pleiotropy there is a hope that, one day, crossing-over will separate the two genes and a polled line can be formed which no longer throws the so-called 'hermaphrodites'. As things are, the only way to avoid them is to use a genetically horned male (he can be dehorned surgically); he will produce heterozygous offspring which will be polled but sexually normal.

Two Swiss breeds are horned and black but otherwise of the same general type as the horned breeds. These are the Grisons Striped of Graubunden which has a white face stripe and white points and the all-black Verzasca of Ticino.

The seventh Swiss breed, the Valais Blackneck, is of quite a different type. It has long, slightly twisted horns and long hair. It is restricted to the canton of Valais. The colour is striking – front half black, back half white. A semi-feral herd of the same type and colour used to graze the park of Blithfield Hall in Staffordshire, England, and may be descended from Valais goats imported in the 14th century. They are named Bagot, the name of the owner of the park. This herd is now dispersed into several small private herds maintained by individuals anxious to conserve this ancient rare breed.

Spain is one of the few countries in Europe which has retained its own characteristic breeds without admixture of Swiss blood. There are two famous dairy breeds in the south. Coat is short, ears medium to short, facial profile straight. Both breeds carry tassels, and sabre horns may or may not be present. The *Malaguena* breed in Malaga and adjoining provinces is light red. The *Murciana-Granadina* of Murcia and Granada is mahogony or black; it is now encountered in nearly all the southern half of the country.

The remaining goats in Spain are less primitive in conformation but also less clearly separated into breeds and less productive. The populations in the Pyrenees, the Meseta and Extremadura have developed convex profiles and horizontal ears while retaining sabre horns. The Mountain (*Serrana*) or Spanish White breed of Castille and Andalucia has long, spirally twisted horns in addition to convex profile and horizontal (or slightly pendent) ears.

It is something of a paradox that the primitive breeds of Europe must now be sought in those with twisted horns which have, in fact, already diverged considerably from the wild type. Typical were the old breeds of the British Isles. They have been given geographical designations according to countries but Werner (1977) suggests that it may be better to classify them as highland or lowland types. The highland would include those called Old Irish, Old Welsh, Scottish Highland and Hebridean. It was also found on the hills of northern England. It was small and long-haired. The lowland variety was larger and short-haired: it would include the Scottish Lowland and English varieties. Horns were variable in both types but often strongly twisted. Colours were also variable but in the lowland type white was commoner and in the highland, black. These local breeds have been eliminated by the introduction of Swiss and oriental breeds starting at the end of the 19th century. The crosses were at first called Anglo-Nubian-Swiss and later 'British'. The first Swiss breed to be imported was the Toggenburg (1890s). This was followed by the Saanen (1903). These breeds were used to up-grade the British and in 1921 the British Toggenburg and British Saanen were recognized. Herdbooks for the two new breeds were opened in 1925. The history of the British Alpine is similar. It is descended from the 'Swiss Mountain' (probably the Grisons Striped) imported in 1903.

The Carpathian and Balkan breeds of southeast Europe are very variable but long hair is the rule and twisted horns are common. In France and Italy the majority of goats belong to extremely mixed populations which are not divided into recognized breeds but among which the *prisca* horn type is common. In Italy two very striking local breeds have been developed by selection on the horns. The *Garganica* in the Gargano peninsula of Apulia has long flattened horns which make a half turn in an open spiral. The coat is long and chestnut coloured. The *Girgentana* breed of the Agrigento province of Sicily has closely twisted vertical corkscrew horns. The coat is white with brown spots on head and neck. Varieties with similar corkscrew horns have also been described in Greece and Albania. It is impossible to say whether this similarity is due to independent parallel selection or to introductions from one place to the others.

Goats on the island of Malta have been intensively selected for milk production. They are hornless animals with semi-lop ears and straight facial profiles. In spite of their reputation as milkers they are neglected on the island and now number less than a thousand. Fortunately they have been exported to other parts of the Mediterranean to improve milk yield – especially to Italy, Greece, Libya and the Maghreb countries. The Maltese is now an important milk breed in Sicily. The origin of the breed is not known but it probably derives from an early union of stock from Italy and North Africa.

Near East and North Africa

The typical breed of the countries between Iraq and Egypt belongs to the black long-haired type known as Baladi in Egypt and Jordan, Mamber or Syrian Mountain in Syria and Israel, Iraqi in Iraq and Anatolian Black in Turkey. Ears are long and lop and often have white markings. Facial profile is straight or moderately convex. Horns are long and spirally twisted in the male; in the female they are shorter and less twisted, or sickle-shaped. It is tempting to think that this goat may be a direct descendant of the screw-horn goats depicted in the art of ancient Iraq and Egypt. The most uniform population is in Syria and away from this centre animals tend to have shorter ears and more variable colours.

The Syrian type of goat is also found in Libya but ears are not always pendulous and colours are very variable. Further west, in the Maghreb, the goat shows clear evidence of being a mixture of small short-eared goats and the lop-eared Syrian type. The Dwarf Hejazi goat of Arabia and the Sudanese Nubian goats also belong to the same group of mixed origin.

Among the lop-eared goats of the Middle East there is one type which sticks out because of its extremely Roman nose and high milk yield. The Nubian goat is not so much a breed as an idea. It should be leggy, short-haired, polled, with a convex facial profile and a high milk yield. Black or red is the characteristic colour. Goats of this general type are found as town dairy goats among the general population of range goats of Syrian Mountain type. In fact the Nubian could in theory be derived from the general Middle Eastern type by selecting for polledness, for Roman nose, for short hair and for milk production. Nubia is the region which includes northern Sudan and southern Egypt and the nearest thing to an ideal Nubian goat is the Zaraibi goat of Upper Egypt. There are no recent descriptions of this goat from Egypt and it appears to be rare. The Sudanese

Nubian is Nubian only in name. Although it is found in the area which was the south of ancient Nubia it is more like the Syrian Mountain in appearance. The Damascus or Shami breed of Syria is more like the Nubian type but it is often horned and the coat is long.

However, the best-known example of the Nubian type is the breed called Nubian in America which entirely derives from the Anglo-Nubian of Great Britain. This last breed was formed in the late 19th century by grading up native English goats with lop-eared goats from the east. The breeds used were apparently the Zaraibi and the Jamnapari and a third breed identified only by the name of its place of origin – Chitral in the extreme north of Pakistan. The Anglo-Nubian is a tall animal with lop ears, sleek coat and Roman nose. Colours are various and milk yield is good. Horns are commonly absent. The breed is used for improving the milk yield of local breeds in the tropics.

Sub-Saharan Africa

The goats of Africa south of the Sahara are predominantly small, short-eared, short-haired and short-horned. There is a wide variety of colours and combinations. They presumably derive from the goats which spread south from Egypt at an early date. In the equatorial zone of West Africa and the Sudan the goats are less than 50 cm high at the shoulder and less than 25 kg in weight and can be defined as dwarfs. This is believed to be due to natural selection in an unfavourable environment. In addition the West African Dwarf often has short bowed legs, i.e. it is achrondroplastic. It is possible that this characteristic was deliberately selected by breeders just for its oddness. These goats have very short horns which may nevertheless show a distinct twist.

The dwarf type extends southwards through central Africa as far as Zaire, Angola and the north of Namibia. From Cameroon it was exported to Germany at the beginning of the 20th century and is now commonly found in European zoos. From Germany it was taken to the United States for use as an experimental animal and is now bred in laboratories and as a pet under the name African Pygmy. There is now a breed society.

The goats of eastern Africa are small without being dwarf. In Somalia they are predominantly white, sometimes with coloured spots on head and forequarters. This is the Galla goat of north-east Kenya.

In Kenya, Uganda and Tanzania the short-eared goats have been called the Small East African but the same type is also widespread in Rwanda, Burundi, Malawi, Zimbabwe and Botswana. Naturally there is considerable vari-ation over so wide an area but few specific types or colours seem to have been selected.

In the north of West Africa the goats are larger and have longer legs. They also have long twisted horns. This Sahel or Sudan goat is found in the whole zone between the desert and the forest and between Chad in the east and Senegal in the west. Ears are usually short but they may also be longer and hanging down. There are many local names. One of the best-known local varieties is the Sokoto (Nigeria) or Maradi (Niger) which is famous for the skin used in the production of Morocco leather. Its colour is bright red and the coat is short and sleek. Milk production is poor but it is a good meat animal. There is an official improvement programme in Niger. This involves the castration of inferior males (usually because they are not of the approved colour). Selected males are distributed to improve the goats in sur-rounding districts.

The goat with pendulous ears already noted in ancient Egypt has spread into northeast Africa. The Sudanese Desert goat of the northern Sudan has pendulous ears of moderate size and the male has long twisted horns which extend sideways. In Somalia the lop-eared type of goat is represented by the Benadir goat of the Webbe Shibele valley in the south. It is larger than the short-eared goat of the north and, while colours are various, the plain white is rare. Horns curve backwards but are not twisted.

Lop-eared goats do not occur in East Africa but they turn up again in the south. The best-known breed, in fact the only local type which strictly merits this name because it has a breed society and a breed standard, is the Boer goat of South Africa. It derives chiefly from the goat of the northern Hottentots and was noted as early as 1661 but the name Boer was used much later in order to distinguish it from the Angoras which were imported in the 19th century. The early unimproved Boer goats were very variable in type. Coat was long or short and colours were diverse though usually pied or spotted. Horns

were long and twisted. Later there was evidence of imported dairy breeds, e.g. hornless animals. In the 1920s farmers in the east of Cape Province started to select for improved meat conformation and in 1959 a breed society was formed for the 'Improved Boer' goat. The breed standard prefers a white goat with reddish brown head and neck. The horns should be prominent and the ears broad and drooping. The coat should be short to medium in length and with no woolly undercoat. The Boer goat has spread east into Mozambique where it is known as Pafuri.

The lop ears and often convex facial profile of the Boer goat may have arisen from Indian breeds (e.g. Jamnapari) imported during the 18th and 19th centuries. It is not clear what is the origin of the pendulous ears in the lop-eared goats found in other regions of southern Africa, for in-stance in the Gwanda-Tuli area of southwest Zimbabwe, in Swaziland and Zululand (the Nguni goat), in Botswana and in Namibia. (For further information on African goats see Ep-stein, 1971, Mason and Maule, 1960).

South and east Asia

In the more distant and peripheral regions of Asia, i.e. southern India, southeast and east Asia, goats are of the small short-eared type which appears to have been the original one in Europe and Africa. In northern India and Pakis-tan the dominant type has screw horns and lop ears like the Middle Eastern type already de-scribed for southwest Asia. In central Asia occurs the fleece goat of Cashmere type which will be described separately (see p. 97).

India and Pakistan have the highest goat populations in the world. While milk is their most important product they are also a very important source of meat and many breeds produce hair – either coarse hair or down. The best-known of the small short-eared breeds is probably the Barbari which is the common breed of the towns of Punjab (both east and west), Uttar Pradesh and Haryana. As a town goat it is usually stall fed. The other short-eared breeds include the Chapper, Damani and Kajli of Pakistan and the Berari, Surti, Deccani and Bengali of India (and Bangladesh). They all have straight facial profiles and short horns (in the Damani they are twisted) but the Chapper and the Kajli have long coats. They may represent the primitive substratum or they may be more recent arrivals. Certainly this is the tradition in regard to the Barbari which is said to have come from the town of Berbera in Somalia and the Surti which is believed to derive from a small Arabian milking goat.

Although goats were known in India by 100 B.C. it is suggested that they did not reach southeast Asia until the 1st millennium A.D. and possibly not until after the Arab invasion of Indonesia in the 14th century. The type of goat was the small short-eared type now known as Kambing Kachang (= bean goat). From Indonesia they spread north via the Philippines and Taiwan to reach the southwestern islands of Japan in the 15th century. They did not reach the main islands of Japan where goats were unknown until the introduction of European breeds at the end of the 19th century. The same small goat spread north from Indonesia through Malaysia, Thailand and Burma to Bangladesh and China. (Equally well it may have spread south by land into Indonesia instead of coming by sea.) The South China goat from Yunnan and Guandong is similar to the Kambing Kachang except that it has disproportionately short legs. Like the Malayan breed it is usually black. There are many other breeds of small short-eared goats in southern and eastern China which differ in colour and coat and horn form.

Superimposed on this stratum of small short-eared goats in Indo-Pakistan there is another quite different type of tall, lop-eared goat with convex facial profile. The best known of this group is the Jamnapari which is native to the Etawah district in the state of Uttar Pradesh. Its ears are long and pendulous and folded-up at the tip. The coat is short except on the thighs which carry long hair. The Jamnapari has been extensively used in other tropical regions – especially southeast Asia – to improve the milk yield of the local goats. The Beetal breed of Punjab (east and west) is similar but smaller. The colour is usually pied and the male has long twisted horns.

There is a group of breeds in northwest India and Pakistan which is reminiscent of the Near Eastern type (see p. 94). They all have long black hair and spiral horns but, unlike the Near Eastern type, they have conspicuously convex facial profiles. These include the various types in south Rajasthan, north Gujarat, and neighbouring parts of Madhya Pradesh which can

collectively be called 'Gujarati'; the Bikaneri of Rajasthan and adjacent districts of Pakistan; and the Daira Din Panah of Muzzafagarh district in Punjab, Pakistan. The other Pakistan breeds – Khurasani, Lehri, Sind Desi and Kamori have various combinations of black hair, spiral horns and Roman nose.

The Americas

Goats were first taken to America from Spain at the beginning of the 16th century. What breeds were taken is not clear but as a result of natural selection they have developed into a hardy but low productive type called 'criollo' in Spanish America. Colour and conformation are very variable but the majority of these goats are small, short-eared and shorthaired.

In Brazil the corresponding type is called 'crioulo'. There has recently been extensive crossing with oriental lop-eared breeds such as the Anglo-Nubian and Bhuj (Gujarati).

In the Antilles the goats are very similar to the small short-eared goats of West Africa and southeast Asia. They must have come originally from Europe and/or West Africa but they do not show the achondroplastic dwarfism of the West African goats.

In the southwestern United States there are three types of goat – dairy goats of European breeds, Angoras, and so-called Spanish goats which are meat-type goats derived from the Mexican Criollo. Some of the latter lack external ears or have very small pinnae. They have been bred to dairy goats (with the retention of the vestigial ear character which seems to be due to a single gene) to produce a new breed called the American La Mancha.

Angora

As one of the two types of fleece-bearing goats the Angora or mohair goat is of special interest. It is a lop-eared spiral-horned goat whose fleece has been selected in a manner parallel to that of the Merino sheep, i.e. by an increase in the number of secondary fibres and a reduction in the diameter and degree of medullation of the primary fibres. Fleece is long, white and wavy. It averages 2.5 kg in weight and 30 μm in fibre diameter.

The breed is native to the province of Ankara

in Turkey from which it gets its name and there seems to be no truth in the tradition that it came originally from further east. In fact Greek and Roman writers record the presence of fleece goats with fine fleeces in Phrygia and Cilicia (central Anatolia) as early as the 5th century B.C.

The breed was not known in Europe until 1541 when a pair were imported as curiosities. Not until 1838 was an attempt made outside Turkey to exploit their unique characteristics; in this year they were first imported into South Africa and several more importations were made between then and 1896. A grading-up programme has produced a large population in the dry karoo in the south-central districts of Cape Province. They are also important in Lesotho where about half the goats are Angoras. Angoras were imported into the U.S.A. from Turkey during 1848–80 and from South Africa in 1925–26. They were tried in many states but are now almost confined to the Edwards plateau in Texas. In the 1930s grading-up of local coarse-haired goats in the Central Asian republics of the U.S.S.R. with American Angoras produced the Soviet Mohair breed. Some importations were made into Australia between 1856 and 1873 but a mohair industry never developed. Likewise imports into New Zealand were also a failure (because of susceptibility to foot-rot). Nevertheless Angora characteristics have survived in the feral populations and in both countries feral goats are being used as foundation stock for up-grading to Angoras in a new attempt to found a mohair industry.

World mohair production is almost equally divided between South Africa (including Lesotho), the U.S.A. and Turkey, but Turkey has the highest Angora goat population.

Pashmina or Cashmere goats

The other type of fleece goat is quite different from the Angora. It retains the coarse hairy outer coat and the commercial fibre is made only from the fine woolly undercoat which has to be combed out from the long hairs. Only 100–200 g of cashmere is obtained from a single individual. The Cashmere goat has erect or semi-lop ears and strong flattened horns. In the female the horns are usually homonymously twisted; more rarely they are straight or sickle-shaped. The male has longer horns which are either homonymously or heteronymously twisted. It was these heteronymously twisted horns which gave rise to the theory that the Cashmere goat is a descendant of the markhor (see p. 90).

The Cashmere goat is found in all the mountainous regions of central Asia. The preferred altitude is 3000–5000 m and the three chief areas of distribution are: 1. the Tibetan plateau including Tibet, Qinghai and northeast Kashmir; 2. the Tien Shan of Xinjiang (China) and Kirgizia (U.S.S.R.); 3. the mountains of Mongolia. With this distribution a better name for this goat would be Central Asian Down goat or Pashmina goat; *pashm* is the Persian for wool and 'pashmina' means woollen. Down is the name used in Russia for the woolly undercoat. But Europeans first encountered the fibre in Kashmir and so this name (in the old spelling) has remained both for the goat and the fibre. Down-producing goats are also found in Iran (which country is one of the major producers of cashmere), Afghanistan and European Russia – the Don breed of the Ukraine.

It is not known precisely where the Cashmere goat originated nor to what extent its fine undercoat is due to natural selection in a cold environment and to what extent it has deliberately been selected by man. Certainly there has not been much increase in weight of down. Small flocks were imported into France and into England during the 19th century but they have not survived. For future experiments of this type a high mountainous environment should be chosen.

(For more information on breeds, and references, see Mason, 1981).

Feral populations

The same hardiness which enabled the goat to survive the original trauma of domestication has enabled it to take easily once again to life in the wild. On many occasions goats have been deliberately liberated or they have escaped from captivity and feral populations have resulted. Such populations exist for instance in New Zealand, in Australia, on many Pacific Islands, in the British Isles, in Mexico, on some Caribbean islands and on Cyprus and some Aegean islands.

The goats in the Pacific are descendants of those brought by early western navigators and left as a source of food for later sailors, e.g. to New Zealand by Captain Cook in 1777, to Hawaii by

Captain Cook in 1778 and by Captain Vancouver in 1792, to Ogasawara (Bonin) islands by Admiral Perry in 1853.

In Britain the first references to 'wild goats' date from the 17th century. Some of these were probably descendants of goats which were depastured in forests; others strays from the herds which were driven from Ireland to England (via Wales) and from the lowlands of Scotland towards the north. Others were abandoned at the time of the depopulation of the Scottish Highlands in the 19th century. In some cases goats were deliberately introduced as quarry for hunters or as attractions at a beauty spot. Most of these populations would derive from the old Welsh, Irish or Scottish breeds but since the introduction of Swiss breeds at the end of the 19th century their blood would also contribute. Most of these goats have long shaggy coats and long sabre-shaped or twisted horns. There is no fixed colour. The majority are found in north and northwest Scotland, in the islands off the west coast, and in the southwest of Scotland, and populations of some thousands are found in each of these areas. There are smaller populations in the borders of England and Scotland, in north Wales and in Ireland (Whitehead 1972).

In New Zealand goats are widespread on the lower parts of the forested ranges in both North and South Islands (Rudge 1976). They were at one time present on twenty-two off-shore islands; because of the damage they cause to vegetation and to soil stability they have been exterminated from all but six or seven. The remaining island populations could be exterminated if necessary, and indeed this is the policy for islands which are flora or fauna reserves, but on the mainland they are present in such large numbers that there is no chance of their being eliminated.

Future prospects

In many parts of Latin America, Africa and southern Asia the goat is the principal source of meat and milk for the subsistence farmer. In the drier areas of Brazil, North Africa, the Near East and Indo-Pakistan it is the companion of sheep as a range animal able to live and thrive on poor ranges. So the future of the species as a domestic and agricultural animal is not in question. This is in spite of the bad reputation the goat has received for its part in leading to erosion when it is allowed to continue to graze denuded pastures or to prevent the natural regeneration of forests.

A more interesting question is what genetic changes would be desirable in order that the goat should play an even greater part in food production. Up to now it has been of benefit chiefly because of its ability to survive where other species cannot. Sophisticated selection programmes have been applied only for milk yield and in European countries such as Norway, France and Switzerland where the goat is not numerically important. There is a crying need for such programmes to be applied to tropical dairy breeds and particularly to breeds such as the Shami in Syria and the Jamnapari and other dairy breeds in India and Pakistan.

When it comes to tropical meat breeds the situation is even more difficult. Many countries are now beginning to value their goats as an important genetic resource and are looking for breeds to import in order to improve their meat production. Tropical dairy breeders can use the Anglo-Nubian or the Jamnapari and even, if they stick to halfbreds, European breeds such as the Saanen and the Toggenburg. But the only improved meat breed is the Boer goat of South Africa and even it has not been subjected to adequate breed comparison trials in a variety of environments. So there is an urgent need for selection programmes to improve the local meat goats in the tropical countries where goat-meat production is important. This should be easier than improvement of milk yield since meat characters (growth rate and conformation) can be measured in both sexes before the age of breeding. When we survey the tremendous changes which have been effected in the goat both in appearance and productivity since it was first domesticated some 10 000 years ago, it should not be too daunting to tackle the presently needed continuation of this path.

References

Bökönyi, S. (1974) *History of Domestic Mammals in Central and Eastern Europe*. Akadémiai Kiadó: Budapest

Bökönyi, S. (1976) Development of early stock rearing in the Near East. *Nature*, **264**: 19–23

Corbet, G. B. (1978) *The Mammals of the Palaearctic*

Region: a taxonomic review. British Museum (Natural History): London

Corbet, G. B. and **Hill, J. E.** (1980) *A World List of Mammalian Species*. British Museum (Natural History): London

Devendra, C. and **Burns, M.** (1983) *Goat Production in the Tropics* (2nd edn). Commonwealth Agricultrual Bureaux: Farnham Royal, Bucks, England

Devendra C. and **Nozawa, K.** (1976) Goats in South East Asia – their status and production. *Zeitschrift für Tierzüchtung und Züchtungsbiologie*, **93**: 101–120

Epstein, H. (1969) *Domestic Animals of China*. Commonwealth Agricultural Bureaux: Farnham Royal, Bucks, England

Epstein, H. (1971) *The Origin of the Domestic Animals of Africa*. Africana Publishing Corporation: New York

Gray, A. P. (1972) *Mammalian Hybrids. A check-list with bibliography*. Commonwealth Agricultural Bureaux: Farnham Royal, Bucks, England

Harris, D. R. (1962) The distribution and ancestry of the domestic goat. *Proceedings of the Linnaean Society of London*, **173**: 79–91

Mason, I. L. (1981) Breeds. In: *Goat Production*, ed. C. Gall, pp. 57–110. Academic Press: London.

Mason, I. L and **Maule, J. P.** (1960) *The Indigenous Livestock of Eastern and Southern Africa*. Commonwealth Agricultural Bureaux: Farnham Royal, Bucks, England

Reed, C. A. (1969) The pattern of animal domestication in the prehistoric Near East. In: *The Domestication and Exploitation of Plants and Animals*, ed. P. J. Ucko and G. W. Dimbleby, pp. 361–80. Duckworth: London

Roberts, T. (1969) A note on *Capra falconeri* – (Wagner, 1839). *Zeitschrift für Säugetierkunde*, **34**: 238–49

Rudge, M. R. (1976) Feral goats in New Zealand. In: *The Value of Feral Farm Mammals in New Zealand*, ed. A. H. Whitaker and M. R. Rudge, pp. 15–21. New Zealand Department of Lands and Survey: Wellington

Schaffer, W. M. and **Reed, C. A.** (1972) The co-evolution of social behavior and cranial morphology in sheep and goats (Bovidae, Caprini). *Fieldiana Zoology*, **61** (1): 1–88

Schaller, G. B. (1977) *Mountain Monarchs: wild sheep and goats of the Himalaya*. University of Chicago Press: Chicago

Schaller, G. B. and **Khan, S. A.** (1975) Distribution and status of markhor (*Capra falconeri*). *Biological Conservation*, **7**: 185–98

Uerpmann, H. –P. (1979) Probleme der Neolithisierung des Mittelmeerraums. *Beihefte zum Tübinger Atlas des Vorderen Orient*, Reihe B, Nr. 28. Ludwig Reichert Verlag: Wiesbaden

Werner, A. R. (1977) Observations on the English goat. *The Ark*, **4** (3): 79–82

Whitehead, G. K. (1972) *The Wild Goats of Britain and Ireland*. David and Charles: Newton Abbot, England

Zeuner, F. E. (1963) *A History of Domesticated Animals*. Hutchinson: London

11

Musk ox

Anne Gunn

*Northwest Territories Wildlife Service,
Canada*

Wild musk oxen

The earliest common and scientific names of
musk oxen (*Ovibos moschatus*) emphasized the
apparent similarity to *Bos*, and the musky odour.
Later studies placed the single species in the tribe
Ovibosini (subfamily Caprinae).

Eight months of snowbound ground, unre-
lenting cold and dryness, with only a brief, cool
interlude of summer, characterize the tundra
home of musk oxen. The outstanding features of
their compact appearance are the long coat and
sharp, curved horns with a prominent boss in
adult bulls. The guard hairs, which are dark
brown, long and coarse, overlie fine brown
wool – qiviut. The wool, which is the finest of any
mammal, has fibres that average about 10 cm in
length and about 15 μm in diameter.

Musk oxen are medium-sized ungulates with
bulls standing about 135 cm at the shoulder.
Adult bulls weigh about 340 kg on the arctic
mainland of Canada (Tener 1965) and adult
females weigh about half as much.

On the arctic islands, musk oxen are usually
associated with meadows where they forage on
sedges, forbs and willow. On the mainland, musk
oxen are found in meadows and where willow and
dwarf birch shrubs grow north of the treeline. In
winter, musk oxen push snow aside with their
muzzles or dig feeding craters with their front
feet. Seasonal movements are usually less than
50 km and, in some areas, musk ox herds show
strong attachment to their home range.

Musk oxen form relatively stable herds usually
of ten to thirty animals of all ages and both sexes.
Bulls also form small bachelor groups or roam as
solitaries until the rut when they attempt to
displace the dominant bull of a herd. A herd of
musk oxen close up into a compact defence
formation when faced by danger such as a wolf.
Cows are probably seasonally polyoestrous and
bear a calf in March or April after a gestation of
240–250 days. A cow can bear a calf every year
as the calf is usually weaned by early to mid-
winter. Under poor conditions after a severe
winter, cows breed in alternate years or not for
several years. Normally, cows do not breed until
3 or 4 years of age but breeding of 2-year-olds or
even yearlings is known under good nutritional
conditions. Male musk oxen tend to grow slightly
faster than females and reach their adult weight
in their sixth or seventh summer – a year later
than the females (Hubert 1974). The longevity of
musk oxen is about 20 years.

Overhunting, partially the result of arctic
whaling activities led to the extinction of musk
oxen in Alaska by 1860, and to serious declines
in Canada. Those declines led to a declaration of
protection in 1917 and by the 1970s the Canadian
musk oxen had recovered sufficiently to allow
limited hunting on a quota system. The
40 000–50 000 Canadian musk oxen are found
almost entirely in the Northwest Territories
(Gunn 1982). The 1500 musk oxen in Alaska
(1978) are descendants of 34 animals introduced
from Greenland in the 1930s. Musk oxen in
Greenland have fluctuated in numbers but were
estimated at 15 000 in 1978. Small herds of musk
oxen also have been introduced to Norway and
Russia which were part of the prehistoric range.

The low Canadian hunting quota (252 in 1979;
Gunn 1982) limits the economic importance of
musk oxen. Similarly, in Alaska and Greenland,
hunting is closely restricted. There is limited sale
of horns and hides as souvenirs; most meat is
locally consumed and a small proportion of the

quota is used for non-native sport-hunting which, if further developed, has substantial local economic potential.

Attempts at domestication

An adult bull will produce about 3 kg of qiviut annually which is worth approximately $U.S. 5000. Since the 1950s, the potential for a cottage industry knitting and weaving qiviut has been evaluated from intensively managed herds in Vermont, Quebec, and Alaska (Wilkinson 1972). The efforts in Vermont and Quebec were ended and the Alaskan experiment requires subsidization to meet the high costs of supplementing the diet.

Musk oxen have been associated with mankind since the end of the Middle Palaeolithic. The cultures were, however, nomadic hunters in a harsh environment and they left no evidence of domesticating their prey species but used musk oxen for meat, robes and horn implements.

Wilkinson (1972) traces the European recognition of the potential domestication of musk oxen for wool. To date, the only endeavours have been experimental and largely unsuccessful. A hunting culture still dominates most Eskimo lifestyles. It will require outside influences including government to foster musk ox domestication, and there are difficulties in transforming a hunting culture to a herding culture.

Northern economics are in a period of rapid change, and a need to encourage ventures to break the boom-or-bust cycle of non-renewable resource development is recognized. Such a climate is favourable to instigation of cottage industries including production of luxury woollens. Also, the need to supplement meat from the declining caribou populations raises a possibility of domesticating musk oxen for meat. The pace of domesticating musk oxen may gain momentum despite the lack of success of earlier attempts. However, high initial capital costs of fencing, etc. and then high maintenance costs for supplemental feeding, and so on, will likely continue to deter all but heavily subsidized ventures. Diseases such as contagious eczema have caused problems under close herding conditions. It is therefore questionable whether the efforts at raising musk oxen on farms will be sufficiently widespread and sustained to lead to their domestication.

References

Gunn, A. (1982) Muskox. In: *Wild Mammals of North America*, ed. J. A. Chapman and G. A. Feldhamer, Ch. 51. John Hopkins Press: Baltimore

Hubert, B. (1974) Estimated productivity of muskox (*Ovibos moschatus*) on north eastern Devon Island, N.W.T. M.S. Thesis, University of Manitoba: Winnipeg

Tener, J. S. (1965) *Muskoxen in Canada: a biological and taxonomic review*. Canadian Wildlife Service: Ottawa

Wilkinson, P. F. (1972) Oomingmak: a model for man-animal relationship in prehistory. *Current Anthropology*, **13** (1): 23–44

12

Potential domesticants: Bovidae

C. R. Field

UNEP-MAB Integrated Project in Arid Lands, Kenya

Introduction

This chapter deals with those Bovidae of the Old World which have recently been the subject of planned domestication enterprises designed to show whether intensive production of domesticated wildlife could justify its continued existence by complementing conventional livestock and paying its way.

These Bovidae occur wild in Africa south of the Sahara where there is the world's largest mammalian resource in terms of species diversity, genetic variability, number, density and biomass.

Six species have been examined for their suitability for domestication. They are: Thomson's and Grant's gazelles (*Gazella thomsoni* and *G. granti*), wildebeest (*Connochaetes taurinus*), oryx (*Oryx gazella*), eland (*Taurotragus oryx*) and buffalo (*Syncerus caffer*). The last three species have shown the greatest potential and will be the subject of this chapter.

Wild species

These three species are members of the family Bovidae. The eland, kudu and bushbuck, which are mainly browsers of bushland and forest, are placed in the tribe Tragelaphini in the subfamily Bovinae. Also in this family are the cattle and buffaloes in the tribe Bovini. The oryx, roan, sable and addax antelopes belong to the tribe Hippotragini in the subfamily Hippotraginae.

The range of the oryx is restricted to arid parts of Africa and Arabia. The eland occurs in subhumid areas and the buffalo lives in both subhumid and humid forested regions of Africa.

Oryx

There are three species of oryx: *O.gazella*, *O. dammah* (or *O. tao*), the scimitar-horned oryx, and *O. leucoryx*, the Arabian oryx. Since they apparently give fertile hybrids when crossed some authors consider them to be conspecific. Others again separate *O. gazella* into *O. beisa*, the beisa of eastern Africa, and *O. gazella*, the gemsbok of southern Africa.

The scimitar-horned oryx lives in the Sahel and semi-desert regions of Chad, Niger, Algeria, Mali and Mauretania. Its range formerly covered most of northern Africa but is now much reduced. The beisa oryx has two races *O.g. beisa* which ranges from Eritrea to the Tana river in Kenya and *O.g. callotis* which occurs south of the Tana into Tanzania. The latter bears a fringe of hairs on its ears which is the main distinguishing characteristic.

The gemsbok was formerly distributed throughout South Africa, Namibia and Botswana. It is now restricted to the Kalahari and Namib deserts although it has been reintroduced to several ranches in South Africa.

The Arabian oryx formerly occurred from southern Arabia to Mesopotamia and the Syrian desert. With the discovery of oil and the introduction of motorized desert transport it was decimated by Arab hunting parties in the 1950s until only small pockets remained along the southern edge of the Rub'al Khali or Great Sand desert (Stewart 1963). It was undoubtedly saved from extinction by conservationists who established captive herds in the arid southwest of the United States where they bred successfully. The process of reintroduction of this oryx into Oman and other selected areas has now begun (Stanley-Price, personal communication).

The adult beisa stands 118 cm at the shoulder

and weighs 131–205 kg. It is greyish fawn in colour with black markings on the head, spine, forelegs and flank. It has long straight horns in both sexes. Oryx prefer semi-desert areas of annual grassland and dwarf shrubs. They occur in herds of up to 200 animals (Dorst and Dandelot 1970).

Diet consists mostly of annual grasses supplemented with protein and energy-rich parts of drought-tolerant perennial grasses (Field 1975). They are classified as bulk-and-roughage dry-region grazers (Hofmann 1973). Oryx have low rates of body water turnover (30–124 ml per kg per day) and low metabolic rates, both adaptations to desert conditions. They drink daily if water is available, but where succulent food occurs they may survive without water for several weeks (King 1979). Oryx first conceive at about 23 months. They have a calving interval of 300 days. Annual calving rate is 102 per cent (King and Heath 1975).

Eland

Eland occur in three distinct populations. The western giant eland (*Taurotragus derbianus derbianus*) is now reduced to scattered herds from Senegal to Nigeria. The central African giant eland (*T. d. gigas*) is more widespread in Cameroon, Chad, Sudan and Zaire while the common or Cape eland (*T.oryx*) is found from Natal to Angola and through to East Africa and Ethiopia. The giant eland may hybridize with the common eland.

Adult common eland may reach 175 cm at the shoulder and are the largest of the antelopes. They weigh 590–909 kg, the male being more massive. They are tawny in colour turning greyish with age. There is a prominent dewlap beneath the neck. Black markings occur on the dewlap, spine, forelegs, tail and muzzle. There is faint white barring across the back. Eland prefer sub-humid areas of open woodland. They are gregarious, occurring in herds from a few individuals up to 200 (Dorst and Dandelot 1970).

Eland are principally browsers requiring a high protein intake (Field 1975). They are classified as intermediate feeders by Hofmann (1973). Eland have a water turnover more than twice that of oryx and drink daily if water if available (King 1979). They seek shade readily.

They first breed at 28.7 ± 2.3 months. Oestrus

occurs every 3 weeks and lasts 3 days. Cows may conceive at the first *postpartum* oestrus which on good food occurs 15–30 days after calving. Breeding is year round. Gestation is 271 ± 3 days and the calving interval is 354 ± 20 days. Annual calving rate is 83 per cent. Birth weights are 30 ± 1.3 kg for males and 25.5 ± 0.7 kg for females (Skinner 1973).

Buffalo

Buffaloes are widespread throughout West, Central and East Africa. Their range is limited in the north by the Sahel and in the southwest by the Kalahari and Namib deserts. There are two races, the dwarf forest buffalo, *Syncerus caffer nanus*, which is found in West and Central African forests and is smaller, reddish in colour and has less massive horns than the Cape buffalo, *S.c. caffer*, which occurs over the rest of the range.

Adult buffaloes reach 100–168 cm at the shoulder and weigh 318–818 kg depending upon the race. They have a bovine appearance being heavily built with short limbs. The colour varies from black to reddish brown. Both sexes have horns with a massive boss along the midline. Buffalo range from dense forest to open sub-humid grassland plains and occur in herds of up to 2000 depending on the habitat (Dorst and Dandelot 1970).

Buffalo are grazers of moist perennial grassland being classified as bulk roughage feeders (Hofmann 1973). Buffalo have a water turnover 2.6 times greater than that of oryx but do not differ significantly from cattle (King 1979). They readily resort to wallowing to attain thermostability.

Buffalo first conceive at 54–58 months and the gestation length is 340 days. Calves weigh 45 kg at birth. The calving interval is 18–25 months and the annual fertility rate is 50–85 per cent depending on the area and the year (Grimsdell 1973; Sinclair 1974).

In all three species twinning is very rare, a single calf being the rule.

Early domestication

Despite a prolonged period of coexistence there is no evidence that man attempted to control oryx, eland or buffalo in any way other than through hunting until the time of dynastic Egypt about 5000 years ago. Together with many other species, both wild and domestic, the oryx and buffalo appear as reliefs in the tombs of Saqqara in Egypt. The purpose of their apparent captivity is speculative but linked to the development of animal cult worship. The scenes illustrate tethered oryx being fed in fenced paddocks and it is thought that they were being fattened for sacrifice. Both buffalo and oryx were hunted by royalty and nobility and were sacrificed as enemies (Smith 1969).

The link between keeping animals for cult worship and domestication is tenuous but arguable. It has been suggested that these cults developed among hunter gatherers and pastoralists in order to propitiate the spirits inherent in each animal. Man tried to approach the spirits through the living animals. It may be that the methods used for propitiation were similar to those used in the process of domestication (Smith 1969).

That no further evidence for domestication of these wild animals has been found in subsequent remains does not invalidate the theory, since much of the ancient Egyptian civilization died with the culture.

Current domestication attempts

Little can be said about the current economic importance of these three species as the process of domestication is in its infancy. Five countries are involved – Zimbabwe, Kenya, South Africa, Namibia and the Soviet Union. However, it has been shown in Zimbabwe (Dasmann 1964) and elsewhere that the utilization of wildlife on ranchland may be a more profitable means of using the land than through stocking conventional livestock. At Galana ranch in Kenya the sale value of cattle, domesticated oryx and sheep has been compared and was in the ratio 10 : 7.5 : 1. However, the ranch area required to raise one cow could support two oryx or five sheep. The value of the oryx would be 50 per cent more than that of the cattle and three times that of the sheep. Apart from growth rate (oryx: 0.18 kg per day, cattle: 0.3 kg per day) oryx outscored cattle and eland from the point of view of water consumption, maturation, calving rate, age at slaughter, dressing percentage and value to the

ranch owner (King and Heath 1975).

Despite this apparent potential the oryx enterprise has not achieved commercial viability. Constraints requiring solutions include low recruitment rates through poor calf survival, high rates of loss through straying and government restrictions on marketing (King, personal communication).

The wild conspecifics of these new domesticants, in particular the buffalo, may play important economic roles from both positive and negative points of view. In Zambia, Marks (1973) recorded that traditional and control hunting contributed over 90 kg of wildlife meat per adult per year in one village under observation and that the buffalo was the most important species taken. Conversely, depredation to cultivation and human lives by buffaloes is often considerable, in particular where man settles the traditional range of wildlife.

Since all three species have only recently been resubjected to intensive domestication efforts, they are still able to breed with their wild conspecifics. Thus it is reported that wild eland bulls have mated successfully with females undergoing domestication and that cow buffaloes have sought wild bulls with whom to mate and then returned to captivity to calve (King *et al.* 1977).

The only differences in distributional ranges between the wild genetic pool and recently domesticated conspecifics is found in zoo animals and others kept under more extensive conditions. Notable among these are the eland herd at Askaniya Nova, U.S.S.R., which began to show signs of inbreeding (Treus and Kravchenko 1968), and the Arabian oryx at Phoenix and San Diego in the U.S.A. Some of the latter are now undergoing repatriation.

There may be good reason for eliminating the wild conspecifics of domesticated species since the wild entice away the domestic. This is a major cause of losses of oryx in the Galana Ranch Project in Kenya. Therefore, it may indeed be advisable to develop game domestication projects outside the areas where the wild species still live (King *et al.* 1977).

Future prospects

The economic potential of these new domesticants has yet to be realized and depends upon physical, physiological and behavioural characteristics which distinguish them from conventional livestock (Field 1979). These attributes result from prolonged exposure to the selection pressures of the African environment. By contrast the zebu cow was only introduced in numbers to sub-Saharan Africa about 1300 years ago.

Wildlife has diversified in relation to a food resource which it is able to utilize more fully than conventional livestock. The eland occupies a food niche only partly used by existing domesticants while the food habits of the oryx are complementary to those of camels, the only domestic species of comparable physiological adaptation to arid conditions.

Eland and oryx breed faster than domestic animals of similar size. Wildlife in general have a higher carcass percentage and a leaner carcass than domestic stock.

Oryx, eland and buffaloes tolerate a number of endemic African diseases. With regard to trypanosomiasis and theileriosis buffaloes may act as silent carriers for the infection of livestock. The *Theileria* carried by eland is also transmissable to livestock but does not produce a serious disease (Karstad *et al.* 1978). When attacked by tsetse flies oryx take evasive action and are bitten less frequently than cattle. Attempts to transmit trypanosomiasis by feeding infected flies on oryx have so far failed (Karstad, personal communication). Many wildlife diseases (e.g. foot and mouth in buffaloes), are closely related to those of domestic stock but attempts at cross-infection have failed. Conversely, rinderpest was introduced into Africa in cattle imported from Europe in the 19th century. At the beginning it killed many buffalo and eland. However, it could not persist in these species when the reservoir of infection in cattle had been eliminated. Once the latter were protected by vaccination the disease vanished from wildlife also (Karstad *et al.* 1978). These findings indicate that the danger to cattle from disease is much less than had been expected. They should counteract the prejudice many farmers have against allowing wildlife to share their ranchland with cattle.

In conclusion, these new domesticants should not be considered as replacements for conventional livestock or vice versa. Each has its own advantages and disadvantages. It is likely that wildlife will be gradually squeezed out of high-

rainfall areas except where the tsetse fly cannot be eradicated. In the latter areas buffalo with their tolerance of trypanosomiasis have an important role to play. In the expanding arid lands oryx may be of considerable importance as meat producers entirely complementary to camels. In medium-potential areas of small-scale ranch or farmland, overgrazing by cattle often inhibits fire and encourages a proliferation of bush which is the perfect food niche in which eland thrive (Field 1979).

References

Dasmann, R. F. (1964) *African Game Ranching*. Pergamon: Oxford

Dorst, J. and **Dandelot, P.** (1970) *A Field Guide to the Larger Mammals of Africa*. Collins: London

Field, C. R. (1975) Climate and the food habits of ungulates on Galana Ranch. *East African Wildlife Journal*, **13**: 203–20

Field, C. R. (1979) Game ranching in Africa. *Applied Biology*, **4**: 63–101

Grimsdell, J. J. R. (1973) Reproduction in the African buffalo, *Syncerus caffer*, in western Uganda. *Journal of Reproduction and Fertility*, Supplement, **19**: 303–18

Hofmann, R. R. (1973) *The Ruminant Stomach*. East African Literature Bureau: Nairobi

Karstad, L., Grootenhuis, J. G. and **Mushi, E. Z.** (1978) Research on wildlife diseases in Kenya 1967–1978. *The Kenya Veterinarian*, **2** (2): 29–32

King, J. M. (1979) Game domestication for animal production in Kenya: field studies of the body-water turnover of game and livestock. *Journal of Agricultural Science*, **93**: 71–9

King, J. M. and **Heath, B. R.** (1975) Game domestication for animal production in Africa: experiences at the Galana Ranch. *World Animal Review*, No. 16: 23–30

King, J. M., Heath, B. R. and **Hill, R. E.** (1977) Game domestication for animal production in Kenya: theory and practice. *Journal of Agricultural Science*, **89**: 445–57

Marks, S. A. (1973) Prey selection and annual harvest of game in a rural Zambian community. *East African Wildlife Journal*, **11**: 113–28

Sinclair, A. R. E. (1974) The natural regulation of buffalo populations in East Africa. II. Reproduction, recruitment and growth. *East African Wildlife Journal*, **12**: 169–84

Skinner, J. D. (1973) An appraisal of the status of certain antelope for game farming in South Africa. *Zeitschrift für Tierzüchtung und Züchtungsbiologie*, **90**: 263–77

Smith, H. S. (1969) Animal domestication and animal cult in dynastic Egypt. In: *The Domestication and Exploitation of Plants and Animals*, ed. P. J. Ucko and G. W. Dimbleby, pp. 307–16. Duckworth: London

Stewart, D. R. M. (1963) The Arabian oryx (*Oryx leucoryx* Pallas). *East African Wildlife Journal*, **1**: 103–18

Treus, V. and **Kravchenko, D.** (1968) Methods of rearing and economic utilization of eland in the Askaniya-Nova Zoological Park. *Symposia of the Zoological Society of London*, No. 20: 395–411

13
Camels

I. L. Mason

formerly Animal Breeding Officer,
FAO, Rome, Italy

Introduction

Nomenclature

In the Old World there are two types of camel, the one-humped and the two-humped. Linnaeus placed these in separate species, *Camelus dromedarius* and *C. bactrianus*, or dromedary camel and Bactrian camel. In this article the terms one-humped and two-humped will be used, for several reasons.

To retain the Latin specific epithets is not appropriate since the two types apparently cross readily and the crosses are fertile in both sexes. It is true that Gray (1972) writes 'According to Wender, male hybrids are sterile.' However, Epstein (1971) and Bulliet (1975), while confirming hybrid vigour, quote evidence from the Soviet Union and elsewhere that hybrids of both sexes are completely fertile. Zeuner (1963) also accepted the fact of hybrid fertility but did not on that account think it necessary to put in the same species types which differ so strikingly and have been separate for so long.

Apart from the question of one species or two it is questionable whether the Linnaean binomial is appropriate to a species which is domestic rather than wild. (The one-humped is known only as a domestic animal, the two-humped almost entirely so.) The terms *C. ferus* forma *dromedarius* and *C. ferus* forma *bactriana* have been used but the description *ferus* seems inappropriate to tame animals.

The traditional terms dromedary and Bactrian camel are also unsatisfactory. The word dromedary comes from the Greek *dromos* (= road) and is correctly applied only to the riding or racing camel. The two-humped type was called 'Bactrian' when it was believed to come from Baktria (Balkh). But this area (the Oxus river valley in northern Afghanistan) was certainly not the place of origin of the two-humped camel and now is inhabited entirely by one-humped.

General biology

Like the Bovidae and the Cervidae, the Camelidae are ruminating animals. However, the camelids are usually separated from the Ruminantia into the group Tylopoda (= pad-footed) because they walk on pads at the end of the third and fourth digits instead of on the sole of the hoof. (The hooves are reduced to claws projecting beyond the pad.) The Camelidae also differ from other ruminants in the morphology of their stomachs and in not having horns or antlers. Indeed it is suggested that rumination was independently evolved in the two groups. Camelids are unique among mammals in the oval shape of their red blood corpuscles. Their amazing water economy is associated with their distribution in arid and semi-arid environments. A one-humped camel does not need to drink in winter or when eating succulent fodder; in summer it can go 3–7 days without water, depending on temperature and vegetation.

In addition to the obvious difference the two-humped variety differs from the one-humped in being woollier, shorter in the leg and darker in colour. It is also adapted to the low winter temperature of central Asia while the one-humped type is the typical animal of the deserts of North Africa and the Middle East.

Well-fed camels can reach maturity at 2 years of age but they are not normally bred till the age of 3 or 4 years. Under normal feeding conditions breeding age is higher (4–6 years). They are seasonally polyoestrous but the season depends on climate, breed and management. The oestrous cycle is 20–28 days in length and oestrus lasts 1–7 days. Copulation is necessary to induce ovulation. Twins do not occur. Interval between calvings is at least 2 years. Gestation period is $13–13\frac{1}{4}$ months in two-humped camels. Exact figures for one-humped camels are few; recorded pregnancy lengths vary from 12 to $13\frac{1}{4}$ months.

All camelids have the same chromosome number, $2n = 74$ (Hsu and Benirschke 1967, 1974).

Numbers and distribution

Although there are only about 17 million camels in the world they are vital to the economy of northern Africa, the Middle East and central Asia. Their distribution is shown in Fig. 13.1 (p. 112) and Table 13.1 (p. 113).

Wild ancestor

The fossil record shows that the early evolution of the camel family took place entirely in North America. Remains of animals of increasing size from early periods have been allocated to a number of genera culminating in the appearance of *Camelus* during the Pleistocene. During one of the Ice Ages a solid bridge between Alaska and Siberia closed the Bering Straits and enabled the early camels to spread into Asia. Fossils of different species are found across the dry centre of the continent and into eastern Europe (southern Russia and Romania). Some of these early camels migrated through the Middle East and across North Africa as far west as the Atlantic, and as far south as northern Tanzania.

In the meantime the camel family died out completely in North America. However, some camelids had migrated to South America and became the ancestors of the guanaco and vicuña. These species, of course, have no humps and this fact raises the question – did the early camels which moved from North America to Asia have humps and, if so, of what type? In theory it is possible that the camel's fat hump, like that of the zebu and like the fat store of the fat-tailed sheep, was a product of domestication and was not present in the wild animal. However, a recent discovery by A. P. Okladnikov (mentioned by Friese 1975), of rock drawings in Choit-Zenker cave, Mongolia, shows a typical two-humped camel from Magdalenian times (10 000–5000 B.C.) long before domestication could have taken place. If this drawing is correctly dated it indicates that the wild camel had two humps. Nevertheless it is still tempting to suggest that the double hump looks much more like a man-made trait than a natural adaptation to some specific environmental condition.

It seems possible then that the first camels in Asia and Africa were two-humped and that the single hump was developed when the camel was becoming adapted to the hot dry conditions of the Middle East and North Africa. Possibly the single hump offered a smaller surface area for water loss by sweating (Bulliet 1975).

Alternatively the first *Camelus* may have had no hump, or a single flat layer of fat along the back and from this the two types of humped camel may have developed independently. The palaeontological record in North Africa gives no clue as to the nature of the hump although Zeuner (1963) lists several sites in the Maghreb where remains of wild 'dromedary' (or a closely related species, *C. thomasi*) are found in association with Palaeolithic man in Acheulian and Mousterian times (80 000–15 000 B.C..) Wild camels appear to have survived in North Africa into the Neolithic period. However, their complete absence from the early Saharan rock drawings and from the writings and tomb and temple paintings of dynastic Egypt indicates that by historical times wild camels had died out in North Africa. Bulliet (1975) suggests that the prehistoric camel lived in a bushy semi-arid environment, that it had no defence against predators (such as lions) and was unable to hide from them in the emerging desert. It became extinct in the Sahara and its surroundings before 3000 B.C.

The wild camel was also becoming rare in the Middle East at this time. However, as late as the early Christian era wild camels were being hunted in south Arabia (where the higher rainfall and absence of predators aided their survival). This is the most probable area of their domestication.

There is much less evidence available about the history of the wild camel in central Asia – the ancestor of the two-humped domestic type. Probably at one time it was present from southeast Russia to northern China. Today the wild two-humped camel occurs only in one small area in the trans-Altai Gobi desert on the border of Mongolia and China (Bannikov 1975). Wild camels were first reported by Przewalski in 1876 when they had a continuous distribution between the Tarim basin and Lake Lop Nor in central Xinjiang, and the Mongolian Altai. Since then their numbers have declined and FAO (1977) estimated the number in Mongolia at only 400–700. A reserve was established in 1975.

These wild camels differ from the neighbouring domestic ones in their more slender build, shorter hair, smaller feet and smaller, more erect humps. (A domestic two-humped camel often has flaccid humps flopping to one side.) Doubts have been cast on the status of these camels as wild animals; it has been suggested that, in fact, they may be feral. However, the general view now is that they are true wild animals although they may have been joined by domestic escapees. The rock drawings of Okladnikov confirm the presence of similar wild camels in the area prior to domestication, and as late as the 1st century B.C. the hunting of wild camels is reported from the northern frontier of China (Epstein 1971).

Domestication and early history

Two-humped

Although the wild two-humped camel now occurs only in Mongolia (and Xinjiang) it was formerly distributed as far east as southern Russia and as far south as Iran and Afghanistan. It was not necessarily first domesticated in its present habitat. In fact it would be expected to be domesticated where it had most contact with man and to survive where it had least. Furthermore the camel was not known in China until the end of the 4th century B.C. which seems to preclude a domestication site on her borders.

Zeuner (1963) and Epstein (1971) report that camel remains have been found in Iran and southern Turkmenistan dating from about 3000 B.C. but there is no absolute proof that those remains were from two-humped domesticated animals. However, Bulliet (1975) quotes recent Russian work, including unpublished papers by Sarianidi, which makes it clear that domesticated camels were known in this area as early as 2500 B.C. Remains dated from various periods between 2500–1600 B.C. have been found in Turkmenistan

(U.S.S.R.) and Sistan (Iran); they include camel bones, clay models of camels pulling wagons, a clay jar filled with camel dung and material made from camel hair. Traces of domestic camels datable to before 1500 B.C. have not been found elsewhere in central Asia so there is no indication that they reached Turkmenistan from the north or east. Bulliet (1975) therefore concludes that the two-humped camel was domesticated on the borders of Turkmenistan and Iran some time prior to 2500 B.C. More recent excavations in Sistan (eastern Iran) have revealed camel bones, dung and woven hair from 2700–2500 B.C. which Compagnoni and Tosi (1978) believe to represent an early stage in the slow process of the domestication of the two-humped camel.

The above-mentioned models do not exhibit humps but the geographical, linguistic and social evidence makes it certain that two-humped animals were involved. To explain the paucity of camel bones at archaeological sites, Bulliet (1975) suggests that at the time of its domestication the two-humped camel was a rare species on the brink of extinction. It may have been domesticated because its large size and strength made it attractive as a draught animal.

From its place of first domestication it had spread to the south Urals and northern Kazakhstan by 1700–1200 B.C., to western Siberia by the 10th century B.C., to the Ukraine by the 9th century B.C. and to China by about 300 B.C. It spread over all parts of the Iranian plateau except the Caspian shores and the Indian Ocean coasts. It was known sporadically in Mesopotamia as early as the 2nd millenium B.C. It also spread to north, east and west Afghanistan and it entered the Indus valley with the Aryans.

The silk route from China through central Asia and northern Iran to Baghdad was first operated with two-humped camels. In Mesopotamia they encountered one-humped camels and were crossed. The hybrids proved superior to the two-humped animals and replaced them on the silk route, at least as far north as the Oxus river. This crossbreeding started in Parthian times (2nd and 1st centuries B.C.) and was given further impetus by the spread of the Arabs into Iran and central Asia. At some stage a long-haired, cold-resistant breed of one-humped camel developed which was adapted to the Iranian plateau. It was therefore able to displace the two-humped camel there as well as in Turkmenistan and Afghanistan. The one-humped camel also replaced the two-humped in Anatolia and in Pakistan and India, and spread, alongside the two-humped, into the North Caucasus and southern Russia and into Soviet Central Asia as far north as the Aral Sea. There are also a few one-humped camels as far east as Xinjiang.

One-humped

As we have seen (p. 108), southern Arabia is the last place in which wild one-humped camels were observed and it is the most likely place for their domestication. The time was about 3000 B.C. Zeuner (1963) and Epstein (1971) give central Arabia as an alternative but Bulliet (1975) points out that the central desert would have been much too dry for the survival of prospective herders before they had the domestic camel at their disposal. He suggests that domestication was effected by an unknown pre-Semitic hunting and fishing people who tamed the camel for milk during the period 3000–2500 B.C. in the region of Hadhramaut, Mahrah and Dhufar (now the northeast of the Yemen Democratic Republic and the southwest of Oman).

The evidence for the domestication of the one-humped camel in Arabia (whether southern or central) is circumstantial rather than direct. No camel fossils have been found in Arabia and the archaeological record, until recently, did not go back further than the 9th century B.C. Kohler (1981) suggests reasons why camel bones are so rarely found on archaeological sites. Camels are slow-breeding and valuable animals so they were slaughtered only when absolutely necessary. Furthermore camel breeding was confined to the nomadic peoples whose traces are rarely found in archaeological excavations. Even caravan animals were apparently never stabled inside cities but kept outside the walls. In the rock art of central Arabia camels are represented as game animals, but the first direct evidence for the domestic camel in south Arabia (Hadhramaut) is in the 6th century B.C.

The circumstantial evidence is, nevertheless, very strong and is summarized by Epstein (1971). The earliest archaeological records of the domesticated one-humped camel come from Egypt, Mesopotamia and Palestine, Arabia's immediate neighbours. The camel was used in the incense trade of southern Arabia. Camel nomadism was introduced from Arabia into Palestine, Syria and the eastern desert of Egypt. In Sumerian cuneiform the one-humped camel was called 'beast of the sea (land)' which suggests the Gulf coast of Arabia. The Assyrians used the Semitic word *gammalu* for the one-humped camel to distinguish it from the two-humped. Camels were sacrificial animals in pre-Islamic Arabia and camel sacrifices continued in some parts of Arabia into Islamic times. In practically all early historical records of one-humped camels (whether biblical, Assyrian, Greek or Roman), the animal is connected with Semitic and especially Arab tribes.

The Danish archaeological work in the Gulf States (Kuwait, Bahrain, Qatar, Abu Dhabi) and the Gulf coast area of Saudi Arabia, which began in 1953, has now shown that a major civilization existed in this area in the 3rd and 2nd millennia B.C., contemporary with that of Sumeria (Bibby 1970). On Umm an-Nar, an island close to Abu Dhabi city, an incised figure of a camel was found on the stone flanking the entrance to a tomb dated to 2800 B.C. The neighbouring settlement of the same date yielded large quantities of bones. Eighty per cent belonged to the dugong. Of the land mammals the bones of camel and gazelle were most frequent but there were a few of domestic goats, sheep and cattle. It is not known for certain whether these camels were domestic or wild. The camel figure on the tomb was accompanied by an oryx (wild) and a humpless bull (domestic). However, Hoch (1979) presents circumstantial evidence for a domestic status. There was no room for wild camels on the island of Umm an-Nar and only if the dromedaries were brought there alive would they be represented in the refuse heaps by such useless and heavy parts as their distal limb bones.

It is now clear that the camel was common and was exploited in southern Arabia a thousand years before it was known in the north. Indeed Bibby (1970) suggests that it may have been precisely here, in southeast Arabia, that the camel was first domesticated.

However, it must again be emphasized that some authorities still consider that there is no firm evidence for camel domestication in the Arabian peninsula prior to 1000 B.C. (Zarins 1978). This author suggests that the idea of using camels was transmitted from Iran to the Arabian coast in the 3rd millennium B.C. and it spread slowly across the peninsula in the succeeding millennia.

From southern Arabia Bulliet (1975) considers that camels entered the Horn of Africa (present-day Somalia) in connection with the sea-borne incense trade during the millennium from 2500 to 1500 B.C. This was at the same time as the short-horned humpless cattle but before the zebu.

There is almost no archaeological evidence of tame camels in northern Arabia, Palestine, Syria or Egypt before 1100 B.C. There are occasional references to camels, a few seals, pots and figurines (some very doubtfully camels) and a famous camel-hair cord found in the Faiyum oasis of Egypt, from the period 2500–1400 B.C., but all these can be explained as introductions of animals or artifacts by traders from the south. There is no evidence that camels were bred in northern Arabia or north thereof during this period.

The references to camels in the Bible stories of Abraham and his immediate descendants (Isaac, Jacob and Joseph) have often been quoted to prove the regular use of camels in northern Arabia during the 19th and 18th centuries B.C. While the reputed gift of camels from Pharaoh to Abraham must be a later insertion, the other references show only that camels were present in the Palestine–Jordan area in small numbers long before they were bred or herded there. This is confirmed by the presence of camel bones at various archaeological sites in Palestine dated to the 18th–16th centuries B.C. In Mesopotamia very occasional representations of one-humped camels (as well as two-humped) are known from an early period (3000–1200 B.C.).

However, the camel was rare in these countries until the Aramaean invasion from central Arabia between the 16th and 13th centuries B.C. But the first great invasion of camel nomads was that of the Midianites (= Ishmaelites) who poured into Palestine and Syria from northwest Arabia about 1100 B.C. '. . . both they and their camels were without number: . . .' (Judges 6: 5).

In Arabia and the Middle East camels were thus available in quantity, certainly by 700 B.C. But the camel breeders were nomads, more or less outcasts, and the camel did not become an important animal until their breeders acquired military, political and economic power, i.e. until the rise of the Arabs. Bulliet (1975) followed Dostal in attributing this transformation largely to the invention of the 'north Arabian' saddle in the period 500–100 B.C. This has saddle bows in front and behind the hump connected by wooden crosspieces along the sides. This arrangement enables the rider to sit on top of the hump while being supported firmly on the animal's backbone. (The 'south Arabian' saddle, by contrast, was behind the hump.) The camel-borne warrior seated on a north Arabian saddle was able to seize control of the caravan trade and hence the camel breeders became integrated into the economy.

The change in status of the camel herders led to the realization that the pack camel was economically more efficient than the mule or the ox cart. As a result, the use of wheeled vehicles was abandoned during the period A.D. 300–600 over almost the whole area from Morocco to Afghanistan. Vehicular transport returned only with the improved cart and carriage technology, better roads, and mechanization of the 19th and 20th centuries.

Turning now to North Africa, there are two possible routes of entry – along the coast from Egypt or along the southern fringe of the desert from the Sudan. Bulliet (1975) presents convincing evidence that the latter was the more important route. The cultivated delta of Egypt would be a biological and sociological barrier to a desert animal and anyway the camel was almost unknown in ancient Egypt. The southern route across the Red Sea is more likely and indeed there were trade routes from the Red Sea to the Nile in the 3rd century B.C. From the eastern desert camels moved south to the Sudan and from Upper Egypt and northern Sudan they probably spread westwards via the south Saharan highlands – Ennedi – Tibesti – Tassili – Ahaggar – Adrar of the Ifoghas. Camels are represented on Saharan petroglyphs of the late or cameline period. These are difficult to date but Zeuner (1963) suggested 1500–500 B.C. They then slowly worked their way northwards and were encountered by the Romans in Tripolitania and Tunisia in the 2nd century B.C. There the camel was changed from a pack to a draught animal.

In late Roman times (after the 3rd century A.D.) large numbers of camels were introduced into North Africa through Egypt but it was not until the Moslem Arab conquest of Egypt in the 7th century and the great invasions of the 11th century, that the camel in North Africa reached the numbers and importance that it has maintained until recently.

Recent history

The present-day distribution of one- and two-humped types resulting from these historical processes is shown in Fig. 13.1. Although the exact boundaries are only approximate, especially in the U.S.S.R., it is striking how closely the limit of the camel-breeding area follows the isohyet for 50 cm of rain. It is also clear that two-humped camels are not found where the mean annual temperature is above 21 °C.

Of course camels are not uniformly distributed over the areas shaded. Table 13.1 gives some idea of their density in different regions. Because of the notorious unreliability of livestock statistics, and of camel statistics in particular, country figures have been rounded to the nearest 10 000.

In spite of the inadequacy of the figures there can be no doubt that the majority of the world's camels are to be found in northeast Africa and the next largest concentration is in northwest India–Pakistan–Afghanistan. In both these regions, and also in West Africa, numbers appear to be increasing. The countries where numbers have decreased most dramatically are: Iraq, Saudi Arabia, Iran, Syria and Turkey. In the last 30 years total numbers in these countries have declined from over 1.5 million to less than 500 000 but even this decline has not significantly affected the world totals.

As for two-humped camels it is safe to assume that numbers are decreasing slowly in all three countries which they inhabit.

History has had a large hand in moulding the uses to which camels are put in different regions. The two-humped camel was apparently first domesticated as a draught animal and it is still used as a draught (and pack) animal throughout its range. The one-humped, on the other hand, is primarily a pack and riding animal and it is used for draught only in four widely separated areas (Bulliet 1975). These are: 1. Indus valley (Pakistan) and northwest India; 2. Cape Bon peninsula, northeast Tunisia; 3. Mazagan (El-Jadida) on the Atlantic coast of Morocco; and 4. Aden in the Yemen Democratic Republic. In India and Pakistan the two-humped camel pre-dated the one-humped and the function of pulling carts and ploughs was transferred directly from the one to the other when the one-humped camel started entering in numbers after A.D. 1000 (with the Baluchis). The idea of draught camels was intro-

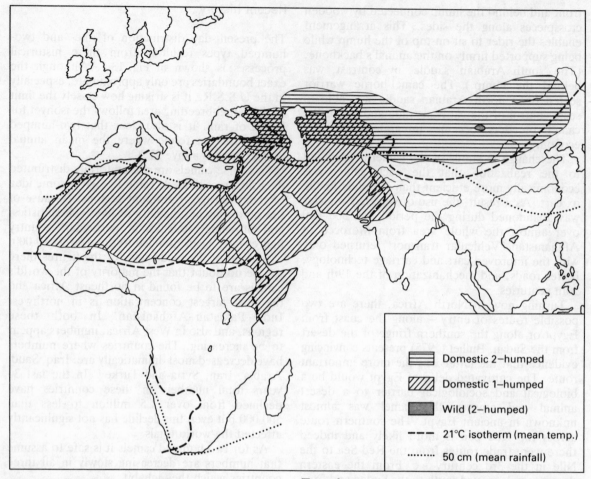

Fig. 13.1 Distribution of camels

Legend:
- Domestic 2–humped
- Domestic 1–humped
- Wild (2–humped)
- — — — 21°C isotherm (mean temp.)
- ········· 50 cm (mean rainfall)

duced from British India to British Aden. The use in Tunisia (to which may be related that in Morocco) is apparently a lineal descendant from Roman times. Tunisia (like Turkey) is thus an area where the wheel never disappeared.

One-humped camels may be divided, according to habitat, into mountain and lowland types. The former is small, compact and coarse-boned. The sole of the foot is very hard; a long fur may develop in winter. It is a pack animal. The lowland camel is larger and the foot tends to be softer. The lowland group is subdivided into the slender desert or riding breeds and the larger more compact riverine or baggage camels. Alternatively the one-humped camel may be divided into baggage and riding types and the former subdivided into hill and plains sub-types.

Two famous breeds of riding camel are the Mehari of the Sahara and the Mahri of Pakistan.

In many countries camels are milked, and in Somalia and the eastern lowlands of Ethiopia their milk forms the staple diet of the population. These camels are used as pack animals, but rarely for riding.

Camels are usually slaughtered for meat at the end of their working lives (or on ceremonial occasions) but in some places, e.g. Somalia, Sudan and northern Kenya, many camels are now kept solely for meat. Much of this production is exported to Egypt, Libya and other oil-producing countries.

Camel flesh is taboo among various non-Moslem peoples of the Near East. These include the Jews, the Zoroastrians of Iran, the Mandaeans of Iraq and Iran, the Nosairis of Syria and the Christian Copts of Egypt and Ethiopia (Simoons 1961)

Table 13.1 Numbers of camels by region and by country (millions)

One humped	1981	1948/52
Northeast Africa		
Somalia	5.55	2.48
Sudan	2.54	1.55
Ethiopia	0.99	0.60
Kenya	0.61	(0.40)
Djibuti	0.03	—
Total	9.72	5.03
West Africa		
Mauritania	0.74	0.10
Chad	0.42	0.07
Niger	0.37	0.16
Mali	0.17	0.08
Western Sahara, Nigeria, Senegal and Upper Volta	0.13	0.08
Total	1.83	0.49
North Africa		
Morocco	0.23	0.19
Tunisia	0.17	0.19
Algeria	0.15	0.14
Libya	0.13	0.14
Egypt	0.08	(0.10)
Total	0.76	0.76
Asia		
India	1.15	0.64
Pakistan	0.87	0.46
Afghanistan	0.27	0.29
Iraq	0.24	0.50
Saudi Arabia	0.16	0.26
Iran	0.03	0.60
South Arabian and Gulf States	0.30	(0.20)
Eastern Mediterranean countries	0.04	0.22
Total	3.06	3.17

Two-humped	1981	1969/71
China	0.61	0.48
Mongolia	0.59	0.64
U.S.S.R.*	0.20	0.25
Total two-humped	1.40	
Total one-humped	15.37	
Total camels	16.77	

* Some camels in the Soviet Union are one-humped but the majority are two-humped. Figures in brackets are estimated. From: FAO (1963, 1982)

Since camels became known to Europeans many attempts have been made to implant them in regions outside their original range (see Bulliet 1975). Only two of these introductions – to the Canary Islands and to Australia – met with any lasting success. The Canary Island population was introduced from Morocco after the islands had been conquered for Spain in the 15th century. They are still bred there and are used in agriculture and as beasts of burden.

Nothing now remains of the camel herd maintained near Pisa in Italy from 1622 until the Second World War, nor of the experimental importations into various European and American countries (including Texas, U.S.A.) during the 19th century, nor yet of the attempts of the British, Germans, and Portuguese to establish breeding herds in the semi-arid parts of their colonies in southern Africa at the beginning of the 20th century.

The big camel success story was in Australia which contains the second largest expanse of arid and semi-arid lands in the world. The story is well told by McKnight (1969). The first successful imports were in 1860 and 1866 and to South Australia. These camels were brought from British India to aid in the early exploration of the dry interior and the success of the venture was attributed to the Baluchi camel drivers (called 'Afghans') who accompanied the animals. Soon camels were being used to carry supplies to outback mining communities and remote pastoral stations as well as helping in the construction of telegraph lines and railways. They reached a peak population of some 20 000 about 1920. During the 1920s and 1930s mechanized transport made the camels redundant; many were turned loose and took readily to a feral existence. When McKnight wrote (1969) there were still a few hundred domestic camels – chiefly used by Aborigines as pack and transport animals. However, he estimated the number of feral animals at 15 000–20 000 ranging over an extensive area in the centre and west of the continent (see Fig. 13.2) with the greatest concentration in and around the Simpson Desert. Although they feed mostly on trees and shrubs and therefore do not compete with cattle for feed, they are considered to be notorious fence destroyers and are persecuted by pastoralists and vermin-control officers. Numbers are therefore declining.

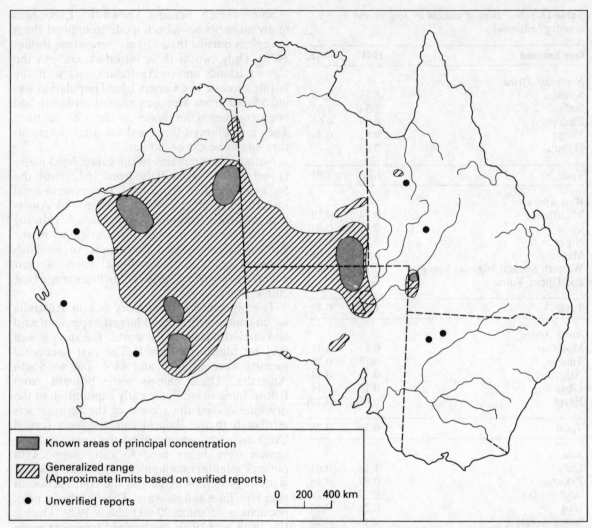

Fig. 13.2 Extent of feral camel range in Australia, 1966. (From McKnight 1969)

Future prospects

The camel is not perfectly adapted to a desert habitat – it needs the aid of man to provide it with water. Nevertheless its ability to go for long periods without water and to live on a thorny and high-fibre diet makes it one of the few animals which can exploit a semi-desert environment. In this habitat it can manage very well by itself – as is shown by its performance in Australia. McKnight (1969) points out that over most of its range in Australia the feral camel is the only large animal present. Even in drought years camels are able to maintain their condition when other ungulates suffer and often die. There is a lesson here for the countries in the original home of the camel, Arabia, and in Mediterranean countries where mechanization has deprived it of its original role as a transport animal. It is the ideal animal to exploit the semi-desert areas of these countries for meat production, provided an easy-care system of management can be developed. In Australia the feral buffaloes of the Northern Territory are regularly mustered and shot for meat. The same system should be applicable to feral camels. In the Near East camels are probably still too valuable to be allowed to run completely wild. A system may be developed whereby the breeding animals run free and the young are periodically rounded up for fattening under more intensive conditions.

There may be other areas of the world where they can be successfully introduced. For instance into the thorn forest (*caatinga*) of northeast Brazil which at present is virtually wasted except for a few wandering goats.

There is also a case for investigating the efficiency of the camel as a milk producer under intensive conditions. Some high yields have been reported in Pakistan especially and, if a suitable selection programme could be implemented, it might produce an animal which could compete with cattle in a hot arid climate and under poor feeding conditions.

Nevertheless it is clear that the major importance of the camel will remain in those areas where its numbers are high and increasing – namely northeast Africa and the northwest of the Indian subcontinent. In the latter it is invaluable as an efficient work animal which can also produce milk and can finally be eaten.

References

Bannikov, A. G. (1975) [The wild camel or Khavtagai]. *Priroda*, No. 2, 62–9. (In Russian)

Bibby, G. (1970) *Looking for Dilmun*. Collins: London

Bulliet, R. W. (1975) *The Camel and the Wheel*. Harvard University Press: Cambridge, Massachusetts

Compagnoni, B. and **Tosi, M.** (1978) The camel: its distribution and state of domestication in the Middle East during the third millenium B.C. in the light of finds from Shahr-i Sokhta. *Peabody Museum Bulletin* No. 2: 91–103. Peabody Museum of Archaeology and Ethnology, Harvard University

Epstein, H. (1971) *The Origin of the Domestic Animals of Africa*, Vol. 2. Africana Publishing Corporation: New York. Edition Leipzig: Leipzig

FAO (1963) *FAO Production Yearbook 1962*, Vol. 16. FAO: Rome

FAO (1977) FAO Project, Mongolia/68/002. Forestry Development and Wildlife Management. (Recommendations of terminal report. Summary in English translated from Russian).

FAO (1982) *FAO Production Yearbook 1981*, Vol. 35. FAO: Rome

Friese, F. (1975) Elefanten, Laufvogel und ein Kamel. *Kosmos*, **71** (8): 331–4

Gauthier-Pilters, H. and **Dagg, A. I.** (1981) *The Camel: its evolution, ecology, behaviour and relationship to man*. Chicago University Press.

Gray, A. P. (1972) *Mammalian Hybrids. A check-list with bibliography* (2nd ed). Commonwealth Agricultural Bureaux: Farnham Royal, Bucks, England

Hoch, Ella (1979) Reflections on prehistoric life at Umm an-Nar (Trucial Oman) based on faunal remains from the third millenium B.C. In: *South Asian Archaeology 1977*. Papers from the Fourth International Conference of South Asian Archaeologists in Western Europe, Naples, pp. 589–638

Hsu, T. C. and **Benirschke, K.** (1967, 1974) *An Atlas of Mammalian Chromosomes*, Vol. 1, Folio 40 and Vol. 8, Folio 389. Springer-Verlag: New York

Kohler, Ilse (1981) *Zur Domestikation des Kamels*. Inaugural-Dissertation, Tierärztliche Hochschule Hannover.

McKnight, T. L. (1969) *The Camel in Australia*. Melbourne University Press: Melbourne

Simoons, F. J. (1961) *Eat Not This Flesh. Food avoidances in the Old World*. University of Wisconsin Press: Madison

Zarins, J. (1978) The camel in ancient Arabia: a further note. *Antiquity*, **52**: 44–6

Zeuner, F. E. (1963) *A History of Domesticated Animals*. Hutchinson: London

14

Lama and alpaca

C. Novoa

Instituto Veterinario de Investigaciones
Tropicales y de Altura
Lima, Peru

and

Jane C. Wheeler

The Florida State Museum,
Gainesville, U.S.A.

Nomenclature

In 1758, Linnaeus designated the two domestic New World camelids, the llama and alpaca, as *Camelus glama* and *Camelus pacos*, placing them in a single genus together with the Old World camels. The two remaining New World species, the wild guanaco and vicuña, were subsequently described and named *Camelus guanicoe* by Müller in 1776 and *Camelus vicugna* by Molina in 1782. In 1775, Frisch correctly proposed that all four New World camelids be classified in the separate genus *Lama*, but in 1950 the International Commission on Zoological Nomenclature declared this work invalid. In 1800, Cuvier also utilized the term *Lama*, yet failed to list or describe the species included in this genus and so is not a valid author under the conditions of Article 12 of the International Code of Zoological Nomenclature. Consequently, according to Article 23, the law of priority, the correct genus designation for the New World camelids should be *Lacma* Tiedemann 1804; however this term must be considered an unused senior synonym (Article 23 a–b) and eliminated since it is completely unknown in the literature. In 1811, Illiger proposed the invalid name *Auchenia* (nec Thunberg 1789), which is frequently seen in print, but it apparently was not until 1827 that Lesson published the first taxonomically correct separate genus designation for the New World Camelidae, *Lama*. In 1924, Miller assigned the vicuña to a separate genus *Vicugna*.

To date, no general consensus concerning the systematic classification of the New World Camelidae has been reached. While some authors classify the llama, alpaca, guanaco and vicuña as separate species of the genus *Lama*, others separate the vicuña, describing it as the only species of the genus *Vicugna*. Additionally, the four forms have also been described both as subspecies of *L. glama*, and varieties of *L. g. glama*, incorrectly utilizing *forma domestica* for the llama and alpaca. One of the factors which has contributed to this confusion is the commonly held, and frequently cited, belief that crosses between the four forms either do not occur, or that when they do, the offspring are sterile. This is not in fact the case. Abundant evidence is available to document both the facility with which such crosses occur and the fecundity of the offspring produced by all possible matches of both pure and hybrid parents (Gray 1972, Fernández Baca 1971). A native folk taxonomy of terms for describing the animals produced by some of these crosses exists among the Quechua speaking pastoralists of the Peruvian Andes (Flores Ochoa 1977).

The above evidence, combined with the fact that all four New World camelids have the same 2n = 74 karyotype (Hsu and Benirschke 1967, 1974), and new osteological evidence indicating a historical relationship between the alpaca and the vicuña (Wheeler, in press), indicate that we are dealing with only one genus, *Lama*. Sufficient systematic research is not yet available, however, to determine if the correct classification of the llama, alpaca, guanaco and vicuña is at the species or subspecies level. In this article the terms 'llama' (*Lama glama*), 'alpaca' (*L. pacos*), 'guanaco' (*L. guanicoe*) and 'vicuña' (*L. vicugna*) will be used.

General biology

The New and Old World Camelidae exhibit the basic processes of ruminant digestion. However, they differ from the infraorder Pecora (or so-called advanced ruminants) in stomach morphology, absence of horns or antlers, and the replacement of hooves with callous pads ending in claws, and are thereby classified in the infraorder Tylopoda (= pad footed). Both lamoids and cameloids manifest specialized adaptations to arid and semi-arid environments, but only the latter can go for long periods without water. The lamoids have also adjusted to the rigours of existence at high elevations in the Andes. They exhibit specialized dentition and digestion characteristics which make possible a highly efficient utilization of sparse and fibrous vegetation. The haemoglobin of llamas and alpacas has a higher affinity for oxygen at high altitude than at sea-level, a feature which facilitates the concentration of oxygen in the blood under conditions of chronic hypoxia. They also have tissue characteristics which allow more efficient diffusion and/or utilization of oxygen. Well-fed llamas and alpacas can first be mated at 1 year of age, but they are not normally bred until the age of 2 or 3 years (Fernández Baca 1971). Females exhibit acyclic oestrus, remaining in heat in the absence of coitus, with ovulation induced by copulation (Novoa 1970, Fernández Baca 1971). Approximately 20 per cent of mated females fail to ovulate, but fertilization of shed ova is high. Embryonic mortality averages 40 per cent during the first month of gestation. Females who lose their embryos return to heat and can conceive if remated. Implantation of the embryo is always in the left horn of the uterus and twinning does not occur. The gestation period is 348 days in the llama, 343 days in the alpaca, and approximately 11 months in both the guanaco and the vicuña. When males and females are herded in continuous association, parturition is seasonal from December to March and the young are thus born during the rainy season when better quality pasturage is available. Under periodic association, births can occur throughout the year. Birth takes place between sunrise and early afternoon, an adaptation which increases survival of the newborn. *Post partum* heat occurs immediately after parturition, but fertile mating only takes place 10 days later.

Numbers and distribution

According to available census data, the current population of llamas and alpacas numbers 6 885 900 animals. They are distributed along the

Table 14.1 Estimated population of South American Camelidae (thousands)

Country	Domestic		Wild	
	llamas	alpacas	vicuñas	guanacos
Peru	900.0	3020.0	50.0	5.0
Bolivia	2500.0	300.0	2.0	0.2
Argentina	75.0	0.2	2.0	109.0
Chile	85.0	0.5	1.0	13.0
Colombia	0.2			
Ecuador	2.0			
U.S.A.	3.0			
Total	3565.2	3320.7	55.0	127.2

Andean cordillera with the zone of greatest productivity located between 11° and 21° S (Table 14.1). Llamas are kept at elevations between 3000 and 5000 m above sea-level, and are found from Colombia to Chile and Argentina (Fig. 14.1). Alpaca rearing is restricted to elevations of 4000 m and above in Peru and Bolivia (Fig. 14.2). Despite the relatively small population of llamas and alpacas, they play a role of primary importance in the economy of the high Andean regions of Peru, Bolivia, Chile and Argentina.

Wild species

The ancestral forms of the New and Old World Camelidae, *Pliauchenia* (early and middle Pliocene) and *Hemiauchenia* (middle Pliocene to late Pleistocene), evolved in North America (Webb 1974). During the early Pleistocene, *Hemiauchenia* extended its range into South America, where it rapidly adapted to the rugged relief and food resources of the Andes, and radiated into *Palaeolama* and *Lama* by middle Pleistocene times. From this Andean origin, *Palaeolama* expanded westward to the Pacific coast and as far north as Texas and Florida. *Lama* expanded towards the east and south where fossil forms of llama, alpaca, guanaco and vicuña are found (Webb 1974) in middle and late Pleistocene deposits (Figs 14.3 and 14.4). The Old World camels, *Camelus*, likewise evolved during the Pleistocene as the result of a Pliocene migration across the Bering Straights into Asia. Both *Hemiauchenia* and *Palaeolama* became extinct at

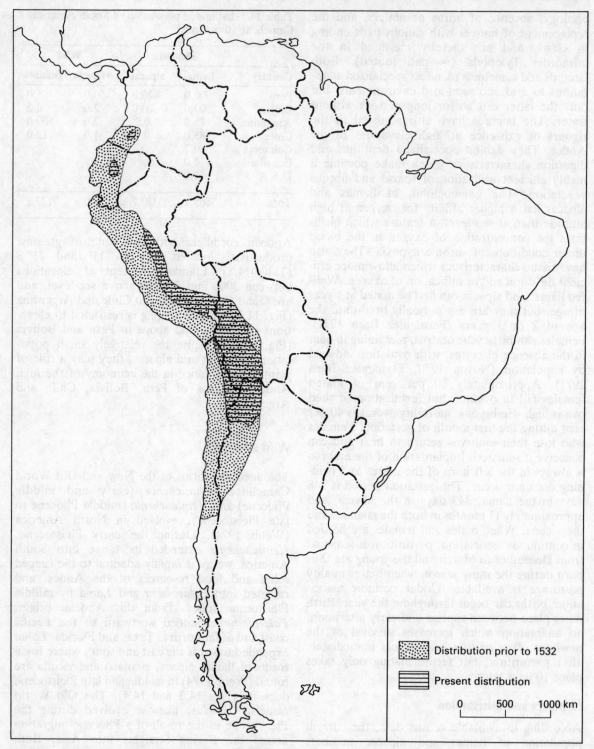

Fig. 14.1 Past and present distribution of llamas
(*Lama glama*)

Fig. 14.2 Past and present distribution of alpacas
(*Lama pacos*)

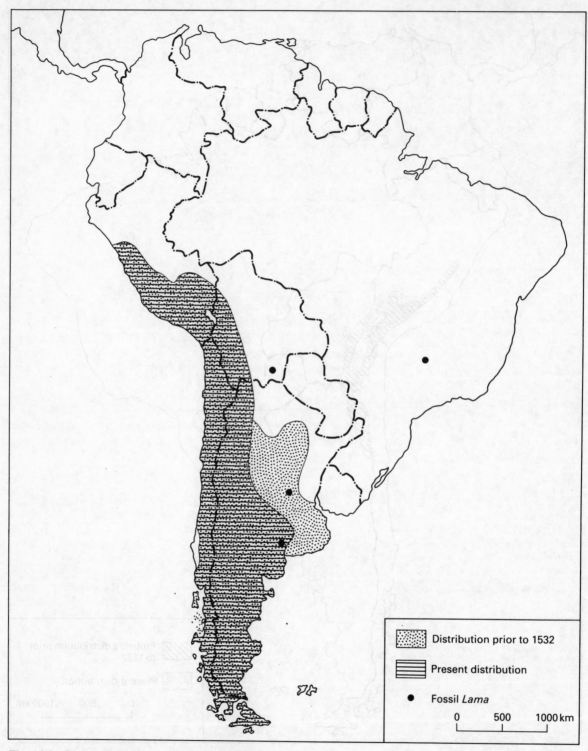

Fig. 14.3 Past and present distribution of guanacos
(*Lama guanicoe*) and fossil *Lama*

Distribution prior to 1532

Present distribution

Fossil *Lama*

0   500   1000 km

the end of the Pleistocene, leaving *Lama*, represented by the wild guanaco and vicuña, and the domestic llama and alpaca, as the only surviving Camelidae in the New World.

Of the four South American camelids, the guanaco presently exhibits the widest range of ecological adaptation, occurring both in shrublands and warm and cold grasslands at elevations from sea-level up to 4250 m (Franklin 1982) or 4600 m (Raedeke 1979) in the Andes. Its northernmost distribution is at present, and appears to have always been, located at about 8° S in the department of La Libertad, Peru. From there its range extends south along the Andean Cordillera to Navarino Island in Tierra del Fuego, east across Patagonia and as far north as the sierras of Curámalal and la Ventana in the province of Buenos Aires, Argentina. In the past, this range probably extended as far as the Paraguay river (Fig. 14.3).

Four geographic subspecies of guanaco have been described. The first, *Lama guanicoe guanicoe* is found in Patagonia, Tierra del Fuego and Argentina south of 35° S; the second, *L. g. huanacus* in Chile; the third, *L. g. cacsilensis* in the *punas* (Quechua for 'high land') of southern Peru and Bolivia; and the fourth, *L. g. voglii* in Argentina north of 32° S. The southern subspecies are the largest wild ungulates in South America, but are still smaller than domestic llamas (Franklin 1982), although the high altitude *cacsilensis* form falls within the size range of the next smallest camelid species, the domestic alpaca. All four subspecies exhibit the same dark brown colouration, white underparts and blackish face.

Guanacos live in both migratory and sedentary groups (Raedeke 1979, Franklin 1982). Family bands consist of 1 adult male and 5–6 adult females with their young, and some occupy permanent territories which the male defends against all other guanacos. Young males unite into migratory troops of up to 50 individuals, but 8 per cent of the total population remain solitary, challenging the dominant males for control of their family groups and territories, especially during the November to February breeding season. They are easily hunted and tamed. The Incas utilized guanacos as pack animals, and the Araucanians distinguished between the tamed *chilihueque* and the wild *luan*. Guanaco populations have been drastically reduced since the arrival of the Spanish, and it is to be lamented that little has been done to protect this endangered species.

Distribution of the vicuña is limited to the *puna* life-zone of the Andes, where they are most common at elevations of 4200–4800 m above sea-level, with a lower limit of 3700 m (Koford 1957). The northernmost distribution of the vicuña, both past and present, is 9°30'S in the department of Ancash, Peru. The southernmost limit is presently 29°0' in the province of Atacama, Chile, but previously extended to the province of Coquimbo (Fig. 14.4).

Two geographic subspecies of vicuña have been described. The first, *L. vicugna vicugna*, predominates in the south, and apparently differs from the second, northern form, *L. v. mensalis*, in its greater size (90 v. 70 cm at the withers), longer molar series (57 v. 45 cm), paler colouration (pale fawn v. fulvous) and lack of a tuft of long white hairs on the chest. However, study of a larger series from different regions is needed to validate these subspecies. Both are characterized by a short growth of extremely fine cinnamon-coloured wool, white underparts, similar facial colouration and their gracile form. Vicuñas live in family groups consisting of an adult male and 4 to 7 females with their young of the year (Franklin 1982). The male establishes and maintains a permanent year-round territory which normally contains a high-ground sleeping area, a lower-elevation grazing area and a water source. The size of the territory depends upon the quality of the pasturage contained therein. Ritual defaecation at communal dung piles provides short-term intra-group orientation in territorial demarcation. Juvenile males and females are expelled from the group prior to the January onset of the birth season. The young males join non-territorial troops of 20–30 animals, and the females join other family groups. Mature males separate from the troops and remain solitary until they establish a territory. Vicuñas are easily hunted, but are very intractable and difficult to tame. The Incas practised rational utilization of this species by harvesting only males and old and infirm animals, and shearing and releasing the rest, at communal drives or *chacos* held approximately every 3 years. Reduction of the vicuña to endangered species status is a post-Spanish phenomenon. Significant progress in reversing this process has been made at the National Vicuña Reserve of

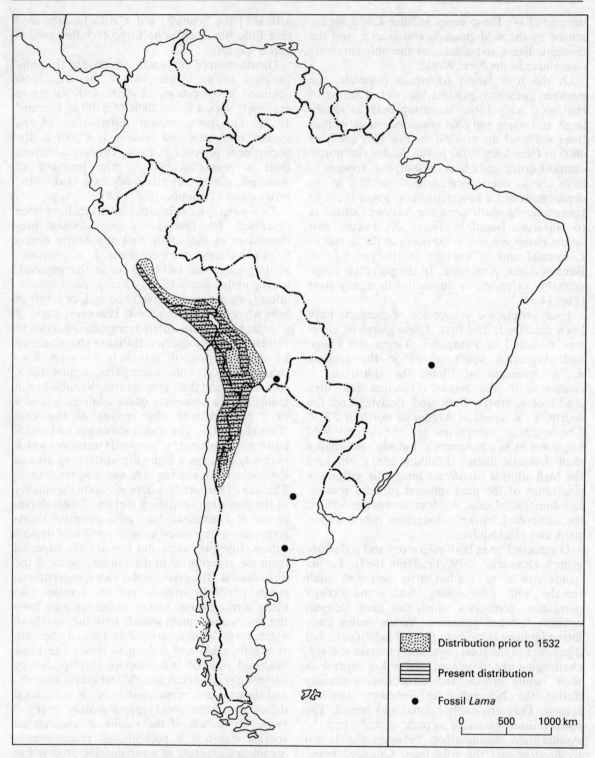

Fig. 14.4 Past and present distribution of vicuñas (*Lama vicugna*) and fossil *Lama*

Distribution prior to 1532

Present distribution

Fossil *Lama*

0   500   1000 km

Pampa Galeras in Peru since its establishment in 1964 (Novoa 1981).

No feral populations of llama or alpaca have been reported, although occasional crossbreeding between wild and domestic Camelidae does occur in nature.

Domestication and early history

Presently available archaeological evidence indicates that llamas and alpacas were domesticated in the Andean *puna* at elevations of 4000–4900 m above sea-level, by 4000 B.C. However, identification of the ancestral form, or forms, from which they were domesticated remains a matter of debate.

Herre (1952) has suggested, on the basis of cranial morphology, that the guanaco is the common ancestor of llama and alpaca. Hemmer (1975) concurs with Herre on the llama, but hypothesizes, principally on the basis of behavioural traits, that the alpaca results from crossing domestic llama with vicuña; while Steinbacher (1953) also uses behaviour to sustain that the alpaca is a domestic vicuña. Finally, there are apparently fossil llamas and alpacas of middle and late Pleistocene date (Webb 1974), which must be considered possible ancestors. Much more archaeological, palaeontological and biological research is needed, however, before this problem can be resolved.

Very few clear-cut osteological characters exist which can be used to identify camelid bones to species level and, with the exception of incisors, these are seldom preserved in archaeological sites. Likewise, osteometric analysis cannot be relied upon for species identification, or to document the domestication process, at archaeological sites located in the *puna*. In descending order, llama, guanaco, alpaca and vicuña all overlap in size, and these four, as well as their intermediate-size hybrids, are native to the zone. Nevertheless, this technique has been successfully used by Wing (1977) to classify domestic camelid remains from non-*puna* sites. For these reasons, the most complete information yet available on camelid domestication has come from the reconstruction of prehistoric animal exploitation strategies at archaeological sites.

The Lake Titicaca basin of southern Peru and western Bolivia is traditionally thought to have been the epicentre of camelid domestication (Murra 1975). This extensive plateau is located at the geographic mid-point in the north–south distribution of the lamoids. It contains the largest contiguous unit of quality natural pasturage resources found at high elevation in the Andes, and was the foremost camelid-producing area discovered by the Spanish conquerors. The importance of wool products exported from the Lake Titicaca basin can be traced back to at least 500 B.C. in textiles with design elements of the Pukara culture found at archaeological sites on the south coast of Peru (e.g. Paracas) and the north coast of Chile (e.g. Alto Ramírez). These textiles are thought to represent the earliest use of alpaca wool in the Andes, and have generated the hypothesis that alpaca domestication was carried out in the basin, by the Pukara culture around 500 B.C., for purposes of wool production. Faunal remains from the type site of Pucará, excavated by Elías Mujica and analysed by Jane C. Wheeler, indicate that domestic camelids played a predominant role in the economy of this early urban centre, but little additional information is available from other sites in the area (Wing, in press) and new evidence of early alpaca domestication in Junín suggests that specialized breeding for wool production may have been the real contribution of the Pukara culture.

Archaeozoological data from Telarmachay rockshelter, located at 4420 m above sea-level in the department of Junín, Peru, provides evidence of *in situ* camelid domestication by 4000 B.C. (Wheeler, in press). To date, 208 664 of an estimated 450 000 animal bones recovered during six seasons of excavation directed by Danièle Lavallée, have been analysed by Jane C. Wheeler. Animal utilization patterns were found to have shifted from generalized hunting of guanaco, vicuña and huemul deer at 7000–5200 B.C., to specialized hunting of guanaco and vicuña at 5200–4000 B.C., to control of early domestic camelids at 4000–3500 B.C., and establishment of a herding economy from 3500 B.C. on at Telarmachay. Identification of the early domestics at 4000–3500 B.C. is based upon the first appearance of alpaca type incisors (which are intermediate in form between those of the llama and guanaco, and the vicuña), together with morphological changes in some lower molars (poorly developed or absent llama buttresses) which may result from controlled breeding, and an increase in the frequency of camelid neonate remains to 56.7 per cent. It cannot be determined if animals with llama type incisors also appear at this time, since these are indistinguishable from guanaco incisors but the presence of both large and small neonates suggests that they probably did. During succeeding periods the incidence of camelid neonates continued to rise, reaching 68.3 per cent for 3500–3000/2500 B.C., and 73.0 per cent for 3000/2500–1800 B.C.. Throughout this sequence, bone distribution indicates that both large and small newborn camelids were brought into the shelter whole, butchered and, in some cases only partially, consumed. A very similar pattern is produced by present-day traditional herders who utilize dead newborn llamas (large) and alpacas (small) for food. The primary cause of this mortality today is enterotoxaemia, a bacterial diarrhoea associated with unsanitary corralling, which kills an average 50% of all young during the first 40 days of life (Fernández Baca 1971). Since this disease does not occur in wild camelids, and such massive hunting of newborn animals is unprecedented, the evidence from Telarmachay suggests that enterotoxaemia, or some similar disease, may have evolved as part of the domestication process and was already starting to affect early domestic alpacas and llamas by 4000 B.C., while the extremely high neonatal mortality rates from 3500 B.C. on probably indicate the use of corrals and the herding of fully domestic animals. Data from other sites in the area (Wheeler Pires-Ferreira *et al*. 1976) indicates that an annual pattern of herd movement from lower-elevation dry-season to higher-elevation wet-season camps located within the *puna* of Junín may have been established by this date. Similar evidence documenting the domestication process should be forthcoming as more archaeological research is undertaken at high elevation sites in the area between Junín and the southern boundary of the Lake Titicaca basin.

Prehispanic llama and alpaca herding spread well beyond the limits of the *puna* ecosystem, where they were first domesticated, and where they always played a primary role in the economy. Guanaco hunting was important in some highland valley sites, such as Jawyamachay in Ayacucho (3400 m above sea-level), prior to the introduction of domestic forms (K. V. Flannery, personal communication), but exploitation never was as intense as in the *puna*. By the end

of the preceramic period, around 2000 B.C., domestic camelids had spread along the highland drainage systems to elevations below 4000 m, where they provided 43.2 per cent of the meat supply at Pikimachay cave in the Ayacucho valley (2850 m) (Wing, in press), 20 per cent at Kotosh on the eastern slope of the Andes (1900 m) (Wing 1977), 84 per cent at Huacaloma in the Cajamarca basin of northern Peru (2700 m) (Shimada 1982) and part of the 76.3 per cent camelid total at Puripica near the Salar de Atacama in northern Chile (3000 m) (Hesse 1982). After this date llama and alpaca remains are found to predominate faunal assemblages from archaeological sites in the highland valleys of the Callejon de Huylas, Ayacucho, Cuzco and the Río Loa; and in the *punas* of the Callejon de Huylas, Junín and the Lake Titicaca basin. In Ecuador, they are common in the valley sites of Sacopampa at A.D. 650–850, and Cochasqui at A.D. 850–1430 (Wing, in press). Ethnohistorical sources record that llamas accompanied Inca armies as far as southern Colombia and central Chile during the period from A.D. 1430–1532.

Evidence for the first spread of domestic camelids to the coast comes from the Caballo Muerto Complex near Trujillo, Peru, where, at 1500–1100 B.C. they provided 15.4 per cent of the food supply (Pozorski 1979). It is not known if these animals, presumably llamas, were being bred at the site, but by the beginning of our era, large herds controlled by the Mochica polity were well established in the area. Domestic camelid remains were also found in burials at Ayalán, in the Santa Elena peninsula of Ecuador at this date (Hesse 1981). On the coast of southern Peru and northern Chile, llama feet and woollen textiles are frequent in burial mounds starting about 500 B.C., but it is not known if herds were present in the area or not. State-owned herds of llamas were maintained on the coast by the Incas until the time of the Spanish conquest.

There is little doubt that llama and alpaca production was rigidly controlled under the Inca empire (Murra 1975). All camelids, both wild and domestic, were technically considered to be State property. State herds were established, largely from animals confiscated during military expansion, to supply the royal armies with pack animals, and for wool production. Shrine herds, divided according to colour, provided sacrificial animals for the State cults. Communally and individually owned herds also existed. Herd records were kept using the *quipu*, a mnemonic device. Textile production and redistribution was largely State controlled, and offerings or sacrifices of cloth accompanied all social, religious, military and political events. A class of hereditary herders, the *yana*, appears to have been emerging prior to 1532. Llama and alpaca distribution reached its maximum Andean extension under the Inca empire, but, with the advent of Spanish rule, and the introduction of European farm animals, camelid production and breeding control began a decline which has never been reversed.

Recent history and present status

Within little more than a century following the Spanish invasion of 1532, llama and alpaca populations had been decimated throughout the Andes (Flores Ochoa 1977). Coastal and highland valley herds were the first to disappear, as their grazing lands were usurped for the production of introduced livestock. In the *puna*, this process was somewhat slower, due in part to the inhospitable climate and the greater wealth of flocks located there (Murra 1975). However, both the prolonged Spanish civil wars, and heavy tribute levies paid either in domestic lamoids or in money obtained from their sale, led to depletion of the herds. In 1567, the crown inspector Garci Diez de San Miguel reported a single, privately owned herd of 50 000 llamas and alpacas during a census of the province of Chucuito, located on the western shore of Lake Titicaca. Five years later, in 1572, Pedro Gutiérrez Flores counted only 159 697 in the entire province. While these figures may not be exact, the trend is evident. Herds continued to decline as tribute payments increased, until by 1561, llamas and alpacas had practically disappeared, even in this, their area of greatest prehispanic concentration (Flores Ochoa 1977).

Under Spanish dominion, the demand for Merino and Churro sheep and their wool far exceeded that for llamas and alpacas. Native herders were pressed to replace their 'sheep of the earth' with sheep of European origin, selling low and buying high (Flores Ochoa 1977). The lamoids were largely destined for slaughter, to provide meat for the forced labourers of the

Potosi mines, or used to transport the ore. Primary competition between the native and introduced ungulates resulted in the disappearance of llamas and alpacas from the greatest part of their range, while the expansion of sheep herding was eventually limited only by poor-quality grazing lands and problems of altitude adaptation. Even today, llama and alpaca herds are primarily relegated to lands located at, or above, the upper limits of agricultural productivity, which are of marginal value for sheep rearing, but to which the lamoids have always been adapted.

Llamas and alpacas have survived within the framework of traditional, non-European, socioeconomic organization because they are an essential element of Andean culture. Since prehispanic time, herd ownership has been considered the primary source of wealth in the Andes, and rituals to assure the increase and well-being of the flocks are still practised. Breeding and herd management procedures are determined by a mosaic of traditional and hispanic techniques, which are not always (scientifically) efficient, but it is in this context that the majority of present-day llama and alpaca production occurs.

Herds of domestic camelids do, in fact, constitute the most reliable human food resource of the *puna* (Thomas 1973). They exhibit greater resistance to drought than sheep, and their mobility assures herd maintenance despite the climatic instability which often adversely affects agriculture in the zone. They convert high cellulose pasture plants, unsuitable for human consumption, into a useful source of stored protein, thus extending productivity into areas where crops cannot be grown. In addition to meat for immediate consumption, or storage as sun-dried *charqui*, the herds provide other goods and services.

The llama is utilized primarily as a pack animal. It can transport loads weighing 25–30 kg over distances of 15–20 km daily (Flores Ochoa 1977), and is employed in both local and long-distance inter-Andean trade for the obtention of non-*puna* food products and other commodities. Its wool is woven into saddle-bags and made into rope. The alpaca, on the other hand, is primarily a wool bearer whose fibre is utilized for the production of finer-quality cloth. The skins, sinews and bones of both animals provide leather products, thongs and weaving tools. Dung, gathered from the communal defaecation piles, is utilized as a primary source of fuel in the treeless *puna*, and as a fertilizer it is essential for effective potato production. Bezoar stones, foetuses, blood and fat all have important ritual uses among Andean pastoralists, but neither animal has ever been used for milk production, ritual or otherwise.

At present, all llamas and 80 per cent of all alpacas in the Andes are under the control of traditional pastoralists (Novoa 1981). These small herd holders maintain flocks of 30–1000 animals on communal grazing lands. The remaining alpacas are kept in large herds which belong to rural cooperatives established in Peru during the agrarian reform programme of the 1970s. In recent years, as the price of alpaca wool has increased on the world market and in relation to the value of sheep wool, the size of these herds has grown.

Peruvian wool began to reach the world market in the 1830s when British-owned export firms were established in Arequipa (Orlove 1977). Trade in alpaca wool was controlled by private deals between brokers until the 1930s, while sheep wool was auctioned in London. Despite the shift to direct sales in the 1940s, these firms have retained their importance as middlemen and England still remains the major distribution centre for Peruvian alpaca and sheep wool. At present, Peru produces an average of 3400 tonnes of alpaca wool annually, which represents 80 per cent of total world production. A total of 2380 tonnes are exported and the remainder is largely destined for local artisan production.

Alpaca wool has a high commercial value because it contains little kemp, has a low felting quality, is very fine, and can be woven into light-weight, soft and lustrous fabric. The price of Peruvian sheep wool is lower, in contrast, because it is coarse and contains large amounts of kemp. Trade in llama wool has not developed due, in part, to its coarseness and colouration irregularities.

Two breeds of alpaca are distinguished on the basis of their wool characteristics. The Suri has long straight fibres, organized in waves which fall to each side of the body in much the same manner as in Lincoln sheep. The Huacaya has shorter, crimped fibres, which give it a spongy appearance similar to that of Corriedale sheep. Animals with wool characteristics of intermediate type exist, but are rare. Crosses between Huacaya and

Huacaya produce a certain percentage of Suri offspring, and crosses of Suri with Suri produce some Huacaya offspring. Although no artificial selection is made, an estimated 90 per cent of all alpacas are Huacayas (Novoa 1981). One possible explanation is that the Suri trait is genetically recessive but no data are available to test this hypothesis. Wool colour varies from white to black and brown, including all the intermediate shades, and tends to vary little across the body.

Two breeds of llama are also recognized but have been little studied. They are the Ch'aku or woolly and the Q'ara or woolless. Colouration is similar to that of the alpaca, but piebald animals are common.

On a national level, pelts of juvenile llamas and alpacas, in great demand for artisan production, are generally obtained from newborn mortalities. Both animals are also used for meat, and in Peru an estimated 8000 tonnes of llama and 10 000 tonnes of alpaca meat are consumed annually. In both Peru and Bolivia, these animals are of primary importance in the popular diet, but their contribution could be greater if breeding practices were improved.

There have been few large-scale attempts to introduce llamas and alpacas into other areas, due in part to a prohibition on the export of live alpacas from Peru enacted in 1843. In 1858, however, Charles Ledger, under commission from the Australian government, landed 274 animals in Sydney after an arduous 4-year trek across the Andes from Peru and a 5-month voyage (Vietmayer 1978). Forty-nine young were born within 6 months of their arrival, and by 1861 the herd numbered 417 animals. Both body size and wool yields greatly increased, despite the fact that they were kept at low elevations. They adapted well to this new environment, and certainly would have continued to thrive, had not increasing revenues from newly introduced Merino sheep cut short interest in the project, which was abandoned by 1864.

Llamas have been imported into the United States since the early 1930s. The total estimated population at present is between 3000 and 5000 animals, and the largest privately owned herd numbers 500. They are well adapted to a variety of environments and elevations across the country, and are bred for wool production and use as pack animals.

Future prospects

The results of biological research carried out over the past 20 years clearly document the productive superiority of llamas and alpacas at high elevation. Introduced species adapt poorly to the environmental stresses of the *puna*, and give reduced yields of inferior quality. Conception and reproductive success rates are low for both sheep and cattle, and altitude-related diseases can still produce significant mortality, even after almost five centuries. They exhibit lower energetic efficiency in regard to both foraging for and metabolizing the native pasture plants than llamas and alpacas, and their sharp hooves and grazing habits have contributed more significantly to the current problems of range degradation. Sheep and cattle rearing also represents a more labour-intensive investment.

While important progress has been made at research centres in Peru, Bolivia and Argentina, much remains to be done before the full productive potential of llamas and alpacas can be realized. Of particular value are efforts to improve and standardize the production qualities of both Suri and Huacaya wool, and Ch'aku and Q'ara meat, and development of the necessary technology for implementing this knowledge. Since the majority of llamas and alpacas remain under the control of poor Andean peasants who utilize traditional herd management techniques which are not always rational, special emphasis must be placed on extension programmes in order to improve production at a national level.

Llamas and alpacas should play an increasingly important role in the economy of Andean nations, given their proven productive superiority over European livestock. They have survived in the marginal high-elevation lands to which recent history has relegated them, but could undoubtedly be more profitably raised on better-quality pastures, and in the lower-elevation zones which they formerly occupied. The successful commercial production of both alpacas and llamas in Australia and the United States indicates that they can be reared in areas beyond the Andes, and their capacity to utilize marginal pasturelands makes them an important, and virtually untapped, potential resource on a world-wide scale.

References

Fernández Baca, S. (1971) La alpaca reproducción y crianza. *Boletín de Divulgación* No. 7. Centro de Investigación Instituto Veterinario de Investigaciones Tropicales y de Altura (IVITA): Lima

Flores Ochoa, J. A. (1977) Pastores de alpacas de los Andes. In: *Pastores de Puna Uywamichiq Punarunakuna*, ed. J. A. Flores Ochoa, pp. 15–22. Instituto de Estudios Andinos: Lima

Franklin, W. L. (1982) Biology, ecology, and relationship to man of the South American camelids. In: *Mammalian Biology in South America*, eds. M. A. Mares and H. H. Genoways, pp. 457–489. The Pymatuning Symposia in Ecology, Pittsburgh

Gray, A. P. (1972) *Mammalian Hybrids*, Commonwealth Agricultural Bureaux, Farnham Royal, Bucks, England

Hemmer, W. (1975) Zur Herkunft des Alpakas. *Zeitschrift des Kölner Zoo*, 18 (2): 59–66

Herre, W. (1952) Studien über die wilden und domestizierten Tylopoden Südamerikas. *Der Zoologische Garten*, 19 (2–4): 70–98

Hesse, B. (1981) The association of animal bones with burial features. *Smithsonian Contributions to Anthropology*, 29: 134–38

Hesse, B. (1982) Archaeological evidence for camelid exploitation in the Chilean Andes. *Säugetierkundliche Mitteilungen*, 3: 201–11

Hsu, T. C. and Benirschke, K. (1967, 1974) *An Atlas of Mammalian Chromosomes*, Vol. 1, Folio 40 and Vol. 8, Folio 389. Springer-Verlag: New York

Koford, C. B. (1957) The vicuña and the puna. *Ecological Monographs*, 27: 153–219

Murra, J. V. (1975) *Formaciones Económicas y Políticas del Mundo Andino*. Instituto de Estudios Andinos: Lima

Novoa, C. (1970) Reproduction in Camelidae. *Journal of Reproduction and Fertility*, 22: 3–20

Novoa, C. (1981) La conservación de especies nativas en América Latina. In: *Animal Genetic Resources Conservation and Management*. FAO Animal Production and Health Paper No. 24: 349–63. FAO: Rome

Orlove, B. S. (1977) *Alpacas, Sheep, and Men*. Academic Press: London and New York

Pozorski, S. G. (1979) Prehistoric diet and subsistence of the Moche valley, Peru. *World Archaeology*, 11 (2): 163–83

Raedeke, K. J. (1979) *Population Dynamics and Socioecology of the Guanaco* (Lama guanicöe) *of Magallanes, Chile*. University Microfilms International: Ann Arbor

Shimda, M. (1982) Zooarchaeology of Huacaloma: behavioral and cultural implications. In: *Excavations at Huacaloma in the Cajamarca Valley, Peru, 1979*, eds. K. Terada and Y.Onuki, pp. 303–36. University of Tokyo Press: Tokyo

Steinbacher, G. (1953) Zur Abstammung des Alpaka, *Lama pacos* (Linné, 1758). *Säugetierkundliche Mitteilungen*, **1**: 78–9.

Thomas, R. B. (1973) *Human Adaptation to a High Andean Energy Flow System*. Occasional Papers in Anthropology, Pennsylvania State University: University Park

Vietmayer, N. D. (1978) Incredible Odyssey of a Victorian peddler. *Smithsonian*, **9**: 91–102

Webb, S. D. (1974) Pleistocene llamas of Florida, with a brief review of the Lamini. In: *Pleistocene Mammals of Florida*. ed. S. D. Webb, pp. 170–259. University Presses of Florida: Gainesville

Wheeler, J. C. (in press) On the origin and early development of camelid pastoralism in the Andes. In: *Animals and Archaeology: husbandry and the emergence of breeds*, eds. J. Clutton-Brock and C. Grigson. British Archaeological Reports (BAR), Oxford.

Wheeler Pires-Ferreira, J., Pires Ferreira, E. and **Kaulicke, P.** (1976) Preceramic animal utilization in the central Peruvian Andes. *Science*, **194** (4264): 483–90

Wing, E. S. (1977) Animal domestication in the Andes. In: *Origins of Agriculture*, ed. C. A. Reed, pp. 837–59. Mouton Publishers: The Hague

Wing, E. S. (In press) The domestication of animals in the high Andes. In: *Adaptations and Evolution in Biota of High Tropical Montane Ecosystems*, eds. M. Monasterio and F. Vuilleumier. Springer-Verlag: New York.

15

Reindeer

S. Skjenneberg

Secretary of the Nordic Committee on Reindeer Research, Harstad, Norway

Introduction

Nomenclature

Despite a great variety of types, *Rangifer* is now considered to be monospecific, *Rangifer tarandus*.

The origin of the term 'reindeer' is not clear, but the same root is found in the Scandinavian languages as *rein* or *ren*, in Old Scandinavian *hreinn* from Germanic *hrán*. The latter may be derived from *kroino* related to the Greek *krios* and West-Germanic *hrind* 'equipped with antlers', a variety of the root in Greek *keras*, Latin *cornu* (Hellquist 1948). In Old French we find the term *rangier* or *rangifer* which may have led to the generic form *Rangifer* (Banfield 1961). Among the reindeer-keeping tribes there is a great variety of names and some people, e.g. the Saami (Lapps) have developed a very rich and specialized vocabulary designating all kinds of reindeer.

Caribou, which is used for the aboriginal reindeer in America, is derived from the Micmac Indian word *xalibu* which means the pawer or shoveller, probably referring to the snow-digging ability of the reindeer (Wright, in Banfield 1961).

General biology

Reindeer belong to the deer family. They look more solidly built than their relatives, especially the small types which inhabit the far north. On the other hand, the environment may also fashion reindeer in an opposite way. Woodland reindeer are tall and slender. In spite of the arctic adaptation, reindeer seem to be capable of adjusting to such very different habitats as Ellesmere Land (84° N) and Sakhalin (46° N) as the most northern and southern extensions. Reindeer never spread naturally to the southern hemisphere.

The coat is very thick and dense. The guard hairs have medulla with air-filled vacuoles. This gives the fur very effective insulating properties. The fur covers even the muzzle and the area between the nostrils. The common colour is greyish-brown with a lighter, often white belly. The summer coat is darker than in the winter and the northern types are lighter than the southern. Moulting occurs between April and July.

It is the only species in the deer family in which both sexes carry antlers. They are shed annually; stags shed them in late autumn after the rutting season, females and calves in spring. Pregnant does keep their antlers until after calving, whereas barren females lose theirs much earlier. Sexual maturity is reached at the age of 18–30 months. The breeding season is late September–November. The oestrous cycle varies in length from 11 to 22 days and oestrus from 1 to 3 days. Gestation length is 225 days (208–238 days). Calving occurs from late April to mid-June. Twins almost never occur but twin foetuses have been observed in slaughtered domestic reindeer. Milk yield is 300–1000 ml per day. The milk is very rich, 10–24 per cent fat content.

The diploid chromosome number is 70.

The diet offers a great variety of plants in summer but conifers are not eaten. Reindeer prefer softer new sprouts and parts of the plants with the highest nutritional value. In winter, lichens, mainly *Cladonia* spp, form the major part of the diet. In woodland, reindeer eat the lichens on the conifers when the snow cover makes it impossible to dig down to the ground. When the snow is soft and loose, reindeer can dig through a layer of 80 cm or even more.

Rangifer is the most gregarious of the deer family. The tundra types are more gregarious than the woodland and the arctic types.

Distribution and numbers

The domestic reindeer area is the northern part of the northern hemisphere. Total numbers approach 3 million.

The reindeer industry is an important part of the economy in the northernmost parts of Fennoscandia, Russia and Siberia, but is still of little importance in Canada and Alaska (Figs 15.1–4, Table 15.1).

Wild reindeer

The great variation in the genus *Rangifer* has led to many attempts to classify it into subspecies and even species. There has been a great confusion of taxonomic systems due to the use of criteria with too large individual variation. e.g. the antlers. Banfield (1961) seems to be justified in saying: 'Reindeer and caribou populations do not readily fit into the classical species and subspecies categories.'

In both Eurasia and America we find four main types of reindeer. The tundra or taiga reindeer occupy the central parts of the reindeer area. In the woodlands the type is larger with a darker colour; there are also differences in the antler shape. The largest reindeer live in the southern forest or mountain regions; they are a larger version of the woodland type. The high arctic regions are occupied by the small, short-legged polar type seen on Svalbard and in the Canadian archipelago.

Reindeer were rare in the mid-Pleistocene, but became a dominant indicator of the cooler phases of the late Pleistocene (Banfield 1961). In the 4-Würm period (70 000–10 000 years ago) the reindeer range was very extensive: Spain, Italy, south Russia, British Isles, Germany and Switzerland (Kurtén 1968, Zeuner 1963). In the late Palaeolithic period reindeer must have been quite abundant as seen from the many discoveries of the remains of many cultures which seem to have relied on reindeer hunting. This is most noticeable in Magdalenian times, 13 000 years ago.

In the New World fossil findings are much scarcer, but *Rangifer* was present in Alaska at

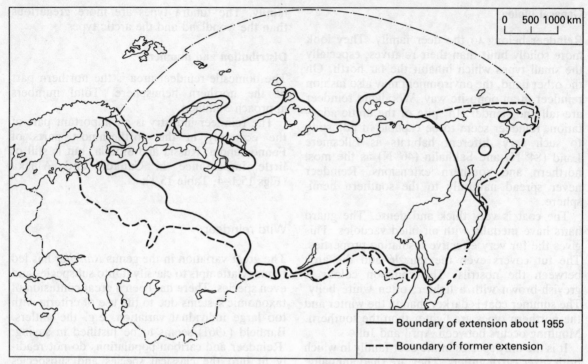

Fig. 15.1 Extension of wild reindeer in the Soviet
Union. Very schematic. (From Heptner
1966)

least 50 000 years ago. A specimen from Fairbanks has been dated back to the Wisconsin
glaciation (30 000–22 000 years ago) when reindeer were distributed in a tundra belt across the
southern edge of the ice sheet (Banfield 1961).

Speculation on how reindeer have spread into
their recent habitat is of little value, as we are not
sure of their original location. At the time when
the Bering Strait was closed, reindeer could easily
roam between the two continents. In Europe,
there are various theories about the reindeer
invasion into Scandinavia, but recent studies by
Siivonen (1975) indicate that the mountain reindeer probably arrived there during the long inter-
glacial period 45 000–35 000 years ago and came
from the south.

Little is known about the type of the wild reindeer which was the origin of the domesticated
reindeer. There is some geographical variation in
the type of the domestic reindeer, so probably
more than one type of wild reindeer was involved.
This question is connected with the history of
reindeer domestication discussed later.

The domestic reindeer in the northern parts of
the U.S.S.R. and almost all the Scandinavian
reindeer are of the tundra type, but domestic
reindeer kept in forest areas are mostly of the
woodland type.

The most noticeable difference between wild
and domestic reindeer is in the colour of the fur.
The wild reindeer are remarkably uniform in their
colour, whereas domestic reindeer display a
greater variety of colours, ranging from almost
black to completely white, often with spots and
larger marks. This is most probably due to selective breeding in order to obtain furs with preferred colours or patterns for clothing.

The distribution of wild reindeer is given in
Figs 15.1, 15.3 and 15.4. The total number is
estimated at approximately 2 million (Table
15.1).

Domestication and early history

It is not surprising that man at an early stage
made use of the special qualities of reindeer,
being as they are so very well fitted to satisfy his
needs in the cold and hostile parts of the world.
In fact, reindeer were the precondition for

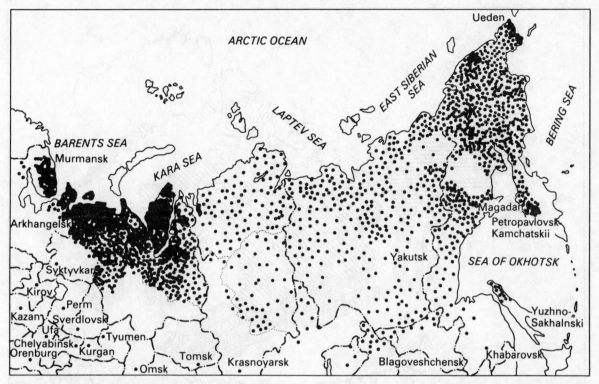

Fig. 15.2 Distribution of domestic reindeer in the Soviet Union. (From Zhigunov 1961)

settling in the extreme north.

The reindeer is gentle and easily tamed. Once tamed, it maintains a certain tameness even when out of contact with man for a year or more. It has a convenient size for use in transportation and may provide man with meat and warm clothes. In addition some tribes utilized the rich milk. It is a gregarious animal, which makes it easy to herd and round up.

Before a discussion about the origin of reindeer domestication, we should define more precisely the term 'domestic' in respect of reindeer. Ingold (1980) has stressed this point. Even if we use the term 'semi-domestic' for the degree of control in which most domestic reindeer are kept, we do not cover all the variations of domestication. Roughly we are able to distinguish between four degrees of reindeer domestication or tameness:

1. Decoy animals in the hunting cultures and reindeer trained for pulling sledges and for riding. Possibly does specially kept for milk should also be placed in this category. The

animals are separated from the herd and are kept in close contact with their owners. They are often dependent upon 'artificial' feeding.
2. Reindeer for the same purposes but kept in captivity for shorter periods and most of the year living in freedom with the herd.
3. Reindeer in smaller herds which are herded every day. They are protected against predators and often guided to adequate pastures. In forest regions they may be given some protection from insects by smoke fires. The reindeer very soon learn to benefit from these advantages.
4. Reindeer in larger herds kept only within certain boundaries but not close-herded. The only contact with man is at the round-ups for earmarking or slaughter. Even then they reach a certain tameness compared with completely wild reindeer.

Reindeer fossil remains have been found associated with human cultures that date as far back as the Chellean–Acheulean epoch (80 000 years ago). Later we know that reindeer were an important prey for Cro-Magnon man and Neanderthal man (25 000 and 15 000 years ago (Banfield 1961). To upper Palaeolithic man rein-

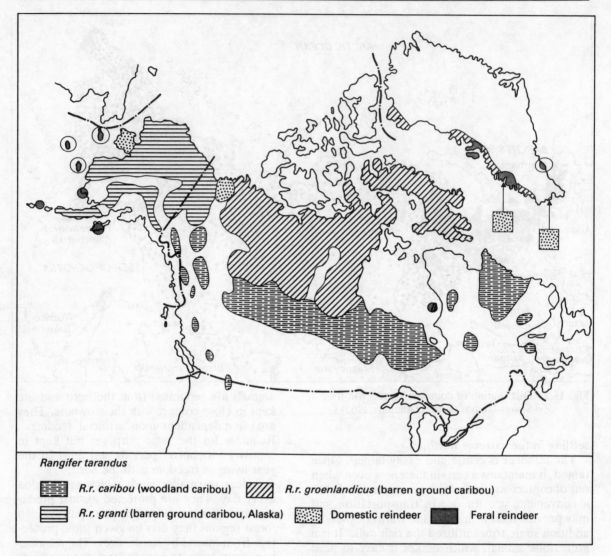

Fig. 15.3 Extension of reindeer and caribou in North America and Greenland

Rangifer tarandus

R.r. caribou (woodland caribou)

R.r. granti (barren ground caribou, Alaska)

R.r. groenlandicus (barren ground caribou)

Domestic reindeer

Feral reindeer

deer must have been one of the most important species hunted, and in the Magdalenian culture reindeer were the dominant prey – in Stellmoor even 100 per cent.

Reliable indications of reindeer domestication are not particularly old. The domestication of reindeer is therefore often considered to be a relatively late event, although not all investigators agree on this. It is natural to assume that the great dependence of prehistoric tribes upon reindeer should have brought about some attempts at domestication. This actually seems to have been

the case with the Saami (Lapps) in Scandinavia some hundred years ago. The origin may have been the use of tamed, decoy animals. Zeuner (1963) points out that even the method of herding combined with nomadism could easily have been used by prehistoric tribes. The old Saami culture has left many remains of stone corrals constructed precisely on the same principles as the corrals built for the round-up of domestic reindeer herds today. Pohlhausen (in Zeuner 1963) suggests that this may have been the case among the tribes at Meiendorf and Stellmoor near Hamburg. In his opinion, two-thirds of the reindeer at these sites must have been herded and only one-third hunted. Sturdy (1975) examined the material from

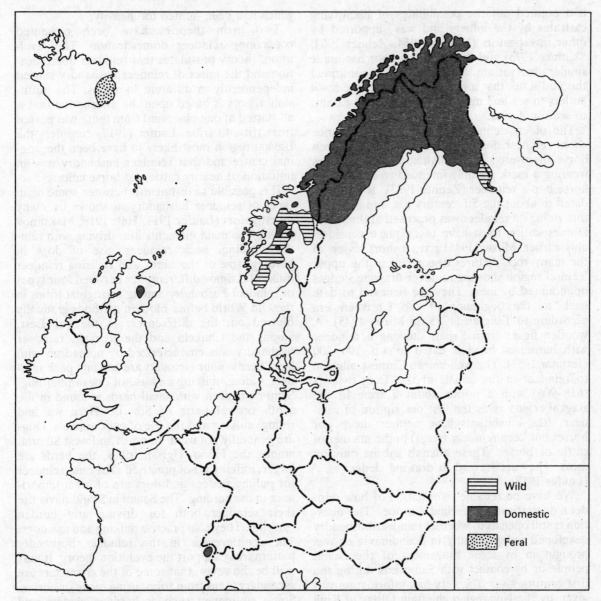

Fig. 15.4 Distribution of reindeer in western
Europe

some of the layers and showed that the ratio of
slaughtered males to females was ten to one and,
moreover, that the animals were killed by axes.
This 'indicates an economy in which reindeer
must have been either under human control or
habituated to man' (Sturdy 1975: 92). If this can
be proved, the earliest time of domestication
should be put back to about 15 000 years ago. On
the other hand, Clark (1980) warns against over-

emphasizing the degree of intimacy in the relationship between man and animal during the late
glacial period. It may, however, be very difficult
to distinguish between site material from selective
hunting and what may be the remains of loose
herding. In the Stellmoor material there were five
or six antlers from castrated reindeer (Zeuner
1963) and this has been interpreted as an indication of domesticated reindeer, but Sturdy (1975)
rejected this idea as the antlers were found in a
group of more than 1000. Already Hatt (1918)

had pointed out the possibilities of identifying castrates by the antlers and was supported by other investigators (Manker 1954, Zeuner 1963, Skuncke 1973) and the present author has made similar observations. Sturdy (1975) has confirmed this and states that it is possible to refer the fossil antlers to sex and age, and also, to some extent, to season of year.

The oldest definitive archaeological evidence about reindeer domestication may be the frozen horse discovered in the Altai, Siberia. It was wearing a mask possibly intended to convert the horse into a reindeer (Zeuner 1963). It has been dated to about the 5th century B.C. and indicates that riding on reindeer was practised at that time. Domestication must have taken place considerably earlier. Mirov (1945) gives a short review of the many rock pictures discovered in the upper Yenisei region showing reindeer drawing sledges or mounted by men. They are believed to date back to the beginning of the Christian era according to Tallgren (1933, in Mirov 1945). A wooden figure from Sajan, showing a reindeer with harnessed head is dated to A.D. 100–200 (Jettmar 1953). The well-known Chinese allusion to reindeer in the annals of the Tang dynasty (618–906) with a notice about a tribe in the Baykal region gives the first description of reindeer: 'the inhabitants have neither sheep nor horses but keep reindeer (stags) in the manner of cattle or horses. These animals subsist only on moss. They are trained to drawing sledges . . .' (Laufer 1917).

We have no reliable information of how reindeer domestication reached Europe. The question is still open as to whether reindeer husbandry developed independently in Scandinavia or was brought in by some movement of the Saami people or by contact with Samoyeds during the first centuries A.D. The very first information was given by the Norwegian chieftain Othar to King Alfred of England when Othar stayed in his court about A.D. 890. He recounted that he had 600 tame, 'unbought' reindeer. Of those, six were decoy animals which were valuable to the Saami for catching wild reindeer (Vorren and Manker 1957). More recent primary sources do not confirm this information on tame reindeer so early, but describe the Saami as reindeer hunters. It is now supposed that reindeer husbandry developed among the Saami some time after 1600, caused by extermination of the wild reindeer

which had been hunted too heavily.

Two main theories have been presented concerning reindeer domestication. The 'evolution' theory postulates that reindeer domestication and the onset of reindeer husbandry started independently in different locations. The 'diffusion' theory is based upon the assumption that it all started at one place and from there was passed from tribe to tribe. Laufer (1917) considers the Baykal region most likely to have been the original centre and that reindeer husbandry was an imitation of nearby cattle and horse cultures.

It is possible to distinguish between some main types of reindeer husbandry as shown by many investigators (Laufer 1917, Hatt 1918, Maksimov 1928). The main elements are: driving with reindeer, riding, pack reindeer, use of dogs in herding, size of the herds and utilizing reindeer milk. Maksimov differentiates between four types of reindeer husbandry among aboriginal tribes in the Old World before modern methods gradually blurred out the differences. In the Far East, among the Chukchi and the Koryak, reindeer husbandry was characterized by nomadism with large herds. Four elements are lacking: pack reindeer, riding, milking and use of dogs in herding. Semi-nomadism with small herds is found in the south central parts of Siberia. Here we find riding, milking and the use of pack animals. Dogs are generally not used. In north and west Siberia, among the Finno–Ugrian tribes, the herds are larger; milking is not practised. They use reindeer for pulling sledges and dogs are of great importance in the herding. The Saami in Scandinavia use their reindeer both for driving and burden carrying. They also practise milking and use dogs. These differences in the reindeer husbandry patterns may support the evolution theory. It may well be, however, that some of the characters are secondary loans from tribes using cattle or horses. Some equipment such as harness, sledges and saddles is evidently such imitation. Linguistic studies give further support to this theory. In Saami there are a lot of terms for driving, pack reindeer and milking which are of Old Norwegian origin. Some attempts have been made to use this for dating reindeer domestication.

The diffusion theory is supported by the fact that the old way of castrating, biting of the testicles, is known all over the reindeer area, as well as the use of lasso and skis (Hatt 1918).

Briefly, we may mention the following factors:

1. Taming of decoy animals may be the earliest type of reindeer domestication.
2. The use of reindeer for transport may be developed from the hunters' need to follow the wild reindeer herds and to bring back the prey, but is more likely an imitation of tribes using horses or dogs for driving.
3. Milking is also an imitation of cultures with cattle or goats.

Two main questions are still unanswered concerning the origin of the domesticated reindeer. 1. Are the specially tamed decoy and transport animals physically handed over from tribe to tribe, or are they animals captured from local, wild reindeer population? 2. Are the herds of 'semi-domestic' reindeer a result of the expansion of a small stock of specially trained animals or developed from a gradually increasing control over wild reindeer populations?

Recent history and present status

Domestic reindeer

Development progressed slowly for many centuries, but during the last decades, the change from barter systems to money economy in the main communities has greatly influenced the original forms of reindeer husbandry. Only parts of the old cultures are left. Mechanization has invaded even reindeer husbandry. But important factors have survived. The domestication or the degree of tameness still justifies the term 'semi-domestic' for herded reindeer. In fact, the development into larger herds reduces the contact between man and animal. Another factor remaining is the migration of the herds between seasonal pastures. Use of reindeer as transport animals is still of importance in the northern parts of Siberia and European Russia, whereas it has been displaced by motor vehicles in the Western countries.

In the Soviet Union reindeer husbandry has adjusted to a common cooperative system. This transition is not surprising. Herding of reindeer among the aboriginal tribes was organized in family groups sharing herding duties.

In China, reindeer are bred by the Evenks in a small area in the northeastern corner of the Inner Mongolian region.

In Fennoscandia the reindeer industry is based upon private ownership and is organized in family groups, or, in Finland and adjacent parts of Sweden, in companies. The herds are relatively large and the reindeer are of the tundra type. In the woodlands of Sweden, we find a more sedentary type of reindeer husbandry with smaller herds of woodland reindeer.

As for the rest of Europe, there is a small reindeer herd in Scotland originating from an import of Swedish reindeer in 1952. A small herd of Norwegian reindeer has recently been established in the French Alps. In the New World the history of reindeer husbandry is very short. It was never started by the Eskimos: the caribou has never been domesticated. Ingold (1980) indicates two reasons for that: the Eskimo way of hunting did not develop the need for decoy animals and the contact on the other side of the Bering Strait was with tribes which used dogs and not reindeer for transport.

Domesticated Siberian reindeer were imported to Alaska in 1891 in an attempt to improve the living standards of the Eskimos when the resources of wild animals for hunting had drastically decreased. The first experiment was far from successful. The Eskimos showed a great aversion to the change from a free hunter's life to more regular duties as reindeer herders. The reindeer went over to a private company which, after some successful years, lost control of the rapidly increasing herds. After 1939 the ownership of reindeer was restricted to the indigenous people. The reindeer industry is concentrated on the Seward peninsula.

Canada's reindeer husbandry was started in 1935 with an import of reindeer from Alaska. It is run in an area of the lower Mackenzie valley and organized on the same principles as in Alaska.

The present-day distribution of domestic reindeer is given in Figs 15.2, 15.3 and 15.4 and Table 15.1.

Breeding of reindeer has in no respect brought about results comparable with other domestic animals. The reasons are obvious: most domestic reindeer live in a semi-domestic or even a half-wild state. The possibilities for controlled breeding are very poor. In the Soviet Union a certain success has been reported in breeding reindeer for better growth. A few trials with artificial insemination are carried out but, for the reasons given above, this method has been of

Table 15.1 Numbers of reindeer and caribou by country and annual harvest*

Country	Number		Annual harvest		Notes
	Domestic	Wild	Domestic	Wild	
China	1 000				Ma Yi-ching (1983)
Soviet Union	2 400 000	800 000	650 000	d	Andreev (1977)
Finland	156 000[a]	500	41 000	0	
Sweden	200 000[a]		43 000		
Norway, mainland	174 000[a]	35 000[e]	55 000	8 000[e]	
Norway, Svalbard		8 000		0	
Iceland		3 600[g]		600–1 000	
UK (Scotland)	80			8	
France	40				
Greenland, reindeer	2 500	500[g]		c.500	
Greenland, caribou		8 500		5 000[f]	
Canada, reindeer	15 200				
Canada, caribou		800 000[b]		21 800[c]	h
U.S.A., reindeer	25 000		d		
U.S.A., caribou		250 000[b]		6 300–11 300	h
South Georgia		2 000[g]		0	1976
Kerguelen		2 000		d	Lesel and Derenne (1975)
Total, reindeer	2 972 820	851 600			
Total, caribou		1 058 500			

* Most information from: Reimers *et al.* (1980)
Notes
a Winter stock
b The Porcupine herd is shared between Canada and the United States.
c Barren ground caribou. No information about woodland caribou harvest
d No information
e Wild and feral
f Information uncertain
g Feral reindeer
h Only official numbers. Extent of 'subsistence hunting' unknown

little importance. Another reason for the slow progress is the lack of control over the most important factor influencing growth, nutrition: reindeer graze exclusively on natural pastures. Until recently, selective breeding also aimed to increase the numbers of animals with special colours popular for fur clothes.

Uncontrolled breeding has created few stable races, although several types are described with differences in respect to size, colour and antler shape. Crossing trials have been carried out, mostly in the Soviet Union, and are reported to have been successful in increasing meat produc-

tion. Polymorphism in the *Rangifer* genetic pool seems to be high, according to the studies of blood serum transferrins. This genetic variability could bring about rapid progress if really effective controlled breeding could be established.

Reindeer are kept mainly for meat and hide production. The output of meat is shown in Table 15.1. The production of milk is of no importance and the use of reindeer in transport is rapidly decreasing. In Asia, an important product is reindeer antlers, usually harvested in the growing phase. They are used in the production of drugs considered to have 'life-stimulating' and aphrodisiac effects. Hides are still used for clothes in the extremely cold climate in the north and sinews are used for sewing fur clothes.

Feral populations

Feral populations of reindeer have developed both naturally and from the wish to introduce reindeer into new areas. The first known experiment was the transfer of domestic reindeer from Norway to Iceland in the years between 1771 and 1787. After the survival of 35 animals, the stock is now approximately 3500. More recently, similar introductions have been carried out to other islands: South Georgia (Norwegian reindeer, 1911–26), St Matthew Island and other islands in the Bering Sea (from Alaska), the Kerguelen Islands in the Indian Ocean (Swedish reindeer, 1955) and to Greenland from Norway, released from the domestic herd established in 1952.

In areas where wild and domestic reindeer are living side-by-side without natural obstacles preventing mixing, there is some interbreeding. This is certainly the situation in south Norway where we find both wild, domestic and intermediate types of reindeer.

Future prospects

Reindeer husbandry is still an important industry in the traditional reindeer areas. As a whole, the population is increasing slowly, but there are definite limits to this increase. There seem to be very few unoccupied areas left and the areas already occupied are threatened by overgrazing in many places. Some progress has been made in feeding reindeer in emergencies and as a supplement to inadequate winter pastures. There are economic limitations to such feeding in a profitable reindeer industry. Used adequately, however, it may give the reindeer owners a more secure income by reducing mortality which is the most important negative factor in the reindeer industry. Losses are due to predators, diseases and undernutrition. Calves constitute the most susceptible group. Losses of up to 10 per cent of the stock are commonly seen, but the calf loss may reach 50 per cent or even more in bad years.

Reindeer research has not yet reached the level of activity that is seen in other domestic animals in spite of the role played by reindeer in man's settlement and exploitation of the vast areas in the north. Regular research was started in the Soviet Union some 50 years ago, and there is now an increasing interest in this field in Western countries.

Interest in the reindeer industry is increasing. This is due to growing interest in the exploitation of the enormous areas in which reindeer are indigenous. The reindeer is here the obvious choice of animal to utilize these natural resources. The U.S.S.R. is planning to increase the domestic reindeer stock from approximately 2.5 million animals to 4 million. But there are many factors limiting this trend. Reindeer ranges must more often give way to other interests. Mining, hydroelectric projects, forestry, roads and railways, tourism, are all factors which both reduce the pasture areas and cause problems in the exploitation of the reindeer.

The biotype is particularly vulnerable. Lichens, the main source of winter nutrition to reindeer, absorb their nutrition through their surface from the atmosphere. This makes them very vulnerable to atmospheric pollution.

When all factors are taken into consideration, however, there are still many areas in the north which can be better utilized by reindeer than they are at present.

References

Andreev, V. N. (1977) [The status of world reindeer breeding and classification]. *Ekologiya*, No. 4: 5–10. (In Russian)

Banfield, A. W. F. (1961) *A Revision of the Reindeer and Caribou, Genus* Rangifer. Bulletin No. 177, National Museums of Canada: Ottawa

Clark, G. (1980) *Mesolithic Prelude: the Palaeolithic-Neolithic transition in Old World prehistory.* Edinburgh University Press

Hatt, G. (1918) Rensdyrnomadismens elementer. *Geografisk Tidskrift (Copenhagen)*, (**7**): 241–69

Hellquist, E. (1948) *Svensk Etymologisk Ordbok.* Gleerups Förlag: Malmö

Heptner, V. G. (1966) In: *Die Säugetiere der Sowjetunion,* Band I: *Paarhufer und Unpaarhufer*, ed. V. G. Heptner and N. P. Naumov. Gustav Fischer Verlag: Jena

Ingold, T. (1980) *Hunters, Pastoralists and Ranchers: reindeer economies and their transformation.* Cambridge University Press: London

Jettmar, K. (1952, 1953) Zu den Anfängen der Rentierzucht. *Anthropos (Freiburg)*, **47, 48** (supplement)

Kurtén, B. (1968) *Pleistocene Mammals in Europe*. Weidenfeld and Nicolson: London

Laufer, B. (1917) The reindeer and its domestication. *Memoirs of the American Anthropological Association*, **4** (2): 91–147

Lesel, R. and **Derenne, Ph.** (1975) Introducing animals to Iles Kerguelen. *Polar Record*, **17** (110): 485–94

Maksimov, A. N. (1928) [Origin of reindeer husbandry]. *Inst. Istorii. Ucheniie Zapiski*, **6**: 3–27. (In Russian)

Ma Yi-ching (1983) Status of reindeer in China. *Acta Zoologica Fennica*, **175**: 157–8

Manker, E. (1954) *Till Frågan om Renskötselns Ålder*. Norrbotten: Luleå

Mirov, N. T. (1945) Notes on the domestication of the reindeer *Memoirs of the American Anthropological Association*, **47** (3): 393–408

Reimers, E., Gaare, E. and **Skjenneberg, S.** (eds) (1980) *Proceedings of the Second International Reindeer/Caribou Symposium, Røros, Norway, 1979*. Direktoratet for Vilt og Fersvannsfisk: Trondheim

Siivonen, L. (1975) New results on the history and taxonomy of the mountain, forest and domestic reindeer in Northern Europe. In: *Proceedings of the First International Reindeer and Caribou Symposium, Fairbanks, Alaska, 1972*. eds, T. R., Luick, P. C., Lent. D. R. Klein and R. G. White. Biological Papers of the University of Alaska, Special Report No. 1, pp. 33–40

Skuncke, F. (1954) *Renen i Urtid och Nutid*. Norsteds: Stockholm

Sturdy, D. A. (1975) Some reindeer economies in prehistoric Europe. In: *Palaeoeconomy*, ed. E. S. Higgs. Cambridge University Press: London

Vorren, Ø. and **Manker, E.** (1957) *Samekulturen*. Tromsø Museums Skrifter, Vol. 5

Zeuner, F. E. (1963) *A History of Domesticated Animals*. Hutchinson: London

Zhigunov, P. S. (1961) Introduction. In: *Reindeer Husbandry*, ed. P. S. Zhigunov. Translated from Russian. Israel Program for Scientific Translations, Jerusalem, 1968.

16
Other deer

T J. Fletcher

*Reedie Hill Farm,
Auchtermuchty,
Scotland*

Introduction

Of the fifty or so species of deer only eleven have attracted man's attention as domesticants. The reindeer has been dealt with in Chapter 15 leaving the musk deer and seven species of the genus *Cervus*, namely the fallow, axis, hog, sambar, rusa, sika and red deer. The two other species are Père David's deer known only in captivity, and the moose known in Europe as elk, each in a monospecific genus. Others, of the genera *Mazama* and *Odocoileus*, are reputed to have been suckled by the South American Maya for either religious purposes or to provide meat; they will not be further considered here. Chromosome numbers are shown in Table 16.1.

Musk deer – *Moschus*

The musk deer are generally placed in a genus. *Moschus*, and a family Moschidae, of their own but the number of species has been somewhat controversial. The most recent revision (Groves 1975) recognizes three: *M. moschiferus* in eastern Siberia, and *M. chrysogaster* and *M. sifanicus* both

Table 16.1 Chromosome numbers of deer

Common name	Species	2n	
Fallow	*Cervus dama*	68	Gustavsson and Sundt (1968)
Axis	*Cervus axis*	66	Hsu and Benirschke (1974)
Hog	*Cervus porcinus*	68	Hsu and Benirschke (1977)
Sambar	*Cervus unicolor*	64–65	Hsu and Benirschke (1973)
Sika	*Cervus nippon*	67	Hsu and Benirschke (1969)
Red	*Cervus elaphus*	68	Hsu and Benirschke (1969)
Père David's	*Elaphurus davidianus*	68	Hsu and Benirschke (1971)
Elk or moose	*Alces alces*	70	Hsu and Benirschke (1969)

extending from the Himalayas to China, the latter at higher altitudes. However, by earlier authors the name *M. moschiferus* has frequently been used for all musk deer or for the Himalayan in contrast to the Siberian species.

Adult musk deer stand only 51–61 cm at the shoulder and weigh up to 15 kg. They resemble the mouse deer or Tragulidae, and differ from other deer, in having no antlers nor any facial or foot glands and in having a gall bladder. The male has tusk-like upper canines. Females are seasonally polyoestrous showing heat at 18–24-day intervals between October and February; oestrus lasts about 24 hours. Twins are usual with birth occurring after a 179–187-day gestation (Bista *et al*. 1979).

Musk deer occur in forested areas of east Siberia, north Mongolia, Manchuria and Korea, south China and the Himalayas. The Himalayan species, *M. sifanicus*, was classed as vulnerable in the 1974 *Red Data Book* of the International Union for the Conservation of Nature and Natural Resources, and this is due to its commercial value as a source of 'musk'. This granular material is secreted from a sac or 'pod' on the ventral surface of the male's abdomen between the umbilicus and the penis. The sac is particularly active during the rut. The musk is used in oriental medicines and as a fixative in western perfumes and that from *M. sifanicus* is the most valuable. In 1972–73 musk made up 8.7 per cent of Nepal's total exports and the trade from that country was valued at more than $ U.S. 1 million per annum (Green 1978)

The increasing value of musk and the declining deer population has caused the Chinese to develop musk-deer farming. A project was begun in 1958 on an island in Foziling reservoir in the Dabie mountains and it has now proved possible to 'milk' males up to fourteen times (Green

1978). Now, however, the extraction of musk is only carried out annually with yields of around 5 g obtainable from each collection; males are productive from 3 to 12 years of age. Several farms have now been established, principally in Sichuan province, some specializing in the collection of the musk and others in the breeding. Deer are housed individually in crates measuring 90 × 90 × 70 cm fitted with mesh floors. Around 90 per cent of females give birth each year and about 60 per cent of the young are successfully reared to maturity (Bista *et al*. 1979).

There would appear to be a strong case for the development of musk farms in Nepal since although synthetic musks have reduced the demand from the perfume industry this is unlikely to influence the medicine trade in the foreseeable future.

Fallow – *Cervus dama*

Fallow are given their own genus (*Dama*) by many authorities but Corbet and Hill (1980) include them in the genus *Cervus*. Two subspecies are recognized, *C. d. dama* and *C. d. mesopotamica*, the latter almost extinct.

Fallow deer usually reach a shoulder height of about 90 cm and a weight of about 90 kg in the buck. There is a conspicuous tassel of hair at the end of the prepuce and the antlers are characteristically palmate; they are cast in the spring and grow through the summer, hardening in time for the autumn rut.

Wild populations are usually spotted with white on a brown ground during the summer but in the winter these spots are less visible. In parks and some introduced populations a number of colour variants exist presumably reflecting man's interference with their breeding. Apart from the spotted form there are all variations between black and near white (Chapman and Chapman 1975).

Fawns, of which fewer than 1 per cent are twins, are born in June and July after a gestation of 229 ± 2.7 days; the does are seasonally polyoestrous (Corbet and Southern 1977).

Cervus d. mesopotamica was, in 1977, reduced to 120–150 individuals in western Iran. The European or common fallow is, however, now widely distributed through the whole of Europe and into parts of Australia, New Zealand and North and South America (Chapman and Chapman 1975)

The sudden appearance of fallow in Holocene sites in Cyprus and in the Bronze Age in Crete, both coinciding with the arrival of other domestic animals, suggests that the deer were imported to these islands by man, presumably as a food animal (Zeuner 1963, Jarman 1976). Fallow deer were probably kept in enclosures or herded by the Hittites and emparked by the Greeks. Columella writing around A.D. 65 described deer in enclosures, specifically referring to fallow deer in Gaul. There is some doubt whether fallow deer existed in Britain after the last glaciation but it seems most likely that they were reintroduced into England by the Normans; they are first recorded in Scottish deer parks in 1290 (Gilbert 1979). By the 16th century they were the most common species in British deer parks which in their heyday may have numbered over a thousand. Whether this was due to their value as food animals or for sporting purposes it is extremely difficult to say; certainly they thrive despite the north European winters. In fact little hunting seems to have been carried on in parks until the 16th century and the emphasis gradually shifted from food production to recreation as techniques for overwintering cattle and sheep improved and as unenclosed deer dwindled.

Fallow deer have been extensively involved in the 'new wave' of deer farming since about 1970, especially in New Zealand, Australia and West Germany. In New Zealand they make up around 15 per cent of the total farm deer population which means that, in 1981, about 20 000 fallow were being farmed there (Anonymous 1981). They are less easy to handle in the yards than red deer and their growing antlers and other by-products are not valued as highly in the Orient, yet their meat commands a premium in the crucial West German market and they appear to be particularly resilient and disease resistant. Within West Germany there are some 200 deer farms and all these use fallow (Reinken 1980). There is a great need for research into the efficiency of grass conversion by all species of deer but anecdotal evidence from Australia and New Zealand suggests that fallow deer may outperform other species of deer as well as cattle and sheep.

Axis deer – *Cervus axis*

The axis deer has a confusing variety of names: it used to have its own genus – *Axis*; in English it is also known as chital, or spotted deer. The confusion has been aggravated by the recent use, especially among deer farmers, of the term 'spotted deer' to describe *C. nippon*, the sika.

The most spotted of all deer, the axis is truly tropical with little or no seasonality of breeding and antler casting or cleaning even after many years in temperate regions. Chital stags grow to about 90 cm at the shoulder and can weigh up to 100 kg. They are readily distinguished from other spotted deer by their white throats and muzzles and the absence of a white rump patch.

Chital are extremely prolific and can reproduce three times in 2 years with a high frequency of twins. Gestation lasts 7–7.5 months. Crosses with hog deer and with red deer produce fertile hybrids. Originally restricted to the Indian subcontinent and Sri Lanka axis deer are now successfully established in north Queensland (Australia) and in Brazil, Hawaii and Argentina.

Chital may have been moved to Europe by the Romans. By the 19th century they were well established in zoos and some parks in England and Continental Europe. Serious attempts to domesticate chital are now being made in Australia where its prolificacy and pale-coloured good-quality meat make up for the difficulties of handling this nervous and active deer.

Hog deer – *Cervus porcinus*

The short-legged and stocky hog deer (also called para) stand only 60–75 cm at the shoulder and yet weigh up to 50 kg. The young are spotted but not the adults. The antlers are shed aseasonally and up to three calf crops may be produced in 2 years although twins are rare.

The hog deer extend from the Indus and Ganges valleys into Burma in a band between the Himalayas and about 24 °N in India, but extending southeast to Thailand, Cambodia and part of Vietnam. They have been established in Australia and Sri Lanka.

Some attempt is being made to farm hog deer in Taiwan but it has not yet attracted attention in Australia where it is restricted to southeast coastal Victoria and some offshore islands.

Sambar – *Cervus unicolor*

Some authorities have included *C. timorensis*, the Timor deer or rusa as a subspecies of sambar but Corbet and Hill (1980) separate the sambar and rusa into two species: *C. unicolor* including *C. mariannus*, the Philippine sambar; and *C. timorensis*. This conveniently fits common usage of the names sambar and rusa respectively.

The various forms of this species show considerable size variation ranging from the largest Asian deer weighing up to 315 kg and measuring up to 150 cm at the shoulder down to the smaller *C. mariannus* of only 40–60 kg and 70 cm shoulder height. They have a coarse and shaggy coat and this is not usually very spotted. The antlers are typically only six-pointed (three on each side) and they are markedly rugose. There is no very obvious white rump patch.

Breeding is usually aseasonal but in some northern areas of the range seasonality seems quite well developed. Twins are born only occasionally.

Cervus unicolor extends throughout India, Burma and eastwards and southwards to include the whole of southeast Asia with the smaller forms occurring in the islands of Indonesia and the Philippines. It has been established in Northern Territory and Victoria, Australia.

Sambar have been hunted for their velvet antlers for a long time and their confinement in small pens and huts has probably been carried on although never as extensively as with sika deer. The Australian sambar are so much prized as a hunting quarry that they are not yet being farmed on a large scale. However they are large and seem likely to gain favour as crossing stock for the more readily available but smaller rusa. In Taiwan and elsewhere they continue to be used as farm animals usually being run in small groups or individually in pens.

Rusa – *Cervus timorensis*

Designated a separate species by Corbet and Hill (1980) the rusa deer are natives of Java, Sulawesi and the Lesser Sundas and differ from their mainland relatives, the sambar, chiefly in their smaller size. They are also called Timor deer or Sunda sambar.

Varying in size between islands, adult male rusa may measure at the largest 110 cm at the shoulder and weigh up to 125 kg, although on farms weights of up to 136 kg have been recorded. Their breeding cycle shows some seasonality at the southern extremity of their range with calves born most frequently between March and April in Australia. Gestation is around 8 months. In appearance rusa deer are similar to sambar deer but their coat is less coarse and their tail less bushy. Rusa are likely to interbreed freely with sambar probably producing fertile hybrids.

In addition to their native range rusa have been introduced very successfully to New South Wales, Australia and to Papua New Guinea as well as to a number of islands including Mauritius.

The rusa is perhaps the most popular farmed deer in Australia although it is outnumbered on farms by the more freely available fallow deer. It finds favour because of its tractability and because it twins more freely than red deer. In addition its antlers are more salable than the fallow. Red deer would no doubt outnumber fallow and rusa were they available.

In the Western province of Papua New Guinea where tens of thousands of rusa exist along the Bensbach river flood plains a deer-farming project has been established by the government, aided by the United Nations, to crop velvet from enclosed rusa for export and eventually to export venison to West Germany. If the project is successful it is intended to assist local villagers to start their own deer-farming ventures.

In Mauritius several sugar estates are attempting to farm rusa for venison as the species has thrived there since its introduction by Dutch settlers in 17th century.

Sika deer – *Cervus nippon*

Some authorities have recognised nearly twenty subspecies of sika deer but as some of these now exist chiefly or only in parks and farms, and since interbreeding of different types has been carried on, the classification is dubious. They are also called spotted deer or 'plum deer'.

Sika deer vary in size depending on race, from the smaller Japanese forms which may only reach 65 cm shoulder height to 109 cm for the largest – Dybowski's sika of Manchuria. They are generally spotted throughout life although these spots may be indistinct in some forms. The white caudal patch is well developed while the throat, neck and face are usually dark.

In the northern hemisphere the calving season is at its height in June; twins are rare. Gestation is about 7.5 months. The antlers are cast in the spring and grow through the summer, hardening in time for the October rut. Details of the oestrous cycle have not been published.

They can form fertile hybrids with red deer.

The natural range of sika is now reduced to Japan and a broad strip up east China to Siberia. Introductions have been numerous and successful, including those to western Europe, North Africa, Madagascar and New Zealand.

In Neolithic sites at Yanshao sika deer bones were second only to pigs in abundance, possibly indicating some control over the deer by man (Watson 1969). The value of the growing or 'velvet' antlers or *panti* as medicine was first suggested during the Qin dynasty (221–207 B.C.) in China (Quan 1981). At what date deer were first specifically raised for production of the velvet antler is unknown.

Cervus nippon is the species of deer most widely farmed in China, Taiwan, Korea and probably the Soviet Union. Supplies of the *panti* and other organs (tails, penises, embryos, sinews, hard antlers, and indeed all parts of the deer) are supplemented by shooting wild deer, but farmed products, including those from reindeer and caribou and from farms in New Zealand, Australia, the Soviet Union and elsewhere are increasingly coming forward to satisfy the Chinese demand. Within China it is estimated that some 270 000 deer are farmed and in 1981 the revenue from the antlers alone was $ U.S. 35 million. Around Peking there are over twenty

farms, two of which carry a thousand deer each and which produce a gross income of $ U.S. 680 per head for an outlay of only $ U.S. 170 each. In Heilongjiang province in northeast China there are well over a hundred deer farms carrying about 70 000 deer; one has 3000 deer on 1000 ha (Quan 1981). The farms vary in intensification from those in which the deer are permanently enclosed in pens to those in which the deer are herded and grazed without fences.

In New Zealand sika are being farmed but represent only 0.5 per cent of the farmed deer, i.e. only around 650. So far they are proving less easy to handle than red deer, and the velvet antlers are proving difficult to market presumably because such small quantities are involved.

Red deer – *Cervus elaphus*

The degree to which the Asiatic subspecies can be considered con-subspecific with the North American wapiti (also called elk) remains controversial. If distinct, then Corbet (1978) favours *C.e. sibiricus* as the name for the Asian forms and *C.e. canadensis* for the Canadian.

Red deer vary in size from the Scottish forms in which the adult stag may weigh only 130 kg, through the large east European red deer where stags reach 300 kg, to the wapiti of North America which can weigh 450 kg. The coat varies from red to grey depending on subspecies and season. The calves are always distinctly spotted but the adults never are. They are pronounced seasonal breeders casting their antlers in March and April, growing them through the summer and hardening them for use as weapons in the rut in October. Hinds are seasonally polyoestrous from October to March with oestrus recurring at intervals of 18.3 ± 1.7 days; gestation in Scottish red deer is 231 ± 4.5 days (Guinness *et al.* 1971). Asdell (1965) gives gestation in the Canadian wapiti as 249–262 days. Twins occur in European red deer at a frequency of only 0.1–1.0 per cent (Mitchell *et al.* 1977).

Red deer range from Europe in a band throughout Asia into North America where they once extended as far south as California and almost as far east as the Alantic. There is a relic population in North Africa. Within Asia some forms are endangered or restricted to farms. European red deer have been established in

Argentina and in New Zealand for hunting and have in each case thrived at the expense of the native flora and fauna.

Red deer were staple in the diet of man of the upper and middle Palaeolithic in southern Europe and in northern Europe in the early Holocene (Jarman 1972). It appears that man moved with the red deer herds and exploited them in an arguably selective way; whether they were herded is not clear. Columella (*c.* A.D. 65) mentions their value as park animals and describes the construction of stone, brick or wooden fences. Certainly red deer were, with fallow deer, the principal occupants of all the mediaeval deer parks of Europe.

In China and probably in parts of the Soviet Union several Asian subspecies of red deer have been enclosed for the production of medicinal products for many years. Wallace encountered red deer tethered in pens in China in 1913 (Whitehead 1972). An estimated 70 000 stags are currently being farmed in the far east of the Soviet Union, the Altai and Kazakhstan, and these produce annually 26 600 kg of growing antler (personal communication, D. K. Belyaev to I. L. Mason).

About 1970, falling sheep prices and rising venison prices stimulated the establishment of deer farming ventures in Scotland and in New Zealand (Blaxter *et al.* 1974). At first these farms developed on poor-quality land but have subsequently tended to become established on better pasture at higher stocking densities. The principal product in both countries was initially venison, destined for West Germany, but rapidly escalating prices for those organs valued in oriental medicine, especially the growing antlers, soon led New Zealand farms to concentrate on antler production and deer farming developed very rapidly with hinds changing hands at auction for up to $ U.S. 2500 (Anderson 1978). In 1980 velvet antler prices fell dramatically and the industry is once more turning to venison production (Yerex 1981). By 1981 there were some 130 000 deer, of which around 85 per cent were red, on about 1600 farms. Wapiti introduced in 1905, into Fiordland on the southwest coast of South Island, New Zealand, have become well established and have hybridized with red deer in that area. Wapiti and wapiti/red hybrids are much valued in New Zealand for their larger carcass, and fetch high prices at auctions.

In Britain deer farming has developed much more slowly with the emphasis always on venison production. Velvet antler cropping was rarely carried out commercially before being made illegal on humanitarian grounds in 1980. By 1981 there were still only around 40 farms carrying about 1500 deer of which red deer are the firm favourites. This was despite acceptance by the government of deer farms as eligible for grant aid on the same level as conventional livestock enterprises.

Père David's deer – *Elaphurus davidianus*

This is distinguished by its long tail reaching almost to the hocks and by its antlers which typically bifurcate some 5–10 cm above the coronet with one beam being carried upwards and backwards. Both beams may then carry several points. The coat is spotted, standing about 115 cm at the shoulder and weighing up to 200 kg (adult stag) the Pére David is slightly larger than the European red deer. The rut occurs in June and July and the single calf is born in April or May. Hybrids with red deer occur occasionally and some at least appear to be fertile.

The Père David does not appear to have existed in the wild for 2000–3000 years and remained unknown to the western world until 1865 when the French missionary Armand David observed it in the Imperial Hunting Park of Nanhaize just south of Beijing. After repeated failures, breeding groups were established in Europe and when, on the destruction of Nanhaize by flood and civil unrest in 1894 and 1900, the Père David became extinct in China, the Duke of Bedford managed to collect enough into Woburn Park in England to form a breeding nucleus from which over thirty zoos and parks – including some in China – have now been stocked. Records of the original purpose for maintaining the deer at Nanhaize do not appear to exist but presumably the Pére David was kept there to supply medicinal products.

The ability of the Père David to flourish in northern Europe and its suitability for park use may mean that it will become of value for venison production. At the moment the scarcity and high price of breeding stock preclude commercial involvement.

Elk or moose – *Alces alces*

The North American moose and the European and Asian elk are considered by Corbet (1978) to be the same species. They differ in that the North American form carries a larger dewlap or 'bell' and is heavier than the European.

The largest living species of deer with a shoulder height of up to 235 cm and a body weight of up to 800 kg, elk are distinguishable by their pendulous upper lip and muzzle, palmate antlers and long legs. The calves which are not spotted are commonly twins, rarely triplets. They are born in late May and June after a 240–250-day gestation. The antlers are cast from November for the oldest bulls until February or later for the youngest; they regrow in the summer hardening for the mid-September rut. Cows are probably seasonally polyoestrous with a 30-day cycle.

It is distributed from Scandinavia in a belt through the centre of the Soviet Union to include parts of Mongolia and Manchuria. In North America, Alaska and Canada carry most of the population although some occur in Wyoming and Utah as well as along the Canadian border of the U.S.A.

The domestication of elk has probably been continuing since at least 200 B.C. An Egypto–Roman plaster cast of a silver vessel of this period originating north of the Black Sea clearly shows a woman milking an elk; the Scythian peoples around the Black Sea certainly represented elk in much of their decoration. In eastern Siberia the Yakuts used elk for riding as late as the 19th century while in Sweden and Estonia they were used for draught and for riding until the 17th century; they are reported to have been faster and more enduring than horses (Zeuner 1963).

The Russians have been running an elk farm in the Urals at the Pechero-Ilych National Park since 1954. There are now around 4000 elk there and these are herded by herdsmen who milk them daily. Selection is reported to have increased the yield to about 8 litres per day and the lactation to several months. The milk has 10 per cent fat and 15 per cent protein. Calves are slaughtered at around 200 kg but adults will usually reach weights of 500 kg. Castrated bulls are used for draught (Anonymous 1974).

References

Anderson, R. (1978) *Gold on Four Feet*. Anderson: Colingwood, Victoria, Australia

Anonymous (1974) *Blue Book for the Veterinary Profession*, **24**: 111

Anonymous (1975) *China Reconstructs*, **24** (3): 48–9

Anonymous (1981) *The Deer Farmer*, (Autumn): p. 29. P.O. Box 11–137, Wellington, New Zealand

Asdell, S. A. (1965) *Patterns of Mammalian Reproduction* (2nd edn). Constable: London

Bista, R. B., Shrestha, M. N., and Kattel, B. (1979) Domestication of the dwarf musk deer (*Moschus berezovskii*) in China. Report submitted to the Nepal Government, National Parks and Wildlife Conservation Office

Blaxter, K. L., Kay, R. N. B., Sharman, G. A. M., Cunningham, J. M. M. and Hamilton, W. J. (1974) *Farming the Red Deer: the first report of an investigation by the Rowett Research Institute and Hill Farming Research Organisation*. Her Majesty's Stationery Office: Edinburgh

Chapman, D. and Chapman, N. (1975) *Fallow Deer, their History, Distribution and Biology*. Terence Dalton: Suffolk

Corbet, G. B. (1978) *The Mammals of the Palaearctic Region*. Cornell University Press: Ithaca, New York

Corbet, G. B. and Hill, J. E. (1980) *A World List of Mammalian Species*. British Museum (Natural History): London

Corbet, G. B. and Southern, H. N. (1977) *The Handbook of British Mammals* (2nd edn). Blackwell Scientific Publications: Oxford

Gibert, J. M. (1979) *Hunting and Hunting Reserves in Medieval Scotland*. John Donald: Edinburgh

Green, M. J. B. (1978) Himalayan musk deer (*M. m. moschiferus*). In: *IUCN. Threatened Deer Programme Proceedings*, pp. 56–64. International Union for Conservation of Nature and Natural Resources: Gland, Switzerland

Groves, C. P. (1975) The taxonomy of *Moschus* with particular reference to the Indian region. *Journal of the Bombay Natural History Society*, **72**: 662–82

Guinness, F. E., Lincoln, G. A. and Short, R. V. (1971) The reproductive cycle of the female red deer, *C. elaphus*, L. *Journal of Reproduction and Fertility*, **27**: 427–38

Gustavsson, I. and Sundt, C. O. (1968) Karyotypes in five species of deer (*Alces alces*, L.; *Capreolus capreolus*, L.; *Cervus elaphus*, L.; *Cervus nippon nippon*, Temm. and *Dama dama*, L.). *Hereditas*, **60**: 233–48

Hsu, T C. and Benirschke, K. (1969, 1971, 1973, 1974, 1977) *An Atlas of Mammalian Chromosomes*. Springer-Verlag: Berlin

Jarman, M. R. (1972) European deer economies and the advent of the Neolithic. In: *Papers in Economic*

Prehistory: studies by members of the British Academy Major Research Project in the Early History of Agriculture, ed. E. S. Higgs, pp. 125–47. Cambridge University Press: London

Jarman, M. R. (1976) Animal husbandry – the early stages. In: *IX Congrés, Union Internationale des Sciences Préhistoriques et Protohistoriques, Colloque 20*, pp. 22–50. UNESCO: Paris

Mitchell, B., Staines, B. W. and Welch, D. (1977) *Ecology of Red Deer: a research review relevant to their managment in Scotland*. Institute of Terrestrial Ecology: Cambridge

Quan, P. (1981) *The Deer Farmer*, (Autumn): p. 4. P.O. Box 11–137, Wellington, New Zealand

Reinken, G. (1980) *Damtierhaltung auf Grün- und Brachland*. Verlag Eugen Ulmer: Stuttgart

Watson, W. (1969) Early animal domestication in China. In: *The Domestication and Exploitation of Plants and Animals*, ed. P. J. Ucko and G. W. Dimbleby. Duckworth: London

Whitehead, G. K. (1972) *Deer of the World*. Constable: London

Yerex, D. (1981) *The Farming of Deer: a review of world trends and modern techniques*. Deer Farming Services: Wellington, New Zealand.

Zeuner, F. E. (1963) *A History of Domesticated Animals*. Hutchinson: London

17
Pig

H. Epstein

Emeritus Professor of Animal Breeding, The Hebrew University of Jerusalem, Israel

and

M. Bichard

Pig Improvement Company, Abingdon, England

Introduction

Nomenclature

The pig belongs to the order Artiodactyla or even-toed ungulates, family Suidae. The word 'pig' is derived from the early Middle English 'pigge', of dubious etymology. The term 'swine', which is of commoner use in the United States than in the British Isles, originates from the Old English *swin*, related to the Dutch *zwijn*, German *Schwein* and Latin *suinus* (sus-ine). The term 'hog', of Middle English derivation, in Britain usually denotes a castrated male pig reared for slaughter, but in America it is used as a general term for the animal, the castrated male being 'barrow'. The word 'boar', from the Old English *bár*, refers to a male uncastrated pig or, as 'wild boar', to both sexes of *Sus scrofa*, the wild pig. The adult female pig is called 'sow' from the Old English *sugu*; the word 'gilt' is used for a female pig until weaning of the first litter, and 'pigling' or 'piglet' for a young pig.

General biology

Pigs have a long mobile snout, and their feet have

two functional and two non-functional digits. Wild pigs have a complete dentition of forty-four teeth, while in some domesticated breeds the number of teeth is reduced. The stomach is two-chambered, simple and non-ruminating. Pigs are omnivorous and in the wild state feed on roots, plant stems, fungi, acorns, chestnuts, beechmast, grain, insect larvae, worms, eggs, frogs, reptiles and carrion. They wallow in mud for hours and travel in bands during the night in search of feed, being most active in the evening and early morning. Adult wild boars tend to remain solitary.

In the larger species of *Sus* the head and body reach a length of up to 180 cm and the height at the shoulder comes to 100 cm; the weight of adult males varies between 75 and 200 kg, and of females between 35 and 150 kg. The thick, hard skin is covered with silver grey to black or brown bristles which form a long upright ridge on the neck and back; in winter a woolly undercoat develops between the bristles. The young are striped.

The litter of wild sows, farrowed 112–120 days after mating, numbers 4–6, and of domestic pigs 8–12; in some domesticated breeds more than 20 live piglings at a birth have been recorded. The sow comes on heat about every 21 days, and the duration of oestrus is 2–3 days. Sexual maturity of wild boar is reached at approximately 18 months, but full growth is only attained at the age of 5 or 6 years; in some modern breeds puberty may be attained by 5–6 months. The life-span is 15–20, occasionally 25 years.

In chromosome number the pig is polymorphic. The Japanese wild boar, *S. scrofa leucomystax*, has a diploid chromosome number of 38 (Muramoto *et al.* 1965), and the same chromosome number has been found in the wild boar of Israel (Wahrman, personal communication) and in domestic pigs; while European wild boar have either $2n = 36$ or 37 chromosomes (McFee *et al.* 1966).

Numbers and distribution

In 1981 the number of pigs in the world was estimated at 779 million, representing an increase of 120 million since 1969–71. Of the total, 370 million were in Asia, 174 million in Europe, 73 million in the Soviet Union, 9 million in Africa,

93 million in North and Central America, 54 million in South America, and 5 million in Oceania.

China, with 310 million, has over 80 per cent of Asia's pigs. In Europe, distribution is more even. West Germany with 23 million has the most and is followed by Poland with 18 million; France, East Germany, The Netherlands, Romania and Spain all have over 10 million. In Africa the largest number of pigs is found in South Africa, followed by Cameroon and Nigeria. In North America the United States has 65 million pigs and is followed by Mexico (13 million) and Canada (10 million). In South America, Brazil has by far the largest number – 35 million (FAO 1982).

Wild species

The ancestors of the domesticated pig are to be found among the wild pigs of the species *Sus scrofa* which ranges throughout Eurasia from Portugal to the western shore of the Pacific Ocean, including southern and eastern Europe, the Soviet Union, and southern and eastern Asia. In North Africa the range comprises the Atlas countries and the Sudan, and until the beginning of the present century it also included Egypt.

Sus scrofa is classed into a large number of subspecies. Before the time when the Mediterranean and northeast Asiatic subspecies of *S. scrofa* were described, which in several cranial characteristics are intermediate between the Indian and European races, *S. s. vittatus* (Malaysia and Indonesia) and *S. s. scrofa* (Europe) were supposed to belong to two distinct species from which different types of domesticated pigs were derived.

The differences between the southeast Asian and the European wild boar mainly concern body size and cranial conformation. *Sus s. vittatus* is smaller than most of the northern races, its height at shoulder being in the range 70–90 cm, and the weight 90–135 kg. The shoulder height of European boar frequently measures more than 100 cm, although its body size is greatly influenced by environmental conditions; weights of adult animals varying between 24 and 350 kg have been recorded. The skull of *S. s. vittatus* is short and high, whereas that of *S. s. scrofa* is elongated.

In the southern European, North African and west, central and northeast Asiatic subspecies of

S. *scrofa* some of the cranial characteristics of
S. s. *scrofa* and S. s. *vittatus* occur in different
combinations. The shape of the skull may approximate to that of S. s. *scrofa*, while that of the
lacrimal bones may be closer to S. s. *vittatus*.

The other species of *Sus* recognized by Corbet
and Hill (1980) are: S. *barbatus* – the bearded pig
of Malaya, Sumatra and Borneo; S. *salvanius* (or
Porcula salvanius) – the pygmy hog of southeast
Nepal and Assam; S. *verrucosus* – the Javan pig
of Java, Sulawesi and the Philippines. Hybrids
between *scrofa* and *barbatus* are fertile in both
sexes.

Domestication and early history

Following the domestication of the dog, that of
other animals in southwest Asia seems to have
occurred only after man had become sedentary.
According to our present knowledge, sheep and
goats were domesticated first, followed by pigs
and cattle. The earliest remains of domesticated
pigs have been excavated at Çayönü in southeast
Anatolia, dated to *c.* 7000 B.C. (Reed 1969, 1977).
Mellaart (1967) noted that at the village site of
Suberbe near the shore of Sugla Gölü in southern
Anatolia, a site dated to the 7th millennium B.C.,
a small race of pigs may have been the only
domesticated animal kept by the villagers. Later,
the domesticated pig occurs at Amouq, near the
Gulf of Iskenderun, in the earliest phase dated to
about 5750 B.C. (Reed 1960). A wall painting in
a temple at Çatal Hüyük, also in southern
Anatolia, ^{14}C-dated to *c.* 5850 B.C., shows the
capture of wild boar by hand (Mellaart 1966).

In northern Anatolia the pig has been found
among the remains of domesticated animals in
Aeneolithic tombs at Alaya Hüyük, northeast of
Boghazköy (2400–2200 B.C.), the pig bones
amounting to 7.3 per cent of the total number of
domestic animal bones.

In northern Mesopotamia the domesticated pig
is represented on painted pottery from Lagash
and in clay figurines from Arpachiyah (Zeuner
1963). At Tell Halaf it is included among a group
of animals carved in stone. Pig bones have been
found also at Hassuna, Nineveh and Tell Aswad
as well as at Gird Banahilk in the mountains of
northern Iraq. The presence of the domesticated
pig at Jarmo in northern Iraq is attested by

numerous teeth and bones from ceramic levels of
the late 6th millennium B.C. In lower Mesopotamia the early Sumerians are believed to have
kept pigs in addition to sheep and cattle. The
domesticated pig probably occurred at Warka,
about 4000 B.C. Clay models of pigs have also
been found in the earliest settlements of the
region, dated to the 5th and 4th millennia B.C. An
ivory pendant from the pre-Sargonic period of
Tell Agrab in Sumer (*c.* 2800–2700 B.C.) has the
form of a domesticated pig with a long head,
upturned snout and pendulous ears (Zeuner
1963). The remains of pigs found at Ur consist
mostly of teeth; nearly all of them are of young
animals and too small for wild boar. The bones
of pigs from Ashnunnak, the modern Tell Asmar,
dated to 2800–2700 B.C. make up 29 per cent of
the bones of domesticated animals.

In the Neolithic oasis village of Sialk in central
Iran the pig is absent during the 6th millennium
B.C. but in the upper Neolithic strata of Belt Cave
it was probably present during the 5th millen-
nium. Bones and statuettes of domesticated pigs
are found at many Iranian chalcolithic and
Bronze Age sites of the 4th and 3rd millennia B.C.
The animal seems to have been introduced into
Turkmeniya from Iran during the Aeneolithic
period.

It is uncertain if the domestic fauna of the early
Indus valley civilization included the pig as has
been claimed by Piggott (1952) for Harappa from
the end of the 3rd millennium B.C. Childe (1935)
remarked that although the pig was present in
large numbers from the earliest times, it was
impossible to determine whether or not it was
actually domesticated or even used as a source of
food by the inhabitants.

In China pig bones occur in great numbers at
the early Neolithic sites of the Painted Pottery
(Yang Shao) culture, dated to the 3rd millennium
B.C. In northern China the pig and the dog are the
only domesticated animals at early Neolithic sites;
both were used for food.

In the whole of Siberia, the region between the
Ural mountains and the Pacific Ocean, the pig
was not bred in prehistoric times. Pig bones are
absent among the domestic animal remains from
the Aeneolithic Afanasyevo culture (2500–
1700 B.C.), the early Bronze Age Adronovo culture (1700–1200 B.C.) and the late Bronze Age
Karasuk culture (1200–700 B.C.), although the

last one was greatly influenced by north China.

In Europe the domesticated pig was first introduced into the Caucasus and Balkans from southwest Asia. It is present in transcaucasian Aeneolithic and Bronze Age burials. At Maikop in the northern Caucasus the pig is found in Copper and Bronze Age kurgans, dating from the 3rd and 2nd millennia B.C. In one of the Bronze Age settlements of the northern Caucasus pig bones amount to 30 per cent of those of all live stock, but part of the pig bones may be from wild boar. In the wooded steppe between the Black Sea and the sources of the Pripet river the pig was the most important domesticated animal during the early Neolithic Age (3000–2700 B.C.) Later, cattle exceeded pigs in importance, and during the Bronze Age pig breeding nearly ceased. From the Caucasus and the wooded steppe north of the Black Sea the domesticated pig was introduced into the east European forest region by way of the Volga and Dnieper valleys.

In southeast Europe domesticated pigs are represented in the second half of the 7th millennium B.C., that is but slightly later than in Anatolia, their presence being attested for pre-pottery Neolithic strata in Greece (Argissa-Magula, Thessaly, and Nea Nikomedeia, Macedonia).

South of Asia Minor, the domesticated pig is absent in Natufian and proto-Neolithic sites in Syria and Palestine. In ancient Syria it ranked as a sacred animal. A scapula from a preceramic level at Munhatta in Palestine may be from a domesticated pig. A votive figurine from a prepottery shrine at Jericho, dated to the 6th millennium B.C., represents a pig (Hitti 1951). In several pottery Neolithic sites in Palestine, dated to the 5th millennium B.C., remains of the domesticated pig are common along with those of wild boar. In 4th millennium Chalcolithic sites all pig remains are from domesticated stock. The statuette of a domesticated pig found in a temple at Jericho dates from the 4th millennium B.C. (Brentjes 1962). At Gezer in Palestine the domesticated pig served for sacrifice during the Chalcolithic Age, the tusk of the boar being considered as a prophylactic against the evil eye. Domesticated pigs continued to be kept in Palestine also at later periods; they are well represented at Bronze Age sites. Down to the time of Isaiah some Jews used to eat pork as a religious rite. The cranium of a sow, about 3 years old at the time of death, found together with the skeleton near the ancient Israel fortress of Hazor (9th century B.C.), is in some respects similar to the skull of S. scrofa from the Hula marshes of northern Palestine.

From Syria and Palestine the domesticated pig reached Egypt where the earliest records of the animal date from the last centuries of the 5th millennium B.C. In the Fayum the bones of pigs and other domesticated animals have been found in an ancient settlement of cultivators along the edge of a large lake that once filled the Fayum depression. A similar domestic fauna occurs in an allied food-producing culture at Merimde on the desert edge of the western Nile delta and at Badari in Middle Egypt during the first half of the 4th millennium B.C. (Childe 1935). At that time domesticated pigs were still absent in Upper Egypt where their presence is first recorded at Toukh in the second half of the 4th millennium B.C. Small clay models of pigs have been found in predynastic graves. At the time of the New Kingdom of Egypt (c. 1500–1000 B.C.) large numbers of pigs were bred in the Nile valley, some herds comprising more than a thousand animals. One of their uses was to tread newly-sown grain into the soil. In ancient times the pig was associated with the Egyptian cult of Set; it was sacrificed and its meat consumed sacramentally, and small figures of sows were worn as amulets.

In Upper Egypt the pig at the Gerzean site of Toukh was very small, resembling the turbary pig of the ancient Swiss lake dwellings. Egyptian pigs from the period of the New Kingdom had a long head with a straight profile, a pointed snout and short erect ears, long neck and flat body, long poorly-muscled legs, a bristly ridge on the neck and back, and a twisted tail. From the colour of the pig of Set in the judgment hall of Osiris, depicted in the Book of Gates, it may be inferred that ancient Egyptian pigs were black. The model of a young pig from an 18th dynasty tomb at Thebes shows that some of the piglings were striped, a feature still encountered in several primitive breeds of pig in Africa and Asia.

South of Egypt domesticated pigs were introduced into Nubia, the region now included in the Sudan and Upper Egypt, at an early time. Prior to the Arab invasion in A.D. 1173 they were common in the Sudan. In one town the Arab invaders killed 700 pigs. The aversion of Moslem

Arabs to the pig is connected with the economy of pastoralists, more especially of those inhabiting semi-arid or arid regions, such as Mongols, Semites and Hamites. Their economy does not include the pig, for the pig cannot easily be herded and does not thrive on dry savannah or desert flora. With some pastoral peoples the absence of the pig among their livestock has later found expression in religious laws that forbid breeding of the animal and consumption of its meat.

Descent of domesticated pigs

The debris of the pile dwellings of the early Neolithic period of Switzerland includes bones of a small domesticated pig, known as turbary. It appears in a definite state without preceding forms of transition between the wild boar of the region and the domesticated type. Hence it is generally accepted that it was not domesticated locally but introduced from somewhere else. This also applies to pigs of turbary type in other Neolithic sites of Europe, e.g. Germany, Hungary, the Balkan countries and European Russia, as well as to those in several Asiatic stations such as Anau and Shah Tépé in Turkmeniya and Mezer in Palestine, and to the turbary pig of Toukh in Upper Egypt.

Since the earliest sites in which pigs occur in a domesticated state are situated in southwest Asia, more especially Anatolia, it may be inferred that the turbary type of pig was derived there from the local wild boar, *S. s. libycus*. Its small size is due to stunting, being paralleled by other animals at an early stage of domestication when malnutrition and close inbreeding were common.

Rütimeyer (1861, 1864), the early student of the domestic fauna of the Neolithic pile-dwellers of Switzerland, first suggested that the turbary pig had a wild turbary ancestor, but subsequently claimed that it came from southeast Asia, because in cranial conformation, and especially in lacrimal shape, it resembled *S. s. vittatus* of Malaysia and Indonesia. This belief was upheld by several later authors who considered lacrimal shape to be a firm basis for distinction between domesticated descendants of European and southeast Asiatic wild boar. Later studies have shown that this view cannot be upheld. The early authors were ignorant of the existence of a series of intermediate forms between European and Indian wild boar

and of the fact that lacrimal shape is not as constant and unaffected by changes in the conformation of other cranial bones as they surmised.

Under domestication the pig has changed in cranial conformation more than any other animal with the exception of the dog. Compared with the crania of wild boar or primitive domesticated breeds the skull in adult pigs of many improved breeds is broader and its anterior part markedly shorter. The brain case is higher and slopes forwards, rendering the break in the profile, originally caused only by the upward slope of the short nasal region, more acute. Owing to their situation on the border between the aboral and oral parts of the skull, the lacrimals must of necessity be affected by far-reaching changes in the form, position and direction of the anterior and posterior cranial parts. Indeed, in the highly improved Berkshire the lacrimals are shorter than the shortest lacrimals known in Indian wild boar.

The skull of young *S. s. scrofa* differs from the adult skull in size, general conformation and proportions, differences which profoundly affect the shape of the lacrimals and the ratio of lacrimal height to length. Lacrimal shape is also affected by sex, the lacrimal indices of European, west Asiatic, Mediterranean and east Asiatic wild sows being considerably smaller than the respective indices of males. The conformation of the crania, not excluding the lacrimals, of wild boar and domesticated pigs is also influenced by body size as well as by diet and mode of life. The head of a pig reared on an insufficient diet is relatively longer, straighter and narrower and its facial part lower than that of well-fed litter-mates in which the head is higher and broader and has a dished profile. Rooting the ground for feed also markedly affects the development of cranial and lacrimal conformation in growing pigs as compared with that of piglings kept on a hard floor and fed from a trough.

Therefore, the shape of lacrimals in domesticated pigs cannot be accepted as a general criterion of racial derivation. Similarity in lacrimal shape between domesticated pigs and any of the geographical races of wild boar may in some instances be due to descent and in others to convergent development. Craniologically there is no necessity to trace long lacrimals in domesticated pigs to the wild boar of Europe, and short lacrimals to the wild boar of southeast Asia. From a craniological point of view it is possible to derive all breeds of domesticated pigs of Asia, Africa and Europe, those with short as well as those with long lacrimals, from the wild boar of Anatolia and adjacent regions. But while this is possible, it may not apply in every instance.

The remains of pigs excavated in many sites of Eurasia have been traced to different local races of wild boar. In particular, pigs larger than the turbary have been claimed to be of such derivation, although in some instances the increase in size may be the result of superior nutrition. There can be no doubt that in several areas local races of wild boar have contributed to the evolution of domesticated breeds. But it is doubtful whether such domestication has occurred independently of the originally domesticated stock, in other words if there have been centres of primary domestication in addition to that in Anatolia. Prima facie, the existence of such independent centres of domestication is unlikely because of the difference, not only of centuries as in the case of Thessaly and Macedonia but of millennia as in China and central Europe, between the time of first domestication in western Asia and the subsequent rise of new breeding centres. It may be assumed that during the long lapse of time between these occurrences domesticated pigs were diffused from the original domestication centre to the confines of Asia and Europe, albeit often in small numbers only. Thus, it is recorded that in the year 3468 B.C. the Emperor Fo-Hi of China ordered his subjects to breed domestic pigs which were imported from the west (Eikamp 1978).

At a primitive state of agriculture pigs are commonly kept in a half-wild condition in the forest, bush and wasteland in the vicinity of the villages. At the beginning of the 3rd millennium B.C. pigs were kept at the early Neolithic stations of the Tripolye culture, between the Black Sea and the sources of the Pripet river, in mixed oak forests. In the northern Highlands of Scotland domesticated pigs were, during the last century, still left almost in a state of nature, searching undisturbed for their feed. In the woods of Norway and Sweden such pigs have gone feral. In southwestern Spain and southern Portugal pigs of the black or red Iberian breed forage in the oak forests which in the mountains of Córdoba and Cordillera Bética abound with wild boar. In conditions of such proximity between domesticated and wild stocks in a natural environment crossbreeding was inevitable, and as the

crossbreds were superior in size to the original turbary type they were used for outbreeding of the latter at many times and places. The increased genetic variability facilitated the evolution of numerous different breeds of domesticated pigs in eastern Asia and Europe.

In Europe the bones of domesticated pigs from the debris of Swiss proto-Neolithic lake dwellings show a small, slender, leggy animal with a small head, flat forehead, shortened jaws, and lacrimals different in conformation from those of European wild boar (Zeuner 1963). Pigs of this turbary type were bred at many European stations, and in some Alpine valleys have survived into recent times. Towards the end of the Neolithic a larger type of pig occurs along with the turbary, and in many Bronze and Iron Age sites of Europe the bones of both types are found in various ratios, and there are also intermediate forms between the two basic ones. In Sweden, at the beginning of the Neolithic, about 3000 B.C., the domesticated pigs of the peasants were considerably smaller than wild boar. Lepiksaar (1973) believes that they had been introduced by peasant immigrants and were subsequently bred without contact with local wild pigs.

In Celtic graves bones and clay statuettes show a relatively uniform leggy pig with a long head, large tusks, a high crest of bristles along the back and a curly tail (Bökönyi 1974). During the period of the Roman conquest domesticated pigs in the occupied territories were larger than the small turbary, but in many places the latter continued to be bred exclusively or along with the larger type. The fattening of pigs also became widespread during this period.

In the Middle Ages the pigs kept in the western regions of Europe were small, in some instances actually dwarfed, while in the eastern part they were larger on average, although the small turbary type was also present in many places. In Denmark these two types seem to have survived until the end of the 18th century, the large type with a long narrow body being represented by the Jutland, and the small type, distinguished by a short back, erect ears and a high crest of bristles along the back, by the Island breed. In Sweden the Forest pig with long legs, erect ears and a longitudinally striped bristly coat was the only domesticated breed kept until the beginning of the 18th century. A similar type was common in Norway. In Finland there were two slightly different unimproved breeds until the middle of the l9th century, the Eastern with short erect ears, and the Western with larger pendulous ears. The Finnish pigs were distinguished from the Swedish Forest type by thicker bones and coarser bristles (Pedersen 1961).

In Germany the early domesticated pigs, characterized by a coarse skeleton, long head and legs, flat ribs, high back and late maturity, remained nearly unchanged from the Neolithic to the 18th century A.D. As in several other parts of Europe they occurred in two different types, a larger one with pendulous ears and a smaller one with short erect ears. These pigs subsisted on forest pasture and lost their usefulness when intensified agriculture began to encroach on their former domains. Remnants survived in the pasture pig of Hanover-Brunswick and the unimproved Bavarian Landrace (Haring 1961).

In Hungary the occurrence of several different breeds has been claimed for the first centuries of the present millennium (Hankó 1954), but this has been denied by Bökönyi (1974) who on the evidence of skulls and extremity bones could distinguish between only two types, a large and a small one. The great variability in the size of pigs during the post-Neolithic periods of Europe may be due not only to the presence and interbreeding of the turbary and locally domesticated boar but also to different conditions of environment, maintenance and feeding at different times and in different places.

Recent history and present status

Development and distribution of breeds in Europe

General Tracing the development of today's pig breeds back to their origins in the types current before 1700 or 1800, even in western Europe, is not as straightforward as in other species of farm animals, – notably cattle. By contrast the pig remained largely a household animal until fairly recently. The relatively low individual value and the rapid breeding rate of the pig meant that keeping track of ancestry of individual pigs was both less important and less easy. In addition the pig has long been unfashionable; those who earned their living from tending it have been given a low social status in many cultures and careful written accounts were not kept.

Pigs have been much less subject to the influences of different local environments than other farm species since they have tended to share the micro-environment which man provided for himself, in his household or farmyard. In more recent times, as specialist pig herds developed which supplied pigmeat as an article of trade, so they became adapted both to the predominant feed supply and to the meat market which favoured a particular type of carcass.

These changes have been produced by two different techniques. First, by selection within herds, where owners have closely observed the performance of individual animals, or their progeny, and chosen certain of them to contribute differentially to the next generation. Second, by controlled selection between herds, often between regions and even countries. Introduction of new stock was often deliberate but at other time's more casual, involving unusual animals brought back from voyages.

A note of caution is necessary when tracing the origin of breeds through recorded pedigrees or contemporary accounts. First, that matings mistakenly attributed to the wrong boar, and piglets to the wrong sow, are quite common even when great care is taken. Second, the innovator who makes important changes in a breed is often in opposition to some prevailing ideas, and may have deliberately concealed the unfashionable or prohibited matings which were the real key to his success. Modern blood-grouping techniques frequently show that a proportion of current declared pedigrees are incorrect.

Pig keeping has continued to be rather unfashionable in Europe to this day, but by contrast this has proved valuable in aiding its evolution. Because pigs are only kept for profit and not for social status (like beef cattle) any new techniques including scientific breeding have been eagerly adopted and great strides have been made in the past 25 years.

Early local breeds in Britain and the Chinese influence The native British pig was of the erect-eared, coarsely-bristled type commonly seen in mediaeval illustrations, but by the mid-18th century many regional types had been developed. Some of these were still very dependent upon scavenging, and in particular upon autumn fattening in the hardwood forests. In other regions, where the forests had already been largely cleared, arable farming and dairy by-products were providing a more reliable though still seasonal feed supply.

It is likely that one of the most powerful forces encouraging such diversity was the alternation of two separate genetic processes. Because pigs were associated with settlements, rather than with nomadic life, and because relatively little movement of people or pigs took place between distant settlements, the mating system tended to produce mild inbreeding. The genetic consequences of this would be greater uniformity within settlements, and consequently diversity between them. Occasional wide outcrosses would occur when a strange animal was introduced and found favour, and this would increase the diversity still further.

While skin and coat colours and markings provide a most convenient method of describing different varieties, not too much emphasis should be placed upon them as indicating basic genetic types. All known colours and markings are caused by relatively few gene pairs. Hence the widespread use in a district of a single new animal and his sons could change the appearance of a majority of the animals within as few as two generations.

Small black or white, short, light-boned and fine-coated pigs were imported from China (Canton district) and Indo-China direct, but also via Naples and Portugal, in 1770–80, and probably also before this time. Trow-Smith (1959) believes that the dark brown colour of the original British pig was made lighter and then changed to white by relatively early importations from an unknown source. The same importations were also probably responsible for the introduction of the lop ear and the maternal qualities associated with it. The black pork types were the results of Chinese, Neapolitan and Portuguese crosses.

Lush (1945) has described the typical stages of breed formation in Britain in the late 18th and 19th centuries.

1. Out of the melting pot came a type which was more productive than the average but not distinctly different in origin.
2. Some of the best animals were gathered into a small number of herds which were then closed to further purchases. Intense inbreeding occurred (not always declared) which then created a distinction in type.
3. If this was successful then the new 'breed'

increased in popularity and more herds were established.
4. The need for a central herdbook arose, so that all breeders could have access to pedigree information.
5. A breed society was established to safeguard breed purity, conduct the herdbook, and promote the breeders' interests.
6. Breeders emphasized the use of their males on other pure breeds, or on commercial stock.

It was not until 1884 that the first pig herdbooks were established in Britain – although the first horse and cattle breeds had their books 90 and 60 years earlier.

Confused though this early period may have been it is clear that the existing regional varieties were crossed in all combinations, and with the imported stock, so that by the mid-19th century several new types had been assembled which subsequently played an important role both in Britain and other parts of the world. Three of these were the Tamworth, Berkshire and Saddleback.

1. The Tamworth seems to have been derived around 1800–50 from the old red types originating in Berkshire. Prick-eared, with a long straight snout, short-bodied and hardy, it was developed for the production of bacon (cured with the skin on) and hence with less backfat than the lard types.
2. The predominantly black Berkshire breed was created out of similar basic stock to the Tamworth, probably by crossing with animals imported from Naples. It had a short head, dished face and prick ears, with a fairly short and deep body. Because this breed was one of the earliest to be recognized as an improved type, it was easily the most important in influencing the development of other breeds in England and many other countries.
3. A black-and-white type was present in Essex which gradually took shape as the Essex Saddleback. This lop-eared pig with fairly straight snout was also developed for the bacon trade. It had a good reputation for hardiness, living outside and milking and caring for its litters well. A similar Saddleback breed was evolved in the west and south of England, and became known as the Wessex. These two breeds merged in 1967 to form the British Saddleback.

Development of the Large White and Danish Landrace In addition to the three types which have just been described, the breed which was destined to become the most important single British type was the Large White. Its origins, again, are not clear, but a prick-eared long white-skinned variety emerged in northeast England which by 1851 was being called the Large White, though also by several other names (and outside Britain it is still referred to as the Yorkshire). Middle White and Small White breeds were also recognized – containing a higher percentage of Chinese blood. The Large White has always been quite a variable breed but it is basically a large-framed, long, white, prolific pig with slightly dished face, bred for the production of bacon.

In general, modern breed development took place rather later in most other European countries, based upon the previously evolved local varieties (or land races). As a consequence the newly acclaimed British types were often used in crosses which then formed the improved breeds of Europe – often called Landrace. For example, large numbers of Middle White and Berkshire, but mainly Large White, boars were used on Danish sows in the mid-19th century. In spite of this the Danes claim that the main foundation of today's Danish Landrace (*Dansk Landrace* – first herdbook 1906) was an original land race from Jutland relatively untouched by these Large White imports, and this type was then selected for the export trade in bacon to Britain (Jonsson 1965).

Today's Danish Landrace is white, very long and fine-boned, with lop ears, a very long straight snout, and quite good ham development. It has gradually become the most extreme bacon-type pig in the world by a most effective system of progeny testing boars at central stations. Denmark pioneered scientific pig improvement from 1895 through a combination of government assistance and control from the farmers' unions and cooperative bacon factories. Many other countries have since been less successful because, although they copied the techniques of boar testing, they failed to insist on the exclusive utilization of stock from the best tested herds.

Spread of Large White and Landrace types throughout Europe. In Germany some of the local land races were mated to imported Large Whites in the late 19th century, and from such crosses were derived both the original German Landrace (*Deutsche Landrasse*) and the German Yorkshire (*Deutsches Edelschwein*) breeds, whose herdbooks began in 1904. The German Yorkshire was a rather stronger boned version of the British Large White. The German Landrace was a lop-eared medium-long white type with long straight snout, but much more heavily boned than contemporary Danish pigs. It was bred for a quite different trade, the manufacture of a wide variety of sausage products and fresh and cured pork joints from heavier carcasses. In more recent times the German Landrace (*Deutsches veredeltes Landschwein*) has changed considerably by incorporating Dutch Landrace (*Nederlands veredeld Landvarken*) blood (itself a product of German and Danish Landrace) from 1953, and Belgian Landrace (*Belgisch Landvarken*) blood since 1972.

These four basic breeds – Large White, Danish Landrace, German Yorkshire and German Landrace – had a major influence on breed development in northwest and central Europe in the first half of the 20th century. The other Scandinavian Landraces (Swedish, Norwegian, Finnish – *Svensk Lantras, Norsk Landrase* and *Suomalainen maatiaisrotu*), the Polish and Russian white breeds, and today's erect- and lop-eared white breeds of all the East European countries, owe their origins to one or more of these four, crossed on to largely unimproved local types.

The Soviet Union, with its vast land area containing nearly 10 per cent of all the world's pigs, has a particularly large number of recognized breeds. Nevertheless most of these derive from crosses between local unimproved breeds and imported stock – chiefly Large White (preceded by Berkshire) but with some Danish and German Landrace and German Yorkshire. Recent blood group studies have demonstrated that Russian Landrace (Long-eared White – *Dlinnouhaja belaja*) have more affinity with Danish, Swedish and Czech types, and less with German and Dutch.

France, Belgium and The Netherlands also developed important Large White and/or Landrace types by successive importations of these four, incorporated into base populations of local sows.

Since 1945, when Denmark decided to restrict the export of breeding pigs (to try to preserve a leadership position), Sweden took over this role and supplied Swedish Landrace to many coun-

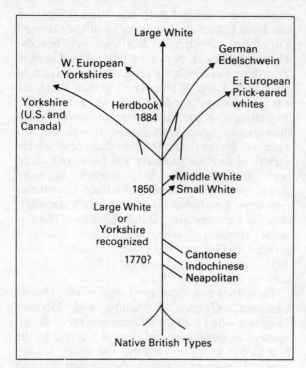

Fig. 17.1 Development and influence of the Large White breed of pig

tries, including the United Kingdom (from 1949), which had not previously had a Landrace breed of its own.

Tracing the exact origin of today's European white breeds has again become complex, since after 1945 the demand for lean meat became universal, breeding aims became much more similar in all countries, and a very complex exchange of breeding stock occurred in many directions (see Figs. 17.1 and 17.2). One factor controlling gene flow between countries in this century has been the existence of health control regulations imposed by individual governments. The Scandinavians have had a good record of freedom from several important pig diseases, and so their breeds have had fairly free access everywhere. Countries in mainland Europe have exchanged stock among themselves but have not found it easy to export to Britain or Scandinavia – and this has perhaps prolonged the split between bacon types and the multi-purpose meat pigs of Germany, Holland and Belgium.

The extent to which Large White (Yorkshire) and Landrace types have come to dominate

European pig production in the period since the Second World War is quite remarkable. By the late 1960s in nearly all the major pig-keeping countries these two types contributed at least 75 and up to 100 per cent of the total breeding populations. The Large White types are still most frequent in all countries except Scandinavia, Germany, Belgium and The Netherlands, which were mainly dependent upon their separate Landrace types. One reason has been that as pig keeping has become a specialist activity, so producers have provided better conditions (feed, housing, veterinary care) for their stock, and have been able to justify keeping the most productive genotypes. Another reason has been the general opposition by the meat trade to coloured pigs (association with fat, unattractive coloured skin on pork joints and bacon, and coloured bristle roots for the fastidious urban consumer).

Influential or surviving local breeds in Europe In Yugoslavia, Hungary and Romania the Mangalitsa (*Mangulac*) was important though its numbers are now very small. It was evolved in late 18th-century Yugoslavia from local breeds, and introduced to Hungary and crossed with the primitive breeds early in the 19th century. In Romania it largely superseded the grey prick-eared native breeds. The Mangalitsa is small to medium in size with a short conical head, wide forehead and ears projecting forward. It occurs in several colours: usually yellowish-white, but also black or black with a white underline, and red. The coat consists of long thick curly hair and fine woolly bristles. It is a lard breed, often slaughtered at 150 kg live weight, but also yields lean meat particularly favoured for some types of sausages.

In Czechoslovakia the Přeštice (*Přeštické*) has survived as a native black pied breed until present times, with an infusion of Wessex Saddleback in 1957. The Large Black from Britain has also existed in quite large numbers in both Czechoslovakia and Hungary where it was known as the Cornwall, long after it had ceased to have commercial importance in its original home. Essex and Wessex blood helped form the Swabian-Hall (*Schwäbisch-Hällisches*) and Angeln Saddleback (*Angler Sattelschwein*) breeds in Germany which, as in other countries, were used as outdoor sows.

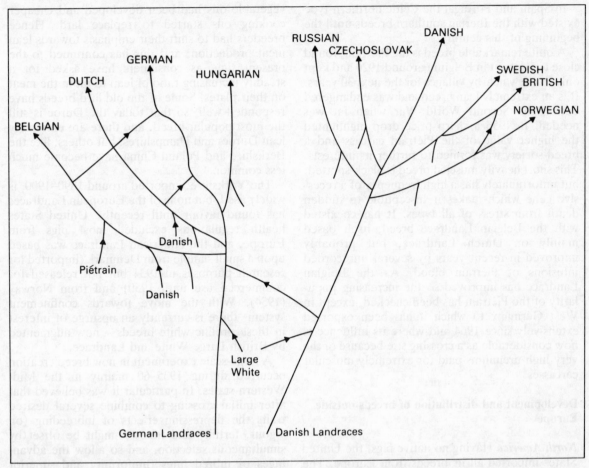

Fig. 17.2 Development of the European Landraces

In Poland two local types have survived after some introduction of British breeds: the Puławy (*Puławska*) is a mainly black-and-white breed originating from a local pied variety with Berkshire blood; the Złotniki (*Złotnicka*), a mainly black spotted breed, was recognized in 1962 based upon primitive strains of long-eared pigs saved from extinction and combined in a modern breeding programme.

In southern Europe some of the original Mediterranean breeds of pig survived until recently in a nearly pure state. In 1966 it was estimated that local breeds still accounted for one-half of the Italian pig population and three-quarters in Spain. In Italy the primitive pasture breeds were late-maturing and usually black-skinned. They mainly had a long head with a pointed snout, straight profile and usually erect

ears, but they have now quite disappeared from commercial production.

In western France a number of unpigmented local breeds could be traced to a common Celtic origin, and were joined into a single breed, the West French White (*Blanche de l'Ouest*), in 1955. The Celtic or northern European type was a late-maturing, leggy, flat-ribbed, yellowish-white pig with long lop ears. In the Pyrenean region the Gascony (*Gasconne*) was a black pig, representative of the Iberian type. These native breeds of the Mediterranean countries had a more compact conformation, a concave head profile with long snout, medium semi-erect ears, and black or reddish brown skin and bristles. A similar breed, the Basque Black Pied (*Pie noire du pays basque*) was found in the chestnut, oak and beech forests of the Basque country, while the Limousin, white with black head and rump, was found in central France.

In Spain and Portugal the Celtic northern type existed with the Iberian southern breeds until the beginning of this century.

A quite remarkable breed is the Piétrain, found close to Brabant in Belgium around 1920 and kept only in a few nearby villages for the next 30 years. It is an extremely lean breed and was endangered during the Second World War when fat was needed! In 1950 a sharp price drop highlighted the higher value of the Piétrain carcass and a breed society was formed to further its interests. This short heavily muscled breed is black-spotted, but unfortunately has a high frequency of a recessive gene which makes it susceptible to sudden death from stress of all types. It has coexisted with the Belgian Landrace breed, itself based mainly on Dutch Landrace, but probably improved in recent years by several unrecorded infusions of Piétrain blood. As the Belgian Landrace has improved so the increasing popularity of the Piétrain has been checked, except in West Germany to which it has been exported extensively since 1964 and where its influence is now considerable as a crossing sire because of the very high premiums paid for extremely muscular carcasses.

Development and distribution of breeds outside Europe

North America Having no native pigs, the United States imported all its breeds from Europe. The early settlers no doubt took pigs for their own support, mainly samples of the types which they kept in their home countries. Then the Berkshire breed, being the first to achieve national prominence in England, was imported from 1820, and black belted pigs from Wessex also arrived around the same time. To satisfy the new conditions of climate and farming, new breeds were then developed from these imported improved types crossed with local stock already in the country, and this was the origin of the Chester White, Duroc, Hampshire and Poland China.

The opening-up of the Corn Belt states produced vast crops of cereals. Consequently until the First World War the greatest use of the pig was in the conversion of a large share of the corn crop into lard and fat bacon for both the domestic and export markets (West Indies and Europe). But by 1923 a method of hydrogenating vegetable oils had been developed and cheaper cooking oils started to replace lard. Hence breeders had to shift their emphasis towards lean meat production, and this has continued to the present time as consumers have asked for a steadily increasing ratio of lean to fat in the meat on their plates. Some of the old lard breeds have responded well, so that today the Duroc is still the most popular breed, and there are extremely lean Durocs and Hampshires, but others, like the Berkshire and Poland China have become much less common.

The Yorkshire, imported around 1890–1900, is widely used, but none of the European Landraces has found favour until recently. United States health regulations excluded most pigs from Europe, and the American Landrace was based upon a small sample from Denmark (imported for research purposes in 1934 but not released for commercial use until 1950) and from Norway (1954). With the move towards confinement systems there is currently an upsurge of interest in these prolific white breeds – now augmented by British Large White and Landrace.

A large-scale experiment in new breed creation occurred during 1935–60 mainly in the Mid-Western states. In particular it was believed that after initial crossing to combine several desired traits the depressing effects of inbreeding (on vigour, fertility and fitness) might be offset by simultaneous selection, and so allow the advantages of inbred lines (uniformity and superior crossing performance) to be exploited. Many different new breeds were created from a variety of crossbred bases and named mainly after the Agricultural Experiment Stations which were involved (Montana, Maryland, Beltsville, Minnesota). None of them has had any lasting impact. It is not clear whether their failure was due to their lack of true merit, or to the fact that the United States pig industry was not yet ready for such lines.

Canada imported improved Berkshires, Yorkshires and Tamworths in the 19th century and the Berkshire was the dominant breed until 1900. After grading standards were introduced, which emphasized white hair colour and specific bacon characteristics, the Yorkshire expanded rapidly and the Berkshire declined to insignificance. Danish Landrace were imported from the U.S.A. in 1950 and these, subsequently augmented by imports from Sweden and Norway, have expanded to second place, in part at the expense of the Yorkshire.

A very interesting example of new breed formation took place at the Lacombe research station in Alberta during 1947–57. The aim was to produce a white breed specifically adapted to Canadian conditions and suitable for crossing with the Yorkshire to produce bacon pigs. The contributions were roughly one-half Danish Landrace, one-quarter Chester White, and one-quarter Berkshire; the emphasis during development was on selection, not inbreeding. Within 10 years of its release the Lacombe breed provided 12 per cent of all pedigree registrations in Canada and still does so.

Latin America All the pigs in this region were brought in from other countries, but some of them arrived so long ago that they are distinguished as 'native', by comparison with modern breeds imported in this century from Europe or North America. Pigs were perhaps first introduced from China in the 15th century when there was regular sea traffic between Mexico and the Far East. Other types, both Celtic and Iberian, came from Europe with Spanish and Portuguese colonists from the 16th century onwards. For the most part the pig in South and Central America is part of a peasant survival economy, playing the role of scavenger. There have been few efforts to improve native types, whatever their adaptation to climate, disease or nutrition, except in Brazil. Where commercial production has been developed (e.g. central and northwest Mexico) it relies almost entirely on imported breeds. Native pig types are not breeds in the sense of having pedigrees or herdbook societies and their origins are frequently unknown.

One racial group is the black hairless type, which is part of household economies in tropical lands from Mexico (*Pélon Tabasqueño*) to Colombia (*Zungo Costeño*). It is prominent in the Gulf of Mexico and Yucatan and in Salvador and Costa Rica (*Pélon de Cartago*). A lard breed, its mature weight is around 80–90 kg and its merit is its ability to survive under very adverse hot climates and to consume bulky feeds and fruit wastes.

A type which was numerous 100 years ago in the corn-growing highland regions of central Mexico was the Cuino or miniature pig. Now nearly extinct, mature males and females may be no more than 12 or 10 kg. It seems to have its origin in the early importations from China. Black, spotted and yellow colours are common; the animals are not swaybacked. It is suggested that such a small animal was well adapted to household life, surviving when the corn crop failed from drought and being easily fattened in better years.

A third Mexican type, *Cerdo Cascate*, is distinguished by wattles, small appendages under the chin. These small pigs, maturing around 60 kg are variable in colour, but very hardy and active for their scavenging life in a dry region.

Another small pig is a native of the central mountains of Honduras. Maturing around 50 kg it is characterized by a long tail 30 cm in length with a hairy switch.

More than half of the pigs in this region live in Brazil. The Canastrão is the largest breed, a lard pig, Celtic in type, red or black in colour, with a few hard bristles in a thick skin. The Canastra is an Iberian type, also a lard pig, black, smaller and with more and softer bristles than the Canastrão.

A third type, much smaller, is known by several names in different regions (Canastrinha, Nilo, Tatu, Macau). It is a small black hairless pig whose origins are Asian, with short legs and head shaped like an armadillo and is often kept in the house.

The Piau pig is found in east central Brazil and is the best and most important national breed. A lard type, it is basically creamy-white with well-defined black spots and probably derived from crosses between Canastra or Canastrão and other imported stock.

In the most southerly states of Brazil commercial pig production is carried out by people many of whom are of German and Italian origin, using mainly the Landrace and Large White breeds often imported from those countries, although the first Landrace came from Sweden and Denmark. Durocs are also popular.

Asia In Asia south of the Soviet Union, pigs are today rare in Moslem countries, while in many parts of India the pig is considered to be an impure animal. In the Far East pigs are much more important, and are the major source of what little meat protein there is in most human diets.

They are particularly associated with the Chinese race, wherever it lives. Epstein (1969) estimated that over a hundred different breeds and varieties occurred when he visited China in 1963, but that many of these were becoming extinct through grading up to other native breeds, or crossing with imported Large White or Berkshire boars.

In central and southern China there are numerous local breeds which are quite highly developed. All of them are lard pigs, adapted to converting water plants and other vegetable feed into fat. The adults are often distinguished by sway backs and by large pendulous bellies, so low that they often scrape the ground, but these may be exaggerated by their bulky diets. They have a wrinkled face, prick or pendulous ears, a straight tail, white, black or variegated skin, and usually sparse, short and soft bristles. Some of them undoubtedly have very high average litter size, with litters of 20 being not uncommon. Typical of these breeds are the Pearl River Delta or Canton (South China type) and the Ningsiang (Epstein 1969). It is from this region that 16th-century traders took pigs to Britain, Naples and Portugal which had an important early influence on the development of European stock.

In the provinces of northern China primitive local breeds with black skins and long coats of hard bristles were widespread until 1950 but many of them have since been graded up to imported European breeds. A similar type is still found in several parts of Korea.

In Tibet and Nepal primitive pigs occur sporadically. They are dwarfed and subsist by scavenging offal and anything they can grub up in the fields. Similar breeds occur in the adjacent provinces of west China where they are essentially grazing animals.

In the villages of Inner Mongolia a few pigs scavenge the market places and streets. They are among the most primitive found in any part of the world and resemble the native pigs of the rain forest in central West Africa in size, colour and conformation. The body and legs are covered with long coarse bristles which form an elongated ridge from the nasal root to the straight tail.

Several primitive types were developed in Taiwan although they are now almost entirely replaced by imported breeds. They are all of the long-eared black type, often heavily wrinkled, and introduced from south China perhaps as recently as this century. There is no longer any trace of two earlier types, the Small Red and Small Black, which probably belonged to the original Taiwanese before the arrival of the Chinese on the island, and may well have come from Malaya.

The South China type extends also into Indo-China, Thailand, Malaysia, Indonesia and New Guinea. In several islands of Indonesia primitive local breeds and occasionally feral pigs are found exclusively, or in addition to the Chinese type.

Africa While accurate statistics on pig keeping in most African countries are difficult to obtain it is quite certain that pigmeat forms an insignificant item of the diet – the reasons are connected with religious taboos, nomadic life-styles, unsuitable climates and serious diseases.

Pigs are bred in the Coptic villages of Egypt and in some of the villages in central and southern Sudan. These animals are primitive, approaching wild boar in appearance. Very few pigs are found in North Africa west of Egypt, other than imported stock kept by Europeans in the vicinity of some of the large towns.

On the southern side of the Saharan barrier the pig is relatively rare. Only in the peripheral provinces of Zaire and the lower Congo and Kasai regions are troupes of pigs found in most villages. From Zaire the pig has spread into Cameroon and Angola, where the majority of tribes living in the tropical rainforest or the highlands possess them. The same types of pigs are widespread in coastal areas of the West African countries, e.g. the Ashanti Dwarf in Ghana. They probably originated from early Portuguese imports (Epstein 1971).

In southern Africa all pigs owned by Africans are descended from stock imported by Europeans. The animals in the Bantu reserves are mainly of two types: Windsnyer – long-nosed, razor-backed pigs of various colours, and Kolbroek – very short, fat, animals of Chinese origin, distinguished by a short snout, dished face and prick ears, and commonly black or brown in colour. European farmers developed commercial production in several former colonies (Kenya, Zambia, Zimbabwe), and used British breeds to supply bacon-type animals, while nearly one-quarter of all African pigs are kept in South Africa: Landrace and Large White types predominate.

Australasia Pigs, imported into Australia and New Zealand by the early British settlers, have never become a major livestock industry because of the concentration on cattle and sheep. But the strict import regulations imposed to keep out diseases have tended to isolate both countries from modern developments – so that, for example, the Berkshire breed has remained relatively important in addition to Large White and Landrace. This tendency to retain older breeds long after they have been replaced in their home countries applies also to the Berkshire in Japan.

Feral populations

Feral pigs are found in the Llanos Orientales of Colombia, in Australia and in both North and South Islands of New Zealand. They are descendants of animals introduced some 100–200 years ago by colonists from Europe. There are also feral types in the United States and several other countries.

Miniature pigs

Miniature strains have been developed for bio-medical research purposes in several countries. The breeding objective has usually been to reduce adult size in order to make handling easier and keep down maintenance costs. In the United States the Vita Vet Lab Minipig had its origin in Florida Swamp pigs in 1948. The Minnesota Miniature was developed at the Hormel Institute from 1949 using several feral strains and imported animals from Guam. The Pitman Moore Miniature was derived from Vita Vet Labs but has been developed separately. Other types used in the United States include the American Essex (Texas), the Ossabaw (Pennsylvania) and the Hanford Miniature (Washington).

In Japan the Ohmini is based upon Chinese miniatures and some imported Minnesota Miniatures, while the Göttingen Miniature in Germany was also formed from the Minnesota strain plus some stock from Vietnam.

Future Prospects

Future evolutionary changes

In the short term, 10–20 years, one can be confi-

dent that recent changes in methods of commercial pig production will continue. Confinement systems and the feeding of balanced rations, based upon grain and protein supplements, will become almost universal in Europe and North America, and make more headway into Latin America and the Far East. Governments of developing countries are already encouraging such specialized systems in order to provide animal protein for their increasingly urban populations.

Primitive pigs are quite capable of surviving on what they can pick up around the household or village, but most of this food is used for maintenance rather than growth or reproduction. Thus it has been estimated that only one-fifth of the total pig inventory in Latin America is slaughtered each year. When pigs are housed, kept relatively free from disease, and fed grain-based diets, the annual kill can equal twice the inventory (e.g. 2000 pigs per year from a 100-sow herd with 1000 growing pigs). This means that a much higher proportion of the feed is utilized to produce new lean tissue (human food) rather than merely to maintain existing animals.

The questions are: can we continue to provide grain to feed our existing pigs in this way, and can we extend the system to the rest of the world's pigs? The first question concerns allocation of existing resources. So long as the present economic system prevails we shall continue to have grain fed to pigs to produce animal protein in developed countries while Third World citizens are short of the food energy and plant protein contained in that grain. This has existed for thousands of years and no solution acceptable to the richer nations has yet been found.

The second question is more concerned with the balance between future demand and supplies. This is impossible to forecast, since we neither know what world population levels will exist (after the present inevitable expansion caused by recent changes in birth and death rates), nor what sustained levels of crop production are possible. Today's levels of grain production are certainly only being achieved by using up non-renewable fossil fuels. But no one knows what alternative energy sources may become available in the future, nor what number of people there will be to feed. Hence it is too soon to forecast how pig production will evolve in the 21st century.

In spite of this uncertainty it must surely be

correct to continue to direct the evolution of our pigs so that they yield the products in demand, with the most efficient use of any grain resources which are allocated to them. At different times and in different places the pig's main or subsidiary products have been: sucking pigs, fresh pork, bacon or ham, manufactured articles (sausages, pies, tinned meats, prepacked foods), sausage casings, lard, skins, bristles and manure (for land, ponds or methane gas). Increasingly in Western countries it is lean meat which is valued today, however it is consumed, but in poorer countries the fat continues to be a valuable commodity. In all countries it must be sensible to make full use of the manure as a fertilizer or energy source.

By-products from food preparation and manufacture can often be utilized by pigs. On the other hand their ability to digest fibrous plant material is very poor since they do not have the symbiotic microflora of the ruminants. Pigs are not grazing animals and are adapted to rooting for underground plant material which is richer in starch and sugars. Sows can certainly cope with cut forage of low fibre content as a source of both protein and energy, because their demands are lower (except in lactation) than in growing pigs.

Many intensive sow herds are now producing over 22 weaned pigs per sow per year in confinement systems based upon an intake of around 1 tonne of concentrated feed per sow. The pigs are growing to 90 kg in around 165 days, with a conversion rate of under 3 kg concentrated feed to 1 kg live weight. They yield a carcass of some 68 kg with a lean content of 50 per cent and some 28 per cent fat and 14 per cent bone. Improving the pig's efficiency of conversion of feed into lean tissue can be attempted in a number of ways:

1. Increase the rate of gain in the growing animal to reduce the maintenance cost per lifetime.
2. Select for lean deposition rather than fat.
3. Select for more efficient digestion and utilization.
4. Increase reproductive rates from sows so that sow maintenance costs are spread over as many offspring as possible.
5. Adopt slaughter weights close to the biological optima.

Today's breeding schemes are tackling most of these approaches.

Changing breeding methods

In Europe and North America considerable genetic progress has been made by breed replacement, as the older types have given way to the new, particularly Large White and Landrace strains. Progress within lines comes from careful testing and selection on test results. In general, performance testing – the assessment of an individual on its own performance – has replaced the older progeny testing system. Both boars and gilts have their rate of gain and subcutaneous fat measured up to around 90 kg, and boars frequently have their individual or group feed efficiency measured also. The trend is to move away from assembling animals from several breeders in a central test station, and instead to conduct the tests on each breeder's farm.

Artificial insemination was first carried out by the Russian Milovanov in 1932 and success was achieved with frozen semen in 1970 by Polge in Britain. The technical problems of getting normal conception rates and litter sizes within a commercial context are gradually being overcome. The fact that several important traits have moderate levels of heritability and can be assessed in the male before puberty, plus the large herd sizes in modern units, suggest that pig artificial insemination may be very slow to achieve anything near to the wide use common with dairy cattle except in special situations. Embryo transplants are quite successful but still involve skilled surgery for both donor and recipient, and as it is not yet possible to preserve embryos by freezing, the technique is not available for gene conservation purposes.

Crossbreeding, utilizing two or more different improved breeds or lines is widely practised. Specialized female lines are being developed, based mainly upon Large White and Landrace strains, and used as crossbred parent sows. These are increasingly mated to specialized males, often crossbred, derived from breeds noted for their rate of gain (like Durocs) or their carcass lean (like Piétrains).

In the future it seems likely that the pace of genetic change will quicken wherever the need for it has been recognized. Not only are the theory and practice of selection and crossbreeding now much better understood, but decision making is concentrated into fewer hands. National schemes and international breeding organizations already exist where millions of slaughter pigs are directly influenced by decisions in a few nucleus herds.

Conservation of genetic variability

If the onward march of the Large White, Landrace, Duroc and Piétrain continues around the world, will this be at the expense of the many existing breeds adapted to their present environments, and will this matter?

The problem here is to decide to which features of their environment these native stocks are adapted. It seems likely that in many cases adaptation will be to feed shortage; or to finding and utilizing a particular type of feed; to locally prevalent diseases; to climatic extremes of heat, cold, dry or wet; or to a local demand for a special product (lard, fat bacon, particular quality of lean). If a new system of pig production becomes possible – because housing and feed are provided, or diseases are controlled – then it may well be more sensible to provide also a new type of pig, already improved, than to set about a selection programme in the native breed.

To this extent the situation may be very different for cattle and sheep, which will continue to exploit the vegetation growing in a particular region, and which may not be able to change so radically to a new more productive genotype.

What must be guarded against is the disposal of local types which possess adaptations that may help in the long-term survival of pig production. Such traits would include ultra-high prolificacy and high teat number even if the piglets currently grow slowly and produce very fat carcasses. Unusual food-seeking habits or other behavioural traits might also be worth preserving; while high activity and strong maternal instincts are valuable in extensive production systems, extreme docility is useful for intensive farming. The ability to cope with large quantities of bulky vegetable diets may well be valuable in the future, at least for sow herds. It seems obvious that every effort should be made to retain and study some of the Chinese breeds. They may provide us with a better understanding of adaptations even if we find it difficult to use their genes directly in other places. While the total number of breeds deserving conservation may not be high, the need for action is urgent since they are disappearing so quickly.

Another aspect of conservation concerns simply the preservation of our heritage. A useful development could be the increasing interest of townspeople in old livestock breeds, and the establishment of farms devoted to their preservation and display. Examples are the Rare Breeds Survival Trust in Britain which has small herds of seven superseded pig breeds and keeps a register of other herds. The Mangalitsa pig in Hungary is reduced to very small numbers, but it has now been rescued and established on three testing and two cooperative farms. For many other breeds this development will probably be too late.

References

Bökönyi, S. (1974) *History of Domestic Mammals in Central and Eastern Europe.* Akadémiai Kiadó: Budapest

Brentjes, B. (1962) Das Schwein als Haustier des Alten Orients. *Ethnographische-Archaeologische Zeitschrift,* **3** (2): 125–38

Childe, V. G. (1935) *New Light on the Most Ancient East.* Kegan Paul: London

Comberg, G. (ed.) (1978) *Schweinezucht* (8th edn). Eugen Ulmer: Stuttgart

Corbet, G. B. and Mill, J. E. (1980) *A World List of Mammalian Species.* British Museum (Natural History): London

Eikamp, H. (1978) Woher stammt das Schwein? *Deutsche Geflügelwirtschaft und Schweineproduktion,* **30** (35): 884–6

Epstein, H. (1969) *Domestic Animals of China.* Commonwealth Agricultural Bureaux: Farnham Royal, Bucks, England

Epstein, H. (1971) *T he Origin of the Domestic Animals of Africa,* Vol. 2. Edition Leipzig: Leipzig; Africana Publishing Corporation: New York

FAO (1971) Report of the 3rd ad hoc Study Group on Animal Genetic Resources (Pig Breeding). FAO: Rome

FAO (1982) *FAO Production Yearbook 1981,* Vol. 35. FAO: Rome

Hankó, B. (1954) [The History of Hungarian Domestic Animals]. Akadémiai Kiadó: Budapest. (In Hungarian)

Haring, F. (1961) Schweinerassen in den übrigen Ländern West- und Südeuropas. In: *Handbuch der Tierzüchtung,* ed. F. Haring, **3** (2): 46–101. Paul Parey: Hamburg and Berlin

Hitti, P. K. (1951) *History of Syria.* Macmillan: London

Jonsson, P. (1965) Analyse af egenskaber hos svin af Dansk Landrace med en historik indledning. *350. beretning fra forsøgslaboratoriet,* Copenhagen

Lantsheere, W. Vergot de, Lejeune, A. and Van Snick, G. (1974) L'élevage du porc en Belgique: amélioration et selection. *Revue de l'Agriculture,* **27,** No. 5

Lepiksaar, J. (1973) Die vorgeschichtlichen Haustiere Schwedens. In: *Domestikationsforschung und Geschichte der Haustiere,* ed. J. Matolcsi, pp. 223–8. Akadémiai Kiadó: Budapest

Lush, J. L. (1945) *Animal Breeding Plans*. Iowa State College Press: Ames

McFee, A. F., Banner, M. W., and Rary, J. M. (1966) Variation in chromosome number among European wild pigs. *Cytogenetics*, **5**: 75–81

Mellaart, J. (1966) The teasing of the great beasts – from Çatal Hüyük. *Illustrated London News*, **11** (6): 24–5

Mellaart, J. (1967) *Çatal Hüyük*. Thames and Hudson: London

Muramoto, J., Makino, S., Ishikawa, T., and Kanagawa, H. (1965) On the chromosomes of the wild boar and the boar-pig hybrids. *Proceedings of the Japanese Academy*, **41** : 236–9

Pedersen, O. K. (1961) Schweinerassen in den nordeuropäischen Ländern. In: *Handbuch der Tierzüchtung*, **3** (2): 30–45. Paul Parey: Hamburg and Berlin

Piggott, S. (1952) *Prehistoric India*. Penguin Books: Harmondsworth, England

Reed, C. A. (1960) A review of the archaeological evidence on animal domestication in the prehistoric Near East. In: *Prehistoric Investigations in Iraqi Kurdistan*, ed. R. J. Braidwood and B. Howe. The Oriental Institute of the University of Chicago, No. 31

Reed, C. A. (1969) The pattern of animal domestication in the prehistoric Near East. In: *The Domestication and Exploitation of Plants and Animals*, ed. P. J. Ucko and G. W. Dimbleby, pp. 361–80. Duckworth: London

Reed, C. A. (ed.) (1977) *Origins of Agriculture*. Mouton Publishers: The Hague and Paris

Rütimeyer, L. (1861) Die Fauna der Pfahlbauten der Schweiz. *Neue Denkschrift der Allgemeinen Schweizerischen Gesellschaft für die Naturwissenschaft, Basel*, **19**

Rütimeyer, L. (1864) Neue Beiträge zur Kenntnis des Torfschweins. *Verhandlung der Naturforschunggesellschaft in Basel*, **4** (1)

Trow-Smith, R. (1959) *A History of British Livestock Husbandry 1700–1900*. Routledge and Kegan Paul: London

Zeuner, F. E. (1963) *A History of Domesticated Animals*. Hutchinson: London

18
Horse

S. Bökönyi
*Archaeological Institute,
Budapest, Hungary*

Introduction

Nomenclature

In spite of their sometimes great morphological differences, all the breeds, groups and types of domestic horses belong to one single species, *Equus caballus*.

An attempt has recently been made to use trinomial nomenclature for domestic animals. In this system the Latin name of the domestic horse would be *Equus przewalskii* forma *caballus*, i.e. the full name of the supposed wild ancestor and an epithet pointing out that it is a domestic form. Apart from the fact that this type of nomenclature has not been accepted by any official zoological body, these names always depend on the current state of the research on the wild forms of the domestic species. The case of the horse is a good example of this: at the time when this suggestion was made, the opinion of the majority of archaeozoologists was that *E. przewalskii* (Przewalski or Przewalsky's horse) was the wild ancestor of the domestic horse. But since then it has been discovered that the earliest name for the wild form of the domestic horse is *E. ferus*, and *E. przewalskii* is only a subspecies of the former. Thus the trinomial name of the domestic horse would be *E. ferus* forma *caballus*. This clearly demonstrates that it is simpler to use the old Linnaean name, in spite of the fact that it is not very logical because domestic horse and wild horse are one and the same species.

General anatomy and biology

The domestic horse belongs to the order Perissodactyla. On each of its extremities there is only one toe, the original third toe; it is therefore very strong and its distal phalanx is covered with the hoof, a modified form of claw which is nail and pad at the same time. The first and fifth toes of the original mammalian foot have completely disappeared but the second and fourth toes sometimes appear as an atavism in a rudimentary form; Bukephalos, the favourite riding horse of Alexander the Great, was such a three-toed animal. On each side of the cannon bones is a splint bone, invisible on the living animal, which is a remainder of the second and fourth rays of the ancient skeletal type.

The reduction in the number of toes was in the interest of speed. Slow-moving animals like bear or man touch the ground with their whole hands (to the wrists) and feet (to the heels). A swifter motion is possible by walking on the toes. The horse and other equids walk (and run) on tiptoe, on their nails. Combined with the relatively short upper leg bones and the long lower leg bones plus the cushioning effect of the lower part of the hooves and the strong springing ligaments, this makes high speed possible. As Simpson (1961) notes, large modern horses 'thanks to the perfection of this mechanism . . . are about as fast as seems mechanically possible for animals of their size'.

Characteristic features of the horse's legs are the chestnuts. They are hairless spots of an oval shape with a thick cornified surface on the inner side of the legs just above the knee and the hock. They are either rudiments of the pads of the side toes or remnants of some kind of gland.

Experts recognize twelve different gaits of the horse but these go back essentially to four motion types: walk, trot, pace and gallop. The walk is a four-beat gait; thus the horse moves each leg separately in sequence. The trot and the pace, both somewhat quicker than the walk, are two-beat gaits: in the trot diagonal legs move together; in the pace the fore- and hindlegs of the same side move together. Pacing is a very comfortable soft motion type for the rider on long trips, and Asiatic nomads appreciate it accordingly. The swiftest motion type of horses is the gallop with a three-beat cadence: beat, pause, quick double-beat; beat, pause, and so on. Depending on the lead, the gallop starts with one hindfoot followed by the other hindfoot and its diagonal forefoot almost at the same time but the hindfoot somewhat earlier, then the other forefoot, and the pause when all four feet are off the ground.

In spite of the fact that the horse lives on vegetable food rich in fibre, it has a simple, one-chambered stomach. The fermentation of cellulose takes place in the rather large caecum, posterior to the small intestine. This system is certainly inferior to the complex stomach and cud-chewing of the ruminants. This means that the feeding of the horse is more expensive which is a serious disadvantage of the species.

The domestic horse reaches sexual maturity at the age of 3.5–4 years. The oestrous cycle is 21 days in length and oestrus lasts 3–7 days. Gestation period is 11–11.5 months. Twins are rare. Length of life can be as much as 40 years.

The chromosome number of all domestic horses is $2n = 64$. That of the Przewalski horse is 66. Crosses between domestic and Przewalski horses give offspring which are fertile in both sexes in spite of their diploid chromosome number of 65. Crosses with asses, zebras and onagers are possible but the hybrids are normally sterile.

Colour

The colouration of domestic horses is rather varied; there is no standard colour for most breeds. As a basic colouration, several authors consider the dun or bayish colour of the Przewalski horse (the only living subspecies of the wild horse) with dark mane, tail, and stockings. The trouble is that the Przewalski horse is only one of several subspecies of the Holocene wild horse. The extinct tarpan of eastern Europe was mouse grey or ash grey with dark mane, tail, back stripe and legs, and one knows nothing about the colouration of the ancient wild horses of north and west Europe.

In the determination of the basic colour of the Asiatic wild horse (Przewalski horse) the *A* ('agouti') gene is the most important. It determines that the peripheral parts of the body be black and the central ones reddish, pale bay or dun. It is typical that the individual hairs show no rings. The black zebra stripes appearing on the elbow and hock regions of wild horses and also some ponies are controlled by an independent gene, though this can be effective in an agouti gene background alone.

In the domestic horses bay is probably the most common colour type. The basic body colour of bay horses is brown, red brown, tan or pale brown, in several shades. Their mane and tail and usually also their stockings, are black. In fact bays of some shades differ from chestnuts only in the presence of these black areas.

The other main colour type is chestnut. The coat of chestnut horses is of various shades of brown without the dark colouration of the aforementioned parts of the bay. The brown can be from flat brown through red brown or golden to liver brown.

The dun varies from yellowish through greyish or brownish. A special type of dun is called isabella. In the skin of isabella horses there is no pigment, and their hooves are therefore waxen yellow and their long hairs are dirty white or light yellow. The skin of real dun horses always contains pigment, so that their hooves are slate grey. Dun horses can be divided into two groups, one with a dark stripe along the spine and long hairs also dark, the other lacking dorsal stripe and with long hairs the same colour as the coat.

Palomino is very close to dun. These horses have golden coats in shades from reddish through dark cream with long hairs flaxen or silver. Some authors put the isabella horses in this colour type, others include the palominos with the duns.

Grey is a colour that first appears as the animal ages. Such horses are born dark brown, reddish brown or black and later develop more and more white hairs. In dapple greys the white hairs are only in rounded spots; in iron greys the white hairs are evenly distributed in a dark coat, silver greys have an overwhelming majority of white hairs in a shiny coat, and in crane greys the white hairs are dominant in a lustreless coat.

The essence of roan is that a coat of any colour except white is sprinkled with white hairs. Blue roan is roaned black, red roan is roaned bay, and strawberry roan is roaned chestnut or sorrel.

So-called white horses are not really white but very silvery animals; some of them are old greys. True albinos probably do not exist among horses, although some breeders refer to horses born very pale or white as albinos.

Black horses have black coat and long hairs. True black horses are extremely rare; some of them are dark bay in the winter and really black in the summer alone.

Piebald is a horse with coloured spots on a white coat. There are three main groups: common piebald (with irregular coloured patches), leopard spotted (with apple-size coloured spots) and agate piebald (around the coloured spots are thin frames in a lighter shade). Among American horses Pinto and Appaloosa are special piebald types.

These are only the main colour types; breeders use many more terms, and hundreds of expressions for the variations. As an example, an ethnographer collected from Hungarian herdsmen 321 expressions describing the colour of horses of which 12 referred just to the colour variants of bay horses.

For an account of the genetics of colour the reader is referred to Jones and Bogart (1971) or Searle (1968).

Geographical distribution and economic importance

The domesticated horse reached many parts of Europe, Asia and Africa shortly after its first domestication, but arrived in the Americas and Oceania much later with the European conquistadors and explorers. Today it lives all over the world except in arctic and periarctic regions.

As Table 18.1 shows, South America is now the real home of domesticated horses not only in regard to their absolute number but also their density and the ratio to the human population. This ratio is also high in North America and in Oceania. Europe is intermediate in both absolute and relative numbers of horses while Asia and Africa are the two continents poorest in horses. Though the absolute number of horses in Asia is greater than in Europe, relative to the vast size of the continent and the large human population, horses are rare.

Today horses serve mostly as work horses in poor countries and as race and pleasure horses in

Table 18.1 The horse population of the world (thousands)

	Average of					Horses per
	1948–52	1969–71	1977	1979	1981	1000 humans
Europe	16 900	7 714	6 000	5 595	5 306	10.9
U.S.S.R.	12 800	7 641	5 996	5 700	5 700	21.3
Asia	11 900	16 360	13 515	17 755	17 580	6.7
Africa	3 000	3 723	3 723	3 646	3 637	7.5
U.S.A. + Canada	6 930	8 029	9 425	9 747	9 928	39.0
Latin America	23 170	20 235	22 530	21 504	23 394	62.7
Oceania	1 300	583	567	622	647	28.2
World	75 800	64 284	61 755	64 569	66 192	14.7

From: FAO (1963, 1980, 1982)

rich ones. Overall their meat has little importance. Mare's milk is drunk fresh, or in a lightly fermented form, as kumys (Soviet Union) but it is important only among horse nomads.

Wild horses

There is perhaps no mammalian family whose evolution is so well known as that of the Equidae. The family originated in the Palaeocene, some 60 million years ago and its evolution took place in both Eurasia and America – at times in parallel and at other times at different stages. The most conspicuous evolutionary changes were: 1. increase in size; 2. reduction in the number of toes; and 3. switch-over from browsing to grazing.

According to Nobis (1971) large- and medium-sized wild horses lived side-by-side during the successive phases of the Pleistocene. By the end of the period the overspecialized large forms died out whereas medium-sized ones survived the drastic climatic change that signalled the end of the period. However no Holocene wild horses existed in the Carpathian basin and on the Iberian, Appennine or Balkan peninsulas. The surviving horses are placed in a single species, *E. ferus*. This lived in a very large area in Europe and Asia and could be divided into at least two clearly distinct subspecies, the tarpan (*E. f. ferus* or *E. gmelini*) and the taki, or Przewalski horse, (*E. f. przewalskii*). Wild horses were still present in America when the first Amerindians arrived but they became extinct about 10 000 B.C.

Compared to modern horses the tarpan was a small (130 cm high) but strongly built animal with short head, broad, flat forehead, a ram nose and small, pointed ears. Its colour was mouse grey or ash grey with a black stripe along the spine. Its mane was short and erect, its tail was short, covered with long, dark hairs. In the autumn some tarpan populations grew white downy hair so that in winter its body was nearly grey, the head, legs, mane and tail remaining dark. This colouration was a very good camouflage in the snow-covered landscape. Because of this characteristic, the tarpan can probably be identified with the white wild horses described by Herodotus in the fourth book of his *Histories* as living beside the marshes at the source of the River Hypanis; this is probably the Bug in the present-day Ukraine, but the district has alternatively been identified as the Pripet marshes of Byelorussia.

In east Europe, particularly in the Ukrainian steppes, the tarpan was fairly numerous. It lived in small groups consisting of mares with their offspring and a stallion. As the steppes became populated after the Russo–Turkish War in the late 18th century, the tarpan gradually decreased in numbers. Horse breeders of the region refused to put up with tarpan stallions which abducted mares from the domestic herds. An unbridled hunt of the tarpan led to the extinction of the east European wild horse by the end of the 19th century; the last one died in captivity in 1918. There remain only one complete skeleton and a skull in Soviet zoological collections.

The other wild subspecies, the taki or Przewalski horse still exists, but only in zoological gardens. The taki is somewhat larger than the tarpan, its withers height measuring 138–145 cm for stallions and 125–140 for mares. It is a stocky, long-muzzled horse with a large head in proportion to its trunk. The neck is thick and the withers protrude only slightly. Its basic colour types are dark bay and dun with reddish and faded brownish tints. It also has a dark stripe along its back, often accompanied by a cross-stripe across the shoulders. The colour is quite light around the mouth. The short, erect mane, the tail and the feet are dark. The lower half of the tail is covered by long hairs; the hairs of the upper part are short.

Originally Przewalski horses populated large areas of central and western Asia where they lived in small groups of eight to fifteen individuals with a single stallion. Decimated by hunting and expanding nomadic herders, they became restricted to deserts and remote mountainous regions in the west of Mongolia and northwest China (Xinjiang). According to a recent FAO report they are now extinct. As at 1 January 1980 there were 388 Przewalski horses in captivity, 154 stallions and 234 mares (Volf 1980).

Some wild horses also survived the end of the Pleistocene in other regions of Europe. Nevertheless, they certainly were far less numerous than the tarpan or the Przewalski horse which means that they surely played a much less important part in horse domestication. Due to the lack of proper bone remains they are not well known.

The chromosome number of the Przewalski horse is $2n = 66$, i.e. two (one pair) more than that of the domestic horse.

Feral horses exist in all continents. Among them the mustangs of America and the brumbies of Australia are the best known, though their numbers have now very much decreased. According to recent observations, with each generation feral horses more closely resemble the Mongolian wild horse simply because their characteristics give a better chance for survival under difficult conditions (Mennick 1979).

All wild and feral horses produce fertile hybrids with domestic horses.

Domestication and early history

The domestication of the horse

The horse was not one of the earliest species to be domesticated; its domestication followed that of the last Neolithic species by about 3000 years.

The domestication of the horse differs in an essential point from that of the earlier domestic animals: dog, sheep, goat, pig and cattle. Whereas the domestication of these five species was without precedent and presumably for the purpose of securing living food reserves, the domestication of the horse happened when man was already aware of further possible uses of his domesticates (milk, wool, draught power). It is plausible, therefore, that in addition to domesticating the horse for its meat, another motive was to utilize it as a working animal.

The main problem is, however, that it is extremely difficult to distinguish remains of domestic horses from those of wild ones, and thus to determine if the horse population of a given prehistoric site is wild or domestic. The sort of morphological changes which occur quite early in other domesticated species cannot be found in early domesticated horses. Therefore, besides developing new morphological methods (e.g. the analysis of the skull characteristics and the variations of the enamel pattern of the cheek teeth) non-morphological methods based on the herd structure, kill-off pattern and especially finds connected with horse keeping and artistic representations had to be worked out.

The earliest domestication of the horse so far known was in the Aeneolithic of east Europe, more exactly in the southern Ukraine (Bibikova 1967). The first site where there is positive evidence for domestic horses is Dereivka, a settlement of the Srednij Stog culture, on the right bank of the Dnieper river, about 70 km from the town of Kremenchug. Its radiocarbon (uncalibrated) date is about 3500 B.C. The inhabitants of the Dereivka settlement originally carried out a specialized hunting of wild horses for their meat, and this gradually developed over time into domestication. At that date, the very high proportion (60 %) of horse bones alone would have suggested a local domestication. But the domesticated nature of the Dereivka horses was also indicated by the careful morphological analysis of a complete skull, of the variation of the enamel pattern of hundreds of teeth and of the very numerous extremity bones. The final proof that these horses were domesticated was provided by six bits made of antler found at the site (Telegin 1973). Since the Dereivka site lies in the middle of the distribution area of the tarpan, there is little doubt that this is the wild form from which these first domesticated horses of Europe derived.

Horse domestication may well have also occurred on the steppes of western and central Asia because the Przewalski horse, the Asiatic wild horse subspecies, lived in abundance there. Unfortunately, due to lack of bone samples studied, we have no direct evidence of this, and the difference in chromosome number between the Przewalski horse and all known domestic horses, including Mongolian ponies, suggests that we ought to be cautious in this matter.

There is not much known about horse domestication in western Europe either. One can probably suppose some horse domestication there too, however, a far less important one than that of east Europe simply because of the rarity of wild horses in the early Holocene. Lundholm (1949) however, demonstrated continuity between the postglacial Swedish wild horses and the domestic horses of the Swedish Bronze Age.

From the area of the first domestication, domesticated horses quickly spread in all directions. First they 'inundated' the Ukrainian and the Russian plains stimulating local domestication wherever wild horse stock was available. Then still in the second half of the 4th millennium B.C. they reached Moldavia, the Carpathian basin and Moravia in the west, probably Bulgaria to the south, the Caucasus, Transcaucasia and possibly even eastern Anatolia in the southeast (Bökönyi 1978, 1980a).

Nevertheless, whereas domestic horses immediately developed a considerable economic importance in the area of domestication, the members of this first wave seem to have been little more than curiosities or status symbols in their new homelands and never started real horse breeding there. It was the same with the second wave which reached the above regions and also the western part of Europe in the second half of the 3rd millennium B.C. This wave probably took domestic horses also to Mesopotamia and other regions of the Near East. Horses in quantity appeared in the central, southeastern and western parts of Europe by the beginning of the Bronze Age around the turn of the 3rd and 2nd millennia B.C. It was this wave which took the domestic horse to Greece around or soon after 1900 B.C. With these numerous horses serious horse

breeding started in these regions too, sometimes reaching the same importance as in the domestication centre.

The early domesticated horses

To the breeder of today the early domesticated horses were small animals, with the size of a pony but with the proportions of a true horse. The withers height of the east European early domestic horses was between 127–145 cm, with an average of about 137 cm. The west European ones were about 10 cm smaller on average. They were strongly built animals with thick legs, somewhat more slender than the wild ones, although throughout horse history they were the closest to their wild form in this respect. Their hooves were narrow with well-arched soles perfectly adapted to hard, dry, steppe soil. The skull was rather large in comparison to the trunk. It had a spacious, vaulted brain-case and a comparatively long, rather wide facial part. The forehead was flat, not very broad and the profile was straight. The teeth were comparatively large showing a rather variable enamel pattern.

Nothing is known about their colouration but we do know that their mane was long and hanging and not short and erect like that of wild horses. This is the result of a mutation occurring after domestication which halted the annual shedding of these hairs. Such long and hanging manes are shown on several representations from the Bronze Age.

As for the use of early domestic horses, there is no doubt their meat was eaten as evidenced by limb bones broken for the marrow and brain-cases opened for the brain; these were delicacies even in prehistoric times. Horsemeat consumption declined in the Roman imperial period and was more or less completely given up in Europe after the conversion to Christianity. Nevertheless, the main exploitation of the horse, even at the earliest stage of horse breeding, was not its meat. This is clearly demonstrated by the ratios of the different age groups of slaughtered animals (kill-off pattern) in prehistoric sites. Whereas species kept exclusively for their meat were mostly killed when immature (presumably not just because their meat was tastier at that time, but because the fodder shortage in winter made it uneconomic to feed any but the breeding stock), animals producing milk, wool or draught power were mostly killed when mature since production reached full development only with maturity. Thus in Hungarian Bronze Age sites the proportion of mature horses was very similar to that of adult cattle kept mainly for their milk and draught power and only secondarily for their meat.

The six bits (cheek-pieces) made of antler found in Dereivka clearly reveal the nature of the horse's main use right after its first domestication: it served as a work animal. The presence of bits excludes or at least minimizes the role of the horse as a pack animal. For a pack horse one needs no bit; a simple halter will do. The question remains whether the horse was first used as a riding or draught animal. Some kind of riding must have existed immediately after the first domestication because without it a horse herd would be impossible to handle and keep together. Nevertheless, formal driving most likely preceded formal riding (Littauer, personal communication). There is no direct evidence to support this assumption but the case can be argued as follows: 1. using a domestic species (cattle) to pull a cart was already known at the time of the first horse domestication; 2. the early bits because of their soft mouthpieces were appropriate only for draught horses and not for riding horses; 3. psychologically and physically (Littauer and Crouwel 1979) it was more tolerable for a horse to be put to a cart than to have a rider on its back; and 4. the representations of horse carts *en masse* precede those of horse riders.

The horse was undoubtedly a better draught animal than cattle, above all because of its swiftness. Whereas a slow ox cart covers up to 24 km in a day (this was the daily coverage of the six-ox wagons of the American West in the last century), a horse team with a cart easily makes 50–60 km a day, even on unpaved roads. The use of the horse as a draught animal brought a revolutionary change in land transport: for the first in his history man was able to accelerate transport to beyond his own speed (Bökönyi 1980b). Horse-drawn carts reached great importance particularly from the Bronze Age on when the invention of the spoked wheel made carts lighter and quicker. At the same time also more precious wares (bronzes and probably textiles) appeared which were worth transporting over long distances. Thus horse-drawn carts revolutionized trade, extending its action radius enormously.

The horse as a draught animal found a direct military use with the introduction of the chariot in the Bronze Age. The chariot gave the armies not just more strategic mobility but also a new weapon, essentially an 'elevated mobile firing platform' (Littauer and Crouwel 1979).

Horse riding appeared in Mesopotamia before 2000 B.C. (Moorey 1970) in the Altai-Sayan region around 1500 B.C. (Hančar 1956), and in Europe sometime in the 2nd millennium B.C. In the Near East riders even reached a tactical importance from the 13th century B.C. (Salonen 1955). In fact, in prehistoric times, riding played a more important role in warfare than in transport.

The last wave of steppe horses to Europe

In the Iron Age the last wave of eastern horses reached Europe. The Scythians thrust as far as the Carpathian basin to the west and brought a great number of fine steppe horses which, through trade or as booty, got to Austria, Italy and Greece.

The Scythian horses are well known both osteologically and through artistic representations (Bökönyi 1968). They were very good horses even to the eyes of today's breeders. They were mostly reminiscent of Arabs but somewhat smaller. Their average withers height was somewhat above 136 cm. There is no doubt that the Scythian horses were excellent, being closely related to the Persian horses. As far as is known conscious horse breeding started first in Persia in the 1st millennium B.C. resulting in an enormous improvement of the horse stock. Several breeds, among them dwarf and large, light and heavy, came into being; this also exerted an influence on the horse populations of the neighbouring peoples.

At the same time, another horse group was living in Europe, west of a line stretching from about Vienna to Venice. They derived from the local Bronze Age horses, and their typical representative was the Celtic horse. The horses of this western group were considerably smaller than the eastern ones having an average withers height of about 126 cm (Bökönyi 1968).

Thus in prehistoric times, the eastern horses were larger than the western ones which is the direct opposite of the present situation (Fig. 18.1). It was the same with their body proportions – the eastern group had thicker, and the western one

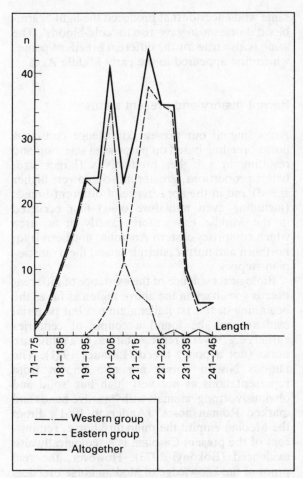

Fig. 18.1 Size variation of Iron Age horses in central and eastern Europe based on the metacarpal length (in cm). (After Bökönyi 1974a)

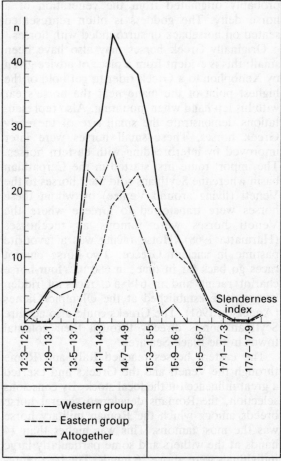

Fig. 18.2 Variation in slenderness index of Iron Age horse metacarpals in central and eastern Europe (in cm). Slenderness index is the minimum width divided by the maximum length and multiplied by 100. (After Bökönyi 1974a)

slenderer, extremities (Fig. 18.2).

From the breeder's point of view, the eastern horses were the better ones; because they were larger they were able to carry heavier loads, to move faster with a rider, to carry more easily a rider wearing armour and to cover longer distances. All these qualities provided reasons why the peoples living in western Europe were anxious to acquire the eastern horses. This, of course, could be achieved by outstanding personages alone who held these horses in such high esteem that they also had them put into their graves (with the original eastern gear).

We must therefore refute the romantic supposition that the Celts were the best horse breeders of their epoch and their horses were of superior quality. Just the opposite, Celtic horses were very small; remains of several individuals with a withers height below 1 m have been found in Celtic sites. Nevertheless, there is no doubt that the west European equestrian traditions originated from the Celts. They held horses in esteem and also granted them a place in their mythology, for the veneration of Epona, goddess of fertility, probably originated from the veneration of a horse deity. The goddess is often represented seated on horseback or surrounded with horses.

Originally Greek horses may also have been small; this is evident from a piece of advice given by Xenophon to a Greek rider to get hold of the highest point of the mane near the horse's ear with his left hand when mounting. Also representations demonstrate the small size of the early Greek horses. These small horses were later improved by interbreeding with eastern horses. The import route first started in the Carpathian basin where the Scythians sold their horses to the Veneti (living around Venice) by whom these horses were transferred to Greece where the Veneti horses were famous as racehorses (Harmatta 1968). Horse racing was a favourite pastime in ancient Greece. Two-horse chariot races go back far in time; in 680 B.C. four-horse chariot racing and in 648 B.C. races of ridden horses were established at the Olympic Games (Anderson 1961). The Greeks could also acquire Scythian horses directly through their colonial towns in the Black Sea area.

The eastern horses reached Italy and Rome through the Veneti and the Greeks and exerted a great influence on the local stock. By conscious selection, the Romans developed several horse breeds among which the Roman military horse was the most famous. This was higher than 14 hands at the withers and some particularly large individuals were above 16 hands. The best representation of this breed is the horse of the statue of Emperor Marcus Aurelius on the Capitoline Hill in Rome.

In the course of the 1st millennium A.D. the last wave of the eastern horse arrived in Europe with the peoples of the Migration Period. These horses, along with the Roman breeds, transformed the European horse stock to such an extent that by the end of this period size and body proportions were similar in eastern and western horses although they could still be distinguished on the basis of the skull form. By the end of the Migration Period and the beginning of the Middle Ages, the heavy 'cold-blood' horses appeared probably as a result of conscious breeding for knights wearing heavy armour. Some authorities (Ebhardt 1958) derive heavy horses from a large wild species, *E. woldrichi* or *E. antoniusi* but this species seems to have become extinct at the end of the Pleistocene and it is more likely that the same wild ancestor that produced the light 'warm-blood' horses also gave rise to 'cold-bloods'. The same is also true for the different breeds of ponies which first appeared in the early Middle Ages.

Recent history and present status

According to our present knowledge conscious horse breeding based on purposeful selection and resulting in a higher productivity (larger size, better proportions, greater muscle power, higher speed) and in the appearance of different breeds (including even miniature ones) first occurred in the Middle East, more exactly in an area which comprises eastern Anatolia, northern Iraq, northern and northeastern Iran and the Transcaspian steppes.

Biological evidence of the existence of different breeds goes back in the above region as far as the beginning of the 1st millennium B.C. but pictorial evidence can be found a couple of centuries earlier, e.g. several representations of a miniature horse (not a pony) breed (Littauer 1971). The famous Nisean horses are also known from representations as not very high but 'solid and obviously strong' animals with 'massive heads and marked Roman noses' (Anderson 1961). From the Median empire the miniature horse, reminiscent of the present Caspian, is osteologically also evidenced (Bökönyi 1973). However, the real proof of the knowledge of Median horse breeders is the existence of the heavy charger that is found not only on reliefs but also osteologically in Median sites. This horse, which was very similar to the great 'cold-blood' horse of the European Middle Ages, was produced on the lush vegetation of the Kopet Dag in Turkmeniya and served as the warhorse of the royal bodyguards, the 'Immortals'.

Today's horse breeds are usually divided into two not always separable groups, the oriental ('warm-blood') and occidental ('cold-blood') horses. The oriental horses are light; their extremities are slender; the hairs of their manes and tails are thin and straight; their temper is lively and their motion fast. The occidental horses are large and heavy; their limbs are thick; their manes and tails are bushy; their disposition is docile and their motion slow. In fact, neither 'oriental' and 'occidental' nor 'warm-blood' and 'cold-blood' are proper denominations for the two groups because: 1. among oriental horses there are breeds of western or mixed origin and vice versa; and 2. 'warm-blood' and 'cold-blood' have nothing to do with the temperature of the blood but with the temperament of the animal. It would be simpler to call the first group 'light horses' and the second 'heavy horses'.

Among the recent horse breeds the Arab or Arabian is the oldest, with a history that goes back to Mohammed's time, into the 7th century A.D. or even somewhat earlier. This is rather strange if one takes into consideration that the Arabs did not get their first horses until around the beginning of the Christian era, before which (e.g. in the Persian wars) they rode on camel back. Arabian horse breeding was promoted by no less a person than the Prophet himself when he discovered the decisive importance of a good cavalry in war. He allegedly had four mares, and at least one strain of pure Arabians derives from them.

The Arabian horse is the result of the long influence of an extremely harsh natural environment and human selection combined with long inbreeding, tough endurance tests and careful keeping and feeding methods. The head is very fine, characterized by its concave profile, well developed forehead, large protruding eyes, narrow fine muzzle and wide nostrils. The neck is long, slender and arched, the chest round and well muscled. The back is straight and comparatively short, the rump long. These features make the Arab an excellent riding horse. The legs are long and slender but well muscled and strong-boned.

The skin of purebred Arab horses is always dark, a kind of antimony grey (*kuhl* in Arabic) but their hairs show a wide colour variation from brown to white. Spotted Arabs are not known and the different shades of dun – palomino, cream, isabella or yellow – are not favoured colours either.

In fact, purebred Arab horses are very rare even in Arabia itself; according to modern estimates their number does not exceed a couple of thousand. In Arabia pedigrees are kept – often only by word of mouth – through the mare alone, and descent is often more highly valued than outward appearance or performance.

Since the Middle Ages Arabian horses – as top representatives of the species – have been imported first by neighbouring areas and then by more distant countries. Everywhere they exerted a great influence on the local stock. Large-scale imports of Arabian horses took place in the last century and since then they have been kept in purebred form outside Arabia.

A breed which bears much resemblance to the Arab is the Barb of North Africa. In body proportions Barbs are very close to Arabs, but they lack the very elegant skull form of the latter.

Another very attractive, light horse is the Turkoman of the Transcaspian steppes. It is a very slender steppe horse and a good racer and jumper. A role in the evolution of the English Thoroughbred is extremely possible.

The lord of the racetrack and the prime example of successful horse breeding is the English Thoroughbred. It has a long history whose beginning is not completely clear. Its ancestors were Arabian, Barb, Turkish, Persian and Egyptian stallions and English mares with occidental background – the fifteen to twenty so-called 'royal mares' bought by Sir John Fenwick, Master of Horse to Charles II on the continent, among whom, according to Wrangel (1928) there were several Hungarian mares. The three most famous stallions which can be found in the ancestry of every English Thoroughbred are the Byerly Turk, the Darley Arabian and the Godolphin Barb. The first was brought to England in 1689, the second shortly after 1700 and the third about 1730. They have a somewhat obscure origin, the Byerly Turk was probably an Arabian, the Darley Arabian could easily have been a Turkoman, and the Godolphin Barb was either a Barb or an Arabian bred in North Africa.

Thus the English Thoroughbred is a mixture (which modern breed is not ?) but a very successful one. As Simpson (1961) points out, Thoroughbreds were developed to be running machines, and every part of their body serves this purpose from their slender nose to their flowing tail. Besides this they are also graceful, attractive creatures, one of the most beautiful breeds of horse. Actually they have no strictly determined outer characteristics; their main feature is their great speed and they are Thoroughbreds because they originate from recognized Thoroughbreds registered in the English Stud Book. They are tall horses, being generally 16 hands or more at the withers. Most of them are black, bay or chestnut; other colours occur extremely rarely.

Thoroughbreds are the only horses which can

run at 64 km per hour over short distances. The fastest time over 1.25 miles (2 km) was also run by a Thoroughbred (King Pellimore) in 1 minute 57.6 seconds in 1976.

A very interesting European light-horse breed is the Lipizzaner. The stud farm of Lipizza near Trieste was founded by Archduke Charles in 1580 in order to produce good-looking, quiet horses for the Imperial Court of Vienna. The original stock consisted of Spanish Andalusian-Barb horses which were subsequently interbred with other Spanish and Neapolitan horses. The result was a breed 157–167 cm high at the withers with a somewhat heavier body, a thicker but well-arched neck and a ram nose. The colour is mainly grey or roan. Its motion is very attractive and since most Lipizzaners are very docile, they have always been the horses of the famous Spanish Riding School at Vienna.

What the Thoroughbred is among race horses, the Standardbred is among trotters. It is an American breed because trotting races developed in U.S.A. from occasional encounters between roadsters. Trotters were of different origin at the start but among modern Standardbred horses, trotters and pacers alike, Thoroughbred descendants quickly took the lead.

Remembering that most modern Standardbred horses derive from Messenger, a Thoroughbred imported from England in 1788, it is not surprising that they are very similar to the Thoroughbred; but they are somewhat smaller, scarcely exceeding 15 hands at the withers, and at the same time a littler longer and sturdier. The trotting record is 1 minute 54.8 seconds by Nevele Pride; the pacer record is 1 minute 49.2 seconds by Niatross, both over a mile (1.6 km).

The cold-blood breeds originate from the heavy military horses of the Middle Ages. The best known are the Percheron from France, the Belgian, and three British breeds, the Shire, Clydesdale and Suffolk. The Percheron is 16–17 hands high and its colour is generally dapple grey but black also occurs. It is an all-purpose horse with a medium-size head, short back and heavy thigh muscles. Its strength is enormous. The Belgian is very similar to the Percheron both in size and conformation, only its colouration is more variable. The Shire is probably the largest breed of this group measuring 17 hands or more at withers and weighing upwards of 1 tonne. In its colouration it is very similar to the Belgian.

The Clydesdale developed around the River Clyde in Scotland in the 18th century. Its colour is black or bay; its height is around 16 hands and its weight is about 800 kg. The Suffolk in the smallest in this group weighing 680–730 kg and reaching about 16 hands. It is a very compact horse with short, thick neck and short back and legs.

Among the ponies the Shetland is worth mentioning. With its maximum height of 10.2 hands it is a strange little creature. In its original home it was a draught animal; it was imported into England to work in coal-mines and as a mount for small children. It is now popular in several European countries and America. For information on other breeds see Glyn (1971), Goodall (1973) or Hope and Jackson (1973).

Future prospects

The power-producing domesticated animals are now on the decline; competition with the internal combustion engine is bringing about a reduction in numbers, especially in developed countries. In this respect, the position of the horse is almost disastrous. The selection commencing very early on, developing the horse exclusively to produce power, as a specialized draught or saddle animal, may have been one of the causes, if not the main one, of the horse's tragedy. The horse has practically no other use but its power. It develops too slowly to be a successful meat producer; being uniparous it is not prolific enough; there are prejudices against its meat; and finally its milk (which is not much anyway) is consumed only by nomads. The horse is still used as a saddle animal in steppe and desert areas but who knows how long this will last?

In western countries the horse for a long time could not compete with the engine. More recently, however, horses have experienced a minor come back, especially the heavy breeds for ploughing on soils unsuited to machinery; the steeply increasing cost of fuel oil has given horses a boost. Whitbread's Brewery in 1971 calculated that it cost £ 500 to run a pair of Shire horses while it would require £ 1000 for a lorry.

According to FAO statistics (see Table 18.1, p. 165), the horse is the only domestic species whose number fell (by 19 %) between 1948–52 and 1977. Since then there has actually been a slight increase in the number of horses. In Europe and the U.S.S.R. there has been a considerable decrease, more than 60 per cent, since 1948–52. On the other hand, in North America the number of horses increased by 43 per cent between 1948–52 and 1981; most of these horses were pleasure horses.

Thus the future of the horse does not look very promising. However, racehorses and pleasure horses will certainly exist for a very long time actually as long as man can afford to keep them, and we seriously hope he can.

References

Anderson, J. K. (1961) *Ancient Greek Horsemanship*. University of California Press: Berkeley and Los Angeles

Bibikova, V. I. (1967) [Studies on ancient domestic horses in Eastern Europe]. *Bjulleten Moscovskogo Obshchestva Ispytatelei Prirody, Otdel Biologicheskii*, **22**: 106–18. (In Russian)

Bökönyi, S. (1968) Data on Iron Age horses of Central and Eastern Europe. *Bulletin of the American School of Prehistoric Research*, No. 25. Peabody Museum: Cambridge, Massachusetts

Bökönyi, S. (1973) The animal remains of Nush-i Jan: a preliminary report. *Iran*, **11**: 9–11

Bökönyi, S. (1974a) *History of Domestic Mammals in Central and Eastern Europe*. Akadémiai Kiadó: Budapest

Bökönyi, S. (1974b) *The Przevalski Horse*. Souvenir Press: London

Bökönyi, S. (1978) The earliest waves of domestic horses in East Europe. *Journal of Indo-European Studies*, **6** (1/2): 17–76

Bökönyi, S. (1980a) La domestication du cheval. *La Recherche*, **11** (114): 919–26

Bökönyi, S. (1980b) The importance of horse domestication in economy and transport. In: *Transport Technology and Social Change*, ed P. Sërbom, pp. 15–21. Tekniska Museet Symposia No. 2, 1979. Tekniska Museet: Stockholm

Ebhardt, H. (1958) Verhaltensweisen verschiedener Pferdeformen. *Säugetierkundliche Mitteilungen*, **6**: 1–9

FAO (1963, 1980, 1982) *FAO Production Yearbook*, 1962, 1979, 1981, Vol. 16, 33, 35. FAO: Rome

Glyn, R. (ed.) (1971) *The World's Finest Horses and Ponies*. Harrap: London

Goodall, D. M. (1973) *Horses of the World* (2nd edn). David and Charles: Newton Abbot, England

Hančar, Fr. (1956) *Das Pferd in prähistorischer und früher historischer Zeit*. Wiener Beiträge zur Kulturgeschichte und Linguistik, No. 11. Herold: Wien - München

Harmatta, J. (1968) Früheisenzeitliche Beziehungen zwischen dem Karpatenbecken, Oberitalien und Griechenland. *Acta Archaeologica Academiae Scientiarum Hungaricae*, **20**: 153–157

Hope, C. E. G. and **Jackson, G. N.** (1973) *The Encyclopedia of the Horse*. Ebury Press and Pelham Books: London

Jones, W. E. and **Bogart, R.** (1971) *The Genetics of the Horse*. Caballus Publishers: East Lansing, Michigan

Littauer, M. A. (1971) The figured evidence for a small pony in the ancient Near East. *Iraq*, **33**: 24–30

Littauer, M. A. and **Crouwel, J. H.** (1979) *Wheeled Vehicles and Ridden Animals in the Ancient Near East*. E. J. Brill: Leiden and Cologne

Lundholm, B. (1949) Abstammung und Domestikation des Hauspferdes. *Zoologiske Bidragen Uppsala*, **27**: 1–288

Mennick, P. (1979) Observations of the wild horse. *Horse News*, Mar.: 4–6, 43–6

Mohr, E. (1971) *The Asiatic Wild Horse*. Allen: London

Moorey, P. R. S. (1970) Pictorial evidence for the history of horse-riding in Iraq before the Kassite Period. *Iraq*, **32**: 36–50

Nobis, G. (1971) *Vom Wildferd zum Hauspferd*. Fundamenta, **6**. Köln/Böhlau: Cologne

Salonen, A. (1955) Hippologica Accadica. *Annales Academiae Scientiarum Fennicae*, Series B, **100**, Helsinki

Searle, A. G. (1968) *Comparative Genetics of Coat Colour in Mammals*. Logos Press: London

Simpson, G. G. (1961) *Horses*. The American Museum of Natural History: New York

Telegin, D. J. (1973) Über einen der ältesten Pferdezuchtherde in Europa. *Actes du VIIIe Congrès International des Sciences Préhistoriques et Protohistoriques*, II, Beograd, pp. 324–7

Volf, J. (1980) *General Pedigree Book of the Przewalski Horse*. Zoological Garden: Prague

Wrangel, D. G. (1928) *Das Buch vom Pferde*. Schickhardt and Ebner: Stuttgart

19

Ass, mule and onager

H. Epstein

Emeritus Professor of Animal Breeding
The Hebrew University, Jerusalem

Ass

Introduction

The ass, *Equus asinus*, is a herbivorous animal of the order Perissodactyla, family Equidae. The genus *Equus* comprises the horse, ass, onager and kiang, zebra and quagga.

The name ass is derived from the Greek *onos* and Latin *asinus*, believed to be akin to the Hebrew term *athon* for the she-ass. Colloquially the animal is known as donkey, originally dunkin, with reference to the commonly greyish brown colouration of the coat.

Wild asses are predominantly grazers in their uninhabited semi-desert range. They require drinking water every second or third day.

In accordance with their origin in a hot steppe region, domesticated asses are happiest in a warm dry climate; they are less capable of enduring cold and moisture than horses. In their nutritional demands asses are very modest; they will feed on coarse pasture, thistles, thorn bush, and if nothing else is available, on paper and dry horse manure. However, they refuse to drink unclean water. In its surefootedness the ass differs markedly from the horse.

Oestrus occurs in spring. The oestrous cycle lasts 19–24 days, in the Somali wild ass 25–28 days (Lang and Von Lehmann 1972). There is a pro-oestrus period of 2–3 days, followed by full oestrus for 3–7 days, with the most favourable time for fertilization between the fifth and seventh days. The gestation period is 11 months. The foal is suckled for about 6 months but will follow its dam until the next foaling season. At the age of 2 years it nearly reaches adult size but full strength is attained only in the third year. The life-span of hard work can be more than 30 years.

Wild and domesticated asses have the same chromosome number, $2n = 62$ (Hsu and Benirschke 1967).

In 1981 the number of asses in the world was 38.6 million, with no major change since 1969/71. Of the total number, 36.7 million were in the developing countries and 1.9 million in the developed countries; 11.9 million were in Africa, 17.2 million in Asia, 1.3 million in Europe, 0.4 million in the Soviet Union, and 7.8 million in America.

In Africa, Ethiopia with nearly 4 million asses has the largest number, followed by Egypt with 1.7 million and Morocco with 1.5 million. In Asia the largest number, i.e. 7.4 million, is in China. In Pakistan (2.5 million) the number has greatly increased since 1969/71, while in Turkey (1.3 million) there has been a large reduction during this period. At the same time the number of asses in Europe decreased by 28 per cent; here the largest numbers are found in the Balkan and Mediterranean countries. In America, Mexico (3.2 million) holds the record in numbers, Brazil coming second (1.4 million). In Oceania there are only about 5000 asses altogether (FAO 1982).

The number of horses in the world exceeds that of asses by almost 70 per cent. However in Africa there are over three times as many asses as there are horses and in Asia their numbers are equal. In Europe and America, on the other hand, the respective numbers of horses are about four times those of asses.

Wild species

Before their near-extinction wild asses occurred throughout the northeast and north of Africa. The two recent subspecies are the Somali, *E.*

africanus somaliensis, and the Nubian, *E. a. africanus*. The boundary between them runs through northern Ethiopia. Survivors number some 500 in all.

The Somali ass formerly ranged over the stony deserts of Somalia and Ethiopia east of the highlands, extending along the coast of the Red Sea from Danakil and the Webbe Shibeli towards the high plateau of northern Somalia. At the present time it is restricted to a region south of Tendaho in Ethiopia and the Los Anod area of Somalia where it may sporadically be encountered on low-lying rocky hills and among sand dunes.

The Somali is a magnificent beast, standing 120–130 cm at the shoulder. The coat is of a pinky fawn, buffy brown or silver grey ground colour. The belly, limbs and a wide ring surrounding the eye are white. The relatively long hair of the mane is black with grey tips, and the brush of the tail is also black. There are about ten thin transverse black bars on each leg, a faint discontinuous dorsal stripe which is sometimes absent, and a brownish patch at the front of the feet above the hooves. A distinguishing feature of the Somali ass from Somalia is the absence of a shoulder stripe, but in animals from the Danakil country a long thin stripe may occur.

The Nubian wild ass ranged over the mountainous semi-deserts between the Nile, the Red Sea and the highlands of Ethiopia, extending into Egypt where it was still hunted by Tutenkhamun and Rameses III during the period of the New Kingdom (*c.* 1500–1000 B.C.). At the present time it is nearly or quite extinct. Wild asses similar to the Nubian are believed to have occurred in Ahaggar and Tibesti in former times. A few isolated herds have been reported still to exist in northern Chad and the Libyan desert, but these may actually be feral or of mixed wild and feral stocks.

The Bisharin, extending from the Egyptian desert into the northern part of the Republic of the Sudan, would sometimes ride down wild Nubian foals or, when the level of the River Atbara fell, catch young asses that had stuck in the mud. They found it impossible to tame a full-grown one and those they caught young would never be entrusted with the water-skins, as they were inclined to go wild (Murray 1935).

The Nubian ass, standing 115–122 cm at the shoulder, is slightly smaller than the Somali; it has a large head with longer ears and rather narrower hooves. The ground colour of the coat is reddish or brownish grey. The muzzle, underparts of the body and insides of the legs have the whitish desert type of colour. The lips are grey, and there is a light-coloured ring around each eye. The mane, tuft of tail, the usually continuous though narrow dorsal stripe, a patch on each side of the front fetlocks, and the distinct short shoulder cross, are black. There are no barrings on the legs.

In the contact area of the Nubian and Somali asses there existed an intermediate type which has been called *E. a. taeniopus* (van Bemmel 1972). It ranged over a small area in northern Eritrea, extending as far as Suakin in the Red Sea hills. Typically its coat showed a combination of barrings on the legs with a dorsal stripe and long thin shoulder cross, but the latter was often absent (Groves *et al.* 1966).

In earlier times the range of the wild ass extended far to the west. Rock drawings of wild asses are known from the Saharan Atlas, and fossils ascribed to wild asses under the name *E. a. atlanticus* have been unearthed in the Atlas mountains. The animal seems to have survived there into Roman times. A Roman mosaic at Hippo Regius represents two wild asses and a number of Barbary sheep, oryx antelopes and ostriches in a hunting scene. The asses depicted show distinct dark markings on the limbs like the Somali race in addition to long bold shoulder stripes which distinguish them from the Somali ass; one of the asses is represented with two shoulder bars. An ass with a shoulder cross and distinct barrings on the legs is shown in a rock engraving from Enfouss (el Hamra) in the Saharan Atlas.

Wild asses proper do not occur in Asia. Ducos (1970) claimed that an extinct species of ass existed in Palestine and Syria between 10 000–8000 B.C., and that this may have been ancestral to the domesticated animal. However, subsequently he conceded that the bones from Mureybit, which he had referred to *E. a. palestinae*, may have belonged to the Syrian onager (Boessneck and van den Driesch 1978). The geographer and historian Strabo (*c.* 63 B.C.–after A.D. 21) noted the abundance of wild camels and onagers (*hemionoi*) in Arabia, but he did not mention wild asses. Therefore, the great majority of authors on this subject are in agreement that the true ass was restricted to Africa in the Holocene, that it was domesticated there, and was thence introduced in its domesticated form into southwest Asia (Littauer and Crouwel 1979).

In the late Pleistocene the range of the ass extended from North Africa into southern Europe. This species is known as *E. a. hydrun-tinus*. Its remains are particularly frequent in Italy but have also been reported from Germany, Britain, Romania, Hungary, Yugoslavia and the Crimea. It became extinct long before domestication of the ass.

Domestication and early history

Nearly all authors on the domestication of the ass are in agreement that this took place in northeast Africa. The wild race that contributed the major share to the domesticated stock was probably the Nubian to which the domesticated asses on a slate palette from ancient Egypt approximate in their long ears, distinct shoulder bar and unstriped legs. These characteristic markings are general in asses with a light grey coat. But the Nubian was apparently not the sole ancestor of domesticated asses, many of which have barred legs, a feature absent in the Nubian but present in the Somali, the Atlas race, and *E. a. taeniopus*. Any of these may therefore have contributed the leg bars to the domesticated ass.

The shoulder cross and dorsal stripe show considerable variability in the wild races. Thus, in the Nubian the cross is short, in *E. a. taeniopus* either absent or long and thin; the typical Somali ass is devoid of a shoulder bar, but in the Somali race of the Danakil area a long, very thin shoulder-bar is fairly frequent. Hybrids derived from wild Nubian and Somali asses display both the shoulder cross and barred legs. The wild Atlas ass had a long, strongly marked shoulder cross, and some individuals had two stripes on the shoulder. The dorsal line is narrow and usually continuous in the Nubian, but faint and either discontinuous or absent in the Somali. It is probable that during the six millennia or more that have passed since the ass was first domesticated the genetic composition of the domesticated breeds has in various degrees been affected by the different wild races.

There are many Egyptian records of domesticated asses dating from the beginning of the 4th millennium B.C. Skeletal remains of asses from this early period have been found in Egypt in association with those of other domesticated animals. A rock engraving in the western desert of Egypt shows an ass with a burden on its back attached by girths. Asses captured either from Libyans or from inhabitants of the delta are depicted on a predynastic slate palette (before 3400 B.C.). On another palette from Giza domesticated asses are shown in a line between rows of domesticated sheep and cattle. From the middle of the 3rd millennium, which is the earliest date at which tombs were decorated in Egypt, the ass is frequently represented in wall paintings.

Asses were kept in large droves in Egypt. In an inscription in a tomb from the middle of the 3rd millennium a scribe has recorded 760 asses among his master's herds of over 5000 animals. In other tombs from the same period inscriptions mention droves of over a thousand asses belonging to individual owners. The trading caravans which came to Nubia and Egypt from such remote places in the south which are now Darfur and Kordofan, and from the country of the Blue Nile, used asses as beasts of burden. Towards the end of the 3rd millennium the caravan master of King Meren-Re started on the return journey from his third expedition to Nubia with 300 asses laden with various products of the region. An expedition to Nubia was accompanied by a hundred asses loaded with presents for the local chiefs.

Asses were employed in ancient Egypt as beasts of burden, for treading the seed into the soil, and for threshing. In the early days the ass was not ridden in the manner with which later generations became acquainted; a seat, somewhat similar to a litter, was fastened on the backs of two animals. Riding on the back of a single ass seems to have become common in the course of the 2nd millennium B.C. In the temple of Dêr el-Bahari a limestone relief shows the saddled riding ass of the Queen of Punt (about 1500 B.C.), and in the Sudan a wall relief from the temple of Amen-Re at Sanam, dated to the 7th century B.C., shows Nubian warriors on asses.

Early in the 3rd millennium, or possibly already towards the end of the 4th, domesticated asses appear in southwest Asia. Before the introduction of the camel they were used in Sinai and the Negev of Palestine in the caravan trade (Albright 1961). In the Plain of Esdraelon the figurine of an ass carrying baskets on its back was found in a chalcolithic tomb (Kaplan 1969). Just after 2000 B.C. Assyrian merchants, established round the court of the Hittite prince of Kanes in the Halys basin of central Asia Minor, used caravans of pack asses which regularly crossed the Syrian steppes and Taurus mountains on their way to and from Mesopotamia. This trade is believed to go back to 3000 B.C. (Childe 1942). In India and Pakistan bones of domesticated asses have been excavated at Rangpur, Gujarat (2000–800 B.C.), Rupar, east Punjab (2000–900 B.C.), Nagarjuna-konda, Andhra Pradesh (2nd century A.D.–1200), Taxilla, Pakistan (mid-1st century B.C.–A.D. 1200), Najjain, Madhya Pradesh (750 B.C.–A.D. 1400), and Jaugada, Orissa (400 B.C.–A.D. 200) (Nath 1973).

In Europe the domesticated ass occurs later than in Africa and Asia. A pottery figurine of an ass from the Messara vaulted tombs in Crete dates from the prepalatial period. Being ill-adapted to a wet cold climate, the ass has never been numerous in the northern regions.

Recent history

Recent domesticated asses are for the most part of a generalized type, medium to small in size and grey or dark brown in colour of coat, combined with a light grey to white underline and inside of the legs. The type may vary in size and colour in different regions but few well-defined breeds have been evolved. With few exceptions, asses are beasts of burden and riding animals. More rarely they are employed for work in agriculture, especially ploughing, or for milk production; but in the past the milking of asses used to be fairly widespread. In Asia and Africa the ass is always the leading animal of camel caravans. In a few countries the meat of the ass is eaten.

Africa The importance of the ass to the peoples of Africa greatly exceeds that of the horse. The area of its distribution comprises the entire northern coast belt and the fringe of the Sahara, Egypt and the regions to the south along the east coast. Commonly the ass is found among nomadic peoples, less frequently among the settled population of West and Central Africa. In South Africa neither the Hottentots nor the Bantu peoples were in possession of asses at the time of the European occupation of their territories.

In Egypt the ass is very common and the prin-

cipal riding animal and beast of burden. The ordinary bedouin ass is a small, nimble, mouse-coloured, half-starved animal. It is of general utility and can carry about 50 kg, being particularly useful in getting down full water-skins from high rock pools beyond the ungainly camel's reach. Muscat asses were formerly widespread in Egypt but are now less frequent. The type, named after the capital of Oman in the easternmost part of Arabia where it is believed to have been developed, is of large size, usually exceeding 120 cm in height. The coat is white, or rather silver grey, turning white with increasing age. The skin, similar to that of white Arab horses, is always black. In general, Muscat asses are well-bred, high-spirited, strong and fast and are usually better treated and fed than the common ass, but they lack the endurance and hardiness of the latter.

In Upper Egypt Muscat riding asses, called Saidi or Hassawi after the village of Hassawiya, are still bred in fair numbers. The Hassawi, standing 135–140 cm at the shoulder, is ridden like a horse with saddle and bridle.

Asses are widespread in the dry northern part of the Sudan but infrequent south of latitude 12 °N where the soil becomes too muddy in the rainy season. The majority belong to the common pack type. This is a small slaty grey animal, between 90 and 100 cm high and often weedy and misshapen. The Bisharin employ their asses only for carrying water. Occasionally they tied up their female asses in a revine which a wild stallion was known to frequent, but the offspring of such a mating were considered to be untrustworthy, and it was only the grandchildren of the wild ass that settled down to domestication (Murray 1935). A small percentage of Sudan asses are of the riding type which has been derived from the pack type by selective breeding. This is a larger ass than the pack animal, standing 100–110 cm at the withers, and often dark brown rather than grey in colour. Muscat asses have also been introduced into the Sudan and from there into the northern part of Tanzania; they are used as riding animals and for crossbreeding with the ordinary local asses. In Zanzibar white Muscat asses until recently served as processional mounts for the family of the Sultan (Zeuner 1963).

In Ethiopia asses of large size, known as Sennar, are derived from the Sudanese riding type. Like the Bisharin of the Sudan, the Beni

Amer of Ethiopia were reported to improve their asses by crossing them with the wild asses of their country. In Eritrea three different varieties are distinguished: Abyssinian, Etbai and Kassala. The Abyssinian stands about 97 cm at the shoulder, the Etbai 100–102 cm and the Kassala 112 cm. The asses of Somalia are usually small and weak. They are not ridden, their principal duty being that of carrying water-skins from the wells. The Somali have never attempted to cross the domesticated with the wild ass of their country, which is quite half as big again as a fine specimen of the domesticated Somali ass.

South of Ethiopia the ass is absent only among a very few East African peoples. It is used for carrying grain to the market and dung to the field. The western Masai even milk their asses. Masai asses have leg stripes. They are bred far to the south and east towards the Zanzibar coast. A few of the Eastern Bantu eat the flesh of the ass, and in one tribe asses are specially fattened for this purpose.

In South Africa the ass was imported by the Dutch East India Company from Persia in 1689 and subsequently from Spain.

West of Egypt, asses are bred throughout the northern part of Africa but not in the coast belt of the Atlantic. In Libya a large type of ass is bred near the coast and a small type throughout the country. In Tunisia, Algeria and Morocco asses occur in large numbers, especially in the south towards the desert.

From the Atlas countries the ass has spread along the fringe of the Sahara and the caravan routes to the south. In West Africa it is the companion of the poorer classes of the population but despised by the wealthy. In the Sahel and Niger zone the range of the ass coincides with the major breeding grounds of the horse in this area. White asses of Muscat type are rare, a few being encountered in the Sahel along the course of the Niger. Farther south asses cannot exist owing to trypanosomiasis, except during the dry season when they are used for transport. With the onset of the rains the high mortality forces their owners to take them back to the north. In the northern parts of Zaire asses take the place of horses.

Asia Asses are common throughout the southern and central parts of Asia. In Arabia they are widespread, being used for pack work and riding. White asses of Muscat type are mainly bred in eastern Nejd and Hasa where in the past the wealthier people used to ride them. But Burton (1855–56) reported that the bedouin of Arabia as well as the wealther classes of Arabs in Syria and Jordan were loth to ride the ass.

From Arabia Muscat asses have been introduced into Palestine, Syria and Iraq. Though the Muscat ass is slower than a horse or camel, it will maintain an easy trot and canter for hours without flagging, and always gains on the horse uphill or on broken ground. In Syria, Darwin (1868) noted, there were four breeds of asses: 1. a light and graceful animal with an agreeable gait, used by ladies; 2. an Arab breed reserved exclusively for the saddle; 3. a stouter animal used for ploughing and various purposes; and 4. the Damascus breed with a long body and long ears. The large Syrian asses with a height of 130–140 cm were carefully selected. They received almost as much attention as horses; they were groomed and well fed, always obtaining their share of barley with the equine companions. Asses of this type have been introduced into Cyprus where they are used for mule breeding.

In Turkey, Iraq and Iran the ass is the common pack and riding animal, except in the marshy regions near the Caspian where it is absent. For work in agriculture it is used only occasionally. It occurs in various sizes and is grey or brown in colour. Striped legs are fairly frequent. The Muscat ass is also found in these three countries. In Iran, where it is known as Hamadan or Kashan, its surefootedness makes it preferable to the saddle horse in mountain regions. Iran has been the source of the Muscat asses of Turkmeniya, the Soviet Union, Afghanistan, Pakistan and India. In Turkmeniya it is bred in two local varieties: Mary or Merv and Hamadan.

In the economy of central Asia the ass is generally of little importance, but for the nomadic tribes it is not less important than the camel. It is particularly numerous in the oases and towns where it is mainly employed for pack work over short distances. These asses are usually of small size with a large head, frequently sand coloured with a black dorsal stripe. Under a large load of wood, reeds or bags, the body of the animal is often invisible, and only the head emerges from the load. In many parts of central Asia the riding of asses is considered to be undignified for men, but among women and children it is common.

There are several types of asses in the Indian

subcontinent. The common type is of medium to fairly large size and is often used for ploughing, sometimes in a span with a bullock, buffalo or camel. The grey Mahratta asses of western India and the pigmy donkeys of Sri Lanka are of diminutive size, standing not more than 50–75 cm at the shoulder and, as Darwin (1868)remarked, not being much larger than a Newfoundland dog. In Burma, Thailand, Indo-China and Indonesia the ass is rare, and in Japan it hardly occurs at all.

In Nepal asses are bred in relatively small numbers. They are found mainly in the hilly midlands; few are kept in the tropical Tarai, and very few in the mountains except in the Mustang area where they are more numerous. The majority of Nepalese asses are derived from Tibet. In both countries they are kept in large droves of fifty and more, and are used for pack work, being capable of carrying a load of about 50 kg. The height at withers of Tibetan asses is approximately 80 cm and rarely exceeds 90 cm. In the Mustang district of Nepal, in the heart of the Himalayas, and in the colder parts of Tibet, the asses tend to develop a thick woolly coat. Their predominant colour is dark brown, more rarely dark grey (Epstein 1977).

Asses are bred in the Mongolian People's Republic and in Inner Mongolia, Xinjiang and adjacent areas of China; there are also a few in southern China, especially in the southeastern provinces. The main breeding areas are situated in northeastern and central China. In some provinces with mixed pastoral and farming zones the ass is an important draught animal in agriculture, and in Mongolia it is kept in large droves; some of these consist only of she-asses which are milked.

Chinese asses vary between 80 and 145 cm in height at withers, and may be divided into two main classes according to size: large and small. Between these principal types there are numerous intermediates. The large type of Chinese ass which is bred chiefly in Shaanxi, Henan and Shandong provinces is represented by the Kwanchung and Siumi breeds. The Kwanchung is an excellent draught animal and is widely used in agriculture. Adult males weigh over 350 kg and females over 300 kg. The Siumi is lighter than the Kwanchung, adult males weighing 230 kg and females 210 kg. The Siumi is used mainly for pack work in the mountains, being capable of carrying a load of up to 80 kg. While the Kwanchung ass has a good disposition, the Siumi is exceedingly stubborn.

Asses of medium size, standing 100–115 cm at the withers, are frequently employed in the northern and northeastern regions of China in flour mills and for short-distance transport. They are capable of carrying loads of 60–80 kg a distance of 30–40 km in a day. The small type of Chinese ass, standing 80–100 cm at the withers, is used as beast of burden, saddle animal or for farm work.

Europe Asia Minor seems to have been the starting point for the introduction of the ass into southern Russia, the Ukraine and the Balkan peninsula. Evidence of the animal's presence in the Ukraine dates from the 9th–8th centuries B.C., while for Greece the ass is attested for the last centuries B.C. and for Crete as early as the 3rd millennium B.C. Asses on Attic vases are depicted with leg stripes. In the Greek colonies on the Black Sea bones of domesticated asses have been found in large numbers. From the Balkans the ass spread into Italy, but the Iberian peninsula seems to have been reached by the animal direct from Africa via Gibraltar. Its extension across Western Europe as far as Britain and Gaul followed in the wake of the Roman conquest. In southern Europe the ass was widely used for vineyard cultivation for which it is eminently suitable. In eastern Europe north of the Balkans it did not become common until the Middle Ages, its advance towards the Baltic being attributed to Jewish pedlars who used the animal to convey their goods (Dent 1972). In central Europe and to the northern limits of its occurrence it was employed mainly by millers, gardeners and smallholders. Millers used it for carrying sacks and turning treadmills and the millstone, the ass having to walk blindfolded in a circle. In mediaeval castles it was also employed for turning the wheel that brought water up from the deep wells.

The flesh of the ass was not generally eaten except by the poorest people. In Athens there was a separate market for its sale. In Hungary it was consumed until fairly recent times; the meat from fattened asses was salted and smoked (Bökönyi 1974). In addition to beef and pork it is said to have been an important component of the famous Hungarian salami.

The milk of asses was highly valued. Resembling human milk in composition, it was fed to babies in several parts of Europe as late as the early years of the present century.

In southern Europe the León-Zamora of Spain and the Poitou of France form a group of very large asses with an average weight of 350 kg in adult males and 250–300 kg in females. The height at withers averages 147 cm, but jack asses may reach as much as 155 cm. In general they are strong in build and lively in spirit. The coat is long, coarse and rough, usually black or greyish black with pale eyes, muzzle and underparts of the body. There are still approximately 1500 León-Zamora she-asses on Spanish peasant farms, but selection and marketing of the stock are not organized. A León-Zamora studbook has existed since 1941. A studbook for the Poitou breed was formed in 1885. Only forty-four Poitou asses remained in 1977 but there is now a programme for their conservation.

The second group of this large-sized type includes the Catalan which extended from Spain into Gascony on the French side of the Pyrenees. Male asses of this group reached a height of 145–160 cm and females 135–150 cm. They are generally long and narrow, and have a quiet temperament. The coat is always dark tending to black, more rarely to dark chestnut. The eyes, muzzle and underparts of the body are reddish brown to grey (Aparicio 1961). A herd of twenty to twenty-five Catalan asses is maintained by the Spanish Ministry of Agriculture for purebreeding and semen collection, and there are about 300 females in the whole of Spain. Although a considerable demand for breeding stock exists in South America, the Catalan is in danger of extinction along with the other breeds of large Spanish asses the number of which decreased by 50 per cent during the last decade (Sanchez Belda, personal communication). A studbook for the Catalan ass has existed since 1880.

The Andalusian of southern Spain, which also belongs to this group of long-bodied asses, has a grey coat. With a population of 2500 registered she-asses it is the most numerous of the large Spanish breeds.

In Italy the large type of ass is represented by the Martina Franca or Apulian and the Sicilian or Ragusan which is descended from Catalan, Martina Franca, Pantelleria and native stocks. These varieties are dark grey or brown to nearly black with light underparts. The Martina Franca has had a studbook since 1943. The Ragusan was reduced to about forty breeding females in 1980.

Outside these high-bred asses which are almost extinct, the asses of Europe are for the most part of a similar type to those of Asia and Africa – small, hard-working and modest in their nutritional requirements. Their colour is either grey or dark brown. They are still extensively used in the Mediterranean countries, in particular in the Iberian peninsula, Italy, Albania and Greece where they are mainly employed for packwork; in Greece they also serve for draught and sometimes for ploughing, alone or along with a bullock. In the central and northern parts of Europe few asses are encountered; they are commonly small and degenerate. Only in Ireland the animal is still widely used by smallholders.

America The ass was introduced into America by Spaniards in the 16th century. In North America the number of asses increases from north to south, while horses show the opposite trend. In the southern states of the U.S.A. such as Tennessee, Kentucky and Missouri, the ass was used chiefly for mule breeding for which the large American or Mammoth breed has been evolved from imported Andalusian, Catalan, Majorcan, Maltese and Poitou stocks. Jackasses of this breed reach a height of 160 cm. A breed society for the American ass was formed in 1888 and a studbook in 1891, but the American ass is now nearly extinct. Mexico has the largest number of asses in America; they are usually of very small size as a result of neglect.

In South America the number of asses increased between 1969/71 and 1981 by about 0.4 million, the largest ass population being found in Brazil. North of 25° S, at the foot of the Andes, the ass is frequently used for work; south of this line it is rare save for the northwestern provinces of the Argentine, Catamarca and La Rioja, where it still occurs sporadically.

Feral asses Asses thought to be feral roam the Ahaggar (Hoggar) high plateau in the central Sahara. They are probably descendants of domesticated animals once bred in this area. Occasionally they are hunted by Tuareg for sport; their flesh is not eaten but is used, if at all, as dog feed. The feral asses on Socotra may be descended from domesticated stock introduced by the ancient Egyptians.

In Asia droves of feral asses roam the desert

regions of Arabia where they encroach on the pasture grounds of the camel herds and have become a veritable pest.

In America the decline in the use of asses for transport and other work has led to the abandonment of many of these, so that there are now droves of feral asses in the southern and western states of the U.S.A., especially in the higher mountain ranges where cattle and sheep cannot be kept economically. Feral asses occur also in the Galapagos Islands.

In Australia there are large populations of feral asses in the drier parts of the country.

Mule and hinny

Asses and horses have been crossed since the time when the ass coming from the south and the horse from the north first met in Mesopotamia, northern Syria and Asia Minor. Hybrids of jack-asses and mares are called mules and those of stallions and jennies are hinnies.

In northern Mesopotamia, northern Syria and Asia Minor mules were bred in ancient times on a large scale. From there they followed the ass to Greece and up the Balkan peninsula and by sea to Sicily, southern Italy and the south of France. In Homeric Greece they were bred and widely used for draught and farm work. At Olympia there were harness races for mules in the 5th century B.C. In the 18th century A.D. mule breeding was a flourishing industry in Italy, Spain and France; at Poitou some 50 000 mules were bred every year. In Britain the chief demand for mules was for army service in India. The breeding of hinnies was a speciality in Ireland. In the United States the number of mules doubled during the decade 1850–60; more than 150 000 mules were foaled in 1889 (Savory 1970).

Mules and hinnies both exhibit a high degree of heterosis, adaptability to unfavourable conditions, heat tolerance, disease resistance, sure-footedness and longevity. The mule stands generally closer to the ass in the conformation of the head, length of ears, shape of muzzle, thin mane, short-haired tail and shape of hooves, but it lacks the light-coloured belly of the ass. The hinny more closely resembles the horse in the conformation of the head, muzzle, relatively short ears and prominent orbits. But in many instances distinction between mules and hinnies

is difficult.

Jackasses have to be trained for mating with mares, and stallions for jennies. They are first put to a female of their own kind at the height of oestrus, which is then replaced by one of the other subgenus. Afterwards they are not allowed to serve a female of their own kind. Fertility is generally lower than in purebreeding of horses and asses. Thus, Epstein (1969) reported that in China pregnancy in hinny breeding is obtained in only 50 per cent and in mule breeding 60–70 per cent of matings, while in purebreeding of horses and asses fertility is 90 per cent. The male hybrids are always, and the females nearly always, sterile. In rare instances female mules have foals either from jackasses or stallions but the male second-generation hybrids remain sterile.

In 1981 there were 14.2 million mules and hinnies in the world. Of these 13.6 million were in the developing countries and only 0.6 million in the developed countries; there were 2.2 million in Africa, 4.8 million in Asia, 0.6 million in Europe, and 6.6 million in America. Ethiopia has the largest number of mules and hinnies in Africa, China in Asia, and Mexico and Brazil in America (FAO 1982). In the developed countries the number of mules decreased by one-half between 1969/71 and 1981, whereas in the developing countries it increased by 23 per cent during this period.

Heavy mules are used for draught in town and country, and light mules for draught, pack work and riding, especially in mountain regions. Medium-sized hinnies are employed for different purposes in mountain agriculture, and light hinnies usually for pack work.

Heavy mules are bred from large jackasses with a height of not less than 155 cm, which are selected for strong forequarters, muscular neck, wide chest, large heart girth and strong metatarsal and metacarpal bones. Such asses are found mainly in France, Spain, China and the U.S.A. The mares for the production of heavy mules weigh between 500 and 600 kg. In Poitou in France a special horse breed, the Poitou or Mulassière, was developed for mule-breeding. This horse had a withers height of 165 cm and a rather coarse conformation. One-fourth to one-third of the mares were bred to stallions, while the larger number were used for mule breeding.

For lighter mules large size is also preferred in the male ass, but the mares may be of any type or breed – draught, riding, trotting or pony.

For the production of hinnies of medium weight, stallions of a light to medium type with a weight of 400–500 kg and a withers height of 150–155 cm are used. The stallion should not be much larger than the jenny and be of a quiet disposition but sexually active. In order to facilitate parturition of the ass the stallion's chest and croup must not be too broad (Aparicio 1961). In Europe, Catalan, León-Zamora and Andalusian, in China the Kwanchung and Siumi, and in the United States the Mammoth she-asses are particularly suitable for the production of medium-weight hinnies.

For the production of light hinnies stallions weighing 350–400 kg and strong she-asses are recommended, but in China Mongolian ponies of a lighter weight and small jennies have to serve the purpose, and in Ethiopia and Central and South America hinnies are bred from ponies and asses of any size.

Future prospects for ass and mule breeding

Taking a long view, the breeding of asses and mules is bound to decline with the growing need for food production and progressive mechanization of agriculture throughout the world. In Europe and North America this trend has been fairly rapid in the recent past, and will doubtless continue in variable degrees in accordance with different agricultural and local social conditions. In developing countries with relatively primitive conditions of transport and agriculture, such as are still common in large parts of Africa, Asia and Latin America, the decrease in the number of asses and mules will be slower and may under certain conditions and for a certain time even revert in several countries, as has recently been the case for asses in Pakistan, Algeria, Brazil and Colombia, and for mules in Pakistan, Mexico and Colombia. For work on small peasant farms the ass can still maintain its hold because it is hardy and quiet, cheap to buy and modest in its nutritional demands, intelligent, patient and easy to train for light draught and pack work. The hardy, strong and surefooted mule, again, has no rival for riding and load carrying on broken ground and narrow mountain paths.

Onager

The onager belongs to the family Equidae, genus *Equus*, subgenus *Hemionus*. The name 'onager' is derived from the classical Greek *onagros*, the combination of *onos* and *agrios*, meaning wild ass, while the term *hemionus* (*hemionos*) means half-ass.

The range of the onager extends through the dry steppe belt of Asia from Syria in the west to Mongolia in the east. In earlier times it also included western Siberia, the European Russia and the southern steppes of eastern Europe.

The hemiones have been placed by Groves and Mazák (1969) in two species, *E. hemionus* and *E. kiang*, which differ in colour, rump conformation and other features. *Equus hemionus* has six geographical races. In the west the Syrian onager or akhdari (*E. h. hemippus*), now probably extinct, is the smallest race. Eastwards the Iranian onager and transcaspian kulan, and the ghor-khar of Afghanistan, Pakistan and India, called khur in the Rann of Kutch, grow larger, while the typical dziggetai of northern Mongolia and the border regions of the Soviet Union and Xinjiang and the desert dziggetai of the Gobi are the largest races. The kiang of Tibet and Ladak comprises three geographical races, a small southern, larger western, and the largest eastern.

The diploid chromosome number of the Persian onager (*E. h. onager*) is $2n=56$, that of the Turkmenian onager or kulan (*E. h. kulan*) is polymorphic (Ryder *et al.* 1978), while those of the dziggetai and kiang are still unknown (Benirschke 1969).

The earliest record of the onager in a domestic context is represented by two wall paintings of tamed onagers, each held by a man, in a temple at Çatal Hüyük V in southern Anatolia, dated to the 6th millennium B.C. At Qalat Jarmo, near Kirkuk in Iraq, onager bones and those of sheep, goats, cattle, pigs and dogs, nearly all of yearling stock, have been found in strata dated to the 5th millennium B.C. Similar bones from the same period have been excavated at the Caspian foreshore in northwest Iran. It is possible, though unproved, that the onager was domesticated in this region at this early period. Whether its meat was eaten or its strength exploited is unknown.

Onager bones from Tell Aswad in the Balikh valley, Tell Chagar Bazar in Mesopotamia,

Megiddo and Beersheba in Palestine and from the northern kurgan of Sialk II in Iran, where they occur together with those of other domesticated animals, are dated to the second half of the 4th millennium B.C. Onager bones from Tell Mefresh in the Balikh valley are dated slightly later in the 4th millennium. On a clay tablet from Susa, Elam, on the eastern mountain fringe of Iraq dated to 3250–3000 B.C., four onagers are represented in a row in association with several men. On another clay tablet from Elam (3000–2800 B.C.) there is an engraving of nineteen onager heads and necks with different facial profiles and manes.

At Tell Halaf remnants of a chariot model and of a painting of a chariot and charioteer on a cup, dated to 3000–2800 B.C., are the first indications of the use of onager-drawn waggons in this region. A potsherd from the same site also shows onagers. At Tell Asmar in Sumer onager bones from the period 2800–2700 B.C. comprise 9 per cent of the total animal bones, the majority of which belong to other domesticated animals. Again, the equine bones recovered from the royal graves at Kish, dated to 2600 B.C., are from onagers. Many works of art from the early dynastic period of Mesopotamia (2700–2350 B.C.) confirm the use of onagers for draught. At a later period (*c.* 2060 B.C.) an inscription on a Sumerian seal cylinder describes the chariot of a city god, drawn by four onagers (*ansu*) from a famous stud. A royal stable list at Ur (2050–1940 B.C.) refers to seventy-eight teams of four onagers each.

The occurrence of onagers among the domesticated fauna was not restricted to early Mesopotamia. Bones or representations of the onager have also been found in Iran, Pakistan and India (Epstein 1971).

From these records it appears that the akhdari was domesticated at a very early time, probably during the 6th millennium B.C., and was employed over a large area in southwest Asia as a draught animal. With the widening of the sphere of the onager's use the Iranian and transcaspian races were probably incorporated with the domesticated akhdari stock. The Sumerians seem to have imported also the larger and stronger Baluchi ghor-khar, which they called 'mountain onager', from Pars.

Several authors have suggested that the onagers employed for draught in western Asia before the introduction of the domesticated ass

and horse were not bred but tamed animals. This would imply that throughout the wide area of their use, extending from Sumer and Elam to Iran and Pakistan, Turkestan, Uzbekistan, the Ukraine and Anatolia, onagers were for more than 3000 years continuously captured, the male foals reared and the adult stallions broken-in. It would further imply that the captured mares and fillies were always and everywhere killed off before they had progeny in captivity. The assumption of such a general, widespread and long-continued waste in different areas, at different times and by different peoples who were successful breeders of sheep, goats, cattle, pigs and dogs seems most unlikely. It is also contradicted by the occurrence in several sites, e.g. Tell Asmar and Qalat Jarmo in Iraq, Belt Cave and Sialk in Iran, and Shah Tepé in Turkestan, of onager bones along with those of other domesticated stock.

From the beginning of the 2nd millennium on, the domesticated onager was replaced by the horse, but wild onagers continued to be hunted, as shown in an Assyrian relief from the palace of Ashurbanipal (668–626 B.C.). More recently, in the Bikaner desert, mounted Baluchis used to ride down ghor-khar foals during the foaling season; about two-thirds of the young animals died prematurely in captivity, but the survivors were easily tamed.

In 480 B.C., as recorded by Herodotus, the Indians in Xerxes' army still fought from war chariots some of which were drawn by onagers, and over 400 years later Strabo (*c.* 63 – after A.D. 21) mentioned the use of onager-drawn chariots in Iran. This accords with the presence of the domesticated onager at Kaunchi-Tepe, Uzbekistan, from the 2nd century B.C. to the 1st century A.D.(Gromova 1949). Pliny (A.D. 23–79) wrote that the onager was easily tamed and that the Romans occasionally produced hybrids from male onagers and female horses, while Columella (*fl.* 1st century A.D.) related that male onagers were crossed with female asses.

References

Albright, W. F. (1961) Abram the Hebrew. *Bulletin of the American School of Oriental Research*, **163**: 36–54.
Aparicio, G. (1961) Eselrassen und – Kreuzungen. In:

Rassenkunde, Handbuch der Tierzüchtung, ed. F. Haring, **3** (1): 199–206. Paul Parey: Hamburg and Berlin

Benirschke, K. (1969) Zoos and the pathologist – a two-way street or cytogenetics on zoo animals. *Acta Zoologica et Pathologica Antverpiensia*, **48**: 29–42

Boessneck, J. and Van Den Driesch, A. (1878) Preliminary analysis of the animal bones from Tell Hesban. *Andrews University Seminary Studies,* **16** (1): 259–87

Bökönyi, S. (1974) *History of Domestic Mammals in Central and Eastern Europe*. Akadémiai Kiadó: Budapest

Burton, R. F. (1855–56) *Personal Narrative of a Pilgrimage to El-Medinah and to Meccah*. Longmans: London

Childe, V. G. (1942) *What Happened in History*. Pelican: Harmondsworth, England

Darwin, C. (1868) *The Variation of Animals and Plants under Domestication*. Murray: London

Dent, A. (1972) *Donkey: the story of the ass from east to west*. Harrap: London

Ducos, F. (1970) The Oriental Institute excavations at Mureybit, Syria: Preliminary report on the 1965 campaign. Part 4. Les restes d'Equidés. *Journal of Near Eastern Studies*, **29**: 273–89

Epstein, H. (1969) *Domestic Animals of China*. Commonwealth Agricultural Bureaux: Farnham Royal, Bucks, England

Epstein, H. (1971) *The Origin of the Domestic Animals of Africa*. Africana Publishing Corporation: New York. Edition Leipzig; Leipzig

Epstein, H. (1977) *Domestic Animals of Nepal*. Holmes and Meier: New York and London

FAO (1982) *FAO Production Yearbook 1981*, Vol. 35. FAO: Rome

Gromova, V. (1949) [The history of horses (genus *Equus*) in the Old World]. *Akademiya Nauk S.S.S.R. Institut Palaeontologii Trudy*, **17** (1): 3–234; (2): 139–61. (In Russian)

Groves, C. P. and Mazák, V. (1969) Some taxonomic problems of Asiatic wild asses; with the description of a new subspecies (Perissodactyla; Equidae). *Zeitschrift für Säugetierkunde*, **32** (6): 321–55

Groves, C. P., Ziccardi, F. and Toschi, A. (1966) Sull' asino selvatico africano. Supplement to *Ricerche di Zoologia applicata alla Caccia*, **5** (1): 1–11

Hsu, T. C. and Benirschke, K. (1967) *An Atlas of Mammalian Chromosomes*, Vol. 1. Springer-Verlag: Berlin

Kaplan, J. (1969) 'Ein el Jarba. Chalcolithic remains in the Plain of Esdraelon. *Bulletin of the American School of Oriental Research*, **194**: 2–39

Lang, E. M. and Von Lehmann, E. (1972) Wildesel in Vergangenheit und Gegenwart. *Zoologischer Garten*, N. F. **41**: 157–67

Littauer, M. A. and Crouwel, J. H. (1979) *Wheeled Vehicles and Ridden Animals in the Ancient Near East*. Brill: Leiden and Köln

Murray, G. W. (1935) *Sons of Ishmael*. George Routledge: London

Nath, B. (1973) Prehistoric fauna excavated from various sites of India with special reference to domestication. In: *Domestikationsforschung und Geschichte der Haustiere*, ed. J. Matolcsi, pp. 213–22

Ryder, O. A., Epel, N. C. and Benirschke, K. (1978) Chromosome banding studies of the Equidae. *Cytogenetics and Cell Genetics*, **20**: 323–50

Savory, T. H. (1970) The Mule. *Scientific American*, **223** (6): 102–109

Van Bemmel, A. C. V. (1972) Some remarks on the African wild ass. *Zoologische Mededelingen*, **47** (21): 261–72

Zeuner, F. E. (1963) *A History of Domesticated Animals*. Hutchinson: London

20

Asian elephant

R. C. D. Olivier

Formerly IUCN–WWF Asian Elephant
Coordinating Centre,
Sri Lanka

Introduction

Nomenclature

Each of the two genera of living elephants contains only one surviving species: *Loxodonta africana*, the African elephant, and *Elephas maximus*, the Asian elephant. The name 'Indian elephant' is too restrictive and the alternative 'Asian' is preferable to 'Asiatic'.

On the unarguable basis of long-term total geographical isolation contemporary elephantologists generally accept three living subspecies. These are *E. m. maximus* in Sri Lanka, *E. m. indicus* on the Asian mainland, and *E. m. sumatranus* on Sumatra. There is dispute as to the length of time the elephants of Borneo have existed there in geographical isolation, but these potentially represent a fourth subspecies *E. m. borneensis*.

General biology

The two surviving species of elephant are readily distinguishable from each other not only with regard to their geographical distribution and detailed anatomy, but also in their general appearance (Carrington 1962: 19–28). The reader is further referred to Sikes (1971) for an excellent description of the general biology of both species. Dittrich (1967) summarizes most of the non-behavioural reproductive data accumulated on the Asian elephant during the last 70 years. Eisenberg *et al*. (1971) believe that sexual maturity, i.e. the ability to produce fertile gametes, may be as young as 6–7 years in females depending on nutritional status and from 7 years on in males, but young elephants, particularly males, may attain sexual maturity some years before they are permitted to mate. Sexual activity continues in both sexes to extreme old age.

External manifestation of oestrus by a female is almost absent. Oestrus has a periodicity of approximately 3 weeks (range 18–26 days), and lasts 4 days (Jainudeen *et al*. 1971, Chappel and Schmidt 1979). However, the results of long-term research at Portland Zoo suggest that Asian elephants come into oestrus only about four times a year in a regular, if unique pattern (M. J. Schmidt, personal communication).

The elephant is polyoestrous, apparently monovular but possibly polyovular as well, and undergoes a number of sterile oestrous cycles before eventually conceiving. Gestation lengths cited in the literature vary from as short as 17 months to as long as 25 months (see Dittrich 1967 for review). The majority of recorded gestations, however, approximate 22 months and Eisenberg *et al*. (1971) concluded that this is the true gestation period and that all reported gestations shorter than 20 months are probably the result of observing mating behaviour during an oestrus period following conception.

Birth, usually of one calf (although twins and even triplets have been recorded), is followed by a period of anoestrus, which is terminated by the series of oestrous cycles that may precede the next pregnancy. The duration of lactational anoestrus is probably variable, there being evidence that it is curtailed if mother and calf are separated soon after birth. This variability in turn gives a variable calving interval of anywhere between 2 and 4 years (Dittrich 1967). The long period of postnatal care accorded to the young, together with the well-developed social life, are consistent with the ethological evidence that elephants show a high degree of intelligence, learned behaviour and adaptability.

Studies by Eisenberg *et al*. (1971), Jainudeen

et al. (1972 a and b), and Scheurmann and Jain-udeen (1972) suggest that male Asian elephants exhibit the vestige of a truly seasonal rut which existed in the early evolutionary history of *Elephas*.

This is known as 'musth', and in the domesticated bull elephant is characterized by aggressive behaviour and temporal gland secretion. In Sri Lanka it was found that musth never appears in males under 10 years of age and appears sporadically in young males from approximately 14 to 20 years of age.

Almost all males over 30 years old show an annual musth cycle. Its duration varies between 2 weeks and 5 months, with an average of 2–3 months. Not all males in a population are in musth simultaneously; instead musth periods are distributed throughout the year although there may be some correlation between peaks of musth and maximum rainfall periods.

Musth is associated with, and possibly caused by, a marked increase in testicular testosterone secretion and there is some indication that it could also be induced by sexual activity. Apparently the aggressive behaviour is androgen-dependent. Factors resulting in a loss of condition suppress musth.

The period of musth appears to aid the male in overcoming the aggressiveness of older cows while at the same time a musth bull can generally dominate other, non-musth bulls. Although a bull can breed whether in or out of musth, the condition of musth is clearly related to the male's ability to achieve a high dominance status on an annual basis. It does not necessarily follow that a male is unable to breed when not in musth. He can indeed breed a female; however, his potency may be somewhat reduced and in a field situation perhaps reproductive success of the male is enhanced during the musth period.

From the above it seems likely that there is complex interaction between non-musth males, musth males, non-oestrous females; and oestrous females. Sike's (1971: 165) observation that the only zoos where African elephant calves had recently been conceived, born, and successfully reared in captivity were those where two rival bulls had simultaneous access to the cows during courtship, may therefore be relevant.

Distribution and numbers in the wild

The past and present distribution of the wild Asian elephant is shown in Fig. 20.1. The species once ranged from the Tigris and Euphrates (45 °E) in the west, east through Asia south of the Himalaya, and north into China at least as far as the Chang Jiang (Yangtze Kiang) (30 °N) and probably further. It may also have existed on Java. Wild elephants are still found in Sri Lanka, India (including the Andaman Islands), Nepal, Bhutan, Bangladesh, Burma, China, Vietnam, Laos, Kampuchea, Thailand, Malaysia (Malaya and Sabah), and Indonesia.

The present distribution is restricted largely to the hilly and mountainous regions that resist human development longest. Modern technology and machinery now make it feasible and profit-able to develop these remnant habitats, and they are currently threatened. Consequently elephant populations continue to undergo habitat encroachment and fragmentation, and in the planning stages of development programmes little consideration is given to elephants. Historical references to the former vast stocks of tame elephants, reflecting former wild populations of a size far greater than those of today, provide the most dramatic illustration of the speed and severity of the man-induced attrition of the Asian elephant.

The numbers of wild elephants estimated to remain in Asia are given in Table 20.1. From these estimates it appears that only about 36 000 Asian elephants remain today in the scattered distribution described above. In view of these small numbers and the continuing exponential escalation of the human activities responsible for their reduction, as well as certain relevant charac-

Table 20.1 Wild elephant populations

Bangladesh	205–222
Burma	5 200–7 200
China	100
India and Bhutan	14 880–16 690
Indo-China (Kampuchea, Laos, Vietnam)	3 500–5 000
Malaya	700–900
Nepal	22–34
Sabah	2 000
Sri Lanka	2 000–4 000
Sumatra	300
Thailand	2 600–4 450
Total	31 507–40 896

From IUCN (1978, 1980)

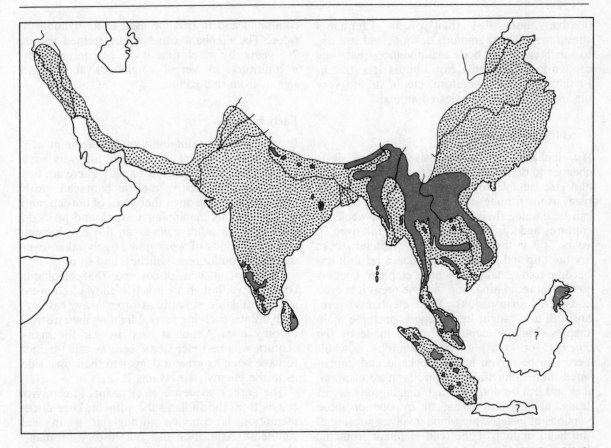

Fig. 20.1 The approximate past (stippled) and present (black) distributions of the Asian elephant. The areas in which the prehistoric existence of *Elephas maximus* remains uncertain are indicated by question marks

teristics of large, long-lived animals, the Asian elephant is now officially considered an endangered species (IUCN 1978).

Domestication and early history

Throughout the range described above the elephant was used traditionally for work, war, and ceremonial, and with the exception of Indonesia and Malaysia (and of war) this is still the case today, albeit to a far lesser extent than formerly. But even where it no longer exists the elephant has passed indelibly into the culture of Asian peoples, through art, legend and religion.

Despite a long history in the service of man, it is debatable whether the Asian elephant can properly be considered a domesticated animal, because until relatively recently there has been practically no breeding under control, and what little there has been has certainly not entailed any selection away from the wild type. Since the use of tame elephants began, the vast majority of births among them have been unplanned, and have been the result of wild males mating hobbled females turned loose at night to forage for themselves in the forest.

In many respects the elephant is very unsuited to domestication, and this is largely responsible for the minimal effort at controlled breeding. To be viable as a full domesticant, an animal must breed and reach full productivity quickly. Elephants do not reach breeding maturity until they are over 7 years old; do not conceive more than once in 2–4 years; carry their young nearly 2 years; generally bear only a single calf which is dependent on the mother for at least 5 years, so that her own work is hampered for almost 7 years; the work potential of the calf itself is negligible for over 15 years during which time ownership involves more loss than profit. Elephants consume such vast amounts of food, and still use so much energy in body maintenance, that they can work hard for only 2 or 3 hours at a stretch. In these respects, therefore, elephants are very unsatisfactory as domesticated animals.

Catching and training

As elephants are so expensive to keep, it is cheaper to catch wild elephants already at an age and size suitable for work, than to breed your own, as the former are easily tamed and trained, can be earning their keep within a few weeks of capture, and can continue to do so for over 40 years. It is in this way that tame elephant stocks are built up and maintained, and as a result it has become traditional among most elephant keepers to discourage mating, and in some areas it is even considered inauspicious. Wild elephants were, and still are, caught by variations on three main themes. Multiple captures can be made by the *khedda* (or *kraal*) method, whereby a whole herd can be driven into a stockade and imprisoned there. Individual capture is more economical and can be done through digging pits across trails, as in south India, or by one or more versions of *melashikar*, which involves noosing of the head or foot of the wild elephant from the back of a tame one.

Females are preferred as working animals because they are more predictable than males. When on musth, males can become dangerous and impossible to handle, and have to be tethered securely at all times. Because they can be out of commission for weeks at a time during musth, adult males are much more of a financial liability than females (non-pregnant females, that is). On the other hand, as the only sex bearing tusks, males are preferred for ceremonial and prestige functions. In Sri Lanka it is estimated that only 7 per cent of males bear tusks; the rest are tuskless males known as *mucknas*.

Whatever the method of capture, or sex of the animal, the procedure involved in breaking and taming is fairly standard. The new capture is securely tied, usually with its feet widely separated so that it cannot apply its strength effectively. It is starved for a few days to weaken it and then training begins, based on reward and punishment. Later, the elephant is tied between two monitor elephants to learn the basic words of command and to become used to a 'mahout' or rider. The elephant can be fully trained within 2–3 years, by which time it can be controlled by a multitude of verbal commands and tactile signals from the mahout.

Early history

In view of the unsuitability of the elephant as a domesticated animal, one might ask why man bothered to tame and train it at all? There are two main answers. Firstly, the elephant can move heavier loads than any other beast of burden, and by virtue of its manipulative trunk and tusks do work that no other animal can do, e.g. logging and other kinds of work now largely taken over by cranes, bulldozers, winches, and so on.

Secondly, it was discovered that elephants could be (although in practice they often were not) formidable in battle as super-heavy cavalry, each animal carrying several fighting men instead of only one. Although used in war for many centuries, tamed elephants seem as often as not to have been as great a danger to their own side as to the enemy (see Hyams 1972).

The earliest evidence of elephants in captivity comes from the Indian subcontinent; later documentation is common throughout south and southeast Asia. For the full history of man–elephant relationships in Asia, the reader is referred to Deraniyagala (1955), Rao (1957), Digby (1971), Hyams (1972) and Olivier (1978).

The earliest records of tame elephants are engravings of not later than 2500 B.C. from Mohenjo Daro on the lower Indus. How it all began is pure speculation. Perhaps stray calves taken as pets sparked the idea. Once started it was obviously much easier to find and capture elephants than it is today because of their very large numbers, and thus a cultural relationship with elephants grew and become an inseparable part of local religion, mythology, war and even everyday life.

All this is recorded in many ancient documents, particularly the collection known as the Gaja Sastre, Sanskrit for elephant lore. Other outstanding documents are the Rigveda of the 20th–15th centuries B.C. and the Upanishads of the 9th–6th centuries B.C..

Elephantology became a branch of Asian governments, and care and training of elephants was overseen by State officials. The possession of a large elephant stable, or *pil-khana*, became the most important symbol of royalty and independent power, not only in India, but in Asia generally.

Local wild stocks were insufficient to maintain the vast tame stocks for military ascendancy, and it became easier to capture the tame stock of one's enemies, or to accept wild stock as tribute from places where they remained abundant. Thus a complex trade in elephants was set up very early on, and the picture is confused because the centres, trading from their own abundant wild stocks, also resold animals imported from similar centres elsewhere.

The three major centres for capturing local wild elephants appear to have been Bengal, Ceylon (Sri Lanka) and Pegu (lower Burma). However, Bengal was itself receiving elephants from elsewhere in India including the Deccan, Orissa and Madras and by sea from Pegu. Ceylon, in addition to its own animals, also imported them from Pegu and was itself serving Indian ports in the south and northwest, whence elephants from both areas could reach Delhi. Madras imported elephants not only from Ceylon, but even from Malaya as late as the 18th century, part of a separate Far Eastern trade. By such means the Mogul emperors were able to build up a large *pil-khana* – 1400 war elephants in 1452 and 3000 between 1463 and 1482. The Emperor Jehangir was reputed to have 12 000 with over 40 000 in his whole empire. Though these figures are prone to various sources of error, they are quite possible. A distinction was made between the 'war' elephant, to which most statistics refer, and other untrained, weaker or immature animals also held in captivity, and even in the 18th century a record exists of 1026 elephants, of which 225 were war elephants. Confusion between the two types of reference has produced a wide variation in quoted numbers.

Despite prehistoric paintings and etchings of elephants, there is no evidence that the animal was caught and tamed in Ceylon as long ago as in India, although the records we do have, which all attest to both the abundance of elephants and the skill of trappers on the island, suggest that Ceylon was probably not far behind. Ceylon was exporting elephants by 600 B.C. so skills must have been well established before then. Any lack thereof would have been corrected after Vijaya invaded from India and his Indian bride arrived

in 483 B.C. with elephants as part of her dowry. In the 3rd century B.C. the Greeks knew of trade in elephants from Ceylon, and Digby (1971) mentions their export to Bihar at that time.

According to the ancient Ceylonese chronicle of history, the Mahavamsa, in the 2nd century B.C., but possibly before, the Sinhalese kings established a royal elephant stable, the *ath-panthiya*, equivalent to the Indian *pil-khana*. This was staffed by a complex hierarchy: a *gaja nayake nilame*, usually a prince, was in charge of various overseers, noosers, cutters of lianas, scouts who located wild herds, mahouts, trainers, food and water collectors, an elephant veterinarian and a special caste of ropemakers. The *ath-panthiya* existed right up to 1815, when the Kandyan king finally ceded to the British, but the title of *gaja nayake nilame* still exists. The Mahavamsa describes royal war elephants, nearly all tuskers or those showing depigmentation, and the methods of capturing them. Indeed, it was an integral part of all princes' upbringing to learn the arts of the elephant.

In the 6th century the elephant trade became very large, and elephants from Ceylon were popular in India. Imports included animals from Burma, and although it is not known when this was started, it was well established by 1165, because in that year the Burmese stopped the trade and the Ceylonese responded by invading. Trade with south India continued to be extensive between 1250 and 1450 and is recorded with Gujarat in 1518–20. By this time the elephant had become entrenched as part of royal culture. In the 12th century a king issued one of the first edicts to protect elephants and other game, and as an apparent result 'wild elephants were so abundant that they could be driven into stockades with little difficulty'. Elephant fights in arenas, involving whole herds, were one of the more popular Sinhala sports, called *gaja kelia*, a term still used for 'great event'. Deraniyagala (1955) refers to at least seven different mediaeval documents on types of elephants and their training, control, health and management. In 1505 the king of Portugal ordered a fort to be built in Ceylon, 'where are all the elephants of India'; as if to confirm this, the Sinhalese king mustered 2200 *war* elephants for his seige on Fort Colombo in 1588.

Throughout the 17th century and after the advent of the Dutch in 1638, the trade with India continued, and elephants were imported from Pegu in Burma. Numerous Dutch sources refer to elephants used by the Sinhalese armies, whose kings in Kandy continued to hold large stables.

In Thailand, elephants have been tamed for at least a thousand years, and the animal is referred to frequently in the Buddhist Jataka tales and features in the great national epic poem, the Ramakien. As war animals elephants helped to form the modern state of Thailand, playing particularly important roles in the 13th–16th centuries. From the early days of the Chakri dynasty until 1921 there was a Department of Royal Elephants directly responsible to the king, to whom all elephants belonged by law.

All Burma's ancient rulers also held stables of war elephants. In 1277 the king of Burma kept 2000 elephants, and in 1586 the king of Pegu had an impressive stable that included four white elephants.

Recent history and present status

Trends in the trapping and trade of Asian elephants between the 17th century and 1978, are documented as fully as the data then available permitted, by Olivier (1978). During this period the use of elephants in battle died out, but their other traditional uses have persisted.

The elephant is still used extensively in lumbering operations, as a transport and pack animal, and is the symbol of wealth and prestige most sought after by the temples, princes, and rich landlords of Asia, even today.

Despite this the number of elephants held in captivity, and the number caught per annum have declined dramatically in the last hundred years, although unfortunately data documenting these trends comprehensively are not available from all the countries concerned, notably India. Although a few thousand tame elephants probably exist in India (several hundred are still caught – illegally – per year, mostly in Assam), I am not aware of any recent statistics save for a figure of 700 working elephants in the three southern states of Karnataka, Kerala, and Tamil Nadu. Of these about 400 are with the State Forest Departments, the balance with temples and timber merchants (Gadgil and Nair 1979).

In 1884 there were more than 20 000 domesticated elephants in northern Thailand alone, and at the turn of the century 1000 were used on one trade route between Chiang Mai and Chiang Saen on the Mekong river. From 13 397 domesticated elephants in 1950, their population gradually fell to 11 022 in 1969 and then dropped catastrophically to 8438 in 1972. The latest statistic is of 5737 elephants in 1978 (Lekagul, personal communication to I.L. Mason).

In China, some Mongol sovereigns such as Kublai Khan (A.D. 1214–94), had a reputed 5000 elephants, while the Manchus at the end of the 18th century had about 60. In 1834 there were eight to ten at court in the capital, but by 1901 there were none. According to McNeely (1975) there were once 200 000 war elephants in the Khmer empire at its height, but in his 1975 report he cites only 582 tame elephants in Kampuchea. The same source cites 902 tame animals in Laos, making the country's former name, 'Land of a Million Elephants', seem incongruous. According to a census by Jayasinghe and Jainudeen (1970) there were only 532 tame animals in Sri Lanka in 1969, whereas Deraniyagala (1955) had been able to 'examine' 670, and in 1671 the king of Kandy possessed 300 *tuskers*.

The only country whose tame elephant population has apparently held its own in recent years is Burma, where the important teak industry is still based entirely on log extraction by working elephants. Before the Second World War there were over 6000 elephants in British Burma. According to Hundley (1980) there were 5973 at the end of 1978, of which 2343 were owned by the State Timber Corporation, 102 by the Forest Department, with the remaining 3528 in private hands. However, there are reports of high mortality amongst prime working elephants, and it is feared that in maintaining the captive population, the wild population has suffered a significant and adverse impact.

There appear to be several mutually reinforcing reasons for the declines noted. The most obvious of these, but possibly not the most important, has been the gradual replacement of the elephant by machines. However, perhaps the most significant cause has been the decline in the wild populations from which the tame ones are maintained. The resumés presented for each country by Olivier (1978) show quite clearly that it was with the advent of colonial regimes from Europe that the elephant was put under pressure in areas hitherto immune to development.

Thus technological innovation, together with other factors associated with colonial aims and attitudes, led in most countries to a really drastic crash in wild elephant numbers during the 19th century. In view of the traditional method of maintaining tame elephant stocks and the lack of captive breeding, this crash was bound to have far-reaching repercussions. Wild populations became smaller, more inaccessible, more fragmented, more mobile, and less predictable than formerly. In other words it was no longer easy to trap elephants in the numbers required to show a profit, a trend anyway aggravated by that of global inflation and a dwindling market. As a result, in those recent times and places where they were legal, *khedda* operations (which are extremely costly to run) have died out, and *melashikar* operations often failed to meet permitted quotas.

Today, however, Asian governments have recognized in their wild elephant populations a rapidly declining natural resource of considerable value, and in a last-minute attempt to conserve it, they have all (with the possible exception of the communist countries), either banned elephant capture outright or placed it under strict government control. Although this legislation is defied, even openly, in some areas, the general consequence has been a further reduction in recruitment to tame populations.

The future

As described above, elephants have for centuries played very important economic and cultural roles in Asia, and these roles are by no means diminished today. It is quite evident that there is a continuing demand for elephants in forestry, religious, and touristic activities. Indeed, with ever-increasing prices of petroleum products, it seems likely that elephants will remain a viable alternative to lorries, bulldozers, tractors, and cranes in the Third World, especially in difficult or mountainous terrain.

However, it is also clear that if Asian elephants are to retain their place as the most intelligent and strongest of working animals, adequate steps must be taken for the conservation and scientific management of wild populations, combined with greater efforts to overcome the difficulties of breeding in captivity.

Organizations such as the International Union for the Conservation of Nature and Natural Resources (IUCN) and the World Wildlife Fund (WWF), are currently actively involved in promoting wild elephant conservation by collaborating with governments in the supply of equipment to national parks; increasing trained manpower; designating new parks and reserves; preserving the integrity of existing conserved areas; and constructing elephant barriers. In addition there is international effort to control the ivory trade through the enforcement of the Convention on International Trade in Endangered Species of Wild Fauna and Flora (CITES). Other significant inputs required are in the fields of education and research. Through public awareness and support for conservation, the chances of preserving the integrity of conserved areas is much improved. Research, particularly monitoring of elephant population trends, is essential for the success of most conservation projects, so that the right decisions are made about what needs to be done.

Even if the above conservation measures are adequately implemented, it is certain that, as development proceeds apace, many elephants will continue to become displaced and 'pocketed'. In these circumstances they become a positive threat to human life and property, but ecological considerations may not permit all elephants captured from problem areas to be reintroduced into the safety of wildlife reserves. In such situations it may therefore be both more humane and more profitable to capture 'problem' elephants and add them to the tame population. As we have seen the demand is still there. While still retaining strict control, some governments (e.g. India) may therefore find it useful, both for the management of wild elephants and for the maintenance of tame stocks, to lift existing total bans on capture. Other countries experiencing rapid destruction of their elephant range (e.g. Sri Lanka) are already experiencing a virtually enforced, if welcome, recruitment to the tame population.

Notwithstanding the above, there remains a need to improve captive breeding performance (quite apart from the fact that this would enhance the conservation of wild populations anyway by reducing a drain on them), and many Asian governments are showing an interest in this. Nepal, where elephants remain important in the forestry, tourism and cultural sectors, is a good example. Here the country's wild population is down to a dozen or less, so there is no question of exploiting it. In recent years the Nepalese have bought tame elephants from India, but this is becoming increasingly difficult (through controls) and expensive. Captive breeding is one obvious answer.

Burma and Thailand are two countries still with relatively high demands for tame animals, but whose wild populations are depleted and poorly conserved and managed; they are therefore interested in captive breeding. Even Sri Lanka is interested, where captive breeding is seen as the last defence line in the battle to conserve an extremely cherished national treasure.

The first step in promoting captive breeding is simply, where possible, to encourage mating actively, rather than the opposite as hitherto. In view of contemporary circumstances, this change in attitude seems to be coming about 'naturally' anyway.

The second step would be to continue relevant research, and to disseminate results and methodologies where they would be most useful. For example, the productivity of male working elephants would be maximized if musth could be prevented by inhibiting testicular activity with oestrogen, or with an anti-androgen such as cryproterone (Jainudeen et al. 1972, a and b). A key area appears to be artificial insemination. Now that our knowledge of the elephant oestrous cycle has improved enormously; now that the approximate time of ovulation can be predicted after a luteinizing-hormone-releasing-hormone (LHRH) treatment regimen (Chappel and Schmidt 1979); now that methods of collecting elephant semen and preserving it under refrigeration with good fertility for five days have been developed; and now that satisfactory methods and equipment for the actual artificial insemination are available; this field promises to be one that will make the biggest contribution to elephant captive breeding in the future, and signal further progress in our domestication of the Asian elephant.

References

Carrington, R. (1962) *Elephants. A short account of their natural history, evolution, and influence on mankind.* Penguin: London

Chappel, S. C., and Schmidt, M. (1979) Cyclic release of luteinizing hormone and the effects of luteinizing hormone – releasing hormone injection in Asiatic elephants. *American Journal of Veterinary Research*, 40 (3): 451–3

Deraniyagala, P. E. P. (1955) *Some extinct elephants, their relatives, and the two living species.* Ceylon National Museums Publication: Colombo

Digby, S. E. (1971) *War-horse and Elephant in the Dehli* (sic) *Sultanate.* Orient Monographs: Oxford

Dittrich, L. (1967) Beitrag zur Fortpflanzung und Jugendentwicklung des Indischen Elefanten, *Elephas maximus*, in Gefangenschaft mit einer Übersicht über die Elefantengeburten in europäischen Zoos und Zirkussen. *Der Zoologische Garten*, 34: 56–92

Eisenberg, J. F., McKay, G. M., and Jainudeen, M. R. (1971) Reproductive behavior of the Asiatic elephant (*Elephas maximus maximus* L.). *Behaviour*, 38 (3–4): 193–225

Gadgil, M. and Nair, P. V. (1979) Elephant and man in south India. Mimeo, 40 pp.

Hundley, H. G. (1980) A report on the status of the Asian elephant in Burma. Mimeo, 19 pp.

Hyams, E. (1972) *Animals in the Service of Man: 10 000 years of domestication.* Chapter 11, Elephants, pp. 138–48. Dent: London

IUCN (1978) *Red Data Book*, Vol. 1, *Mammalia*. International Union for Conservation of Nature and Natural Resources: Gland, Switzerland

IUCN (1980) Summarised minutes of the 2nd Meeting, Asian Elephant Specialist Group. Mimeo, 49 pp.

Jainudeen, M. R., Eisenberg, J. F., and Tilakeratne, N. (1977) Oestrous cycle of the Asiatic elephant, *Elephas maximus*, in captivity. *Journal of Reproduction and Fertility*, 27: 321–8

Jainudeen, M. R., Katongole, C. B., and Short, R. V. (1972a) Plasma testosterone levels in relation to musth and sexual activity in the male Asiatic elephant, *Elephas maximus. Journal of Reproduction and Fertility*, 29: 99–103

Jainudeen, M. R., McKay, G. M., and Eisenberg, J. F. (1972b) Obervations on musth in the domesticated Asiatic elephant (*Elephas maximus*). *Mammalia*, 36(2): 247–61

Jayasinghe, J. B., and Jainudeen, M. R. (1970) A census of the tame elephant population in Ceylon with reference to location and distribution. *Ceylon Journal of Science* (*Biological Sciences*), 8(2): 63–8

McNeely, J. A. (1975) Draft report on wildlife and national parks in the lower Mekong basin. Mekong Committee, ECAFE. Economic and Social Commission for Asia and the Pacific (UN): Bangkok

Olivier, R. C. D. (1978) Distribution and status of the Asian elephant. *Oryx*, 14 (4): 379–424

Rao, H. S. (1957) History of our knowledge of the Indian fauna through the ages. *Journal of the Bombay Natural History Society*, 54 (2): 251–80

Scheurmann, E. von and Jainudeen, M. R. (1972) 'Musth' beim Asiatischen Elefanten (*Elephas maximus*). *Der Zoologische Garten*, **42** (3/4): 131–42
Sikes, S. K. (1971) *The Natural History of the African Elephant*. Weidenfeld and Nicolson: London

21

African elephant

I. Douglas-Hamilton

IUCN African Elephant Specialist Group, Nairobi, Kenya

Introduction

Nomenclature

The African elephant can easily be distinguished from the Indian species. It has an arched back and large ears. The head is sharply angled, compared to the Indian's bulbous dome, and is carried at a higher angle. Two subspecies are generally recognized, *Loxodonta africana africana*, the larger bush elephant, and *cyclotis*, the forest elephant. The subspecies can interbreed and hybrid populations exist. The bush elephant is found mainly in savannah in South, East, Central and West Africa, in semi-desert in Namibia, and in the Sahel as far north as the Assaba mountains in Mauritania. The forest elephant lives in forests of West and Central Africa, and some savannah areas such as the Garamba National Park in northern Zaire.

Present distribution and status

Although greatly diminished, the range of both subspecies is still extensive, of the order of 7 million km^2 but often at very low densities (Fig. 21.1) (Douglas-Hamilton 1979). The IUCN

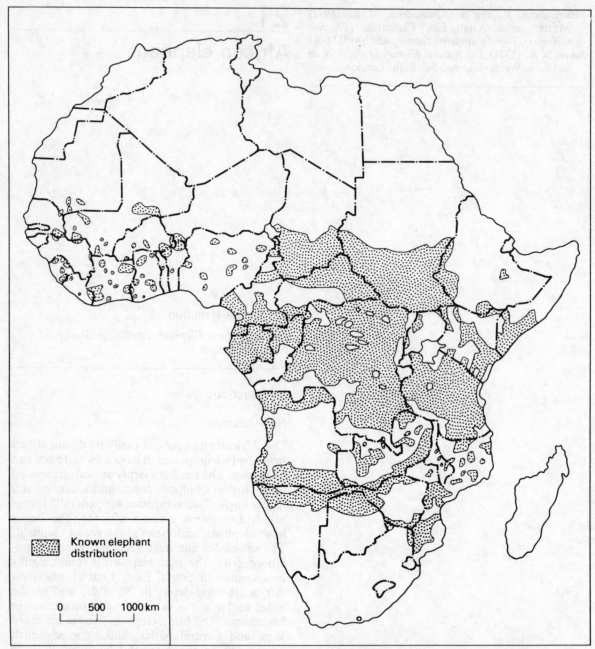

Fig. 21.1 Known distribution of African elephant. (From Douglas-Hamilton 1979)

Elephant Survey estimated a continental minimum of 1.3 million elephant between 1976 and 1979 (IUCN 1980). However, the majority of populations were reported to be decreasing due to the ivory trade and a rapid loss of range in the face of expanding cultivation. The introduction and spread of automatic weapons in Africa is now endangering the elephants in many areas where they were formerly regarded as safe.

Biology

The African elephant usually produces one young

at a time, but twinning occurs at a frequency of approximately 1 per cent. Some populations exhibit a marked breeding season (Hanks 1969, Laws 1969), but mating can occur in any month. Copulation usually takes place after a short chase by the male who mounts the female from the rear and remains up for some 120 seconds.

The gestation period is 22 months after which the young have a prolonged period of dependence on the mother and family unit which may last for 15 years or more, making it uneconomic to domesticate elephants from birth. Elephants, both Asian and African, are usually caught in the wild when already 5–15 years old. There have, therefore, been no attempts at selective breeding to improve the domestic stocks. However, the elephant's potential longevity is 60 years and the working life may be 30 years or more (Laws 1970, Sikes 1971).

Early domestication

Knowledge of tamed elephants reached Egypt at least as early as 1500 B.C. and an engraving of a domestic Asian elephant appears on the tomb of the vizier Rekhmir. However, there is no historical record that African elephants were domesticated until the Ptolemaic dynasty began capturing them for their armies about 270 B.C. When Alexander the Great died (323 B.C.) his generals, including Ptolemy, vied for his empire, and in the battles that followed war elephants originally captured from India were used, playing much the same role as a tank force. Victory often went to the side with the larger number (Gowers 1953). When Ptolemy, in Egypt, was cut off from the Asian supply of war elephants by the Seleucid dynasty, he and his son turned to the African continent to renew the supply. Expeditions sailed down the Red Sea as far as modern Eritrea, and the Horn of Africa. Ptolemy also sent envoys to Meroë, capital of a semi-hellenized civilization of the Upper Nile, in whose culture elephants played an important role. According to Scullard (1974), the Kingdom of Meroë had already domesticated elephants, although whether or not the technique had been learned from the Orient is not known. In any event, African elephants were vital to Ptolemy for his army, and while Indians probably helped with the training, he may also have obtained expert assistance from Meroë.

The only time African elephants were pitted against Asian elephants was at the battle of Raphia (217 B.C.) and contemporary accounts attest the superior size and training of the Asians. There is some possibility that the African elephants belonged to a small North African race now extinct, as is convincingly argued by Gowers (1948).

Use of African elephants in warfare spread from Egypt to Carthage with Hannibal's famous crossing of the Alps and, after the defeat of the Carthaginians, was carried on by the Romans in a relatively minor way, with the last recorded use being made by Mark Anthony in 43 B.C. Thus, for a period of 250 years, the ancients domesticated hundreds, if not thousands, of African elephants, and made use of them in scores of major battles (Scullard 1974). The main use of the elephant was to neutralize the enemy cavalry, break up the line of the infantry and breach walls or fortified encampments, or, if they faced other elephants, to take them on in individual duels. In many cases, cavalry facing elephants for the first time became quite uncontrollable.

The elephant behaviour described in ancient accounts was often no more than a pattern found normally in the wild. For example, a matriarch, if sufficiently aroused, may charge a pride of lions, a group of hyaenas, or a party of human beings on foot, with her family following on a broad front (Douglas-Hamilton 1972). Just as prisoners were trampled to death by the war elephants of the Greeks, the Carthaginians and the Romans, so have lion cubs been observed destroyed by elephants in the Lake Manyara National Park (Makacha and Schaller 1969). Again, leaning against walls until they topple, is much the same as leaning on trees for the same purpose.

What is more remarkable is that the ancients were able to switch on aggression by command, apparently sometimes with the additional stimulus of alcohol. Elephants in the wild, when facing an enemy, are usually finely balanced between fight and flight, and most charges are made as a demonstration, followed by withdrawal. One would not expect an excited elephant to distinguish readily between friend and foe. In many ancient battles the elephants charged the enemy and then retreated through their own lines, causing havoc to friend and foe alike.

There are also ancient accounts of war elephants being made to fight each other to the death. In the wild, bull elephants kill each other only extremely rarely. Presumably, this happens only when the animals are on musth, a hormonal condition of heightened aggression and sexuality, well known in the Asian species, when elephants become dangerous to their keepers, and which has recently been conclusively demonstrated in the African species through intensive behavioural observations (Poole and Moss 1981). Normally, however, bulls know each others' strength and a simple threat display, such as a head-shake or nod in the direction of the rival, is a sufficient signal of aggression to allow the weaker one to escape and the stronger to assert his superiority without any stress.

After the fall of Rome and Carthage, the African elephant disappears from the history of warfare and domestication, apart from fragmentary references to its being used by the Ethiopians in their battles against the Arabs.

Recent domestication

For thirteen centuries, there is no mention of domesticating the African elephant, and the knowledge gained by the ancients was lost. It was not until the end of the 19th century that the question again arose. King Leopold II of Belgium was struck by the intelligence, dexterity and usefulness of Asian elephants which he encountered in Ceylon (Sri Lanka). He financed an expedition in 1879 from Bombay, with four Indian elephants and thirteen mahouts, which arrived on the east coast of Africa, near Dar-er-Salaam, and set off to cross half the continent. Unfortunately, the elephants died one by one, and by the time they reached Karema, on Lake Tanganyika, only one remained. However, the expedition was judged a success in that elephants could be moved over long distances of Africa, despite the fact that the entire expedition was wiped out on the way home by an attack of Ruga-rugas.

Efforts were resumed in the Congo some 20 years later, once again at the instigation of the Belgian king. After studying native methods, the officer in charge of the project, Commandant Laplume, tried using pitfalls covered in branches, one of the time-proven methods used not only in Africa but also by cavemen to trap mammoths in Europe, and at times by the Ptolemys and the Carthaginians. The first elephant that fell into one of the pits was rescued by its family who broke down the retaining walls. The next one apparently died of shock. After these initial reverses, a modification of the Indian *khedda* system, where elephants are driven into stockades, was tried. Unfortunately, although the elephants walked into the traps, they proved to be quite intractable and the Belgians were forced to release them (Laplume 1911, Leplac 1918).

All this early experimenting lacked both Indian mahouts and trained monitor elephants which could be used to quieten new captives. Eventually, the Belgians developed a technique in which the capture teams literally ran the elephants down, splitting up large herds into small units and scaring other elephants away from the selected calf and its mother with blank shots. The commandant would ride out on a horse in front of the mother, distracting her, while the men on foot lassoed the calf by the feet and tied it to trees. Then monitor elephants were brought up to calm the captive and lead it away (Denis 1962). They learned to avoid taking animals that were either too small, as these invariably died, or too large, in which case they were untamable. Heights between 1.5 m and 1.8 m were judged the best (Huffman 1931). The method of capture remained the same for 30 years, but the capture teams acquired such dexterity that eventually they were taking animals that were near adult (Offermann 1930). The record size was 2.13 m. New captives were broken in according to methods brought from India. Local Azande tribesmen were trained as 'cornacs' by professional Indian mahouts, and even adopted a version of one of the Indian elephant songs which they sang to their elephants as they took them down to the river every day. To be a cornac was a matter of pride and status.

The elephants were used in many places and learned to draw carts. They were harnessed to ploughs and handled logs in the forests. An experiment to compare their efficiency with tractors was conducted at Bambesa cotton station in 1928. During the ploughing season, which corresponded with the rains, an elephant could work without stopping in wet soil, whereas the tractor would frequently skid or subside in soft patches. When it came to negotiating obstacles like termite hills, the elephant could surmount them easily whereas the tractor was likely to turn over. Again, when the plough snagged on a rock, the elephant would immediately sense the obstruction and ease off before breaking the tackle. Finally, the elephant needed no imported fuel, oil, or spare parts, nor any mechanical knowledge on the part of its driver. In those days, taking into account the labour conditions of Africa and the frequent breakdown of tractors, the elephant was considered faster than a tractor and certainly much more economical (De Jongh 1929).

By the end of 1930 there were thirty elephants in use at Api, and forty in the new training school of Gangala-na-Bodio both in northeast Congo (now Zaire) (Huffman 1931). Of these, four had been born in captivity. Other countries, however, did not follow suit (Caldwell 1925).

With increased mechanization in the 1950s, and cheaper fuel, the use of elephants even in the Congo began to loss its economic edge over other forms of traction. The school diversified its activities, sold trained elephants to zoos and circuses and, until the eve of Independence in 1960, encouraged their use in scientific research and in the film industry (Lefebvre 1960). In the next often chaotic 20 years, the elephant school survived by a thread. When the station was overrun by Simba rebels, the cornacs hid their elephants in the bush and for 2 years kept them alive, bringing them food secretly, even though they themselves were unpaid. Many of the elephants died, but when I visited them in 1974 there were still nine left. Two of them were monitors, of which one, Wanda, had been born in captivity in 1929. However, the harness had rotted and rusted away, the carts were out of order, and the elephants were kept going more as a tradition than for any possible earnings.

Unfortunately, the great boom in tourism experienced in East Africa during the 1960s never developed in Zaire. Had it done so, the elephants could easily have been used to carry visitors to view the wild animals, as they are in the Kaziranga Park in Nepal. That this would be feasible is proved by the highly trained African elephants at the Basle Zoo which take thousands of visitors every year for rides on their backs, and which I have seen showing the utmost good nature and

consideration for children runing between their legs.

Future prospects

Elephant domestication in Africa could only survive through drive, vision, organization, money and a market. Ironically, now that Zaire and other African countries are experiencing chronic deficiencies in transport, fuel and spare parts, the elephant could once more help with ploughing and carrying of goods. Unfortunately, although African cornacs were depicted on coins in Hannibal's time, 2000 years ago, elephant training has not survived in African culture, and the Belgian experiment was too short-lived to reintroduce the practice. It is unlikely that the domestication of the African elephant will survive the five ageing elephants and their cornacs still alive, in 1981, in the remote northeast of Zaire.

References

Caldwell, K. (1925) Elephant domestication in the Belgian Congo. *Journal of the Society for the Preservation of Fauna of the Empire*, Part 7: 71–82
De Jongh, E. (1929) Emploi des éléphants à la station de selection cotonnière de Bambesa (Uele). *Bulletin Agricole du Congo Belge*, **20**: 283–4
Denis, A. (1963) *On Safari*. Collins: London
Douglas-Hamilton, I. (1972) on the ecology and behaviour of the African elephant. D.Phil. Thesis, Oxford
Douglas-Hamilton, I. (1979) African elephant ivory trade study. Final Report to U.S. Fish and Wildlife Service. (Typescript)
Gowers, W. F. (1948) African elephants and ancient authors. *African Affairs*, **47**: 173–80
Gowers, W. F. (1953) The African elephant in history. In: *The Elephant in Central East Africa*, ed. W. C. O. Hill *et al.*, pp. 143–50. Rowland Ward: London
Hanks, J. (1969) Seasonal breeding of the African elephant in Zambia. *East African Wildlife Journal*, **7**: 167
Huffmann, C. (1931) La domestication de l'éléphant au Congo Belge. *Bulletin Agricole du Congo Belge*, **22** (1): 7–22
IUCN (1980) IUCN *Bulletin* **2** (1/2): 1–16.
Laplume, J. (1911) La domestication des éléphants au Congo. *Bulletin Agricole du Congo Belge*, **2**: 405–18
Laws, R. M. (1969) Aspects of reproduction in the

African elephant *Loxodonta africana. Journal of Reproduction and Fertility*, Supplement **6**: 193–217

Laws, R. M. (1970) The biology of the African elephant. *Science Progress*, **58**: 251–62

Lefebvre, R. (1960) Note concernant le soixantième anniversaire des expériences de domestication de l'éléphant au Congo Belge. *Service des Eaux et Forêts Chasse et Peche. Bulletin*, **7** (25/26): 655–61

Leplac, E. (1918) La domestication de l'éléphant d'Afrique au Congo Belge. *Bulletin Agricole du Congo Belge*, **9**: 37–77

Makacha, S. and Schaller, G. B. (1969) Observations on lions in the Lake Manyara National Park, Tanzania. *East African Wildlife Journal*, **7**: 99–103

Offermann, P. (1930) La capture et la domestication des éléphants dans la province Orientale. *Bulletin Agricole du Congo Belge*, **21**: 384–7

Poole, J. H. and Moss, C. J. (1981) Musth in the African elephant, *Loxodonta africana. Nature*, **292** (5826): 830–1

Sikes, S. K. (1971) *The National History of the African elephant*. Weidenfeld and Nicolson: London

Scullard, H. H. (1974) *The Elephant in the Greek and Roman World*. Thames and Hudson: London

22
Dog

Juliet Clutton-Brock

*Department of Zoology, British Museum (Natural History),
London, England*

Introduction

Nomenclature

Despite their great array of sizes, shapes, and colours, domestic dogs throughout the world are sufficiently interfertile and sufficiently similar in their basic characteristics to be treated as a single domesticated species, the *Canis familiaris* of Linnaeus. In his *Systema Naturae* of 1758 Linnaeus listed the dog as a species separate from the wolf, *C. lupus*, and other wild canids. *Canis familiaris* has subsequently been designated the type of the genus which could cause problems if agreement were to be reached on the removal of domestic animals from formal zoological nomenclature. The question need not be discussed here, however, because in order to be consistent with the rest of the book the vernacular name 'dog' will be used. This term, when applied, together with the names of the breeds, adequately describes the domestic animal in all its variations and distinguishes it from the other members of the genus *Canis*, these being the wolf, coyote, and four species of jackal.

Linnaeus divided domestic dogs into eleven groups based on differences in the carriage of the ears, the length of the coat, size of body, and length of tail. There have been innumerable attempts, before and after the time of Linnaeus, to classify dogs on their physical characteristics, behaviour, or place of origin. Many authors have accorded subspecific status to their 'types' or to individual subfossil specimens recovered from archaeological excavations, but as explained in the Appendix (pp. 434–8), the use of subspecific names for domestic animals should be avoided. The subspecies is not a synonym for the breed and the two terms have quite separate definitions. A subspecies is a distinctive, geographical segment of a species that differs morphologically from the rest of the species and it is always restricted to a geographical area. A breed of dog comprises a group of animals that has been selected by man to possess a uniform appearance that is inheritable and distinguishes it from other groups of dogs. The breed is a product of artificial choice of characters that are not necessarily strategies for survival but are favoured by man for the hunt, or for economic, aesthetic, or ritual reasons. A breed differs from a subspecies of a wild animal in that it is not necessarily restricted to a geographical area although it may be so.

When writing about dogs it is important to understand the characteristics of a breed because there are probably more than 400 breeds of dog in the world today. Many of these may only exist for a short time; they can be created, re-combined into new breeds, or lost, according to need or fashion. Almost all fall within the ten main groups shown in Fig. 22.1.

General biology

Similarities of the dog to other species of Canis

All dogs, whether Great Dane or Pekingese, have the same basic anatomy and physiology, and the same general behaviour patterns. For the following account the European sheepdog or Alsatian will be taken as a generalized example of a domestic dog.

The sheepdog conforms to the genus *Canis* in being a social carnivore with a long-muzzled head, strong jaws, large ears, a long tail, and a body that is adapted for fast running over great distances. The senses of sight, smell, and hearing are all acutely developed. Social behaviour, based on submission and dominance within a hierarchy of individuals, is highly evolved and there are complicated patterns of vocalization,

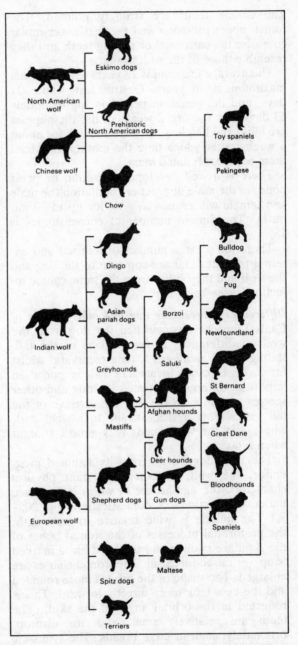

Fig. 22.1 A hypothetical scheme for the relationships between modern breeds of dogs and the four geographical races of wolf from which they are likely to be descended

facial expression, and bodily posture (including tail-wagging) for communication between individuals.

The normal dental formula is $I\frac{3}{3}$, $C\frac{1}{1}$, $P\frac{4}{4}$, $M\frac{2}{3}$.

The canine teeth are strongly pointed. The fourth upper premolar and the first lower molar comprise the carnassial or cutting teeth and they resemble those of the wolf in shape.

The average life-span is 12 years with a normal maximum of 16 years. Oestrus lasts about 21 days, and the gestation period is approximately 63 days. For the first 2 weeks of life the puppies are blind and helpless; they are suckled for about 6 weeks, after which time the mother may feed them with regurgitated meat.

There is a well developed baculum, or penis bone, in the male dog. After copulation the male and female will remain in a tie for up to 30 minutes. The diploid number of chromosomes is 78.

There are a large number of internal and external parasites that are common to the dog and the wolf and they suffer from the same canine infectious diseases.

Differences between the dog and other species of Canis Linnaeus distinguished the dog from the wolf on differences in the carriage of the tail, stating for *C. familiaris* 'cauda recurvata' whilst for *C. lupus* 'cauda incurvata'. This is indeed an important difference between the dog and other species of *Canis*. In the dog the carriage of the tail varies from a sickle shape to a tight curl, whereas in the wild canids it is almost straight when relaxed.

The sheepdog differs from its assumed progenitor, the wolf, in other important physical characters, the significance of which is discussed under the section on domestication (pp. 205–7.) The muzzle is wide relative to its length, the postorbital processes of the frontal bones of the skull are swollen and the head has a marked 'stop', or raised forehead. The frontal sinuses are enlarged. The shape of the orbit is more rounded and the eyes look more directly forward. This is reflected in the orbital angle of the skull. The teeth are relatively small and are disproportionately small in large breeds. The tympanic bullae are reduced and flattened.

The first digit (hallux) on the hind foot is often developed as a dew claw which is almost unknown in wild canids. The gland that wild canids have on the dorsal side of the tail near its root is reduced or absent altogether.

The number of mammary glands is related to litter size. This is more variable in the different breeds of dog than are many other anatomical characteristics. The toy dogs normally bear less than 4 puppies whilst the larger breeds may have litters of 9 or more. The average number of pups in a wolf litter is between 4 and 7 (Mech 1970).

Most dogs become sexually mature before they are a year old, whilst in the wolf maturity is not reached until the animal is about 22 months old. Furthermore, the sheepdog and all other highly domesticated dogs have two oestrus periods a year. All wild canids breed only once a year, as do the dingo and the basenji, and the pups are born at the optimum time when the prey species are most abundant.

Although it is often claimed that dogs differ from other canids, especially wolves, in their propensity to bark, this is a question of degree rather than a real difference. Wolves can and do bark but never so intensely as dogs.

Geographical distribution and economic importance

Dogs are found today in every region of the world that is inhabited by man. As they are such an integral part of every human culture from Eskimo to city dweller it is hard to assess their overall usefulness and economic importance. Perhaps it should just be said that for the last 10 000 years dogs have been an indispensable adjunct to the well-being of humanity.

Wild species

The taxonomic position of the dog as a member of the genus *Canis* is clear and established, the other members of the genus, according to most classifications, being the wolf, *C. lupus*, the coyote, *C. latrans*, and the four species of jackal, *C. aureus*, *C. simensis*, *C. mesomelas*, and *C. adustus* (for their distributions see Fig. 22.2). The wolf is an extremely successful predator whose distribution, after that of man, is probably the most widespread of all large mammals. Until relatively modern times the wolf was found throughout Europe, Asia (including India), and North America, but not in South America, Africa, or Australasia. Geographical variation in the species follows Bergmann's rule in that those wolves living in the cold northern zones are much larger and heavier animals than those that inhabit the hot arid regions of Arabia and India.

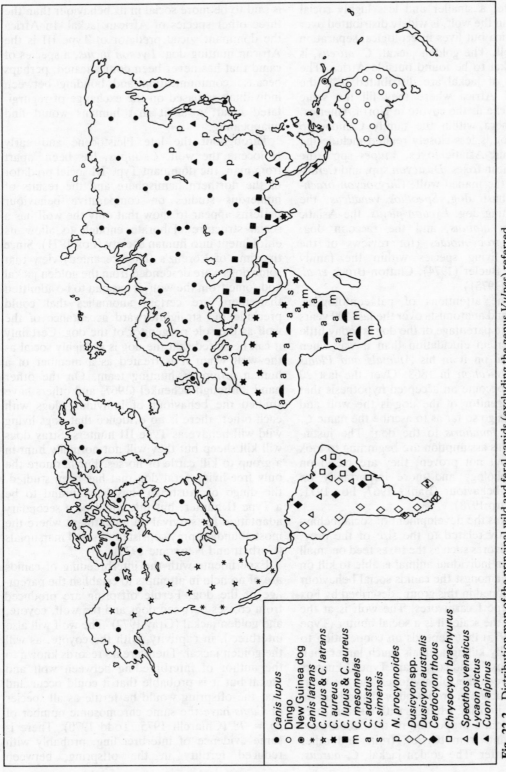

Fig. 22.2 Distribution map of the principal wild and feral canids (excluding the genus *Vulpes*) referred to in Chapter 22. The domestic dog is not included; its distribution is ubiquitous

- ● *Canis lupus*
- ○ Dingo
- ◉ New Guinea dog
- ✳ *Canis latrans*
- ✶ *C. lupus & C. latrans*
- ■ *C. aureus*
- m *C. lupus & C. aureus*
- m *C. mesomelas*
- a *C. adustus*
- s *C. simensis*
- p *N. procyonoides*
- ◇ *Dusicyon* spp.
- ◇ *Dusicyon australis*
- ◆ *Cerdocyon thous*
- □ *Chrysocyon brachyurus*
- △ *Speothos venaticus*
- ◖ *Lycaon pictus*
- ▲ *Cuon alpinus*

The coyote, a smaller and less highly social carnivore than the wolf, is widely distributed over North America but lives in ecological separation from the wolf. The golden jackal, *C. aureus*, is the only jackal to be found outside Africa. The four species of jackal are distributed over the continent of Africa where they fill the same ecological niche as the coyote in North America.

Other genera within the family Canidae to which the dog is less closely related include the European and Asiatic foxes, *Vulpes* spp., the South American foxes, *Dusicyon* spp. and *Cerdocyon thous*, the maned wolf, *Chrysocyon brachyurus*, the bush dog, *Speothos venaticus*, the African hunting dog, *Lycaon pictus*, the Asiatic dhole, *Cuon alpinus*, and the raccoon dog, *Nyctereutes procyonoides* (for reviews of the thirty-seven living species within the family Canidae see Bueler (1974), Clutton-Brock *et al.* (1976), Fox (1975)).

Despite the attentions of palaeontologists, ethologists, and anatomists over the last 100 years the immediate parentage of the dog is today little nearer to certain elucidation than it was when Darwin wrote on it in his *Animals and Plants under Domestication* in 1868. Over the last 20 years it has become an accepted hypothesis that the sole progenitor of the dog is the wolf and many authors go so far as to ascribe the name *C. lupus* forma *familiaris* to the dog. The justifications for this assumption are beginning to look convincing but not proven; they are based on dental morphology and more importantly on comparative behaviour (Banks 1967, Fox 1971, Hall and Sharp 1978).

In carnivores the development of social behaviour is directly related to the size of the prey. Solitary carnivores such as the foxes feed on small prey which an individual animal is able to kill on its own, and amongst the canids social behaviour is least developed in this group, described by Fox (1975) as Type I carnivores. The wolf is at the other end of the scale; it is a social hunter (Type III of Fox 1975) that depends on cooperation to obtain food by killing animals much larger than itself. The jackals and coyote are Type II hunters and fall between the solitary foxes and the wolf. They will associate in groups but will also hunt on their own or in pairs, whilst as with all carnivores they will obtain a proportion of their food from scavenging and from berries, fruits and other vegetable matter. The golden jackal, *C. aureus*,

is said to be more social in its behaviour than the three other species of African jackal. In Africa the dominant social predator of Type III is the African hunting dog, *Lycaon pictus*, a species of canid that has never been domesticated, perhaps because communication and bonding between individuals is based on the exchange of regurgitated meat; a habit that humans would find unappealing.

Throughout the late Pleistocene and early Holocene the wolf, *C. lupus*, has been, apart from man, the dominant Type III social predator of the northern hemisphere and the results of numerous studies on comparative behaviour patterns appear to show that only the wolf has a social structure elaborate enough to allow its enfoldment into human society (Fox 1971). Since the time of Lorenz's (later recanted) view that some dogs were descended from the golden jackal and some from the wolf, it has had to be admitted that there are certain anomalies that could preclude the straightforward acceptance of the wolf as the sole progenitor of the dog. Certainly it can be stated that the dog is as highly social as the wolf when it is treated as a member of a human group or hunting team. On the other hand, although Schenkel (1967) and others have studied the behaviour of individual dogs with each other, there is no evidence that dogs living wild will behave as Type III hunters. Stray dogs will kill sheep but they will not normally hunt in a group to kill cattle or horses. Furthermore the only free-living feral dog that has been studied, the dingo of Australia, has been found to be a Type II hunter, but this could be a secondary adaptation for survival on a continent where the most abundant prey consisted of small marsupials (Corbett and Newsome 1975).

Experiments with the interbreeding of canids are of no help in attempts to establish the parentage of the dog. Fertile offspring are produced from crosses between dogs and the wolf, coyote, and golden jackal (Gray 1972). The wolf will also interbreed, in captivity, with the coyote, as will the golden jackal. There are no records known to the author of interbreeding between wolf and jackal but it is probable that it could occur and that the offspring would be fertile as all species of *Canis* have the same chromosome number of $2n = 78$ (Chiarelli 1975, Todd 1970). There is some evidence of interbreeding, probably with reduced fertility in the offspring, between domestic dogs and the South American foxes belonging to the genera *Dusicyon* and *Cerdocyon thous* (Clutton-Brock 1977, Gray 1972). These foxes also have high chromosome numbers, varying from $2n = 74$ to $2n = 76$ and they are taxonomically closer to *Canis* spp. than are the Eurasian foxes. There are only a few tenuous records of interbreeding between dogs and *Vulpes* spp. which have a much lower chromosome number, e.g. $2n = 34$–38 in *V. vulpes* and $2n = 40$ in *V. rueppelli*.

As the dog, wolf, coyote, and golden jackal can probably all interbreed and have fertile offspring it could be argued that there is little justification for retaining them as separate species. These canids do not normally interbreed in the wild, however, and they are morphologically so distinct that their status as biological species is incontrovertible. A large factor in their reproductive separation is the seasonality of their breeding periods. The adult male domestic dog is in breeding condition all the year round but male coyotes and probably male wolves and jackals have a breeding season lasting only for about 2 months each year. This is coincident with the conspecific females (Mengel 1971), which have only one oestrous cycle a year in wild canids as opposed to the domestic female which has two. According to Mengel, F_1 generation hybrids between coyote and dog retain the annual breeding cycle but there is a shift in the timing so that the hybrids breed approximately three months earlier than wild coyotes. This shift prevents further interbreeding with the wild population and effectively precludes the introgression of dog genes into *C. latrans*. It is probable that there is an equal incompatibility in the breeding seasons of dog/wolf hybrids and dog/jackal hybrids with those of the wild species.

In summary it may be said that there is no conclusive evidence on either morphological or behavioural grounds to prove that all domestic dogs are descended from a single ancestor. On the other hand the wolf does appear to have had the greatest share in their parentage. It is probable that the small western Asiatic wolf, *C. lupus arabs*, was the progenitor of most European and southern Asiatic dogs, including the dingo. Perhaps the golden jackal interbred with these dogs from time to time and with dogs that migrated southwards through Africa with humans at a relatively late period. The small Chinese wolf, *C. l. chanco*, was probably the ancestor of the early Chinese dogs (Olsen and Olsen 1977), whilst the North American wolf was the main progenitor of the Eskimo dogs. The Plains Indians' dogs could have been interbred with coyotes, whilst the dogs of South America were originally locally domesticated members of the genus *Dusicyon* which were later replaced by European dogs and some interbreeding may have occurred (see p. 210).

Although these suppositions are lent some support by the archaeological evidence, with finds of subfossil canids from the prehistoric period, they remain as speculations for which modern biochemical investigations such as work on blood proteins have as yet provided no new clues (Seal 1975).

Domestication and early history

Man and wolf in the late Pleistocene and early Holocene

The association between human hunters and wolves could have begun at least 40 000 years ago, with the emergence of *Homo sapiens*. By the end of the last glaciation, 10 000 years ago, the partnership was fully and irrevocably established. It is not possible to know when man first tamed a wolf puppy any more than it is possible to know when he first killed an animal with a stone tool. The progress from taming to full domestication, and the consequent development of a new kind of animal must be gradual, although it is much more rapid than the evolution of new species in nature.

What evolutionary and ecological conditions promoted the migration of early hominids into the northern hemisphere during the Pleistocene period can only be guessed at but it is probable that *Homo sapiens* evolved in response to an environment where the most readily available food was provided by the vast herds of ungulates. Apart from the large cats, including the sabretooths, the wolf was man's only competitor for this prey.

Both the human hunters and the wolf were wide-ranging, highly intelligent predators whose sophisticated behaviour patterns evolved as an adaptation to the harsh life of the Ice Age landscape, where group hunting was essential for the slaughter of the largest mammals – rhinoceros, mammoth, bison, horse, and reindeer. Perhaps it is improbable that man and wolf cooperated in their hunting during the upper Pleistocene but it may be surmised that wolf puppies were occasionally taken from dens by women and children who fed and played with them. There are ethnographic parallels for the nurturing of young wild animals by people in hunter-gatherer societies, and it has been repeatedly shown in recent years how easy it is to tame young wolf cubs and to rear them in a human community. Unlike other more solitary carnivores the wolf will remain more or less tame as an adult because the hierarchical structure of its social behaviour patterns allows it to accept its human owner as the dominant member of its group or pack (Clutton-Brock 1981).

A tamed wolf is, however, very different from a domesticated dog, and it is not until such animals have been bred in captivity for many generations that the morphological changes associated with domestication would be observable in the fragments of bones and teeth that are usually all that is retrieved from archaeological excavation.

One point that should be remembered when the premise is made that man's partnership with the wolf began in the upper Pleistocene is that during the last glaciation, over almost all of its ubiquitous range, the wolf was a very large animal, as large as the largest tundra wolves of today. Yet a number of claims for finds of 'domesticated' canid skulls have been made from late Pleistocene sites in Europe, western Asia, and North America, and all of these are from small animals, no larger than the small Arabian wolf of the present day. The difference in size between these small canid remains and finds of the large *C. lupus* is so great that it seems very unlikely that they are immediately related, even though one of the most important changes brought about by early domestication is a decrease in size.

At the end of the Pleistocene, around 14 000 years ago, with the amelioration in climate, there was a decrease in size amongst many groups of mammals and this was quite marked in the carnivores of western Asia as shown by their fossil record (Kurtén 1965). It is from this period that there is more substantial evidence for the early domestication of the dog, with finds from archae-

ological sites in North America, northern Europe, and Asia (Lawrence 1967). Perhaps the most interesting of these early finds is the complete skeleton of a young dog that was found buried with a human in the Natufian level of the site of Ein Mallaha in northern Israel (Davis and Valla 1978). This dog is dated to between 12 000 and 10 000 years ago. Other finds of dog have come from the earliest levels of the Tell at Jericho (Clutton-Brock 1979) and from the cave of Palegawra in Iraq (Turnbull and Reed 1974). The remains are from animals that were slightly smaller than the present-day Arabian wolf but larger than the local jackal which also has a smaller cranial capacity and teeth that are morphologically slightly different.

The relationships of man with the first dogs

All the early finds that can be attributed with some certainty to dog come from the cultural period of the Mesolithic (or Natufian as it is called in the Near East). During this period human societies, world-wide, had a subsistence that was still based on broad-spectrum hunting and the gathering of wild cereals. Settled communities and incipient agriculture were beginning, however, in western Asia and clay-lined storage pits have been excavated from Natufian sites in Israel. Furthermore the system of hunting had changed from the short-distance attack on animals with the aid of large stone hand-axes to the long-distance shooting of prey with arrows that were armed with microliths. I should like to put forward the theory here that this change of weapon was associated with different hunting methods resulting from the world-wide spread of the domestic dog as a hunting partner that could track down wounded animals and could retrieve game from difficult terrain such as undergrowth or water. Washburn and Lancaster (1968: quoting Lee) described how one bushman with a trained pack of hunting dogs brought in 75 per cent of the meat of a camp, whilst six other resident hunters in the group, without dogs, brought in only 25 per cent of the meat.

We can only speculate on the functions of the first dogs in human societies but it seems reasonable to extrapolate broad outlines from hunting peoples of the present day. Probably the dogs fulfilled a variety of needs in communities in different parts of the world, depending on the way of life of the people, their hunting methods and traditions, and the climate. Puppies would have been cherished as pets and were probably often eaten as in New Guinea until recent times (Titcomb 1969). Adult dogs would have been valued as scavengers that would help clear the living site of garbage; they would also be necessary partners in the hunting and retrieval of game animals, and would help keep the people warm by sleeping with them at night. They would also have the very important function of acting as guard dogs, against the intrusion of outsiders both human and animal.

During the 7th to the 4th millennia B.C. the usefulness of dogs to man became even greater with the establishment and spread of agriculture and livestock husbandry. The dog must have been indispensable to the early farmers for not only could it be trained to herd flocks of sheep and other livestock, it would guard them from predators, of which its progenitor the wolf was the most important, and it would drive marauding wild ungulates away from valuable crops.

Changes in the dog brought about by domestication

Morphology Darwin, amongst many other authors, believed that because variation in the dog is so extreme there must have been more than one ancestral species. Scott (1967) maintained that only one new trait seemed to distinguish the dog from all wild canids, this being the different carriage of the tail, and that this mutation indicated that domestication had occurred only once from a single progenitor which he believes, on behavioural evidence, to be the wolf. Whichever of these views is nearer the truth there is one aspect of domestication that applies to all animals whose breeding is controlled by man. This is the reproductive isolation of a small number of individuals from the wild population which are interbred amongst themselves and constitute a new 'founder population'. These founder animals will not contain the full genetic diversity of the wild species, although by subsequent interbreeding with wild individuals it may be increased. Genetic variability in the newly domesticated animals will be restricted but at the same time mutations that would be deleterious in wild animals will be preserved and even selected for by man. It is in this way that the great variety of

dogs that we see today have been developed, whether from one ancestor or many, in one locality or in several different parts of the world.

Selection in the earliest tamed canids was probably for animals that were small, docile, and affectionate, whilst at the same time able to survive on the least possible food. Aggressive animals would have been slaughtered, and those that were not extremely robust and adaptable would have died. As with all domesticated animals, neotony (the retention of juvenile characters into adult life, together with associated submissive behaviour) played a part. Dogs that remained dependent on their substitute parent, the human owner, and had the large eyes and appealing form of the puppy would be favoured.

Apart from the different carriage of the tail and perhaps of the ears which may have been selected for at an early stage because it allowed a ready distinction between the wild canids and the dog, the most obvious changes were in a reduction in the size of the body and a change in shape of the skull and the relative sizes of the teeth. It is of course the changes in the skull and skeleton that enable the subfossil remains of canids from prehistoric sites to be identified as dogs. In the skull the palate and maxillary region became shorter and wider, the tympanic bullae became smaller and flatter, and the orbital angle greater. The upper tooth row became more bowed and the angle of the mandible deeper with the ventral edge more convex.

In the earliest finds of dogs the teeth are only marginally smaller than those of the small Asiatic wolf; their decrease in size seems to have lagged behind that of the bone, so that they are often compacted in the jaws. In more advanced domestic dogs the teeth become much smaller, especially the canines and carnassials, and they may be widely spaced in the jaws. The cusps on the premolar teeth may also become fewer and less complex in structure.

Presumably the decrease in size, which happened with the early domestication of most mammals, was a result both of selection for smaller and therefore more manageable animals as well as an adaptation to a less propitious environment with an often inadequate food supply.

Herre and Röhrs (1973) have established that many domesticated animals, including dogs, have a smaller relative brain size and weight than those of the wild progenitor. They maintain that this is because of unconscious selection by man for animals with the smallest brain weights because such animals will be less wary, more docile, and more easily tractable than individuals with larger brains. Hemmer (1976), however, holds that the small size of the brain is a consequence of the ancestral population being of small size. He believes, with other authors, that it was the west Asiatic small wolf that was the mainline progenitor of the dog and that the brain of this wolf does not differ markedly in size from that of primitive dogs such as the dingo.

In more highly domesticated dogs, changes in the shape of the head, e.g. in the bulldogs or greyhounds, or in the ears being held upright or pendent, were presumably the result of deliberate selection, whilst small changes like the presence of the dew claw were probably the result of genetic drift or of a founder effect.

Pelage The change in pelage that occurred in the dog and its possible relationship to changes in temperament are two of the most fascinating aspects of a study of the process of domestication. No experiments have been carried out to the author's knowledge on mutations in the coat colour of tamed wolves but observations have been made on coat colour variants and temperament in foxes, *Vulpes vulpes*, that are bred for their furs.

Keeler (1975) has shown that there is a close correlation between mutations of coat colour from the wild form and the lessening of fear and the enhancement of other characteristics associated with domestication. Thus, amber foxes are relatively odourless, the tail is less bushy, and they are less aggressive than red foxes. They also eat more quietly and grow fatter. In addition it has been recorded that the pituitary gland and the adrenal gland are reduced in size in the colour mutants.

It therefore seems probable that mutations in coat colour could also be associated with behavioural changes in the tamed wolf and that selection for colour variants could, as well as distinguishing the wild from the domesticated animals, also accelerate the process of habituation and breeding under conditions of captivity.

It may be speculated that changes in coat colour with related docility of temperament

occurred early on in the domestication of the dog and that the most favoured mutant form was the all-tan or ochreous body, with white tip to the tail, white on the muzzle and white on the lower limbs. This is the characteristic colouring of the dingo, the New Guinea dogs, many pariah dogs and mongrels, and the African Basenji. The skins of mummified dogs from ancient Egypt also appear to have had a uniform coat colour which was probably ochreous/tan.

Vocalizations There is still much work to be done on vocal communication in the Canidae, especially in the jackals. One of the reasons for assuming that the wolf was the main-line progenitor of the dog is that the vocalization patterns of these two canids are very similar. The coyote, *C. latrans*, and the golden jackal, *C. aureus*, have very complicated repertoires of howling that are markedly different from those of the wolf and dog. It is this evidence that persuaded both Lorenz (1975) and Scott (1967: 377) that the wolf was the only progenitor of the dog.

Until recent years many authors, including Darwin, maintained that wolves do not bark, or that if they do bark in captivity it is because they have learned to do so from dogs. All modern observers, however, agree that wolves do bark as an alarm signal given near the den, whereas howling is the means by which wolves communicate whilst hunting. Human hunters are therefore more likely to hear wolves howling than barking.

It is probable that there has been intensive selection for increased barking in domestic dogs because of its value as a warning of intruders. Dogs may also bark as a learned attention-seeking device and also they may bark in mimicry of the human voice. It is not unusual for a modern house-dog to bark as soon as a telephone bell begins to ring, presumably because of its association with speech.

Behaviour The claim made by J. P. Scott in 1950 that dogs behave in human society in the same way as wolves in wolf society remains, after much further work on the behaviour of canids and humans, substantially true. The studies that have been carried out on the behaviour of humans, dogs, and wolves have done much to elucidate why dog and man are so compatible, whilst detailed work on the behaviour of modern dogs has shown how artificial selection has altered temperament and agonistic behaviour in the different breeds under investigation.

Many social animals depend on a hierarchical system of differentiation between individuals in the maintenance of territory and in reproductive behaviour. It appears, however, that it is only in the social carnivores and in some primates, including man, that there is a rigid structure of personal relationships that are based on submissive and dominant behaviour. It is active submissive behaviour, as defined by Schenkel (1967) that is the clue to the parallel behaviour systems of wolf and man and it has presumably evolved as an inhibiting mechanism to prevent fights-to-the-death between group hunters whose way of life is based on the killing of large animals. Such fights do, of course, occur but it is the pattern of submissive behaviour that maintains a measure of control between animals that are by nature skillful killers.

The form that submissive behaviour will take in the wolf or dog depends on the attitude of the dominant animal. If he responds in a negative or intolerant way the submissive animal will normally retreat but if the superior animal wishes to enter into a communication with the inferior then harmonious social interaction can take place. The patterns of behaviour that control communication between wolves or dogs are ritualized and can be seen to recur time and time again. They have been documented by Schenkel (1967) and by many other workers including Mech (1970). The carriage of the ears and the tail plays an integral role in submissive and dominant behaviour, as does the marking of 'scent posts' by urination. In breeds of domestic dogs, those with pendulous ears and low-slung tails such as the spaniels appear to have permanently submissive or fawning behaviour whilst those with upright ears and a tail that is carried high are dominant and aggressive. This character of a breed can even be artificially enhanced, at least to the human observer, by the clipping of ears and tails as is done in some countries, e.g. with the Dobermann Pinscher and Boxer.

Scott has carried out interesting experiments on submissive and agonistic behaviour of four modern breeds of dog of approximately the same body size to demonstrate their variation and genetic basis. These studies (Scott and Fuller 1965) on puppies of Basenji, Beagle, Cocker Spaniel, and Wire-haired Terrier showed that there are consistent differences in their patterns of activity, aggression, and fearfulness. Perhaps the most interesting, although predictable conclusion, to come out of these observations was that with the Basenji and the Cocker Spaniel they were dealing with two different kinds of animal; the first on which training (essentially a process of taming) had a profound effect and the second which is apparently born tame and on which training has little effect. It can therefore be seen that domestication is a continuously developing process and some breeds have undergone much more intense selection for submissive and affection-seeking behaviour than others.

Recent history and present status

The early development of breeds

The archaeological evidence indicates that local populations of prehistoric dogs differed from each other and that there were separate kinds of dog as early as the 5th millennium B.C.. By 3000 years ago the main lines of dogs that we know today were beginning to be depicted in works of art. There were large, heavy hunting dogs of the mastiff type in Asia, as is well known from the Assyrian and Babylonian friezes, and there were greyhounds and short-legged dogs from Egypt. In Europe from Iron Age times onward there are the remains of dogs in a wide variety of sizes and in China the Pekingese dog is known from at least A.D. 700.

The Romans were probably the first to breed dogs (and other domesticated animals) systematically. They had massive fighting dogs, sheepdogs, guard dogs, and lapdogs, all of which are described with care in the literature as well as being depicted on mosaics and wall paintings, and in statues. Figure 22. 1 shows a hypothetical scheme for the relationships of the main groups of dogs to each other and to the four geographical varieties of wolves from which they are probably descended. None of these dogs, with the exception of the Pekingese, is likely to have been unfamiliar in appearance to the Romans.

Although the main groups of dogs had probably been developed by the early historic period it was not until the time of Darwin, little more than a hundred years ago, that there was any knowledge of evolution or the mechanism of the inheritance of variable characters. There were no set standards to which a breed of dog should conform, except for a few highly-prized kinds of dog such as the Saluki or the Pekingese, but even with these the individual dogs were much more variable than they are today. It was not until the idea grew of holding competitive shows for dogs in order to assess their relative merits that breeds were constrained within the inflexible patterns of size, shape, and colour of today.

In Britain, which has always been a centre of dog-breeding since Roman times, the first competitive dog show was held in Newcastle in 1859, and it was for pointers and setters only. From then onwards shows became increasingly popular and in 1873 the British Kennel Club was established to direct the conditions under which shows should be held and set the standards by which dogs should be judged. The establishment of the Kennel Club meant several old British breeds that were in danger of extinction, e.g. the Deerhound, were preserved, and it also meant the development of new breeds, such as the toy dogs, the Shetland Sheepdog and the Boxer.

The functions and economic history of the dog

The dog became an integral part of the human economic and social system at the time when it joined man as a hunter perhaps 10 000 or more years ago. By enabling humans to obtain more meat, at a period when there was a decline in the vast herds of large ungulates that had lived on the plains of the northern hemisphere during the Pleistocene, the dog became an important element in the progress of the human species towards total dominance of the world. In later periods it is probably true to say that few farmers, throughout the world, flourished without the aid of dogs to guard the homestead and herd the livestock. After the emergence of the first cities and an affluent elite in society the dog became an essential part of many sports from bear-baiting to hunting gazelle, lion or elephant. There is no ancient or modern civilization in which the dog has not played an important role, but this aspect will not be enlarged upon here as there are many excellent publications on the subject.

Until the beginning of the 19th century A.D.

dogs were bred for their use to man, which could include companionship, but they were not bred to any great extent as pets. After the spread of industrial development and the decline of the baiting and hunting sports there was a great increase in the value of dogs, simply as pets, and at the present day dog-breeding has become a world-wide commercial enterprise worth millions of pounds. Unlike livestock animals where economic value lies in their productivity, the value of a dog lies in its status as a 'special animal'. The most rare and beautiful dog, according to the fashion of place and time, will be the most sought-after by the rich. For this reason it is unlikely that genetic diversity within populations of dogs will be lost, although in the most developed and affluent countries the highly-bred humanized dog could altogether displace the common mongrel.

There must be few opportunities for the dog to fulfil any new role in human society but within recent years, especially in the United States, dogs have been used in psychiatric hospitals as part of therapeutic procedure. It has been reported that most patients respond very well to being given a pet with whom they can communicate non-verbally and this leads to a marked improvement in the interactions between patients and staff (Corson *et al.* 1977). Within recent years there has been a great expansion in the training of guidedogs for the blind and of dogs for use in the search for illicit drugs and explosives. There is also an increasing demand for dogs in their traditional role as guardians of property.

The cultural attitudes of people to their dogs vary greatly in different parts of the world and although they seem often to be contradictory and to have little in the way of a rational basis they would probably be found to have good historical and functional explanations if properly investigated. In times of famine dogs have always been killed for food, as a last resource, but it seems that since the early prehistoric period there have been few countries where the dog has been reared primarily as a food animal. In many parts of Asia dogs are considered to be animals that are unclean and untouchable, this being a taboo that has probably developed from awareness of the diseases and parasites that are carried by dogs. No 'civilized' woman would suckle a puppy, any more than she would knowingly eat one, but in the ancient Pacific both practices were common until recent times (Titcomb 1969). Only in Australia, New Zealand, and New Guinea were there indigenous animals that could serve as prey for dogs that were brought by the first people to the ancient Pacific. Imported pigs and dogs were the only sources of meat available for the people of the smaller islands and consequently there was no development of hunting cultures there. The dogs lived on the same vegetable diet as that eaten by the humans and they not only fulfilled the function of suppliers of meat but were also cherished pets.

Feral dogs

The dingo of Australia is a feral dog that has lived wild for so long that it could be argued that it is now as much a wild species as the jackal or the coyote, and indeed all work done on the behaviour of the dingo indicates that this canid is now a Type II hunter. If it were not known that it must have been taken to Australia by prehistoric peoples from Asia the dingo would be classified as a separate species akin to the jackals and coyote. In fact the dingo is probably the only 'purebred' domestic dog in the world and it is most likely to be directly descended from the Indian wolf, *Canis lupus pallipes*. It is a relict that survived as an isolated feral population in Australia, for thousands of years without the introgression of dog genes from any other breed, until the arrival of the first Europeans with their dogs. Furthermore, it is probable that the true dingo is descended from only a very small founder population of dogs that survived the journey from southeast Asia to the continent of Australia.

Dogs could not have been introduced into Australia before approximately 12 000 years ago because they never reached Tasmania and it was at around this date that the links between Australia and Tasmania were broken by flooding of the Bass Strait due to rising sea-level. The almost complete skeleton of a male dog has been excavated from South Australia by Mulvaney *et al.* (1964) and this has been dated by radiocarbon from associated charcoal to *c.* 1000 B.C. It has been claimed that this skeleton is identical to that of a modern dingo of 18 weeks of age. Although dingoes are notorious as killers of sheep and are treated as vermin by most stockmen in Australia, their extermination would be a great loss for they provide a unique relic of the

dogs that must have travelled with man all over southeast Asia in the prehistoric period.

There is another group of feral dogs that is probably also of great antiquity in Papua New Guinea. The dogs from the highlands, known as New Guinea singing dogs because of their eerie communal howling, are probably also of unmixed descent from a prehistoric population.

The feral dogs of India and Africa have a more mixed ancestry than those of Australia and New Guinea but in many areas they live in populations that have been reproductively isolated for a very long period. As the pariah dogs of India are treated as untouchable they are not interfered with and live very much as wild animals that are subject to the harsh struggle for survival in a land of extreme poverty for both human and animal. These dogs fulfil a most useful function as scavengers that will clean up the detritus associated with human habitations, and they are tolerated as a means of waste disposal. This situation is in direct contradiction to that of affluent countries where stray dogs can become a serious problem especially in cities, where they are a threat to traffic and a sad reflection of human thoughtlessness.

The African dogs, of which the Basenji from Zaire is one kind, are not truly feral like the dingo and the pariah dogs. These dogs live as camp-followers to the African hunters and are often trained to retrieve game. Like the true pariah dogs, however, they mostly eke out an existence as scavengers.

Early domesticated canids other than true dogs

Anubis, the jackal god of ancient Egypt

There is no agreement amongst Egyptologists on whether Anubis, the 'jackal god' was supposed to be representative of a jackal or a fox, but it was unlikely to be a true dog. This impressive black canid, with upright ears and an alert expression, was often depicted by the ancient Egyptians and was worshipped as the patron of embalmers. To the present author the evidence suggests that Anubis was modelled on the wolf-jackal of Egypt, *C. aureus lupaster*. This is the largest of all jackals and it is generally classified as a subspecies of the golden jackal. The best-known representation of

Anubis is the black and gold figure from the 'treasury' of Tutankhamun. Mummified remains of jackals have been reputedly recorded from Asyut, the ancient Egyptian city of Anubis, but although there was obviously a very close link between a species of canid and ritual in ancient Egypt it cannot be proved that the animals were truly domesticated, as opposed to being kept in captivity.

Domesticated canids in South America

There seems little doubt, from the descriptions and illustrations of the famous naturalist and traveller, Charles Hamilton-Smith, that before the 19th century arrival of Europeans, the Indians of South America domesticated the local canids belonging to the genus *Dusicyon*, as well as the crab-eating fox, *Cerdocyon thous*. Probably a number of different species of *Dusicyon* were domesticated by the Indians and they fulfilled many of the roles of true domestic dogs but were never as habituated to humans nor were they very companionable. They were therefore replaced by European dogs as soon as these became available (Clutton-Brock 1977).

The Falkland Island 'wolf'

The origins of this little-known extinct canid are unknown. It was not a wolf but looked more like a large variety of the South American 'foxes' of the genus *Dusicyon*, and it is generally named *D. australis*. The South American 'foxes' in their morphology appear to lie between the true foxes of the genus *Vulpes* and the species of *Canis*, and they have the typical pelage and skull characters of wild canids. The pelage and morphology of the Falkland Island wolf, however, carried some characters that are generally associated with domestication and in many ways it must have looked like a dingo. It had a white tip to its tail, and white on its lower limbs, a wide muzzle, large somewhat compacted teeth, especially in the premolar region, and expanded frontal sinuses. There are, however, characters in the teeth that distinguish them immediately from those of true dogs and if this canid was indeed a feral domestic animal then its ancestry must either have been from an indigenous South American canid or from hybridization between South American canids and true dogs (Clutton-Brock *et al.* 1976, Clutton-Brock 1977).

The most peculiar fact about the Falkland Island wolf is that until Europeans reached the Islands it was the only mammal living there and it survived by feeding on sea-birds. It is very unlikely that this large carnivore could have survived as the only species of mammal on East and West Island throughout the periglacial conditions of the Pleistocene, even if at that time the islands were accessible as a result of a lower sea-level. The distance of the Falkland Islands from the mainland is too great (*c.* 480 km) for it to have reached them since the end of the last glaciation and the only remaining explanation is that this canid was taken there as a domestic animal at some time during the last 10 000 years. The Falkland Island wolf was described by Darwin 20 years before its final extinction in about 1880.

References

Banks, E. M. (ed.) (1967) Ecology and behavior of the wolf. *American Zoologist*, **7** (2): 221–381

Bueler, L. E. (1974) *Wild Dogs of the World*. Constable: London

Chiarelli, A. B. (1975) The chromosomes of the Canidae. In: Fox (1975) (op. cit.), pp. 40–53

Clutton-Brock, J. (1977) Man-made dogs. *Science*, **197** (4311): 1340–2

Clutton-Brock, J. (1979) Mammalian remains from the Jericho Tell. *Proceedings of the Prehistoric Society*, **45**: 135–57

Clutton-Brock, J. (1981) *Domesticated Animals from Early Times*. British Museum (Natural History) and William Heinemann: London.

Clutton-Brock, J., Corbet, G. B. and Hills, M. (1976) A review of the family Canidae with a classification by numerical methods. *Bulletin British Museum (Natural History), Zoology*, **29** (3): 119–99

Corbett, L. and Newsome, A. (1975) Dingo society and its maintenance: a preliminary analysis. In: Fox (1975) (op. cit.), pp. 369–79

Corson, S. A., Corson, E. O'L., Gwynne, P. H. and Arnold, L. E. (1977). Pet dogs as nonverbal communication links in hospital psychiatry. *Comprehensive Psychiatry*, **18**(1): 61–72

Davis, S. J. M. and Valla, F. R. (1978) Evidence for domestication of the dog 12 000 years ago in the Natufian of Israel. *Nature*, **276** (5688): 608–10

Fox, M. W. (1971) *Behaviour of Wolves, Dogs, and Related Canids*. Jonathan Cape: London

Fox, M. W. (ed.) (1975) *The Wild Canids; their systematics, behavioral ecology and evolution*. Van Nostrand Reinhold: New York and London

Gray, A. P. (1972) *Mammalian Hybrids*. Commonwealth Agricultural Bureaux: Farnham Royal, Bucks, England

Hall, R. L. and Sharp, H. S. (1978) *Wolf and Man Evolution in Parallel*. Academic Press: New York and London

Hemmer, H. (1967) Man's strategy in domestication – a synthesis of new research trends. *Experientia, Basel*, **32**: 663–6

Herre, W. and Röhrs, M. (1973) *Haustiere – Zoologisch Gesehen*. Gustav Fisher: Stuttgart

Keeler, C. (1975) Genetics of behavior variations in color phases of the red fox. In: Fox (1975) (op. cit.), pp. 399–415

Kurtén, B. (1965). The carnivora of the Palestine caves. *Acta Zoologica Fennica*, **107**: 1–74

Lawrence, B. (1967) Early domestic dogs. *Zeitschrift für Säugetierkunde*, **32** (1): 44–59

Lorenz, K. Z. (1975) Foreword. In: Fox (1975) (op. cit.), pp. vii–viii

Mech, L. D. (1970). *The Wolf: the ecology and behavior of an endangered species*. Natural History Press: New York

Mengel, R. M. (1971). A study of dog-coyote hybrids and implications concerning hybridization in *Canis*. *Journal of Mammalogy*, **52**: 316–36

Mulvaney, D. J., Lawton, G. H. and Twidale, C. R. (1964) Archaeological excavation of rock shelter No. 6 Fromm's Landing. *Proceedings Royal Society Victoria*, **77** (2): 479–94

Olsen, S. J. and Olsen, J. W. (1977) The Chinese wolf, ancestor of New World dogs. *Science*, **197**: 533–5

Scott, J. P. (1967) The evolution of social behavior in dogs and wolves. *American Zoologist*, **7** (2): 373–81

Scott, J. P. and Fuller, J. L. (1965) *Genetics and the Social Behavior of the Dog*. University of Chicago Press: Chicago

Schenkel, R. (1967) Submission: its features and functions in the wolf and dog. *American Zoologist*, **7**: 319–31

Seal, U. S. (1975) Molecular approaches to taxonomic problems in the Canidae. In: Fox (1975), (op. cit.) pp. 27–39

Titcomb, M. (1969) Dog and man in the Ancient Pacific. *Bernice P. Bishop Museum Special Publication*, No. 59: Honolulu, Hawaii

Todd, N. B. (1970) Karyotypic fissioning and canid phylogeny. *Journal of Theoretical Biology*, **26**: 445–80

Turnbull, P. F. and Reed, C. A. (1974) The fauna from the terminal Pleistocene of Palegawra Cave, a Zarzian occupation site in northeastern Iraq. *Fieldiana Anthropology*, **63** (3): 81–146

Washburn, S. and Lancaster, C. S. (1968) The evolution of hunting. In: *Man the Hunter*, ed. R. B. Lee and I. De Vore, pp. 293–303. Aldine: Chicago.

23

Foxes

D. K. Belyaev

Institute of Cytology and Genetics, Novosibirsk, U.S.S.R.

Silver fox

The silver fox is a colour phase of the widespread red fox, *Vulpes vulpes* (also called *V. fulva* in North America), differing from it in having black pigment. There are two types of silver fox, the Canadian (or Standard) and the Alaskan; in the latter the hair is brown in the ears, on the shoulder blades and at the root of the tail. In Canada the Alaskan type is commoner in British Columbia and the Standard in Quebec. There used also to be a few silver foxes in the wild in Eurasia; these were of the Alaskan type. Only the Canadian type is bred on farms.

The black pigment develops under the control of one or other of two independently-inherited incompletely-dominant genes: *Al* – Alaskan, and *R* – Canadian. Matings between the Alaskan and red fox produce 'cross-foxes' in the first generation and crosses between the Canadian black and the red fox produce so-called 'bastards'. The cross-fox is reddish brown; black hairs extending up the legs and across the shoulders meet a spinal extension of the black hairs from the tail to give the 'cross'. The bastards are closer to the red fox in their colour, but have an admixture of black hairs on their extremities. It is not clear whether the difference in colour in the F_1 is connected with some specificity in the *Al* and *R* gene effects, or with the effect of some modifiers (Robinson 1975).

Silver foxes have a white band on their guard hairs which is specially visible on the black background. The concentration of silver hairs on one or other part of the body, the body area covered by the silver hair (i.e. the percentage of silveriness), and the size of the silver band are positively correlated. All these characters are under polygenic control and are inherited quantitatively.

Under breeding conditions in cages, the fox has maintained the same seasonal rhythms as in nature. Reproduction is strictly seasonal; the mating season, irrespective of geographical latitude, begins usually not earlier than the second half of January, and ends in the second half of March. Females have only one oestrus a year, which lasts 3–5 days.

The gestation lasts 51–52 days so the whelping season is from the middle of March to the middle of May. The mean litter size is 5–5.5 cubs, but cases of 13–14 cubs are known. The female's potential fertility, i.e. the number of eggs ovulated, varies between 1 and 15, being 6.3 on the average. Embryonic deaths occur chiefly in the pre-implantation period.

Males on farms mate, on the average, with 5–6 females in one season, but this number is variable. Both males and females become pubescent at 10–11 months.

Characteristic of foxes is the strict seasonal rhythm of fur formation and moulting. The fur becomes mature by the end of November and remains good during December and January. In February or March moulting begins and lasts until the autumn. From the second half of summer, at the same time as the loss of the old hair, the hair forming new winter fur grows. The reproduction and moulting rhythm is controlled by the photoperiod, i.e. the seasonal changes in the length of daylight.

Foxes have a diploid chromosome set of $2n = 34$ but have, in addition, a varying number (1–7) of microchromosomes which seem to be of heterochromatic nature. The number of cell clones with additional microchromosomes is correlated with the type of defensive behaviour in respect to man, i.e. with the degree of domestication. In foxes showing a high degree of domestic behav-

iour, i.e. those resembling the domestic dog, the number of mosaic cell clones is larger than in foxes with a wild behavioural type (Belyaev 1979).

About 365 000 skins of silver foxes are produced annually on farms. The bulk of them, i.e. about 350 000, are produced in the U.S.S.R. and about 15 000 in Finland.

The foxes bred on farms are not domesticated *sensu stricto*. They preserve all the main characteristics of the wild species: behaviour and strict seasonal rhythm of physiological functions. Crosses between farm-bred and wild foxes are fertile. So are the offspring from such crosses. The chromosome number in wild and farm-bred foxes is the same.

The beginning of breeding of silver foxes in captivity goes back to 1892, when Charles Dalton (Canada) obtained offspring from parents caught in the wild. The aim of breeding was to obtain fur which was highly valued on the market. High prices paid for fur, and especially for living breeding animals, as well as Dalton's success, stimulated the formation in Canada and the U.S.A. of special farms, the number of which was counted in hundreds before the First World War. In Europe, silver fox farms began to appear after the war. In the early 1930s there were in the world as many as 11 000 farms on which over 300 000 silver fox skins were produced. In 1940, 1 350 000 skins of silver foxes, the largest number in the history of fox breeding, were sold on the international fur market. The main producers were Canada, the U.S.A., Norway, Sweden, and the U.S.S.R. During the Second World War, and especially shortly after it, the number of foxes on farms was sharply reduced, due to decline in demand and hence the fall of prices for the fur. This was the result of competition from mink fur which was becoming fashionable leading to a sharp increase in its price and an intense development of mink breeding (Kaplin *et al.* 1955).

The main traits for which the fox was selected all over the world were good fur, high fertility and stability of reproduction. Due to this selective breeding, the fur of silver fox is characterized by a good development of hair and an intense black colour. Brown shades have always been undesirable and selection has been made against them. The requirement as to the silveriness of the fur was different at different stages of fox breeding; before the First World War, skins with a low percentage of silveriness were appreciated, i.e. where the silver hair made up no more than 25–50 per cent of the skin surface; later on, skins with a silveriness of 75–100 per cent fetched a higher price.

Breeding has considerably influenced fur quality and reproduction of foxes; however, it has not created any special breeds or races, although it has adapted them to life and reproduction in captivity. In the initial stages of breeding, the reproductive ability of farm-bred foxes was much lower than later on. This was due to the improvement of breeding techniques and to selection.

Several mutations have been described on farms. In U.S.A. and Canada, there was the 'pearl' mutation causing dilution of the main colour. It has been found that the pearl colour is controlled by at least two independent recessive genes having similar phenotypical effects. In the 1930s and 1940s there appeared platinum and white-faced foxes independently on farms in different countries (Canada, Norway, and the Soviet Union). Both of these forms are characterized by a specific piebaldness on the face and a white collar. The white-faced foxes do not differ from the silver fox in the general shade of colour while the platinum ones are much lighter. A genetic analysis has demonstrated that these colours are controlled by allelic dominant genes which are lethal when homozygous (Mohr and Tuff 1939, Cole and Shackelford 1943). In 1944, on a farm in Georgia (U.S.S.R.), a fox of white colour with black spots scattered all over the body was born. This colour was found to be controlled by a dominant gene allelic to the platinum and the white-face genes. When homozygous this gene causes death during the embryonic or postnatal periods (Belyaev *et al*. 1975). Skins of the platinum and white-faced foxes were produced in great numbers in the first years of, and shortly after, the Second World War. Thus in 1946 at the Hudson Bay Company auction alone, 1100 skins of platinum foxes were on sale. Later on their number sharply decreased due to the fall in prices.

Nowadays, fox breeding exists practically only in the Soviet Union. It is difficult to say whether fox breeding will be revived in other countries, and if so, when. In the Soviet Union, the main colour mutations – platinum, white-faced, and the Georgian white – are maintained.

A real domestication of the fox can hardly be expected under the existing breeding conditions on farms. As experiments on domestication of the fox have demonstrated, the key moment of domestication is selection for behaviour, i.e. a hereditary reorganization of behaviour with respect to man. In the course of such breeding the behaviour of the fox becomes closer to that of the domestic dog and, in parallel with this, many other characteristics, beginning with reproduction and moulting, are changed. As a result of such selection foxes acquire the ability to reproduce twice a year, i.e. they lose the seasonal rhythm of reproduction. Among foxes bred for behaviour there appear animals with traits characteristic of dogs (drooping ears, a specific piebaldness, position of tail, etc.) (Belyaev 1979).

Of course, some selection for behaviour is done on farms but it is carried out with low intensity and unconsciously, due to the existing correlation between behaviour and fertility. That is why some domestication-induced changes found in the laboratory may also be expected on farms.

Arctic fox

Blue and white are alternative winter colours of the arctic or polar fox, *Alopex lagopus*, which has a circumpolar distribution in Eurasia and North America. Most wild foxes are white but on farms only blue foxes are bred. There are two types of blue fox; the Alaskan differs from the Greenlandian in its darker colour, which is due to the higher concentration of pigmented hair in its coat. The 'blue' colour of the arctic fox is determined by the fact that among the guard hairs there are both white and pigmented hairs. The blue fox has no white band on its hair. The blue colour develops under the control of one dominant gene but heterozygous animals are somewhat lighter than their blue parents. The colour variation (the degree of silveriness) in the blue fox is under polygenic control (Johansson 1960). In summer the white fox is greyish brown and the blue fox is greyish.

Farm-bred polar foxes have all the properties characteristic of the wild species. They maintain the strict seasonal rhythm of reproduction and moulting and the behaviour of wild animals.

The breeding season in farm-bred polar foxes begins, as a rule, not earlier than the second half of February and ends at the end of April or the beginning of May. Females have one oestrus within the breeding season, which lasts 6–8 days. The length of gestation is 52 days. The whelping season is usually from the middle of April to the end of May. In the Alaskan blue fox the breeding season is on the average 1 week later than in the Greenlandian. The mean litter size is 8–9 cubs, but cases of as many as 25 cubs are known. In Alaskan females the fertility is somewhat higher than in the Greenlandian ones (Johansson 1960). The mean potential litter size is 16–18, and the embryonic mortality rate is about 50 per cent.

A male mates with 5–6 females during the breeding season; the age of puberty in both sexes is 10–11 months. The seasonal rhythm of moulting and fur formation characteristic of the blue fox is similar to that of the silver fox. The photoperiodic control of reproduction and moulting is also present in the blue fox.

The diploid chromosome number varies from 48 to 50, due to variation in pairs 13 and 26. This variation does not affect fertility and viability; 48-, 49- and 50-chromosome polar foxes exist in equal proportions (Wurster and Benirschke 1968). Hybrids between *A. lagopus* and *V. vulpes* are known but they are sterile.

About 2 500 000 polar fox skins are produced on farms annually – 1 250 000 in the Soviet Union, 600 000 in Finland, 350 000 in Poland 250 000 in Norway.

The farm-bred polar fox is not domesticated *sensu stricto*, although it is adapted to life in captivity. Breeding of arctic foxes on farms started at the beginning of the 20th century in Scandinavia (Norway, Sweden) and in the 1920s it began in U.S.A., Canada and the U.S.S.R. Farm breeding of the polar fox thrived especially in the 1930s, in Norway and Sweden. In Norway, in 1939, there were 30 000 polar foxes on farms, and in Sweden 44 000. Later on, because of decrease of skin prices, there began a sharp reduction of the number of farms and of farm-bred polar foxes (Kaplin *et al.* 1955).

Polar foxes have been bred for the density and softness of fur, their purity of colour, i.e. the absence of brown shade, and for high reproductive ability. Selection has resulted in the formation of two types of polar fox: one with a darker colour and a pronounced degree of silveriness (silvery blue fox), and a lighter one (veiled blue fox). The former type is closer in its colour to the Alaskan blue fox, while the latter to the

Greenlandian. As in silver foxes, during farm breeding of arctic foxes there has appeared a platinum mutation characterized by piebaldness on paws, belly and face, and by a general dilution of colour.

Nowadays, the number of polar foxes on farms is greatly reduced, and it is hard to tell whether it will go up again in the near future.

In a real domestication of the polar fox, as in that of the silver fox, the main thing is the hereditary reorganization of behaviour (Belyaev 1979). Selection for behaviour is hardly probable on industrial farms; therefore the domestication-induced changes in the polar fox will develop slowly.

References

Belyaev, D. K. (1979) Destabilizing selection as a factor in domestication. *Journal of Heredity*, **70**: 301–8

Belyaev, D. K., Trut, L. N. and **Ruvinsky, A. O.** (1975) Genetics of the *W* locus in foxes and expression of its lethal effects. *Journal of Heredity*, **66**: 331–8

Cole, L. J. and **Shackelford, R. M.** (1943) White spotting in the fox. *American Naturalist*, **77**: 289–321

Johansson, I. (1960) Inheritance of the colour phases in ranch bred blue foxes. *Hereditas*, **46**: 753–66

Kaplin, A., Ivanov, V. and **Pastushenko, M.** (1955) [*Fur.*] Foreign Trade Publishing House: Moscow. (In Russian)

Mohr, J. L. and **Tuff, P.** (1939) The Norwegian platinum fox. *Journal of Heredity*, **30**: 227–34

Robinson, R. (1975) The red fox (*Vulpes vulpes* L.) In: *Handbook of Genetics*, ed. R. C. King, Vol 4, pp. 399–409. Plenum Press: New York and London

Wurster, D. and **Benirschke, K.** (1968) Comparative cytogenetic studies in the order Carnivora. *Chromosoma*, **24** (3): 336–82

24

Raccoon dog

M. H. Valtonen

*College of Veterinary Medicine,
Helsinki, Finland*

The wild animal

The English name, raccoon dog, for *Nyctereutes procyonoides* arises from its resemblance to the raccoon, *Procyon lotor*. The most commonly used local vernacular name, *tanuki*, comes from Japan. In the fur trade the raccoon dog is known as Ussurian raccoon, Russian raccoon, Finnraccoon, Japanese fox or sea-fox.

The raccoon dog is taxonomically closely related to the dog and the fox, but phylogenetically was early differentiated from the other members of the family Canidae (Wurster 1969). It is quite fox-like, differing in smaller ears and shorter legs and tail. It is brownish grey above and yellowish brown beneath. The large black cheek patches resemble the mask of the raccoon, hence the name. The underfur is thick and soft with long and coarse guard hairs.

The raccoon dog is primarily nocturnal. It inhabits forests and thickets near the sea, lakes and rivers. It is omnivorous, feeding on both small prey and vegetation like corn and berries. Body weight increases greatly in the autumn and in cold areas this canid may sleep in burrows through most of the winter, emerging only on warm days.

Raccoon dogs are seasonally monoestrous, pairing up in the autumn in the wild although mating takes place only the following February or March, depending on the climate. Polygamous mating is successfully practised in captivity. Pro-oestrus lasts 2–14 days and oestrus 3–4 days; the gestation period is 59–64 days and litters are large: 6–7 cubs (Valtonen *et al.* 1977).

The diploid chromosome number of the Japanese raccoon dog is 42 (Wurster 1969). Finnish raccoon dogs originating in Russia have $2n=56$ (Makinen 1974). Future studies will establish whether there is in fact more than one raccoon dog species.

The raccoon dog is native to eastern Siberia, Korea, eastern China, northern Indo-China and Japan (Bueler 1973). In the 20th century the distribution of the raccoon dog has been expanded far beyond the original range through extensive introductions (Fig. 24.1). During the years 1928–58 almost 100 000 individuals were introduced into western Russia, the Caucasus, Kazakhstan and Siberia (Heptner and Naumov 1974). The raccoon dog is now well established in European Russia and has spread to Scandinavia and central Europe. Nowak and Pielowski (1964) found no natural restraints which might prevent raccoon dogs from occupying the whole of western Europe. The introductions to Asiatic Russia met with little success; in this region the species is still rare.

Breeding in captivity

Widely hunted for its fur, the raccoon dog grew rare in its native countries. The skins were used for several purposes: making fur jackets, collars, caps, even bellows and decorations for drums (Kroll and Franke 1977).

In the Far East raccoon dogs are known to have been managed in captivity already in ancient times. No information, however, is available about their first domestication.

At the beginning of the 20th century fur-animal breeding was spreading from North America to other countries. Thousands of state-supported farms started to breed raccoon dogs in Japan, but with only a few breeding animals in each farm production stayed low. Today Japan has few raccoon dog farms left, but 30 000–40 000 wild animals are hunted yearly.

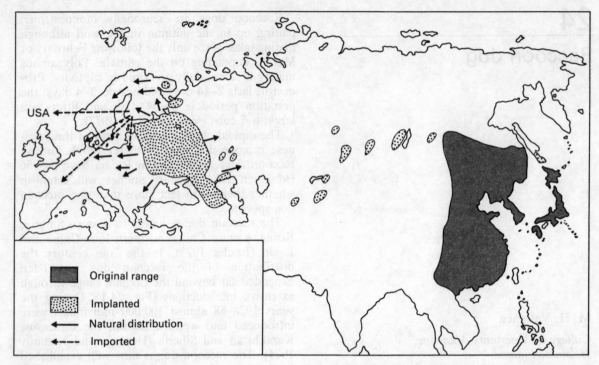

Fig. 24.1 Distribution of raccoon dogs

Increasing demand for raccoon-dog fur stimulated the start of farming in Russia in 1928. In 1934 there were fifteen state-owned fur farms breeding mostly raccoon dogs, but the silver fox gained ground and raccoon-dog farming ceased in the U.S.S.R. in 1945. In the 1930s farming was tried also in Sweden and Germany but with little success. In Germany most of the breeding animals died of an unknown disease in 1937 and the rest, including valuable black and white mutations, were killed and pelted (Schmidt 1973).

The successful introduction of raccoon dogs considerably increased the importance of this species in the fur trade. It was being hunted more than formerly in the Far East. According to Kroll and Franke (1977) the total catch in 1950 was about 500 000 raccoon dog pelts.

In Finland farming began in the early 1970s. The stock originated from a wild population from Russia. Each farm was started with one or more captured couples. By 1980 the number of farms was about 500 with 15 000 females producing 55 000 young.

Though cautious, the raccoon dog is easy to handle and feed in captivity. Due to this good adaptability to domestication raccoon-dog breeding is both successful and profitable. Although the fur industry is easily affected by changes in fashion, the demand for long-haired fur is likely to increase. Interest in raccoon-dog farming has been awakened in several European countries. Finnish raccoon dogs have been exported to Denmark, France and Poland, some animals even to the United States and Canada. Interest in raccoon-dog farming is also on the increase in Japan and the U.S.S.R.

References

Bueler, L. E. (1973) *Wild Dogs of the World*, pp. 217–23. Constable: London

Heptner, V. G. and **Naumov, N. P.** (1974) *Die Säugetiere der Sowjetunion*, Vol 2, pp. 70–97, Seekuhe und Raubtiere. Fischer: Jena. (Translated from the Russian edn, 1967)

Kroll, J. and **Franke, C.** (1977) *Jury Frankels Rauchwaren-Handbuch*, pp. 162–4, Tier und Fellkunde. Rifra-Verlag: Murrhardt

Makinen, A. (1974) Exceptional karyotype in a raccoon dog. *Hereditas*, **78**: 150–2

Nowak, E. and **Pielowski, Z.** (1964) Die Verbreitung des Marderhundes in Polen im Zusammenhang mit seiner Einburgerung und Ausbreitung in Europa. *Acta Theriologica*, **9**: 81–110

Schmidt, F. (1973) Vom Marderhund (*Nyctereutes procyonoides* G.). *Deutsche Pelztierzuchter*, **47**: 91–2

Valtonen, M. H., Rajakoski, E. J. and Mäkelä, J. I. (1977) Reproductive features in the female raccoon dog (*Nyctereutes procyonoides*). *Journal of Reproduction and Fertility*, **51**: 517–8

Wurster, D. H. (1969) Cytogenetic and phylogenetic studies in Carnivora. In: *Comparative Mammalian Cytogenetics*, ed. K. Benirschke, pp. 310–29. Springer-Verlag: New York

25
Cat

Roy Robinson

St Stephens Road Nursery,
London, England

Introduction

The domestic cat rivals the dog as the most widely kept household pet, observing neither regional nor national boundaries. The cat may be kept as a companion or as an economic animal. In most ordinary households, the cat is maintained as a pet while on farms and the like, the animal is kept also to deter rodent pests. Many commercial properties have a cat for the same (official) purpose.

Nomenclature

The domestic cat is frequently referred to as *Felis catus* (or *F. domesticus*). The first name is due to Linnaeus but he applied the designation to the blotched tabby which is merely a mutant variant of the wild striped or mackerel tabby. The domestic mackerel tabby was subsequently named *F. torquata,* a designation not in common usage.

Two English vernacular words meaning cat are 'puss' and 'tabby', frequently linked to cat as in pussy-cat and tabby-cat. The word 'cat' in various forms occurs in several modern European languages and is thought to have derived from earlier use by the Romans. Zeuner (1963) suggests that it can be traced to Berber languages, where it occurs in various forms. The Arabic word for cat is *quttah,* for instance. He also suggests that 'puss' may be derived from the Egyptian goddess Pasht or Bast but this is dubious because puss is an ancient English word for a she-cat, later used to mean cats of both sexes. The word is commonly employed to call the animal. Tabby is thought to be one of a series of corruptions of the Arabic word *attabiya,* a particular striped silk fabric originating in Baghdad about the 12th century and called after an Umayyid prince named Attab. Thus, only the word 'cat' is sufficiently ancient to afford a clue to the place of origin of the animal and this is by no means conclusive.

Reproductive biology

For a carnivore, the cat is notably prolific. Under optimum conditions a female will attain sexual maturity at 7–12 months, depending upon the season of birth and breed. The male usually lags behind the female by 1 or 2 months. The cat is sometimes referred to as a seasonal breeder but this is not so in any rigid sense. Under a long-day light regime, as is often the case for individuals residing in a household with artificial illumination, the cat will mate and litter throughout the year. However, there is a tendency for litters to be born in the early spring and again in the late summer.

The female is facultatively polyoestrous in that several consecutive heat periods will occur in the absence of a male but none after a successful copulation. Ovulation is induced by the complicated fore-play and act of coitus, and is almost unknown without it. In the presence of a male, the actual heat period lasts about 4 days but, in the absence of a male, the period may be as long as 9–10 days, followed by a recurrence within 2–3 weeks.

A sterile copulation results in a pseudopregnancy which lasts about 36 days. A successful pregnancy lasts about 65 days with little variation after allowing for uncertainty as to the actual day of heat in which the eggs were fertilized. The mean litter size is 4 kittens with typical variation of 3–10. The young are old enough to be weaned at about 8 weeks of age. The secondary sex ratio is 104 males: 100 females. A *postpartum* oestrus is absent but the female is capable of coming into heat some 3–4 weeks after the litter is weaned. Maximum litter size is reached by the third litter and remains at this level for as many as 15 litters (Robinson and Cox, 1970). A female may reproduce for as long as 8–10 years but the last litters of her fertile life are usually smaller than usual and tend to be erratic.

Karyotype

The domestic cat has 19 pairs of chromosomes. In general, three types may be recognized as follows: metacentrics (2 large, 3 small), submetacentrics (7 large, 4 small) and acrocentrics (2 small). The sex chromosomes are metacentrics, the X medium-sized and the Y small. The wild subspecies *F. silvestris libyca, F. s. ornata* and *F. s. silvestris* have karyotypes identical with the above. This is one of the primary reasons for postulating that these are geographically subspecies or races of a single widely-distributed species or species complex.

Wild species

The domestic cat is almost certainly derived from the *Felis silvestris–libyca* species complex. This complex consists of a number of races of small cats inhabiting the whole of Europe and the larger offshore islands (e.g. Britain, Corsica, Sardinia, Sicily, Crete, etc.), Asia, except the far north, and North Africa. Some have been classified as distinct species but it is doubtful if this is warranted biologically. Those so far examined have identical karyotypes and crosses between them produce fertile offspring. However, considering the vast area involved, it is not surprising to find regional variation of phenotypes due to adaptation to local conditions. These are now distinguished as subspecies or races, rather than as species (Weigel 1961). There is little reason to dispute Weigel's designation of *F. silvestris* for the whole complex.

Felis s. silvestris is commonly called the European wild cat and *F. s. libyca,* the African wild cat. The two forms are very similar, the main differences being that the latter has a more lissom body, is lighter in colour and has a less well-defined tabby pattern. These two subspecies are specially mentioned because both have been suggested as the ancestor of the domestic cat. The subspecies *F. s. ornata* (steppe cat) and *F. s. ocreata* (dun cat) have also been thought to be involved. This now seems improbable; it is almost certain that the domestic descended from *F. s. libyca* and any contribution from other subspecies has been negligible. Hemmer (1976a, 1978) has refuted some of the wilder suggestions for the polyspecific origin of the animal. The likelihood of *F. margarita* (sand cat). *F. manul* (Pallas's cat) or *F. temmincki* (Temminck's cat) being involved in the evolution of the domestic cat is extremely small. (For distribution of these species, see Fig. 25.1.)

The only species other than *F. s. libyca,* which could have made some contribution to the early domestic cat is *F. chaus*. While the majority of the mummified Egyptian skulls examined by Morrison-Scott (1952) could be attributed to *libyca*, a very small proportion could have been *chaus*. However, this finding merely means that *chaus* was kept in captivity, not necessarily that it was domesticated nor that it contributed significantly to the domestic cat genome. This latter aspect has been discussed by Robinson (1977). *Felis chaus* has only a vestigial tabby pattern, as has a breed of domestic cat known as Abyssinian. However, the Abyssinian pattern is due to a dominant allele at the tabby locus and almost certainly arose by mutation, rather than by hybridization with *chaus*. *Felis chaus* is a larger beast than *libyca* and the two species would probably not normally copulate. Under domestic conditions, they can be induced to mate and the hybrids are fertile (Gray 1972). The fertility cannot be taken as indicative of ancient hybridization because many of the small cats have very similar, if not identical, karyotypes (Robinson 1979) and fertile hybrids are possible for many of the species.

Although, initially, the domestic cat may have been a prized possession and was carefully guarded, it is notoriously difficult to confine a female in heat. As the animal became more numerous these restrictions were presumably relaxed. It is easy to imagine that some crossing with local wild cats occurred. Most of this crossing would have taken place during the early days, before the wild cat retreated in the face of expanding civilization. It is difficult to believe that such crosses would have contributed to the domestic population, for the 'hybrids' would almost certainly be wilder than the domestic parent. The tamer may have stayed around but the wilder would have become part of the wild population.

The above reasoning may have prompted Suminski (1962, 1977) to suggest that no 'pure' races of *F. s. silvestris* exist in Europe at this time. He considers that all populations have some admixture of domestic genes. However, it must be pointed out that Kratochvil and Kratochvil (1970) could find no evidence of variation of cranial capacity in a sample of *F. s. silvestris* from the west Carpathians which could be attributed to domestic cat genes. This implies that the influx of domestic genes to wild populations is trivial or that natural selection against the genes is severe. Popular accounts of *F. s. silvestris* being initially timid, yet ultimately tamable, should be regarded with some disbelief. The alleged wild kittens could be feral domestics because even kittens of household females must become accustomed to humans at an early age if they are not to display some fearful behaviour as adults. Even if the kittens are *F. s. silvestris,* it is possible for them to have some domestic ancestry, thus explaining their amenability.

Hillaby (1968) has cited a report that *F. s. libyca* is common in Zimbabwe and can be domesticated, perhaps more easily today than in the past. These animals roam about in the semi-urban areas, scavenging for food and interbreeding with the domestic. Such populations would be slowly evolving a new form of cat, probably less fearful of man than the aboriginal *F. s. libyca* and less intimidated by the new environment in which it finds itself. Quite possibly, it would not require much introgression of domestic genes to evolve such a type.

Domestication

The exact origin of domestication is unknown but there is reason to think that the process occurred during the rise of the flourishing civilizations in the Fertile Crescent of the Middle East. More specifically, it would be related to the early days of agriculture which ushered in settled farming and stock keeping. This implied houses, granaries and barns. A new environment was created which was quickly exploited by the house mouse, as judged by the masses of skeletons recovered from the

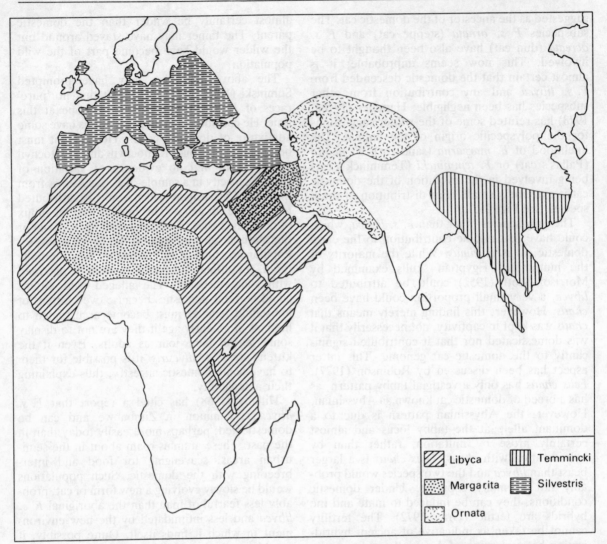

Fig. 25.1 (a) and (b) Distribution of *Felis* species and subspecies mentioned in text

Legend:
- Libyca
- Margarita
- Ornata
- Temmincki
- Silvestris

basements of dwellings in Middle East archaeological sites. It has been proposed that the cat was domesticated in order to combat the ravages of mice. This is a possibility, of course, but it probably implies more purposeful behaviour on the part of human beings than that which actually occurred. It is not unreasonable to propose that the cat followed the mouse into the farms and villages partly as a scavenger and partly to prey upon the abundant mice.

However, a development now occurred which completely changed the situation and may have accelerated the domestication process. The Egyp-tians had emerged as an influential people in the Middle East and were keenly interested in all forms of animal life. The Egyptians associated specific animals with gods. In particular, the male cat was regarded as sacred to the potent sun god Ra and the female cat to the fertility goddess Bast. For these reasons, the cat was kept in temples as a religious symbol. Eventually, the animal itself became deified. It was bred by the well-to-do, revered and interred in special cemeteries after death. These latter became so large that visitors marvelled at their size. The Egyptians jealously guarded their cats, would carry off any animal they saw outside Egypt and would be vengeful towards anyone caught ill-treating the animal.

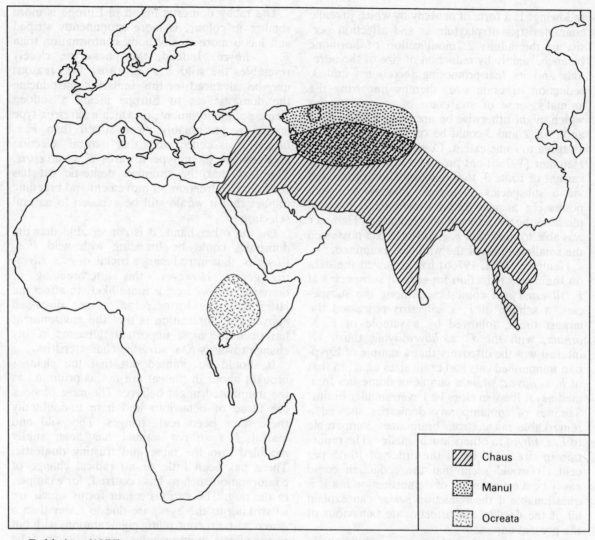

Chaus
Manul
Ocreata

Baldwin (1975) speculated on the likely progression of events leading up to the adoption of the cat by the Egyptians as follows:

1. Period of competition (prior to *c.* 7000 B.C.) characterized by wild cats competing with man for small mammals and birds.
2. Period of commensality (*c.* 7000–4000 B.C.). Characterized by 'half-wild' cats scavenging in, and feeding upon, vermin within the early settlements.
3. Period of early domestication (*c.* 2000–1000 B.C.). Characterized by the confinement of cats in temples for religious purposes.
4. Period of full domestication (*c.* 1000 B.C. onwards). Characterized by secularization of cat keeping in Egypt and the earliest diffusion of the animal from the country.

Attention may be directed to the considerable timespan involved in the domestication process. Cats mature in roughly a year, hence the citation of differences in number of years for each step is an approximation of the number of generations. Therefore, the periods can be reckoned in terms of thousands of generations, ample time for the observed changes of behaviour to occur. Evidently, the cat population was sufficiently variable genetically for some individuals to adapt to confined quarters, tolerate close proxity to man and other cats, and to conceive and rear kittens in such an environment.

The biological route by which the cat was domesticated could have been one or more of the following: 1. a form of neoteny by which juvenile characteristics of playfulness and affection persist in the adult; 2. modification of hormone balance, mainly by reduction of size of the adrenals and its fear-promoting secretions; and 3. reduction of brain size, thereby impairing the animal's sense of awareness of an environmnt which might otherwise be uncongenial. Routes 1 and 3 or 2 and 3 could be correlated; as indeed all three to some extent. Comparatively recently, Hemmer (1976b) has proposed a fourth route, a variant of route 3, that domestication proceeded via a subspecies of the *F. silvestris* complex possessing a small brain which made it more pliant to the processes of domestication. Hemmer was able to show that *F. s. libyca* in fact possessed the smallest brain of the wild cats examined.

Hemmer (1972, 1976b) has reviewed the data on brain-size reduction for several subspecies of *F. silvestris* and domestics. Among the subspecies, a sample of *F. s. silvestris* possessed the largest brain, followed by a sample of *F. s. ornata*, with the *F. s. libyca* lying third. Of interest was the discovery that a sample of Egyptian mummified cats had brain sizes equal to that of *F. s. libyca*; while a sample of domestics from mediaeval Russian cities had even smaller brains. Samples of contemporary domestics showed a remarkable range, from brain sizes comparable to *F. s. libyca* to others much smaller. The reduction in size could be of the order of 10–15 per cent. It would seem that the reduction could easily be a concomitant of domestication but it is questionable if the reduction *per se* can explain all of the docility and affectionate behaviour of the present-day cat.

No studies on the relative sizes of the main endocrine organs, comparable to those of Richter (1954) on the rat (see Ch. 39 of this volume), have been accomplished for the cat. In the rat, docility or absence of fear appears to have been developed through a marked reduction in the size and activity of the adrenals. On the other hand, the ovaries mature earlier and function more actively in the domestic animal. In the cat, the only evidence for modification of the endocrine glands is an almost complete absence of seasonal breeding behaviour. It is probable that the sexes mature earlier than in the wild cat but this sexual precocity requires confirmation. Mean litter size does not differ between the domestic and wild cat.

The tabby domestic found in Europe is more sombre in colour, is more prominently striped and has a more stocky body conformation than *F. s. libyca*. Indeed, the phenotype closely resembles the wild *F. s. silvestris*. Two reasons may be advanced for the similarity. Introducing the domestic cat to Europe meant a sudden change of environment, to which a *silvestris*-type animal is presumably more suited than *F. s. libyca*. It is conceivable that natural selection changed the phenotype towards *F. s. silvestris*. The fact that the ordinary domestic cat has considerable freedom of movement and breeding implies that it would still be exposed to natural selection.

On the other hand, it is conceivable that the domestics could be breeding with wild *F. s. silvestris*, thus introducing a trickle of *F. s. silvestris* genes. However, this interbreeding of domestic and wild cat is more likely to affect the latter than the former, for reasons discussed above. The implication is that the environment has been the most important influence in any change towards *F. s. silvestris* characteristics.

It should be pointed out that the changes brought about in the cat are not as profound as the uninitiated might believe. The most obvious are those of behaviour and here undoubtedly there have been real changes. The wild and fearful *F. s. libyca* animal has been subtly modified into the tame and trusting domestic. There has been little or no radical change of conformation such as has occurred, for example, in the dog. The various colour forms which are so striking to the eyes, are due to fewer than a dozen mutants controlling pigmentation, with but minor effects on morphology or physiology. The domestic cat can become feral very easily and often successfully. The supplying of food by humans has two effects; it ensures that the individual remains in its home range and it maintains a high population density. This latter aspect could be an important factor in maintaining tameness. The wilder individuals are those most likely to become the free-ranging feral animals in suitable environments but also the pitiful strays which can be a nuisance in towns and cities, where the possibilities for a successful feral life are poor.

History

The keeping of cats became so popular that

evidence for its domestication began to appear in other countries, for instance, Crete (100 B.C.), Greece (500 B.C.), Libya (500 B.C.), India (300 B.C.), and China (200 B.C.). Evidently, the cat was spreading slowly throughout those regions of the world which had attained a level of prosperity and culture which made it possible to support an aesthetically pleasing household pet. The sub-duing of Egypt by the Romans and the rise of Christianity dissociated the cat from religion. The Romans are credited with introducing the domestic cat to Europe, e.g. Italy (A.D. 100), Switzerland (A.D. 200), Germany (A.D. 1000) and Britain (A.D. 400). The evidence points to an inex-orable migration, almost certainly promoted by man. All dates are approximate and drawn from Zeuner (1963).

Whatever popularity the cat may have had in Europe under the Romans began to wane in the early Middle Ages. Cats were associated with witchcraft and satanism. They were hunted and killed, thrown into bonfires or roasted in cages on long poles, in order to drive out the devil. Later, mummified or dried cats were buried in founda-tions or entombed in walls of houses as good-luck charms or rat scares. Some of these interments may have been the result of accidents but some cats were entombed with birds, mice or rats in their mouths. These latter had probably been prepared beforehand as set pieces to strengthen their potency against vermin (Zeuner 1963).

The practice of keeping cats aboard ships dates from Egyptian times and was continued by many other cultures, both ancient and modern. The cat acted as a charm for a safe voyage, as a fasci-nating companion and as a destroyer of vermin. In this manner man unwittingly aided the cat in its diffusion throughout the world. The spread of the animal throughout Eurasia was doubtless by both land and sea. Subsequently, the cat was transported to the American continent by successive waves of British and Spanish colo-nizers. Still later, it was taken to Australia and New Zealand by the early British settlers.

Population genetics

Over the last two decades the world-wide census of coat colours and length which has been in progress has led to a number of interesting conclusions. Certain coat colour mutants are known all over the world and this would indicate that they are of ancient origin. These are black, blue dilution, blotched tabby, orange (sex-linked), white spotting, yellow- or blue-eyed white and long hair. Man is probably the main agent in preserving the mutants. Most breeders have an appreciative regard for the unusual which is rein-forced by the belief that colour variation is a hall-mark of domestication.

Many of the mutants may be common but this does not imply that they have equal frequencies in all regions of the world. Three mutants which have comparatively uniform frequencies are blue dilution, white spotting and yellow/blue-eyed white. Blue occurs at a low frequency in most regions while spotting is moderately common. In contrast, all-white is very rare. This is interesting since all-white cats seem to be held in esteem by most people. The explanation lies in the adverse side-effects of the gene. These are deafness, indifferent mothering ability and elevated mortality. The only area where white cats are common is in a part of Turkey, near Lake Van, where they are bred as a commercial venture for tourists, but there is confusion between extreme white-spotted and the true all-white individual.

Black is remarkable in that the frequency is high for most regions. The precise reason for this is unknown but the phenotype would not be a disadvantage for a largely crepuscular or nocturnal animal. A more subtle speculation is that the causative gene makes the individual less fearful or less disturbed by external stimuli (Robinson 1980). A mutant gene which confers a degree of placidity would possess an advantage under conditions of domestication particularly if it arose early in the process.

A similar argument of less fearful, or placid, behaviour may also apply to blotched tabby but with a small difference. This colour is very common in some regions but rare in others. The areas of higher frequency are those of high population density of man and cats, an environ-ment which would favour placidity colours. However, regions of high frequency may also be explained as the point of origin of the mutant. In the case of blotched tabby, this is Britain, and there is evidence that British colonists carried the mutant to America and Australia at moderate to high frequencies (Todd 1977).

Orange cats occur more frequently in southeast Asia and Japan than in the rest of Eurasia. It has been suggested that the mutant arose in this region and has spread westwards into Europe. This is plausible since orange is an attractive colour. Long hair also displays regional variation, reaching its highest levels in the Middle East and in the southwest of the Soviet Unicn. It is probably not a coincidence that the first long-haired cats imported to Britain were referred to as 'Angoras' or 'Persians'.

Two colours which have a restricted regional occurrence are Abyssinian and Siamese. The former occurs at a low frequency in parts of the Soviet Union and in southeast Asia; while the latter occurs at somewhat higher frequencies in southeast Asia. Both occur very infrequently or not at all elsewhere. It seems probable that both mutants have arisen comparatively recently, perhaps within the last few hundred years.

Breeds

The concept of breeds in cats dates from the middle to late 19th century (notably in Britain), at which time names were adopted and standards of excellence were drawn up. Breeds of cat fall into two broad categories, the British, European or American (according to country) and the Foreign. These categories refer in the main to head shape, body conformation and coat quality. The British type is of more stocky appearance and often has a heavier coat than the more lissom Foreign. The early cat fanciers had to make do with the stock to hand, choosing likely specimens of the basic colour phenotypes of the general populations. The native breeds of Britain were of the compactly built 'cold climate' conformation. The Manx is classed as a British breed. The alleged origin from the Isle of Man is fanciful for the dominant gene producing the characteristic anury is known to occur in many parts of the world.

About this time, a few individuals bearing the Siamese mutant were imported into Britain from Siam. These were of the slender 'warm climate' body conformation and they caused a sensation. Henceforth, all non-native breeds had to conform to the same sinuous body type under the general title of Foreign breeds. Later, the Abyssinian (origin unknown) and Burmese (from Burma) mutants were added to the list of Foreign breeds. This is how the two breed categories came into existence. Another basis of classification is hair length, the former Persian breeds are now known as Long-hairs, as opposed to the Short-hairs. The difference in this case is due to a single gene, the long coat being a recessive characteristic.

At one time, most cat breeds owed their origin to single genes or a few combinations. These could be termed the basic breeds. Of later years, there has been a more exploitative approach to finding new breeds, by combining the mutants so that many of the basic breeds are bred in numerous colours. A few of the gene combinations have been given distinctive breed names. The outcome is that the distinction between breed and variety has become blurred. The most recent mutants to be recognized as breeds are the Cornish and Devon Rexes (in Britain) and the Wire-hair (in America). All three modify the hair, the Rexes producing a short, often wavy, coat and the Wire-hair a more rough-textured pelage. All are bred in the usual colours.

The newest development is the creation of a Foreign type breed named Oriental in a range of new colours, some of which are not covered by existing breeds. It is traditional for long-haired breeds to be of British type but efforts are being made to introduce long-haired animals with Foreign body type. Some of the new breeds have been given absurd names, such as Somali for the long-haired Abyssinian, although the 'breed' has no connection with Somalia. Cymric, for the long-haired Manx, although the 'breed' has no connection with Wales. There is, in fact, little scope left for really distinctive new breeds, unless a new mutant appears with a novel phenotype which can be promoted *per se* or in suitable combinations with existing mutants. A full account of breeds and varieties may be found in Robinson (1977).

References

Baldwin, J. A. (1975) Notes and speculations on the domestication of the cat in Egypt. *Anthropos*, **70**: 428–48

Bökönyi, S. (1976) Development of early stock rearing in the Near East. *Nature*, **264**: 19–23

Gray, A P. (1972) *Mammalian Hybrids*. Commonwealth Agricultural Bureaux, Farnham Royal, Bucks, England

Hemmer, H. (1972) Hirngrossen Variation im *Felis silvestris* Kreis. *Experientia*, **28**: 271–2

Hemmer, H. (1976a) Zur Abstammung der Hauskatze:

sind Siamkatzen und Perserkatzen polyphyletischen Ursprungs? *Säugetierkundliche Mitteilungen*, **24**: 184–92

Hemmer, H. (1976b) Man's strategy in domestication: a synthesis of new research trends. *Experientia*, **32**: 663–6

Hemmer, H. (1978) Were the leopard cat and the sand cat among the ancestry of the domestic cat races? *Carnivore*, **1** (2): 106–108

Hillaby, J. (1968) Ancestors of the tabby. *New Scientist*, **38**: 404–405

Kratochvil, J. and Kratochvil, Z. (1970) Die Unterscheidung von Individual der Population *Felis s. silvestris* aus den Westkarpaten von *Felis s. f. catus*. *Zoologicke Listy*, **19**: 293–302

Morrison-Scott, T. C. S. (1952) The mummified cats of ancient Egypt. *Proceedings of the Zoological Society of London*, **121**: 861–7

Richter, C. P. (1954) The effects of domestication and selection on the behaviour of the Norway rat. *Journal of the National Cancer Institute*, **15**: 727–38

Robinson, R. (1977) *Genetics for Cat Breeders*. Pergamon Press: Oxford

Robinson, R. (1979) Cytogenetics of the Felidae. *Carnivore*, **1** (*Supplement*): 63–8

Robinson, R. (1980) Evolution of the domestic cat. *Carnivore Genetics Newsletter*, **4**: 46–56

Robinson, R. and Cox, H. W. (1970) Reproductive performance in a cat colony over a 10-year period. *Laboratory Animals*, **4**: 99–112

Robinson, R. and Manchenko, G. (1981) Cat gene frequencies in cities of the U.S.S.R. *Genetica*, **55**: 41–46

Suminsky, P. (1962) Les caractères de la forme pure du chat sauvage *Felis silvestris*. *Archives des Sciences, Genève*, **15**: 277–96

Suminski, P. (1977) Zur Problematik der Unterschiede zwischen der Wildkatze, *Felis silvestris*, und der Hauskatze. *Säugetierkundliche Mitteilungen*, **25**: 236–8

Todd, N. B. (1977) Cats and commerce. *Scientific American*, **237**: 100–107

Weigel, I. (1961) Das Fellmuster der Wildelebenden Katzenarten und der Hauskatze in vergleichender und stammesgeschichtliche Hinsicht. *Säugetierkundliche Mitteilungen*, **9** (Supplement): 1–20

Zeuner, F. E. (1963) *A History of Domesticated Animals*. Hutchinson: London

26

Ferret

Clifford Owen

Formerly Leicestershire Museums, Leicester, England

Introduction

The ferret was named by Linnaeus in 1758 *Mustela furo*. Its name prior to that had been, in Latin, first *viverra*, later *furectus*, *furetus* and *furo*. In French it is *furet*, German *Frettchen* and in Spanish *hurón*. In England it is commonly called ferret but the dark form, which resembles the polecat, is sometimes called 'fitch-ferret' or 'polecat-ferret' to distinguish it from the white-coated, pink-eyed form. Some think the dark form is a cross between the wild polecat and the white ferret but there is no evidence for this. Males are called hobs or bucks and females jills or does.

The Mustelinae, to which the ferret belongs, are highly successful hunters with long bodies and short legs: lithe in their movements, they readily pursue their prey in holes and burrows. The well-developed canines are often used in a bite which penetrates the skull and the hold is retained until the victim ceases to move. Prey is frequently stored near the nest. Strong-smelling anal glands are used to scent mark territory and in some species emptied violently when the animal is attacked. The ferret does this. Ferrets are either polecat coloured (whitish body hair and dark guard hairs), erythristic or white; they have less prominent eyes than *M. putorius*.

Their breeding begins in the year after birth. The testes, which are partially withdrawn during the non-breeding period, enlarge and descend into the scrotum by March and the female shows a greatly enlarged vulva at this time. The whitish body fur of the male becomes orange tinted from secretions of the skin glands. Mating is protracted, sometimes lasting several hours. Gestation takes 42 days. Towards the end of this period the female makes a nest from any soft material available, shredded and mixed with moulted hair. Birth weight is about 10 g and the number in a litter varies from 1 to 15, averaging 7.5. The young are born naked and blind. They begin to chew meat, brought to the nest by the mother, when about 3 weeks old. The eyes open at about 4 weeks and they explore the environs of the nest. This appears to cause the mother anxiety and she frequently carries or drags them back into the nest. Litters are weaned at between 6–8 weeks. About 2 weeks after the young are weaned the female comes into oestrus. Females with small litters may oestrate whilst suckling (Rowlands 1967) but this only occurs when the number of corporea lutea or suckling young is fewer than five. A second litter is common: three litters can be got by increasing the day length with artificial light in early spring but failure to produce milk and the occurrence of pseudo-pregnancies are frequent. If well cared for ferrets live up to 8 years and are active until a few weeks of their deaths.

The chromosome number is $2n = 40$ (Matthey 1958). Ferrets are kept for hunting in Europe and North Africa. They are mainly used for catching rabbits (*Oryctolagus cuniculus* L.) for food: less often for killing rats (*Rattus norvegicus* Berkenhout). Rabbits were domesticated and also kept in captivity in a semi-wild state in warrens (large enclosures) from which they moved out. As the rabbit spread so did the ferret and it is now used in most of the countries where the rabbit is common. Ferrets were introduced into New Zealand, where they are now feral. They are kept as pets and in zoological gardens in North America.

Wild species

There are three wild Palaearctic polecat species – the polecat (*Mustela putorius*), the steppe or asiatic polecat (*M. eversmanni*) and the North American black-footed ferret (*M. nigripes*). The validity of the first two species has long been debated and the question as to which of them is the ancestor of the domestic form has been the subject of a number of studies.

The distributions of the two species overlap in eastern Europe, where steppe conditions occur. The ferret has a post-orbital constriction (Ashton 1955, Ashton and Thompson 1955) usually narrower than 15 mm which appears to place it nearer to *M. eversmanni* than to *M. putorius*. However, by comparing the measurements of the skulls of polecats reared in captivity, ferrets reared under wild conditions and F_1 hybrids between ferrets and *M. putorius* and those of *M. eversmanni*, it appears that any similarity between the ferret and *M. eversmanni* is coincidental. The differences between *M. putorius* and *M. eversmanni* may be explained by their different ways of life. *Mustela eversmanni* is a steppe animal while *M. putorius* lives in more wooded country; the eye position in the former is suitable for scanning the sky for predators. The technique of discriminant analysis used by Rempe (1970) indicates that the ferret is a domesticated form of *M. putorius*, and the skull differences are characteristic of domestication – the reduction of the brain, bulla, orbits and dentition. There are several races or subspecies of *M. putorius* and the coat colour of the 'polecat-ferret' may well be that of the race originally domesticated. Skins examined at a furrier's in Athens by the author and said to be of wild polecats resembled those of ferrets; as there were no skins available in Greek museum collections the question of a wild polecat with pelage like that of a ferret from that area awaits study.

The present distribution of *M. putorius* is not accurately mapped nor is that of *M. eversmanni*. In 1930 Cabrera claimed that ferrets existed as a truly wild animal in North Africa. He saw the ferret being used to hunt rabbits by Berbers in the Rif mountains of Morocco. The earliest reliable account of the use of the ferret, by Strabo (*fl.c.*63 B.C.–A.D. 24) gave its origin as Libya, and this predisposed Cabrera to believe the hunters when they said that the young were taken wild. At this time the Spanish authorities could not move freely off the main roads because of the disturbed state of the country. Cabrera left Morocco for the Argentine and did not carry out any search in the field. His informants listed areas where the animal lived in the wild. Investigation in 1962 in these areas showed that the Beni Ahmed, Beni Khaled, Beni Ersin and the Uaringa knew nothing of wild ferrets. Indeed the Ruafa had bred them as domestic animals from time immemorial, keeping them in their houses, tethered in a hanging length of the bark of the cork-oak. The reason that Cabrera and his countrymen were told that the ferrets they saw were taken young from the nest in the wild was the severity of the Spanish game laws, in force in Morocco. These placed very severe restrictions on the ownership and use of ferrets. Moroccan ferrets were brought to Britain by the author and bred for some years. Their colour and behaviour were similar to those of British ferrets, except that no white specimens were bred.

Ferrets and *M. putorius* interbreed readily if kept in suitable conditions. Difficulty may be encountered with polecats caught when adult but it has long been the custom for hunters to cross their ferrets with wild polecats to increase their vigour: ferrets escape in areas where wild polecats exist, so that there has been gene exchange over long periods. There may well be an appreciable amount of ferret in the polecats in Wales. Interbreeding between ferrets and *M. eversmanni* has been recorded but the author considers that these records may not be valid. Interbreeding between ferrets and stoats (*M. erminea*) is recorded by Millais (1904) but this cannot be verified as no material was preserved. Recent work in the Soviet Union states that while *M. mustela* and the ferret have chromosome number $2n = 40$, *M. eversmanni* has $2n = 38$, and that they have not been able to interbreed *M. eversmanni* with the others (Volobuev *et al.* 1974).

Domestication and early history

Archaeological evidence is scanty although the increasing attention paid to bones in excavations is producing more material. However, to date no finds antedate the historical evidence. Although Aristophanes, *c.*450 B.C., and Aristotle *c.*350 B.C., both mention animals which could have

been ferrets, the first convincing account is that of Strabo who, describing their use in the Balearic Isles to control a plague of rabbits, attributes their origin to Libya.

As the use of the ferret is associated with the exploitation of the rabbit it is reasonable to assume that it was first domesticated somewhere in the region originally inhabited by the rabbit, i.e. northwest Africa and Iberia. Rabbits were moved by man northwards through Europe as a domesticated animal, and also kept in confinement on islands, in enclosures or in warrens. Pliny (A.D. 23–79) mentions the rabbit and the ferret, as does Isidore of Seville in A.D. 600. Mediaeval manuscripts mostly rehash Aristotle, until the great hunting manuscripts appear. There is a reference to the use of ferrets by the Emperor Frederick II in Germany in 1245. It is said that Genghis Khan used them in a hunting circle at Termed in 1221. The best manuscript is the Livre de Chasse of Gaston Phebus, Comte de Foix, who reigned over two principalities in southern France and northern Spain. He wrote it between May 1387 and his death about four years later. This shows ferreting with a muzzled ferret and purse nets. In Queen Mary's Psalter of about 1340 two well-dressed women are shown ferreting. The Normans introduced the rabbit to Britain at the end of the 11th or the beginning of the 12th centuries and the first mention of the ferret in Britain is in A.D. 1223. The time of the appearance of the white, pink-eyed ferret is unknown: the first reference is by implication in *c.* A.D. 1421 in a translation of a French poem into English as *The Siege of Thebes* by John of Lydgate (Owen 1969). The ferret is described in 1551 as being of the colour of wool stained by urine (Gesner 1551).

Present status and future prospects

Ferreting provides food, sport and the control of rabbits and rats. For the poor a little bread, milk and olive oil will yield a return in meat.

The use of the white and dark forms arises from the preference of the hunters and the differing terrain and plant cover occupied by the rabbits. On open grassland the dark form is preferred as it is more vigorous and active in moving from burrow to burrow above ground. Its vision may be better than the pink-eyed form, enabling it to move more rapidly. In close country with thick hedges the white form is often preferred as it is more easily seen and does not travel away so quickly. In Morocco in 1961 only the dark form was seen except for one white specimen taken from a poacher who said he bought it from a sailor.

In European ferreting nets are placed over the holes to catch the rabbits. It is easy for a ferret to slip under or through a net and get lost. The hunter stands back from the holes as if the rabbit sees him it will not bolt into the net. Instead it will tend to turn back and be killed by the ferret underground. This means digging to recover the rabbit, and perhaps also the ferret, which may have fed and fallen asleep. The Berbers in the Rif of Morocco use a different technique; they push springy wads of vegetation into the holes and stand close with hands ready to seize the rabbit when they see the filling move. Others stand ready with sticks in case the catcher fumbles. The ferret cannot depart unobserved. There is no advantage to be had in using white ferrets, which may be why there are none.

Myxomatosis, which arrived in England in 1953–54, reduced the rabbit population of possibly 60–100 million by about 99 per cent (Sheail 1971): a consequent even greater reduction in the number of ferrets took place, as when rabbits are few and scattered ferreting is uneconomic. Some ferrets were released and many killed.

Apart from major introductions as in New Zealand, feral ferrets appear to succeed in establishing breeding colonies on smallish islands (Mull, Man and others) more readily than they do on the mainland; although such feral colonies are reported from time to time in Britain, no detailed accounts have been published. In Morocco the author could find no evidence that they had established themselves in the wild, even in apparently suitable country.

Breeding for hunting is on a very small scale. In Great Britain a newly formed society aims to improve breeding methods, and increasing leisure is leading to a resurgence of interest in the ferret. The emergence of mild forms of myxomatosis has allowed the rabbit to increase again and ferrets are increasing in number. Large colonies are kept for use in physiological research and these are selected for docility and ease of maintenance.

References

Ashton, E. H. (1955) Some characters of the skulls of the European polecat, the Asiatic polecat and the domestic ferret. *Proceedings of the Zoological Society of London*, **125**: 807–9 (Addendum)

Ashton, E. H. and Thompson, A. D. D. (1955) Some characters of the skulls of the European polecat, the Asiatic polecat and the domestic ferret. *Proceedings of the Zoological Society of London*, **125**: 317–33

Cabrera, A. (1932) Los mamiferos de Marruecos. *Trabajos del Museo Nacional de Ciencias Naturales. Serie Zoologica*, **57**, Madrid

Gesner, C. (1551) *Historiae Animalium*. Zurich

Matthey, R. (1958) Les chromosomes des mammifères eutheriens. Liste critique et essai sur l'évolution chromosomique. *Archiv der Julius Klaus-Stiftung*, **33**: 253–97

Millais, J. G. (1904) *The Mammals of Great Britain and Ireland*. Longmans Green: London

Owen, C. E. (1969) The domestication of the ferret. In: *The Domestication and Exploitation of Plants and Animals*, ed. P. J. Ucko and G. W. Dimbleby, pp. 489–93. Duckworth: London

Rempe, U. (1970) Morphometrische Untersuchungen an Iltisschädeln zur Klärung der Verwandtschaft von Steppeniltis, Waldiltis und Frettchen. *Zeitschrift für wissenschaftliche Zoologie*, **180** (3/4): 185–367

Rowlands, I. W. (1967) The ferret. In: *The UFAW Handbook on the Care and Management of Laboratory Animals*, pp. 581–93. Livingstone: London

Sheail, J. (1971) *Rabbits and their History*. David and Charles: Newton Abbot, England

Volobuev, V. T., Ternovski, D. V. and Grafodatski, A. S. (1974) [The taxonomic status of the white African polecat or ferret in the light of caryological data. *Zoologicheskii Zhurnal*, **53**: 1738–40. (In Russian)].

27

American mink

R. M. Shackelford

*University of Wisconsin,
Madison, U.S.A.*

Introduction

Nomenclature

Mink, weasels, martens, otters, skunks, badgers and wolverines are placed in a single family, the Mustelidae. They are medium to small carnivores with slender bodies, short legs, digitigrade feet, partially webbed toes, dense pelage and well-developed scent-glands. Systematists have included all North American mink in one species, *Mustela vison*; Anthony (1928) lists eleven subspecies. Except in French-speaking countries where 'vison' is used, mink is the normal term employed.

General biology

A solitary and territorial species in nature, the mink tolerates others of its kind only at mating and during the raising of the young by the female. Individual pens are provided for adults, the only exception being the occasional breeder who allows two littermate juveniles to remain together until pelting (November). Sexual dimorphism in body size is characteristic of the species, although more pronounced in some breeding isolates than in others. Males are approximately twice as heavy as females, ranging in weight from 1.4 to 4.5 kg. Both the twice-yearly moult and the reproductive cycle follow a circadian pattern.

Mink require higher-quality protein diets than are satisfactory for foxes. Early breeders fed meat and fish almost exclusively; nowadays ground mixtures of animal by-products (meat, fish, poultry), cottage cheese, eggs and cooked cereals fortified with appropriate vitamins and trace minerals are in general use. Dry pelleted rations are in limited use with good success and probably represent the food of choice for the future.

Reproductive physiology

Breeding is limited to a single period extending from the first days of March through the first week of April (in the northern hemisphere). Neither behaviour nor appearance of the external genitalia is a reliable guide in ascertaining the mating response to be expected of the female: if she accepts the male when exposed, she is in heat! Ovulation follows copulation after 36–40 hours; fertilization and descent of developing eggs into the uterine horns are comparable to the pattern in other mammals. Implantation usually does not take place immediately, the blastocysts remaining free in the uterus: implantation occurs in approximately inverse relation to the lateness of the season with the result that some females mated during the last days of March bear their litters of 1–10 (5 is a good average) young (kits) at the same time as others mated some 3 weeks previously. Most litters (95 per cent) are born during the first two weeks of May, although the mating spread may have been five weeks. Gestation length ranges from 38 to 76 days, the average being about 50 days.

A majority of females will accept the male a second time 6–9 days post coitus and subsequently ovulate. Most litters derived from such double matings will be sired exclusively by the last-used male, a much smaller number entirely by the first male, and occasionally the litter will consist of kits resulting from eggs fertilized at both matings. In the case of 'split' litters, all young are born at the same time and in the same stage of development: with a week's interval between matings, 90 per cent of the kits will be sired by the last-used male.

The mink female can best be classified as seasonally polyoestrous, with three or four sets of eggs maturing and capable of being ovulated at intervals of 6–9 days. Along with the rabbit and cat, the mink had long been listed in texts as an example of a non-spontaneous ovulator. The work of Hansson (1947) not only demonstrated that mating or rough sex play was necessary for ovulation, his detailed investigations laid the basis for the complete elucidation of reproduction in this species. Experiments by Shackelford (1952) confirmed and extended Hansson's supposition that superfoetation could occur as a result of delayed implantation. Mink are polygamous: commercial breeders find one male to five females satisfactory. Both sexes reach adult size by 6 months, are sexually mature the following March, and are seasonally sexually active. Productive females are customarily retained for three breeding seasons, rarely for longer than 5 years; select males have been used as long as 10 years.

Karyotype

The chromosome number of mink was first reported (Shackelford and Wipf 1947) to be 14 pairs, one pair having satellites; later investigations by at least two workers using tissue culture materials have demonstrated that 15 pairs is the correct number.

Distribution and economic importance

Mink fur has always held an exalted position among the finest furs, and consumption keeps pace with increasing production. The best estimate of world pelt production is presented in Table 27.1. Considering 4 to be the average litter size, approximately 7 million females are presently in service world-wide. The vagaries of the industry in the United States from 1969 to 1978 can be seen from Fig. 27.1.

Wild species

Mink are native to North America, Asia and Europe but were extinct in at least the northern part of Europe by the time the first ranches were established. The Eurasian mink is *Mustela lutreola*. A feral population derived from

Table 27.1 Estimated 1980 world pelt production of ranched mink

Soviet Union	10 000 000	East Europe	650 000
Denmark	3 950 000	Japan	550 000
Finland	3 700 000	United	
United States	3 250 000	Kingdom	220 000
Canada	1 250 000	West Germany	180 000
Sweden	1 200 000	France	170 000
Netherlands	800 000	Ireland	100 000
China	750 000	Belgium	75 000
Norway	700 000	All Others	500 000
Total		28 045 000	

From: Hudson's Bay (1980)

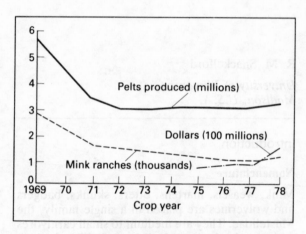

Fig. 27.1 United States mink ranches, pelt numbers and values. (From Wisconsin Department of Agriculture, Trade and Consumer Protection 1979)

domestic escapees of *M. vison* has reintroduced the genus *Mustela* to the Nordic countries and to the British Isles; it is considered a pest by game breeders in Britain. Commercially produced colour-phase mink of the world have been derived almost exclusively from importations of breeding stocks from Canada and the United States although native stocks are reported to have been utilized to some extent.

So far as is known, domestic mink are completely interfertile with their wild counterparts wherever found. Escapees, at least in the United States have no difficulty in joining the wild population which still occupies its historic range

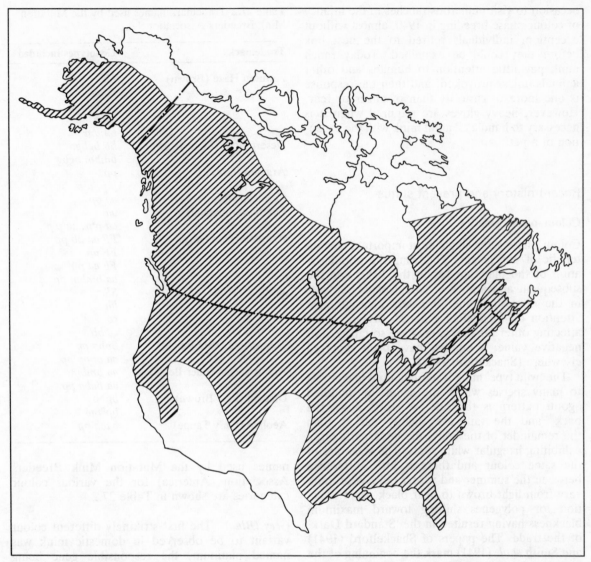

Fig. 27.2 Distribution of wild mink (Canada and United States)

(sometimes in urban areas). Trappers not infrequently garner pelts of dominant colour phases normally found only on ranches.

Domestication and early history

Domestic mink have been derived from the various geographic races indigenous to North America (Fig. 27.2). The time and place mink were first bred in captivity is not as precisely

known as is the case for the fox (*Vulpes vulpes*); the date 1866 cited by Caras (1967) for North America appears to be sufficiently accurate. Pioneer breeders produced pelts that did not compete strongly with those taken in the wild and consequently the future possibilities for the species were not appreciated at the time. The sharp rise in production followed the introduction of colour-phase breeding.

Despite a combative nature, wild mink breed satisfactorily in captivity. The behaviour of domestic mink in close proximity to humans, especially strangers, has changed markedly during the past decades. When this writer had occasion to visit numerous ranches in the infancy of colour-phase breeding in 1940, almost without exception, individuals retired to the nest box before they could be examined. Today ranch mink pay little attention to humans and other animals unless provoked, and then the response is one more of curiosity than of anger or fear. However, heavy gloves are required when it is necessary to handle all adult mink with the exception of a pet.

Recent history and present status

Colour-phase genetics

Colour phase is the single most important characteristic of any pelage. Following the popularization of the platinum fox in the 1930s and the subsequent appearance of colour-phase mutants in the domestic mink, breeders focused most attention on Mendelian genetics. Mutant genes affecting other fur characters have so far been of negative value; a number have been reported elsewhere (Shackelford 1957).

The 'wild type' mink is self-coloured in contrast to many species where the agouti or modified agouti pattern is the rule. A stripe down the back, and the tail, are slightly darker than the remainder of the body, the underparts often exhibiting irregular white spots. The sexes are of the same colour and there is little difference between the summer and winter pelt. Wild mink vary from light brown to near black, rigid selection for polygenes shifting toward maximum blackness having resulted in the 'Standard Dark' of the trade. The papers of Shackelford (1941) and Smith *et al.* (1941) mark the beginning of the genetic reports appearing in technical journals relative to this species.

Twenty mutant loci affecting colour phase are known in domestic mink. Several separate loci produce similar phenotypes; a number of genes have pleiotropisms that range from unimportance to severe limitations; and a few genes have three, four or five mutant alleles. Either singly substituted or in appropriate combinations, genes are available to produce any nuance of colour biologically possible for the species. A brief summary of the most important genes and their effects on commercial pelt production (historically and in present use) will be attempted here. Trade-mark

Table 27.2 Trademark names used by the Mutation Mink Breeders Association

Trademarks	Genotypes included
Autumn Haze (Brown)	*bb*
	bgbg
	bb bmbm
	bsbs
	bmbm
Desert Gold (Light Brown)	*bb bgbg*
	bmbm bgbg
Argenta (Grey)	*pp*
	p^{s}p, p^{s}p^s
Cerulean (Blue)	*aa pp*
Lutetia (Gunmetal)	*aa*
	aa p^{s}p, aa p^{s}p^s
Azurene (Pale Grey)	*FF aa bb pp*
	FF pp
	FF aa pp
	aa bmbm pp
Jasmine (White)	*SS*
	hh
	cc
	cc bb
Tourmaline (Pale Beige)	*bpbp pp*
	aa bpbp pp
Arcturus (Lavender Beige)	*aa bmbm*
	aa baba pp
Diadem (Pale Brown)	*bpbp*
	bp^{p}bp^p
Aeolian (Grey Taupe)	*baba pp*

names used by the Mutation Mink Breeders Association (America) for the various colour categories are shown in Table 27.2.

Grey (Blue) The first strikingly different colour variant to be observed in domestic mink was named platinum, the responsible gene being symbolized by *p*: the main effect of this gene is pigment clumping. An allele, *p$_S$*, produces a darker grey called steelblu. A phenotype similar to steelblu is obtained by substitution of a gene at a second locus (*co*). A phenotype identical to platinum appeared in the northwest, and was called Imperial platinum (*ipip*).

The most useful gene for combination with other mutant genes in the production of many of the pale shades is *a* (or *al*). The resulting aleutian colour phase is a soft gunmetal grey which is very beautiful. Mink of *aa* genotype are difficult to raise as a result of 'Aleutian disease', caused by a pleiotropic effect of *a* rendering them incapable of dealing satisfactorily with a ubiquitous virus that is not usually pathogenic to non-aleutian individuals.

Light brown One or another of the several genes for 'brown' may have been circulating undetected in dark mink before selection for maximum darkness became effective. Not so strikingly different from wild type as platinum, the brown phase called Royal pastel was recognized at about the same time with *b* assigned as an appropriate symbol. The 'screwneck', a pleiotropic effect of *b*, was discussed by Shackelford and Cole (1947); Erway and Mitchell (1973) have reported that by manganese supplementation in pregnant females this condition can be avoided.

Mutations at four additional loci bring about brown colour phases similar to Royal pastel: 1. green-eyed pastel (*bgbg*) so-named because a pleiotropic effect of this gene diminishes the eye pigmentation; 2. Moyle buff (*bmbm*) is somewhat lighter than the preceding types in unselected stocks; 3. ambergold (*baba*), a reddish brown unless selected for modifiers to suppress the redness; 4. Socklot (*bsbs*) a brown first recognized in Finland – there are three additional alleles at this locus, with $bs^S bs^S$ a lighter brown than Socklot, one still paler that is of an identical genotype ($bs^M bs^M$) and phenotype to the American Jenz palomino, the lowest in the series ($bs^A bs^A$) producing a pink-eyed white (Nordic albino) identical in appearance to the albino (*cc*). The palomino colour phase (*bpbp*) is a pale yellowish brown with pink eyes.

White Albino (*cc*) was the first white recognized in the mink. It is not completely white, the extremities being lightly pigmented. The Hedlund (*hh*) white is free of pigmentation except for the eyes which are black; a pleiotropic effect of *h* is deafness which is believed to be the basis for mating difficulties with *hh* females. Several genes are known for white spotting, but none are as satisfactory as *h* for production of completely white pelts.

Patterns All genes enumerated to this point have been discussed in the context of their relation to the wild type, the effects being manifest over the entire body. Several mutant genes produce patterns: only one of them, *F*, which results in the blufrost of the trade, is of more than passing interest. In *Ff* individuals ventral white spotting is increased, and the feet and tail tip may be white; varying numbers of white guard hairs are scattered over the body; the underfur varies from medium grey to almost white which brings about the desired enhancement of contrast between guard hair and underfur. Survivability of *FF* individuals ranges from 0 to 100 per cent depending on modifiers in a particular breeding group, and *FF* males are invariably sterile. Notwithstanding these difficulties, *F* has been one of the most useful genes in mutant pelt production: *Ff* mink that are simultaneously homozygous for one or more of the recessives comprise the Breath-of-Spring series, and pelts of surviving *FF* individuals are placed in a category called 'homozygous types' which consistently command high prices.

Combining genes to obtain established colour types has almost ceased, consequently colour-phase groups are tending to become breeding isolates. Mink ranchers keep unusually accurate records of their breeding operations, this practice having been a requisite to rational colour-phase production; however, herdbooks are unknown for domestic mink.

Future prospects

Mink pelts traditionally have been held in high esteem. They are durable, have good thermal qualities and as a result of their light weight lend themselves superbly to the art of the designer. As we have seen practically any natural colour possible in the fur of any species is in production: genes have been identified affecting other fur characteristics such as long hair, curly hair and absence of guard hair (the rex of other species). In a word, the mink pelage offers a versatility second to none.

The proportion of wild mink pelts in the United States market has steadily declined since a dependable supply of domestic pelts became available, particularly with the introduction of the mutant colour phases. It is estimated that more than 90 per cent of the mink pelts sold at auction are produced on ranches, and that pelts from wild-caught mink will continue to decline as a result of diminishing numbers of trappers and the disappearance of suitable habitats caused by human population pressure.

Except for the fitch (*Mustela putorius*), the mink is the only Mustelid to be domesticated successfully. Little space is required per individual and the animal protein requirements are met by the use of packing-house by-products, poultry offal and fish scrap, which removes the species from competition with humans in this respect. Most diseases to which other livestock fall heir are not a problem, and individual pens suspended above the ground provide the maximum in hygienic conditions. The three diseases that present health problems – distemper, botulism and viral enteritis – are avoided by the prophylactic use of vaccines developed over the years. It would seem probable that this newly domesticated species will be with us so long as high-fashion furs are in demand.

References

Anthony, H. E. (1928) *Field Book of North American Mammals*. Putnam: London and New York.

Caras, R. A. (1967) *North American Mammals – Furbearing Animals of the United States and Canada*. Meredith Press: New York

Erway, L. C. and **Mitchell, S. E.** (1973) Prevention of otolith defect in pastel mink by manganese supplementation. *Journal of Heredity*, **64** (3): 111–19

Hansson, A. (1947) The physiology of reproduction in mink with special reference to delayed implantation. *Acta Zoologica*, **28**: 1–136

Hudson's Bay (1980) Estimated 1980 world pelt production of ranched mink. *Hudson's Bay Post (New York)*, **9** (1): 4

Shackelford, R. M. (1941) Mutations in mink. *Transactions of the Wisconsin Academy of Science, Arts and Letters*, **34**: 45

Shackelford, R. M. (1952) Superfetation in the ranch mink. *American Naturalist*, **86**: 311–19

Shackelford, R. M. (1957) *Genetics of the Ranch Mink*. Hoffman: New York

Shackelford, R. M. and **Cole, L. J.** (1947) 'Screw-Neck' in the pastel color phase of the ranch-bred mink. *Journal of Heredity*, **38**: 203–9

Shackelford, R. M. and **Wipf, L.** (1947) Chromosomes of the mink. *Proceedings of the National Academy of Science*, **33**: 44–6

Smith S. E., Whitaker, C. H., Davids, L. F. and **Noble, P. V.** (1941) The inheritance of three coat color mutations in ranch-raised mink. *Journal of Heredity*, **32**: 173–6

Wisconsin Department of Agriculture, Trade and Consumer Protection (1979) Mink production, 1978/79. Wisconsin Agriculture Reporting Service: Madison

28
Mongoose

Elizabeth Mason
Edinburgh, Scotland

Nomenclature and biology

The word 'mongoose', plural 'mongooses', is of Dravidian origin, through Prakrit *manguso*. It is applied to the keen-eyed and highly agile mammals belonging to the Herpestinae, a subfamily of the Viverridae, widely distributed throughout Africa and southern Asia and with two species in southern Europe. Although there are 10 genera, 30 species and about 170 subspecies, only 3 concern us here – the 3 which, since ancient times, have been famous for their skill in killing snakes. These are the ichneumon of Africa, *Herpestes ichneumon* named by the Greeks 'the tracker', the Indian grey mongoose, *H. edwardsi* and the small Indian or golden spotted mongoose, *H. auropunctatus*. Some authorities class the golden spotted mongoose as a subspecies of the Javan mongoose with the name *H. edwardsi* and the small Indian or golden *Herpestes* means 'creeping' in Greek and is cognate with herpes, the human skin diseases; it hardly seems appropriate to such active, leaping animals.

Mongooses are small to medium-sized – the ichneumon is one of the largest – head to rump length 65 cm, tail length 45 cm; the Indian grey is one of the smaller – head to rump length 45–50 cm, tail length 38–41 cm. The body is long in relation to the legs and the tail is long, bushy at the root and tapering towards the tip. The head is ferret-like with sharp snout and short, rounded ears which can be closed to keep out sand or water. Typically, there are five toes, though some species have only four and the claws are not retractile. The soles of the feet are often quite bare and this may contribute to the animal's ability to climb walls and trees. Our three species resemble each other in having long, coarse hair which probably acts as a barrier to snake bite. The senses of sight and smell are very acute. Courtship consists in the male overcoming the female's reluctance to receive him. This may take him an hour or more and when he has succeeded he opens his mouth during copulation and keeps pressing it against the female's neck without actually biting her. The gestation period is 8–9 weeks, resulting in 2–4 young which may be born at any time of the year. In captivity, ichneumons have been known to live over 20 years.

Mongooses hunt by day and eat everything they can catch. Prey includes small mammals, birds, snakes, lizards, frogs, insects and worms. They are also prepared to eat fruit and are adept at sucking out eggs by making an opening at one end of the egg. As soon as the prey moves, the mongoose leaps after it killing it on the run by a well-aimed neck or head bite. The swiftness and accuracy of its aim are incredible. Perhaps because mongooses do so much searching for food in such a variety of ways, the practice play of the young involves all sorts of leaping, hiding and searching games which persist into adult life. Sometimes the animals stand up on their hind legs and go through all the motions of observing, chasing and catching an imaginary prey. Sometimes they roll themselves into a furry, impenetrable ball, just as they do in moments of real-life danger.

Domestication and early history

In spite of being fierce and courageous while hunting, mongooses are easily tamed, but only when young. This fact, plus their willingness to kill snakes, rats and mice must have made them welcome in and around the habitations of early man in India, in Java and in Egypt. The ichneumon, or Egyptian mongoose, was allowed the freedom of human dwellings in Old Kingdom times. It was probably domesticated in the sense that cats are, i.e. it lived with humans if and when it chose, but its breeding was not under their control. It soon became a sacred animal which, according to Zeuner (1963), was kept in temples from New Kingdom times onwards when mongoose mummies became frequent. In Ptolemaic and Roman times large numbers were kept. So perhaps they were bred. Roman ladies began to keep them as fashionable pets. The centres of the mongoose cult were at Letopolis in the upper Delta and at Herakleopolis, in Middle Egypt, south of the Faiyum oasis. The reputation which the ichneumon had for devouring crocodile eggs was no doubt one of the causes of hostility which existed between Herakleopolis and Arsinoë, in the Faiyum oasis, where the crocodile was held sacred – indeed Arsinoë's earlier name was Krokodilopolis. It was even said that the wicked little mongoose could pop into the open mouth of a sleeping crocodile, gobble up its insides, make a hole in the belly and escape unharmed. There are many good representations of mongooses on Egyptian painting and Roman mosaics and even a coin from the time of Hadrian on which the mongoose replaces the crocodile as the symbol of Egypt. As human numbers increased and viper numbers decreased, the popularity of the mongoose was eclipsed by that of the cat. Some ichneumons are still kept, outside towns, and foreigners like them as pets but the ordinary Egyptian who breeds poultry (almost unknown in ancient Egypt) regards them as pests.

In India, the oldest known tales about mongooses come from 1000 B.C. in the Mahabharata, the great Indian heroic epic. In the Panchatantra, a group of Sanskrit stories from around 100 B.C., the mongoose is portrayed as hero and helper of man, protecting him from cobras by killing them. Many English readers will be familiar with Rudyard Kipling's story *The Jungle Book* in which a brave little mongoose called Rikki Tikki Tavi is an important character.

Recent history and present status

The status of the mongoose has suffered in recent years as the result of ill-considered introductions, notably into the West Indies and the Pacific islands. In 1872 the small Indian mongoose was introduced into Jamaica in order to kill rats in the sugar plantations. From there it was taken to Fiji (to kill snakes and rats) in 1873 and 1883 and in the latter year it was also taken to the Hawaiian islands (Watkins, personal communication). From Jamaica it spread into most of the other Caribbean islands and to the coast of South America between Guyana and French Guiana. It killed some rats but also exterminated some native species. In Fiji, for instance, only 15 per cent of its diet consists of rats and at the same time several species of frogs and of ground-nesting rail are now rare (Gorman 1975). The Indian grey mongoose has been introduced into Mauritius and the Ryukyu islands.

The mongoose is also a carrier of rabies and of a number of nematodes and other parasites transmissible to farm animals. Indeed a large majority of scientific publications deal with it as a veterinary problem or with campaigns for its control.

In some places, such as India and the French Antilles, however, the mongoose has a special status as a fighting animal, used in a cruel sport popular with male spectators. A mongoose is thrown into a small cage with a snake and prodded with sticks until it finally leaps on the snake and kills it. If its natural reluctance to engage in a fight not of its own choosing cannot be overcome by repeatedly forking it up and dropping it on to the snake, it is taken out of the cage and a 'better' mongoose is put in. Deraniyagala (in Grzimek 1972) reported similar reluctance to fight on the part of both animals when he put a cobra and a golden spotted mongoose together in a cage. After 15 minutes the mongoose was still trying to escape and the cobra just wanted to be left alone. After 50 minutes of intermittent fighting the tired animals were separated.

Mongooses are not immune to cobra toxin, though they can withstand six times the dose that would be lethal to a rabbit. If they do get bitten, they do not find and eat the 'antidote' plant, *Ophiorrhiza mungos*, as popularly supposed. The poison fangs are short and cannot penetrate far into the thick coat of the mongoose. In any case, the cobras used by snake-charmers all over India in mongoose fights staged for tourists have had their poison fangs removed.

The mongoose is difficult to breed in captivity (Watkins, personal communication from Fiji). It

may sometimes be kept as a pet. Its rat-catching ability is no longer exploited. It appears doomed to be treated as a pest – especially where it is an introduction – or to be caught for snake-fighting exhibitions.

References

Gorman, M. L. (1975) The diet of feral *Herpestes auropunctatus* (Carnivora: Viverridae) in the Fijian Islands. *Journal of Zoology*, **175**: 273–278

Grzimek, B.(ed.) (1972) *Grzimek's Animal Life Encyclopedia*, Vol. 12. Van Nostrand Reinhold: New York

Zeuner, F. E. (1963) *A History of Domesticated Animals*. Hutchinson: London

29

Civets

A. L. Berhanu

Formerly Wildlife Biologist, Wildlife Conservation Organization,
Ethiopia

Wild species

There are five species in the subfamily Viverrinae of the family Viverridae from which the civet of commerce is obtained, as follows:
1. *Viverra civetta*, African civet, distributed through most of sub-Saharan Africa except the drier parts of southern Africa.
2. *Viverra zibetha*, large Indian civet, in the mainland of southeast Asia including northeast India and south China.
3. *Viverra tangalunga*, Malay civet, in Malaysia, Philippines, Sumatra and Borneo.
4. *Viverra megaspila*, large-spotted civet, India and Burma to Indo-China and the Malay peninsula. This species is sometimes divided into two species, *V. megaspila* in southeast Asia and *V. civettina*, the Malabar civet, in India.
5. *Viverricula indica* (or *malaccensis*), small Indian or lesser oriental civet, in India, southeast Asia and south China. It has been introduced into Madagascar and Zanzibar.

This classification follows Corbet and Hill (1980). Some taxonomists (e.g. Ellerman and Morrison-Scott 1966), on the other hand, divide *Viverra* into three subgenera, including *civetta* under *Civettictis*, *zibetha* and *tangalunga* under *Viverra*, and *megaspila* under *Moschothera*.

Civets are sometimes called 'civet cats'.

Seven major characteristics distinguish Viverrinae from other Viverrids. They are: 1. perineal gland (situated between the anus and the vulva); 2. digitigrade and subdigitigrade feet; 3. sectoral and tuberculated teeth with reduced molars; 4. presence of marker chromosomes; 5. uniform spotted colouration; 6. carnivorous, insectivorous, piscivorous and omnivorous food habits; 7. vertical pupil.

Civets are unspecialized mammals with poor sight but sharp sense of smell and hearing. Social communication is thus performed by scent-marking and vocal calls such as hissing, snorting, and coughing. They live where vegetation cover is thick and water is available.

Average weight of *Viverra civetta* is 12–18 kg; it is less for other species. Average height is 30–40 cm and body length (head to tail) 60–80 cm for *V.civetta* and 30–60 cm for other species.

For *V.civetta*, Ansell (1978) believes that breeding is seasonal in Zambia, whereas elsewhere it is year-round. Litter size is 2–3 in captivity and commonly 4 in the wild. Gestation period averages 65–75 days. Both sexes perform scent-marking which is done by pressing the perineal gland against a dry object. The marking is performed by anal dragging, in squatting, quadruped or handstand positions depending on the particular behaviour under display. I have observed civets marking and sniffing at various heights from ground level and thus it is difficult to establish its exact role in terms of territorial display, sexual behaviour or establishing dominance.

Wurster and Benirschke (1968) noted that the NF (total number of major chromosome arms in a female complement) in the Viverridae varies between 66 and 72 and the diploid chromosome number is $2n = 34$–52. This family has distinct satellite marker chromosomes and the subfamily Viverrinae has distinct karyotypic features which the other subfamilies lack.

Economic exploitation

The civet's economic importance lies in its musk which has a special persisting intrinsic odour and ability to fix other aromatic products; it is used as a base for perfumes. The musk or civet is produced by the perineal gland; it is yellowish in colour and has the consistency of butter. It has been valued as an important constituent of perfume since biblical times. Alpinus described the keeping of African civets in Cairo in 1580.

The centre of civet production in Africa is Ethiopia. During 1975–78 a total of 5830 tons of musk was officially exported from that country, chiefly to France. Only the males are used in production since it is said that in the female the civet becomes contaminated with urine. They are caught by hunters who track and net them in the wet season or trap them in the dry season. They are then housed individually in long narrow semi-cylindrical cages made of thin branches. One farmer has between 5 and 50 or sometimes as many as 100 animals. They are kept until productivity declines and released if they survive. They are fed with porridge supplemented once or twice a week with meat, offal or eggs. The musk is collected every 9–12 days with a horn spatula. One collection amounts to about 10–15 g per animal.

Viverra civetta can be bred in captivity and has bred at Ghana University and at various zoos (see Mallinson 1969, Ewer and Webber 1974). However, they are extremely shy and retiring and the rearing of young had a low success rate. Because civets are common, easy to catch and, if well cared for, have a potentially long cage-life, there has never been any incentive to try and breed them in captivity for commercial musk production.

In spite of the value of its product there seems little chance that the civet will be domesticated.

Acknowledgement

Information obtained by Dr E. S. E. Galal and Dr M. Alberro is gratefully acknowledged.

References

Ansell, W. F. H. (1978) *The Mammals of Zambia*. Department of Wildlife Physiology: Chilanga, Zambia

Corbet, G. B. and Hill, J. E. (1980) *A World List of Mammalian Species*. British Museum (Natural History): London

Ellerman, J. R. and Morrison-Scott, T. C. S. (1966) *Checklist of Palaearctic and Indian Mammals* (2nd edn). British Museum (Natural History): London

Ethiopia (1973) Improved management practices for civet farming. Imperial Ethiopian Government, Wildlife Conservation Organization. Mimeo.

Ewer, R. F. and Webber, C. (1974) The behaviour in captivity of the African civet, *Civettictis civetta* (Schreber). *Zeitschrift für Tierpsychologie*, 34: 359–94

Mallinson, J. J. C. (1969) Notes on breeding the African civet, *Viverra civetta*, at Jersey zoo. *International Zoo Yearbook*, 9: 92–3

Wurster, D. H. and Benirschke, K. (1968) Comparative cytogenetic studies in the order Carnivora. *Chromosoma*, 24: 336–82

30

Rabbit

Roy Robinson

St Stephens Road Nursery, London, England

Introduction

The rabbit has proved to be one of the most useful of mammals. It is employed by man for four purposes: for meat production, as a laboratory animal, as a domestic pet and for fur and wool. While not being mutually exclusive, the four uses do not overlap very much and somewhat different types of rabbit have been developed for each. The keeping of rabbits as a sporting pursuit has almost entirely ceased although this was an important factor in an early but abortive attempt to domesticate the animal. True domestication only occurred at a later stage and as a direct result of man's search for sources of meat.

Nomenclature

Today the word 'rabbit' is applied to animals of all ages. However, this is a relatively recent usage, formerly 'rabbit' was only used to describe young animals. The more general term was 'coney' or 'cony', which is still used to some extent by gamekeepers and especially in the fur trade to denote rabbit skins dyed to simulate more expensive furs. (The cony of the Bible is, of course, the hyrax.) 'Bunny' is commonly used to describe a young rabbit, mostly by children. The male is known as a buck and the female as a doe. The Romans called the rabbit *cuniculus* which is believed to come from an Iberian word from which are also derived the words for rabbit in several modern European languages, e.g. *conejo*, *conelho*, *coneglio*, *kaninchen*, and, of course, cony.

Reproductive biology

Unlike many rodents (mouse, rat) rabbits cannot be kept in pairs in cages but must be housed separately. The female is always introduced to the male, as this minimizes fighting. A female in heat will crouch and exhibit lordosis so that the male can mount her. An unreceptive female will ignore the male and merely explore the hutch although the male may be making attempts to mount her.

The rabbit reaches sexual maturity at the age of 5–6 months, there being a short period of about 1–2 months during which copulation can occur but no ovulation. The species has no regular oestrous cycle, the Graafian follicles ripen in waves but degenerate after about 7–10 days. A female may remain in heat for several weeks in the absence of a male. Ovulation occurs only following coitus or from strong sexual stimulus, e.g. does copulating with one another. Either a pregnancy or a pseudopregnancy follows, the latter (lasting 16–17 days) if the eggs are not fertilized.

The majority of does will produce litters at any time of the year but there is often a pause during the colder winter months and, more rarely, during the hotter summer months. Copulation may occur but the follicles fail to rupture. Gestation lasts 31–32 days, with little variation, except that does with large numbers of developing foetuses tend to have shorter periods than those with only a few. Litter size is more variable, being dependent upon breed and body size. Small breeds may have a mean of 4, with a range of 1–6, medium-size breeds a mean of 5–6, with a range of 1–8, while a large breed may have a mean of 6–7, with a range of 3–12 young. The first few litters tend to be smaller than those produced during the prime breeding years. A certain amount of foetal loss is common in the rabbit, suggesting that the actual conception rate is higher than that observed. The secondary sex ratio was found to be 113 males to 100 females.

The rabbit has the equivalent of a *postpartum*

oestrus in as much as a batch of follicles are ready to release their eggs and the does will accept coitus almost immediately after littering. However, if the female is suckling a large litter, say over 4–5 young, the fertilized ova do not mature but are resorbed. The female will normally accept mating about a week after the young are weaned. This roughly coincides with the mother returning to her normal healthy condition after suckling the litter, during which her condition may temporarily deteriorate.

Karyotype

The rabbit has 22 pairs of chromosomes. The karyotype is composed of 11 metacentric or submetacentric (4 large, 6 medium and 1 small) and 11 subacrocentric or acrocentric (3 large, 6 medium and 2 small) pairs of autosomes. The X is a medium and the Y is a small submetacentric. In conventional stained preparations only a few of the chromosomes can be individually and consistently identified although several well-defined groups may be recognized. However, in banded preparations, all of the chromosomes can be individually identified. A review of the literature on banded chromosomes may be found in Stock (1976) and Hagelton and Gustavsson (1979). Stock discusses the similarities and dissimilarities between the karyotypes of the rabbit, hares and pikas, and attempts a phylogenetic interpretation.

Wild species

The rabbit is placed in the order Lagomorpha, allied to the Rodentia, which consists of two families, the Ochotonidae (pikas) and the Leporidae (rabbits and hares). The New World has the greatest number of species; both hares and rabbit-like species. The hares of North America are also known as jack-rabbits while the rabbit-like species are known as cottontails. The true rabbit, *Oryctolagus cuniculus*, is abundant throughout Europe except in the colder regions of Scandinavia and the Soviet Union. Two species of hares occur in Europe, the mountain, blue or varying hare, *Lepus timidus* in the mountain regions of northen Europe, which turns white in winter, and the common brown, red or European hare, *L. europaeus*, in almost the whole of Europe except

the extreme north. This is now believed to be conspecific with *L. capensis* of Africa whose name has priority. The rabbit and common hare are often confused for there is a superficial resemblance. However, the hare is the larger animal with longer legs and is more brightly coloured. It lives entirely above ground even when rearing young, whereas the rabbit is an extensive and tireless burrower and rears its young underground.

Several subspecies or geographical races are recognized. Among these are the rabbit of Algeria and Morocco (*O. c. algiris*) and that of Spain and offshore islands (*O. c. huxleyi*). These races are distinctly smaller than the northern Europe rabbit (*O. c. cuniculus*). It is doubtful if any of the other races (e.g. *O. c. brachyotus* of the mouth of the Rhône valley in France) warrant separate designation. Some geographical differentiation is to be expected in consideration of the vast area involved and the virtual isolation of some races.

Despite the claims made in the older literature for hybrids between the rabbit and other species, these have not been substantiated by controlled experiments. The best-known name was the 'leporid', the alleged hybrid between the rabbit and the European hare. It is possible for eggs of one species to be fertilized by sperm of the other but these do not develop beyond the blastocyst stage for rabbit eggs and not beyond the 2-cell stage for hare eggs (Gray 1972). One reason for the persistence of the leporid myth is the existence of a breed of rabbit (Belgian Hare) which mimics the hare in conformation and colour. However, the Belgian Hare is a true rabbit. Attempts to produce hybrids between the rabbit and cottontail species have also failed. Less than 50 per cent of rabbit eggs were fertilized by artificial insemination with cottontail sperm and these did not survive beyond the blastocyst stage.

Initially, it seems that the wild rabbit had a restricted distribution, being confined to the Iberian peninsula. In view of the animals adaptability and reproductive powers it is surprising that its range was so limited but the reason appears to be the dense forests which covered most of Europe after the last glaciation. The rabbit is suited to open country and did not spread very rapidly until much of the forest had been cleared by man. Even then, the spread was seemingly due to further intervention by man because the Pyrenees mountain range separating

Spain from France was an effective block to easy migration.

Domestication and early history

The Phoenicians encountered the rabbit about 1100 B.C., during their journeys to Spain. It is doubtful if the Romans knew or took much interest in the animal until about 100 B.C. when the writer Varro suggested that the rabbit be kept in leporaria. These were walled-in enclosures for keeping captured hares until they were required for the kitchen. They were spacious compounds, complete with cover for the animals. The rabbit apparently thrived and bred in these surroundings. This method of keeping rabbits continued for many centuries. Although the Romans were mainly concerned with the rabbit as a meat animal, the enclosures were later adapted as a means of providing rabbits for game. The animals were hunted with small dogs and with bows and arrows. The sport was considered suitable for women. King Henry IV of France is said to have had a large enclosure or rabbit garden at Clichy, near Paris, by the 16th century.

However, the problem of securing the rabbits in the enclosures must have been tremendous and the practice was partially abandoned in favour of keeping rabbits on small islands. Quite a number of these were established during the Middle Ages. One was founded in the Lake of Schwerin in Mecklenburg in the 15th century. Queen Elizabeth I of England is reputed to have had such islands in the 16th century. Pfaueninsel, near Berlin, was set up as a rabbit island during the late 17th century (Zeuner 1963). Their abandonment coincided with the curtailment of the authority of the nobility and of petty kingdoms.

The keeping of rabbits either in enclosures or upon islands can scarcely be described as domestication. Moreover, the catching or hunting of the animals for meat or sport would not be conducive to the evolution of tame individuals. On the contrary, the most alert or wilder animals were the most likely to survive. Domestication of the rabbit was not achieved by this means. However, the rapid spread of the rabbit throughout Europe is attributed to the practice since the animal would almost certainly be escaping into the countryside from the enclosures. Even the confinement to an island would not be secure since,

although the rabbit may not take readily to water, it is nonetheless a strong swimmer. Thus, while the rabbit could have gradually spread throughout Europe by natural migration, the process was undoubtedly, albeit unwittingly, aided by man.

The process of domestication of rabbits was probably accomplished by mediaeval monks. The need to find a food which could be eaten during Lent led them to adopt an item which was popular among the Romans, to wit, the unborn or newborn young of the rabbit. Surprisingly, this was not regarded as meat. It is known that these were being eaten with relish about A.D. 600 but it is impossible to assess when rabbits were first bred purposely to provide young for this purpose. For some centuries pregnant rabbits would be caught, kept until they littered, and then released.

The above method was exceedingly primitive and, under these circumstances, attempts to breed rabbits in the monasteries must have received serious attention. One account mentions that in 1149, the abbot of the Benedictine monastery of Corvey on the Weser requested the abbot of Solignac in France to give him two pairs of rabbits. By the 16th century, rabbit keeping was well under way. The first coat colour mutants were reported from this time; their appearance is considered by some people to be the hallmark of domestication. At this time, too, selection had been remarkably successful in breeding for increased body size, for an ancient record mentions rabbits in Verona which were four times the normal size (Nachtsheim 1949, Zeuner 1963, Spriggs 1978).

Recent history

Feral populations

It was common practice to carry rabbits upon ships in the Middle Ages and to deposit these upon islands by mariners as a future source of food. Upon some islands, where the conditions were favourable, the rabbits multiplied so quickly that they were detrimental to the native fauna and flora. A well-known case is the rabbits of Porto Santo in the Madeira islands. Here, the animals not only quickly adapted themselves but became quite small (600–800 g) and slightly different from the mainland rabbit.

The rabbit was introduced into Australia in 1859 and several decades later, in the absence of its usual predators, had become a serious pest. A similar process occurred in New Zealand. On the other hand, the rabbit has not been able to establish itself in America, probably because of the presence of predators and of rabbit-like species already occupying the suitable ecological niches. All of these cases are examples of tame (or semi-tame) rabbits successfully becoming feral. Evidently, at that time, captivity had not succeeded in modifying the basic attributes of the species. Even at this time, if the rabbit is not socialized with man at an early age, it will remain nervous and timid as an adult.

Various coat colour mutants occur in wild populations, such as black, blue, yellow or minor white spotting on the nose and paws, mostly as rare variants (Robinson 1958). Of greater significance is when a particular mutant is sufficiently numerous to constitute a polymorphism. One instance is the prevalence of a black phenotype in a northern coastal belt in Tasmania (Barber 1954). The frequency of blacks varied with the amount of rainfall and particularly with areas of rainforest. Berry (1977) noted an agouti versus black polymorphism on the island of Skokholm (Wales). The blacks appear less timid than wild type and spend more time in feeding outside their burrows. When food is limited, as on most small islands, this gives them an advantage for survival in spite of higher risk of predation due to their more conspicuous colour.

Myxomatosis

No article on the rabbit would be complete without a discussion of the myxomatosis epizootics. Myxomatosis was first observed among domestic rabbits in Uruguay in 1889, followed by similar outbreaks in Argentina and Brazil. The disease was eventually traced to a virus but not until 1942 was it discovered that the virus was endemic to the South American wild rabbit (*Sylvilagus brasiliensis*) in which it produces a mild disease. Clearly, the European has no resistance to the disease for the mortality rate is exceeding high (90 per cent or higher). The myxoma virus does not affect other mammals, with the exception of the hare and then only rarely. The transmitting vectors are flea and mosquito species (Thompson and Worden 1956).

The virus was utilized as a biological control of rabbits in Australia. The initial trials were unsuccessful but in 1950 the disease began to spread increasingly rapidly with substantial kills until about 1955–56. The mortality rate was initially high but showed a fall later. This appeared to be due not to an increase in resistance of the rabbit but to the appearance of 'attenuated' or less virulent strains of the virus. The rabbit population was decimated, farm crops and pastures improved, and it was possible to increase sheep-stocking considerably.

In 1952, rabbits infected with myxoma virus were released in France and quickly spread throughout the country and neighbouring countries. Again the rabbit population was practically wiped out. In 1953, the disease was found in Kent, England. The spread of the disease was slower than that observed in Australia but nonetheless persistent until the whole of Britain had its share of infected rabbits. The virus is now endemic in all of these rabbit populations and sporadic minor outbreaks of myxomatosis occur whenever a local population passes a critical level.

Breeds

Nachtsheim (1949) has discussed in some detail the domestication of the rabbit and has given approximate dates for the occurrence of the major colour and coat length mutants. The rabbit is one of the few mammalian species for which it is possible to probe so far back into the past in this manner.

Date	Mutant genes
Before 1700	Albino, blue dilution, brown, Dutch, non-agouti, yellow.
1700–1850	English, Himalayan, long hair.
1850–1900	Harlequin, steel-grey, tan pattern.
1900–1950	Blue-eyed white, dark chinchilla, light chinchilla, rex, satin, waved, wide band.

All of the important breeds are based upon one or a combination of a few of the above mutants. In addition, a silver, in which abundant white hairs are distributed evenly throughout the coat, and lop-eared, in which the ears are enormously developed, were produced quite early in the domestication period. It is not possible to include either of these in the above list because the silvering and allometric growth of the ears are due to polygenes and were developed progressively over many generations by selective breeding.

There are innumerable breeds of rabbit, many of which are known under different names in different countries. This is perhaps inevitable but unfortunate in that it leads to confusion and could give the impression that many more breeds exist than really is the case. The main reason for this state of affairs seems to be a combination of nationalism, geographical isolation and ignorance. One is certainly not justified in concluding that breeds of similar phenotype are distinctive solely because they have different names. Breeders of the past were not above importing stock of a breed and altering the name either to one more congenial to their tongue or out of nationalism. Another pitfall to avoid is that of recognizing as distinctive those breeds which differ only slightly in body conformation. The albino breeders are probably the worst offenders in this respect. Very few breeds of rabbit have been in existence long enough for genetic differences to arise as a consequence of breeding isolation. Such differences may exist but it is doubtful. In this, the rabbit differs from the larger livestock.

The many breeds are based upon the Mendelian recombination of fourteen colour and of three coat mutants. Full details of the phenotypes and genotypes of most breeds have been listed by Robinson (1958) and Fox (1974). Many breeds have the same basic colour or coat genotype but differ in weight or body conformation. The range of body weight is relatively enormous for a small mammal, varying from 1–1.5 kg for the smallest breeds (Hermelin, Netherland Dwarf, Polish) to 6–6.5 kg for the largest (Blanc de Bouscat, Deutsches Riesenkaninchen, Flemish Giant). Only the dog and horse can compare with such a variation of weight within a species. There is also the Belgian Hare, a breed which crudely mimics the European hare, with elongated limbs, ear pinnae and body. The English Lop has huge pendulous ears which drag along the ground. The European Lop has shorter ears.

Rabbit breeds in Britain are divided into two groups, fancy and fur, largely on grounds of tradition. The older breeds were largely bred for exhibition, with the emphasis on fancy markings. The most well-known are Angora, Belgian Hare, Dutch, English, Flemish Giant, Harlequin or Japanese, Himalayan, Lop, Polish, Silver and Tan. However, the rapid development of the rabbit in the years immediately after the First World War, as a meat and fur animal, brought a number of new breeds to the fore in which the quality of the pelt was paramount. The distinction between fancy and fur breeds is trivial, of course, in the sense that most breeds will yield a useful pelt. In any case the distinction has never been strongly upheld outside Britain.

Undoubtedly, the coat quality of the new fur breeds is superior to the old fancy breeds. The fur is denser, more uniform in colour and, because the breeds are heavier, the pelt is larger. Some breeds, such as the Chinchilla which is an imitation of *Chinchilla laniger*, became popular. Certain albino breeds are useful because they produce a white pelt which can be uniformly dyed.

However, breeding rabbits for skins has not been outstandingly successful in more recent years, largely because it is inclined to be labour intensive. Thus, it may be an industry suited for developing countries. The production of top-quality pelts is beset with problems. It is far from easy to mass-produce large numbers of prime, matched skins. The economics of meat production usually means that the rabbit is slaughtered too early in life for the skin to be in perfect condition. Rabbit fur is too fine to be hard-wearing. This has meant that the skins are rarely used for other than cheap garments or small items (e.g. gloves). Rabbit hair has been used extensively by the felting industry, although its inferior wearing quality is evident if used by itself.

The fineness of rabbit hair is an asset in the production of wool which is the plucked or shaven hair of the long-haired Angora breed. Normally this is white from the albino variety. It is usually mixed with fine Merino sheep wool to give it more substance and to improve its wearing quality. Angora rabbit wool features in luxury items.

The most recent breeds to arrive on the scene are those developed for commercial meat production. This is a development since the Second World War and the best of these are superior to the ordinary breeds which have not been expressly bred for the purpose. This situation is to be expected and it is probably wise to go all-out for specific meat carcass breeds rather than attempting to up-grade an existing breed. It is well known in genetic circles that intensive selection for a few desirable characters is more successful when the base generation is the result of crosses between unrelated breeds in the previous generation. The two well-known meat rabbit breeds are the Californian and the New Zealand White, both of which are rapid-maturing medium-heavy to heavy rabbits. Several different strains of each breed are now in existence.

Genetics

In the early days of mammalian genetics, the rabbit was studied in order to confirm the generality of Mendelian heredity and to determine the inheritance of the various mutations preserved by fanciers. This was interesting work and it became apparent that heredity in the rabbit was in no way exceptional. However, genetic studies with the rabbit steadily declined as the house mouse increasingly became the first choice for research in mammalian genetics.

Approximately ninety mutants have been described, a number in considerable detail but others very sketchily indeed (Sawin 1955, Robinson 1958, Fox 1974, Lindsey and Fox 1974). The majority of the colour and coat mutants are preserved in the breeds and colour varieties bred by fanciers but the majority of genetic abnormalities are in constant jeopardy of being lost. A number are being maintained at the Jackson Laboratory, Bar Harbor, Maine, U.S.A., but the majority are maintained only at the whim of current research programmes and the personal interest of research workers. No international or coordinated effort to conserve mutant genes from extinction exists in the rabbit, as in the house mouse.

The rabbit has few long-inbred strains, compared with the house mouse or rat, due in part to the long generation interval for the species. Festing (1979) lists twenty inbred strains (seven being substrains of strain III), all of which are maintained at the Jackson Laboratory. Fortunately, the strains are being steadily and systematically typed for immunological and biochemical variants in the modern manner. A few are being used to conserve mutant genes producing anomalies. The nomenclature for designating inbred strains has been standardized for the house mouse and this should be done for the rabbit. The use of these in publications for concise and accurate

description of experimental material cannot be over-emphasized.

The rabbit has been utilized in biochemical research, where its large size and general robustness can be useful. To facilitate this sort of research, biochemical, biophysiological and morphological data on the normal rabbit are often required. A considerable amount of information has been collected by Weisbroth *et al.* (1974), Altman and Katz (1979) and Kaplan and Timmons (1979). Most of the diseases which afflict the rabbit are discussed by specialists in Weisbroth *et al.* (1974).

Present situation

The keeping of rabbits occurs at several levels. The animal makes an excellent pet for children, for it is large and robust enough to be handled (or mishandled) without serious harm befalling it. Also, it can usually be found in the event of its straying from its hutch. In fact, it can be allowed to run loose in a confined space, even with children present, provided a responsible person is in attendance. This makes the rabbit useful as a recommendable addition to the range of small animals which may be kept in the pet section or miniature zoo of a school or municipal authority. Any of the patterned breeds, such as the Checkered Giant, Dutch, English or Himalayan are suitable because of their contrasting markings.

Rabbit keeping for exhibition is a highly organized undertaking in several countries, especially in Britain and the United States but a little less so in European countries. Indeed, it probably ranks after cage birds, cats and dogs in this respect. It has a big enough following and is sufficiently well organized to have a central governing body in most countries. This activity implies that large numbers of rabbits are bred to universally recognized standards of excellence. These are the traditional breeds which date from the middle of the 19th century. The majority are bred for their aesthetic appeal and even the so-called dual-purpose breeds (meat and fur) are bred mainly for their fancy attributes in this context.

However, it should not be forgotten that the rabbit was originally domesticated as a supplier of meat. Although the monks pioneered the keeping of rabbits in confined quarters and hutches, the practice was adopted by the peasants and smallholders of late mediaeval Europe right up to the present day. At first, rabbit meat was treated as a delicacy eaten by the nobility, who presumably owned the rabbits and upon whose estates the rabbits were reared. This was a carry-over from keeping the animal in rabbit gardens or warrens. However, the rabbit is a prolific breeder and small enough to be easily passed from hand to hand. In later centuries, the rabbit became part of the diet of peasants and the poorer classes in general.

The advantage of the rabbit is that it requires scant space in which to live and feed, compared with larger livestock, such as the goat or pig. Furthermore, it can thrive on almost any edible green food, especially if this can be supplemented by grain from time to time for suckling does and growing young stock. Thus, in many situations, the rabbit can be a worthy competitor to the larger livestock and in other situations as a means of providing variety in the consumption of meat and meat by-products.

During the Second World War there was a tremendous increase in keeping rabbits for meat, a trend which soon waned as far as the man-in-the-street was concerned. However, the war and the lean years which immediately followed had roused interest in the possibilities of the animal for modern meat production. Breeders in the United States were particularly interested and went so far as to develop two new breeds for the purpose. These are the Californian and New Zealand White. They are medium-large animals, of stocky conformation, with plenty of meat on the haunches. The young stock are particularly rapid growers, in order to produce a meaty carcass in the quickest possible time. To achieve the rapid growth, the young are fed specially formulated high-protein diets.

Similar efforts were also under way in Europe, notably in Eastern Europe and the Soviet Union. Here, the approach has been twofold: 1. improvement of existing breeds of meat rabbits which have always been a feature of continental breeding; and 2. development of new breeds by a programme of intercrosses of established breeds, followed by selective breeding. A number of new breeds have been produced but whether or not these are greatly superior to the established

breeds is doubtful. It must be remembered that some of the original breeds, such as the Blanc de Bouscat, Blanc de Termonde and others, had already attained a high level of productivity as meat producers.

Future prospects

There is little doubt that rabbit-meat farming should be included among the options open to developing countries to improve or add variety to their food production. Every effort should be made to exploit the rabbit's ability to digest almost any form of edible greenstuff. The rabbit also thrives on a monotonous diet provided it is wholesome. The ideal situation would be that of a locally available green feed which is in surplus, a by-product of processed food for human consumption or unpalatable to man. It may, of course, be necessary to augment the protein content of the diet if this is exceptionally low, and to include a vitamin and trace-mineral supplement. Large-scale rabbit farming may then be feasible although logic demands that an experimental pilot scheme be tried out initially. Since labour costs should be low, it should be possible to construct relatively simple hutches or pens in order to reduce the capital outlay. The methods of American or European rabbit farming could be employed but suitably, if not radically, modified for local conditions.

An alternative is to foster a large number of more modest units at the village level. These may vary from one-man concerns to cooperatives of a number of interconnected smallholdings. This may mean encouraging and instructing people in the art of keeping an animal which may be strange to them. Once again, the principle is that of utilizing green feed which is locally available, relatively easily obtained and not normally used for human consumption. The rabbit is economically efficient at converting low-quality feeding stuff into high-protein meat with relatively simple equipment, provided the labour cost is low. In feeding a family, a man may not consider his own labour, and this is the situation where the potential of the rabbit should be most useful.

An organization has been formed to further the interests of rabbit farming on a global scale. This is the World Rabbit Science Association. The address of the secretary is Tyring House, Shurdington, Cheltenham, Gloucester GL51 5XF, England. This organization publishes a newsletter of coming events and progress of rabbit farming in most countries of the world. In individual countries, the primary source for information for local conditions governing rabbit keeping is normally the agricultural department of the government. Rabbit keeping is being encouraged even in the most advanced industrial nations.

References

Altman, P. L. and Katz, D. D. (eds) (1979) *Inbred and Genetically Defined Strains of Laboratory Animals.* Part 2 *Hamster, guinea pig, rabbit and chicken.* Federation of American Societies for Experimental Biology: Bethesda, Maryland

Barber, H. N. (1954) Genetic polymorphism in the rabbit in Tasmania. *Nature,* **173**: 1227–9

Berry, R. J. (1977) *Inheritance and Natural Selection.* Collins: London

Festing, M. F. W. (1979) *Inbred Strains in Biomedical Research.* Macmillan: London

Fox, R. R. (1974) Taxonomy and Genetics. In: Weisbroth *et al.* (1974) (op. cit.)

Gray, A. P. (1972) *Mammalian Hybrids.* Commonwealth Agricultural Bureaux: Farnham Royal, Bucks, England

Hagelton, M. and Gustavsson, I. (1979) Identification by banding techniques of the chromosomes of the domestic rabbit. *Hereditas,* **90**: 269–79

Kaplan, H. M. and Timmons, E. H. (1979) *The Rabbit: a model for the principles of mammalian physiology and surgery.* Academic Press: London

Lindsey, J. R. and Fox, R. R. (1974) Inherited diseases and variations. In Weisbroth et al. (1974) *(op. cit.)*

Nachtsheim, H. (1949) *Vom Wildtier zum Haustier.* Paul Parey: Hamburg and Berlin

Robinson, R. (1958) Genetic studies of the rabbit. *Bibliographia Genetica,* **17**: 229–558

Sawin, P. B. (1955) Recent genetics of the domestic rabbit. *Advances in Genetics,* **7**: 183–226

Spriggs, G. (1978) Hopping through history, rabbits over the years. *Country Life,* **163**: 1912–3

Stock, A. D. (1976) Chromosome banding pattern relationships of hares, rabbits and pikas. *Cytogenetics and Cell Genetics,* **17**: 78–88

Thompson, H. V. and Worden, A. N. (1956) *The Rabbit.* Collins: London

Weisbroth, S. H., Flatt, R. E. and Kraus, A. L. (1974) *The Biology of the Laboratory Rabbit.* Academic Press: London and New York

Zeuner, F. E. (1963) *A History of Domesticated Animals.* Hutchinson: London

31
Coypu

L. M. Gosling and J. R. Skinner

*MAFF Coypu Research Laboratory,
Norwich, England*

Introduction

Nomenclature

Coypus (*Myocastor coypus*) are hystricomorph rodents that are sometimes included with the hutias in the Capromyidae and sometimes separated in the Myocastoridae. Osgood (1943) lists five subspecies, including the type form *M. c. coypus* but these are in need of revision. 'Coypu' or 'coipo' is derived from the original Indian name of 'water-sweeper' (Murua *et al.* 1981) and is preferable to the alternative name 'nutria' which is the Spanish for otter.

General biology

Coypus have a superficially rat-like appearance with a scantily haired tapering cylindrical tail. In England fully-grown feral males (over 100 weeks old) average 6.70 kg and females 6.36 kg (Gosling 1977). Average lengths from the nose to tail-tip are 995 mm and 980 mm respectively. The pelage is composed of long glossy guard hair and a soft dense underfur.

Many of the external features are adaptations to a semi-aquatic habit. The nostrils are valvular and the lips can be closed behind the orange-coloured incisors while the animal cuts through submerged food plants. The ears, eyes and nostrils are placed high on the dorsal surface of the head and are exposed as the animal swims or floats. The large hind feet are strongly webbed and propel the animal with powerful alternate thrusts.

The forefeet are strongly clawed and used to excavate both burrows and underground parts of food plants such as roots and rhizomes. Coypus dig complex burrow systems, sometimes up to 6 m into the banks of waterways and these usually contain spherical chambers with large amounts of nest material. Surface nests are also built either concealed in waterside vegetation or in shallow water where they are supported by the stems of emergent vegetation. Freshwater mussels are sometimes eaten but, otherwise, coypus are entirely herbivorous and feed on a broad range of aquatic and semi-aquatic plants (Gosling 1974, Murua *et al.* 1981).

Coypus share many of the unusual features of hystricomorph rodent reproduction. The gestation period is about 19 weeks and mean litter size 5–6 with a range of 1–13 (Newson 1966). There is a *postpartum* oestrus but otherwise the female cycle is irregular, suggesting that ovulation is normally induced by male behaviour (Newson 1966). The mean *postpartum* interval of feral coypus in England is 2.1 weeks (Gosling 1980). There is extensive prenatal mortality with losses from resorption of individual embryos and abortion of entire litters. Coypus breed throughout the year. The young are born fully furred and active and weigh on average about 225 g. They can survive if artificially weaned at 5 days (Newson 1966) but are normally suckled for an average of 7.7 weeks (Gosling 1980). The teats are in 2 rows of 4 or 5 on the dorso-lateral body surface and the young suck while the mother is in a normal sitting position. The median age for maturity is 3–7 months depending on season (Newson 1966). Coypus are gregarious in their social behaviour and have a polygynous mating system (Gosling 1977).

Average potential longevity in captivity is 6.3 years but for a year before this coypus start to lose weight and to show other signs of incipient senility (Gosling and Baker 1981). In feral populations females survive better than males (Newson 1969, Gosling and Baker 1981) but there

have been no studies of survival in undisturbed populations in South America.

Coypus in the wild

South America

The range of coypus in South America includes a wide variety of environmental conditions, from the subtropics of northern Argentina to the severe winters of Patagonia in the south. *Myocastor c. bonariensis* from northern Argentina cannot tolerate cold weather and when introduced to temperate climates, such as that of the British Isles, suffers severe frost lesions and sometimes heavy mortality in the winter. The subspecies native to Patagonia is *M. c. santacruzae* and the account of Ashbrook (1948) suggests that its members are well adapted to icy conditions.

Myocastor c. coypus inhabits the rivers and lakes of central Chile although its numbers are now reduced. The dark-skinned coypus of southern Chile (*M. c. melanops*) swim in both fresh and brackish water (Osgood 1943) and a similar tolerance of saline water is found in *M. c. bonariensis* both in its natural and its introduced range. The identity of the coypus found in the extreme south and in Tierra del Fuego is not known but they were once numerous; 3000 skins were marketed at Punta Arenas in 1939. The coypu of Bolivia, provisionally named as *M. c. popelairi*, similarly remains uninvestigated but Osgood (1943), describing it as 'exceedingly large and dark coloured', believed it to be a distinct form.

Although protective legislation has been introduced in some parts of South America coypus have been severely reduced in numbers and range through over-exploitation (see p. 248). Ashbrook (1948) described the decline of coypus in Argentina and Osgood (1943) and Barlow (1969) mention similar trends in Chile and Uruguay respectively.

Feral populations

Coypus were exported from South America as early as the 19th century but there are no reports of feral populations that originated at this time. In contrast, the world-wide sales of breeding stock that followed the increase in value of coypu fur in the 1920s led to the establishment of numerous wild colonies. There are now feral populations in North America, Europe, the Soviet Union, the Middle East, Africa and Japan. Where it was recorded the type introduced was *M. c. bonariensis* and feral populations of this subspecies certainly exist in North America, Great Britain, The Netherlands, Germany, Italy and the Soviet Union (Ashbrook 1948, Laurie 1946, Litjens 1980, Santini 1980, Vereshchagin 1936, *et al.*).

Most feral populations became established following escapes from inadequate enclosures or when disgruntled farmers released their stock as they encountered the problems of slow breeding and low profitability.

In the Soviet Union there were systematic introductions from 1930 onwards (Vereshchagin 1936). The subspecies was *M. c. bonariensis* and, not surprisingly, they did not survive at the most northerly release points. However, large populations became established in the extensive reed swamps and fens of the south and presumably these are still exploited for fur and meat. Feral coypus appeared in Turkey as immigrants from the Soviet Union and similar international movements occur from West Germany to the Netherlands (Litjens 1980).

Attitudes towards feral coypus vary in relation to their economic exploitation. In North America and the Soviet Union where the populations are cropped, the damage by coypus to crops, drainage systems and natural plant communities is regarded as acceptable or controllable. In other areas, particularly those such as England and The Netherlands, which have drainage systems that are vulnerable to damage by burrowing, coypus are regarded as pests and intensive control exercises are mounted (Norris 1967, Litjens 1980).

The early history of utilization and the start of captive breeding

Wild coypus have been intensively exploited throughout their natural range at least from the mid-19th century and probably much earlier. Ashbrook (1948) mentions their great abundance at that time but for the next 60 years coypus were ruthlessly hunted for their fur and meat. The numbers killed were immense: Samkow and Trubezkoij (1974) record a yield of around 10 million pelts a year from South America up to 1910. This level of exploitation led to near extinction in many areas; by 1924–28 Argentinian exports had declined to an average of 175 600 per year and by 1931 only 200 000 South American pelts appeared on the market. Protective legislation was enacted in a number of countries but it has been only partially successful.

The 1920s also saw a dramatic increase in the value of coypu pelts stimulated partly by fashion and partly by a reduced supply. With depleted wild populations, trappers were forced into other means of livelihood and captive breeding began using wild-caught stock. Prices of $100 for breeding pairs were not unusual. Ashbrook (1948) believed that the first fur farms were established in 1922 although Aliev (1967) mentions a farm near Buenos Aires at the beginning of the last century, established in response to an earlier decline in the numbers of wild coypus.

The first farms consisted of large fenced enclosures around lagoons and other marshland areas. Later these areas were subdivided into functional units including sections for breeding, rearing and for growing coypu food crops such as corn and alfalfa (Ashbrook 1948). Some selection was carried out by removing animals judged to lack desirable pelage characteristics (Samkow and Trubezkoij 1974). Eventually small pens were built for breeding groups, and for single females with dependent young. Weaned young were separated into larger pens before selection for breeding or fur production. All pens contained water for swimming and this was considered essential for high-quality fur production.

The farms established in North America and Europe were similarly variable in the size of enclosures and in the degree of control over breeding. While there were no farms of the fenced lagoon type, some breeders did enclose sections of natural streams or ponds. Many of these farms never became profitable and, both in South America and elsewhere, there was a sharp decline in the industry in the recession years leading up to the Second World War. In some countries, e.g. Great Britain where about fifty farms were established (Laurie 1946), production stopped entirely and was never revived. In other countries, such as Germany and Poland, there was a reduced wartime production; in Germany the pelts were used to line the flying suits of pilots and the carcasses as a source of coupon-free meat (Lang 1970). The end of the war marked the end of the early phase of coypu husbandry and the start of modern fur-production techniques.

The development of intensive husbandry techniques

The exploitation of coypus for fur and meat expanded dramatically after the Second World War. In the United States the cropping of feral populations, particularly in Louisiana, rapidly became more important than captive breeding. In contrast the Polish industry has concentrated on selective breeding in close captivity and the production of valuable colour varieties. The third major producer, Argentina, has a mixed policy of captive breeding and cropping wild animals. These alternatives are pursued on a smaller scale in a number of countries throughout the world: small numbers are bred in captivity in East Germany, Hungary, Czechoslovakia and France and some animals are cropped from wild populations in Chile, Paraguay and Uruguay. The scale of fur production in the Soviet Union is not known in detail although 100 000 pelts were brought by the state in 1973 (Il'ina 1975). There is an extensive Russian literature on husbandry, genetics and selective breeding (e.g. Il'ina 1975, Kuznetsov 1979).

Without doubt Poland is the main producer of coypus bred in close captivity. The first imports were from Argentina in 1926 and by 1939 there were 4000 breeding females. Production collapsed during the war but, with government grants to set up farms, there were around 70 000 breeding females by 1958, some in small private concerns and some in larger state-owned farms. There was reduced production between 1957 and 1962 (Kopański 1962) but, this period apart, breeding stock and yield increased steadily. By 1977 production had reached 1 600 000 pelts (Kukla 1977).

Husbandry techniques in Poland and the Soviet Union vary between year-round confinement in small groups and winter confinement with release into large enclosures during the summer (Aliev 1967, Il'ina 1975). The first practice is more common in breeding valuable colour varieties which requires carefully controlled selective breeding. Coypus are easily kept in pens and are generally fed a mixture of a suitable pelleted concentrate, root crops and green food plants.

In coypu farms in Czechoslovakia the most important cause of death is pseudo-tuberculosis, a disease caused by the organism *Yersinia pseudo-tuberculosis*. Strongyloides infestations and coccidiosis sometimes develop in conditions of poor hygiene (Lang 1970).

Breeding is carried out either with small groups of males and about 10 times the number of females (up to about 40), or more selectively using 1 male and up to 10 females. Sometimes single females are placed with males for mating and then returned to an individual pen. Some mating procedures are based on the erroneous belief (see above p. 247) that female coypus have oestrous cycles of 25–30 days (e.g. Il'ina 1975).

Coypu fur, widely termed 'nutria', is marketed either with or without the guard hair. The stripped underfur was the typical fashion fur of the 1920s, but the entire skin has become more popular in recent years. The underfur of the belly is the most dense and the most valuable. Desirable pelt characteristics are an even, deep and dense underfur – 15 000 hairs per cm^2 of skin constitute a valuable fur – and a large size. Sizes are graded in four stages, from very large (over 75 cm from eyes to base of tail) to small (60–65 cm). Large skins (70–75 cm) can be obtained from animals of 4 kg body weight and it is apparently economic to kill animals when they reach this size at about 8 months old. Pelt quality is assessed using complex systems of 'points' which have varied over the years and in different parts of the world. Furs of the highest quality are obtained in the winter when moult is minimal (Il'ina 1975, Niedźwiadek and Kawińska 1980).

'Standard' pelts, the wild type, have brown or grey underfur and guard hairs of various shades of brown. These pelts still form the majority of the furs produced even though the proportion of other colour varieties has increased substantially in Poland and, presumably, the Soviet Union. Such increases depend largely on planned production since colour variants have occurred spontaneously in inbred captive stock for many years. Ten colour varieties were known in Argentina by 1958 and the first albinos were exhibited in 1936 (Aliev 1967, Bettin 1959). Many varieties similarly appeared in Russia and West Germany during the 1940s and 1950s. However, breeders usually sold these animals for quick profit and production remained low (Lang 1970). In 1973 colour varieties were still rare in Argentina, for example, and the emphasis appeared to be on cropping standard pelts from wild populations or those enclosed in large lagoons.

In contrast, the larger farms in Poland and the Soviet Union have concentrated on the selective breeding of colour varieties and on their large-scale production. These varieties include white, black, golden and beige types and many others. The genetics of coat colour determination is extremely complex, as in other mammals. For example, there are at least four genetically distinct types of white coypu including white Azerbaijan and Severin coypus, produced in Russia, and white Italian (Il'ina 1975, Kukla 1977). The large-scale production of unusual colour types can sometimes be achieved by inbreeding but this often leads to reduced fecundity. For example, inbreeding snow-white nutria or crossing them with golden animals results in a 25 per cent reduction in offspring production (Kuznetsov 1974). Other deleterious consequences of inbreeding include reduced grooming, so that the pelage becomes matted, and decreased resistance to disease (Bettin 1959).

Coypu meat is both edible and palatable and has been eaten since early times in South America. It is also used as a by-product of the fur industry in North America, Europe and the Soviet Union both for human and for animal consumption. The meat contains about 20 per cent protein and 8–10 per cent fat and is sold in a variety of forms including that of sausage and pâté.

Future prospects

Intensive selective breeding of existing colour varieties is likely to continue, especially in Poland and, to a lesser extent in the Soviet Union. Of secondary concern may be attempts to combine the traits of high growth rate and fecundity with desirable fur quality including uniform depth and high density. In the short term these efforts will presumably be directed towards the production of pelts such as the 'Greenland' type, which are currently most valuable, but such preferences are a matter of fashion. Such breeding programmes must depend partly on an improved knowledge of the genetics of coypu coat colour which, compared with contemporary work on the deter-

minants of coat colour in other rodents, particularly mice, is at an early stage of development.

While farm-bred animals have taken a progressively larger share of the international fur market in post-war years, large numbers of wild or feral animals are still caught and sold. For example, in Argentina, only 5 per cent of the 1.2 million skins produced in the 1971/72 season were from captive animals. In North America all of the 1–2 million pelts obtained each year between 1970 and 1976 (Deems and Pursley 1978) were from feral populations, mainly in the wetlands of Louisiana. All of these animals are wild types and their pelts consequently less valuable than those of selectively bred coloured varieties. In 1976 North American pelts averaged $5.25 in value. It seems likely that the trade in 'nutria' will continue to polarize between these relatively cheap skins and the more costly products of intensive selective breeding from Eastern Europe.

Ironically, while feral coypus thrive, the indigenous populations of South America are threatened, in spite of protective legislation. While it is to be hoped that these animals will be preserved because of their intrinsic value as members of the South American wetland fauna, it is also important to retain viable populations as reservoirs of genetic diversity that can be tapped in the future.

References

Aliev, F. F. (1967) Numerical changes and the population structure of the coypu, *Myocastor coypus* (Molina, 1782), in different countries. *Säugetierkundliche Mitteilungen*, **15**: 238–42

Ashbrook, F. G. (1948) Nutrias grow in the United States. *Journal of Wildlife Management*, **12**: 87–95

Barlow, J. C. (1969) Observations on the biology of rodents in Uruguay. *Life Sciences Contributions, Royal Ontario Museum*, **75**: 1–59

Bettin, L. (1959) [Argentina leads the way in nutria breeding]. *Deutsche Pelztierzuchter*, **33**: 77–8. (In German)

Deems, E. F. and Pursley, D. (1978) *North American Furbearers: their management, research and harvest status in 1976*. University of Maryland Press: College Park, Maryland

Gosling, L. M. (1974) The coypu in East Anglia. *Transactions of the Norfolk and Norwich Naturalists Society*, **23**: 49–59

Gosling, L. M. (1977) Coypu, *Myocastor coypus*. In: *The Handbook of British Mammals*, ed. G. B.

Corbet and H. N. Southern, pp. 256–65. Blackwell Scientific Publications: Oxford

Gosling, L. M. (1980) The duration of lactation in feral coypus (*Myocastor coypus*). *Journal of Zoology*, **191**: 461–74

Gosling, L. M. and **Baker, S. J.** (1981) Coypu (*Myocastor coypus*) potential longevity. *Journal of Zoology*, **197**: 285–312

Il'ina, E. D. (1975) [*Animal Breeding*]. Kolos: Moscow. (In Russian)

Kopański, R. (1962) [Breeding fur-bearing animals in Poland]. *Medycyna Weterynaryjna*, **10**: 616–9. (In Polish)

Kukla, F. (1977) [New data on the inheritance of colour in nutria]. *Chovatel*, **16**: 217–8. (In Czech)

Kuznetsov, G. A. (1974) [Genetics of the colour of snow-white nutria]. *Nauchnye Trudy Nauchno-Issledovatel'skogo Pushnogo Zverovodstva i Krolikovodstva*, **13**: 53–9. (In Russian)

Kuznetsov, G. A. (1979) [Breeding nutria]. *Krolikovodstvo i Zverovodstvo*, (2): 31–3. (In Russian)

Lang, H. (1970) [Quo vadis nutria? What is the future of nutria breeding in the Federal Republic?] *Deutsche Peltztierzuchter*, **44**: 128–131. (In German)

Laurie, E. M. O. (1946) The coypu (*Myocastor coypus*) in Great Britain. *Journal of Animal Ecology*, **15**: 22–34

Litjens, B. E. J. (1980) [The coypu, *Myocastor coypus* (Molina), in the Netherlands. I. Fluctuations in the population during the period 1963–1979]. *Lutra*, **23**: 43–53. (In Dutch)

Murua, R., Neuman, O. and **Dropelmann, I.** (1981) Food habits of *Myocastor coypus* in Chile. *Proceedings of the Worldwide Furbearer Conference*, ed. J. A. Chapman and D. Pursley, pp. 544–58

Newson, R. M. (1966) Reproduction of the feral coypu (*Myocastor coypus*). *Symposia of the Zoological Society of London*, No. 15: 323–34

Newson, R. M. (1969) Population dynamics of the coypu, *Myocastor coypus* (Molina), in eastern England. In: *Energy flow through small mammal populations*, ed. K. Petrusewicz and L. Ryszkowski, International Biological Programme Publications

Niedźwiadek, S. and **Kawińska, J.** (1980) The influence of different rearing systems on nutria fur. *Scientifur*, **4**: 7–12

Norris, J. D. (1967) A campaign against feral coypus (Myocastor coypus (Molina) in Great Britain, *Journal of Applied Ecology*, **4**: 191–9

Osgood, W. H. (1943) Mammals of Chile, *Field Museum of Natural History Publications*, *Zoological Series*, **30**: 131–4

Samkow, J. A. and **Trubezkoij, G. W.** (1974) [Coypu breeding outside the U.S.S.R.]. *Deutsche Pelztierzuchter*, **48**: 130–2. (In German)

Santini, L. (1980) La nutria (*Myocastor coypus* Molina) allo stato selvatico in Toscana. *Frustula Entomologica* (new ser.), **1**: 273–88. (English summary)

Vereshchagin, N. K. (1963) [Experiments in acclimatising nutria (*Myocastor coypus bonariensis* Rengger) in the south of U.S.S.R.]. *Trudi Azerbaidzhanskogo Filiala Akademii Nauk U.S.S.R. Zoologicheskaya Seriya*, **29**: 1–66. (In Russian)

32

Guinea-pig or cuy

B. Müller-Haye

Research Development Centre, FAO, Rome, Italy

Introduction

Nomenclature

The common English name for this animal is misleading. It is not a pig and does not come from Guinea; in fact it is a South American rodent. Other European languages also refer to it as pig, e.g. *Meerschweinchen* in German, i.e. a little pig which came across the sea. This leads to confusion and the name should be changed. Simpson (1980) proposes to use the Latin *cavia*. Gade (1967) and Gilmore (1950) adopt the Quechua word *cuy* or *cui* which is the usual one in South America and the one preferred by the present author. From the Galibi (French Guiana) comes *cobiai* which also gives the French *cobaye* and the Spanish *cobayo*. The English fanciers refer to it as 'cavy'. Other names are *acuri, aperca, cochinillo* or *conejillo de las Indias, coic, coyi, cuize, curi, curia, curtela, cuye, prea, quiso, sucuy*.

The nomenclature for *Cavia* species differs among authors as evidenced in Gilmore (1950), Hückinghaus (1961, 1962), Gade (1967) and Weir (1974).

For the purpose of this contribution the nomenclature of Hückinghaus (1961, 1962) is followed

which is the most advanced study published, although it is limited in so far as his research was based exclusively on the comparison of skull characteristics. The consideration of traits such as weight, fertility, colour, coat etc., could assist in further clarifying differences. Weir (1974) suggests studying the genetics of various cavy crosses for greater distinction between species.

General biology

The cuy (*Cavia porcellus*) is a small animal with an inoffensive temperament; it very rarely bites man. It cannot climb or jump and hence escape from its cage. A number of favourable characteristics have led to its widespread use as a meat animal amongst Andean rural populations, as a laboratory animal for health, nutrition and genetic research and, of course, as a pet. It is a harmless animal but it suffers from a nervous disposition and responds to sudden disturbances with fear reactions such as whistling, squealing, grunting, quick moves or timid rigidness. Cavies incessantly communicate among themselves making a variety of noises. Males, although good-natured with other species, often fight fiercely among themselves.

According to Asdell (1964) age at puberty varies from 33 to 134 days with an average of 88 days, but Ibsen (1950) has observed females that were bred as young as 20 days of age. Females reach puberty at an earlier age than males, i.e. 55–70 days under normal conditions and 45–60 days when well fed, while males would be 56–70 days old under the same well-fed conditions (Zaldivar and Chauca 1975). The oestrous cycle lasts about 16 days with great variation in individual animals (12–20 days). The majority of females will come into heat within a period of 2 days of giving birth, but this heat lasts on average only 3.3 hours against the normal 8.3 hours (Asdell 1964). They ovulate spontaneously. The female is normally receptive to the male up to 10 or even 15 hours. Gestation length averages 68 days again with considerable variation, 59–74 days (Ibsen 1950, Paterson 1972). The size of the litter affects gestation length, large litters being carried for a shorter time.

The cuy has two mammae; if there are more these are usually non-functional. Litter size is 2–4, but litters of up to 8 young have been recorded. As a result of the long gestation period compared to other rodents the young are fully developed at birth with fur formed and eyes open. They are herbivorous and start eating grass and other feedstuffs within hours. There is no nesting. The mother permits suckling up to 3 weeks. Zaldivar (1981) points out that good management and improved strains can double production. While the average number of young per dam per year is 5.5 in two parturitions in the Andean region, he could raise this figure to 10.8 in 4.1 births under experimental conditions. In commercial enterprises in temperate zones, however, 5 litters per year are possible, which would mean 12–15 animals weaned per year.

The covering capacity of an adult male is usually 5–8 females, but Zaldivar and Chauca (1975) suggest a sex ratio of 1 to 15. There is no breeding season for the cuy in its natural habitat, but in northern areas it will usually not breed during winter unless heating and artificial daylight are provided.

Where the cuy is kept for meat production, as in the Andes, growth characteristics are important. Birth weights usually range from 70 to 110 g but with improved strains 130 g has been observed (Jara-Almonte and Castro 1975) and single-born animals can be as heavy as 150 g. Animals less than 40–50 g usually die. Weaning is normally carried out at 3–4 weeks of age when the animals weigh about 200 g. However, the same improved material of Jara-Almonte and Castro (1975) reached weights of 365 g for males and 338 g for females at 28 days. At 3 months they had reached 872 and 799 g respectively. Normal mature weight is otherwise 750–800 g for males and 700–750 g for females at about 5 months, but much heavier animals are on record. The cuy will continue to grow until it is 15 months old, but weight gains at this age are mainly due to fat deposition. The life-span of cuys depends largely on the purpose for which they are kept. Females in production should be culled after 18–24 months and males at 12 months to broaden the genetic base. Average age of pet animals is 5–7 years, although they may sometimes live longer. Guzman (1968) analysed cuy meat and found a composition of 20.3 per cent protein, 7.8 per cent fat, 0.8 per cent minerals and 70.6 per cent moisture. The carcass normally dresses out at about 65 per cent and at 67 per cent in castrated animals (Zaldivar and Chauca 1975).

The chromosome number of *Cavia* species is $2n = 64$.

Distribution and importance

The distribution of the cuy as a laboratory animal and pet is world-wide, but as a meat animal it is of importance only in the Andes and there are considerable numbers in Peru, Colombia, Ecuador and Bolivia. Zaldivar (1981) reports that Peru has 67 million cuys which produce annually 17 000 tonnes of meat. Colombia probably has half a million animals which yield annually 390 tonnes of meat. Gade (1967) estimated a total of 2 million animals for Bolivia and Ecuador.

It should be mentioned that the cuy is also used for curative purposes in indigenous medicine and for religious ceremonies. The custom dates back to pre-Incan times, but is still practised today. A detailed description of what is done with patient and animal and how the ceremony is celebrated is provided by the Biblioteca Agropecuaria (1979). Christianization has not eliminated ancient beliefs and deep-rooted conviction that spirits can be indulged by offering them cuy meat and offals. On All Souls Day the dead are offered their ration of cuy meat.

Wild species

The superfamily of cavy-like rodents (Cavioidea) includes three families: cavies (Caviidae), capybara (Hydrochoeridae) and agoutis (Dasyproctidae). Hückinghaus (1961) identified three species in the genus *Cavia*. *Cavia aperea* has the widest distribution. It can be found practically all over South America from Venezuela to Argentina with the exception of the humid tropical lowlands of Brazil, Ecuador and Peru and the cold zones of Chile and Argentina. The species *C. fulgida* is found in mountainous areas of southeastern Brazil and *C. stolida* in northeastern Peru. It is accepted that the cuy, *C. porcellus*, was domesticated from *C. aperea* because they exhibit common characteristics, and of the nine subspecies *C. a. tschudii* of Peru is the most likely ancestor.

Weir (1974) reports on the interrelationship of cavies and reviews the results of crosses between *C. aperea* and *C. porcellus*. The F_1 and F_2 hybrids were fertile but males resulting from F_3 females × *C. porcellus* males were not. George *et al.* (1972) studied the karyotypes of *C. aperea* and of *C. aperea* × *C. porcellus* and found no gross morphological differences between the chromo-

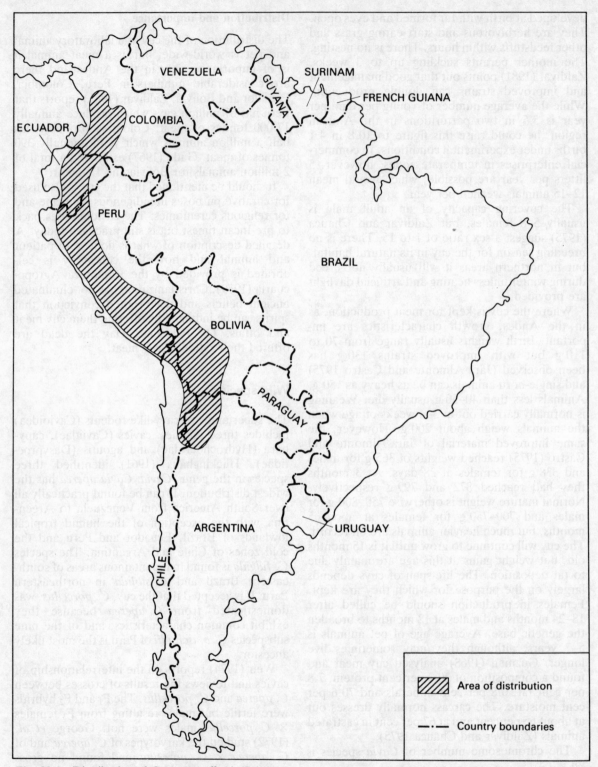

Fig. 32.1 Distribution of *Cavia porcellus* in South America

somes of the two species. There was a slight difference in chromosome 3 between the two species but also a polymorphism within *C. aperea*.

Reciprocal crosses between *C. fulgida* and *C. porcellus* have been obtained, but F_1 males are sterile and females are only occasionally fertile (Gray 1972).

The wild species are smaller than the domestic and of agouti colour. The gestation period is shorter (62–64 days).

Domestication and early history

The cuy is one of the few mammals which has been domesticated in the Americas; the others are the llama and alpaca. Wing (1977) recognizes three stages in the use of animals in the Andes. The first stage she dates between 7000 and 4000 B.C. in early cultural sites in Peru below 4000 m. In some places the cuys make up 40 per cent or more of the fauna. In the *puna*, i.e. over 4000 m, inhabitants lived on hunting deer and camelids. It is not clear if cuys were already domesticated or were captured when living close to human settlements in search of food waste and shelter. The samples do not exhibit an increase in size during the period, which argues against domestication. The second stage, in which domestication took place, dates from 4000 to 1000 B.C. It went hand-in-hand with a shift of the human subsistence from hunting to herding with a sedentary or semi-sedentary way of life. If one observes today's association of man and cuys in Indian huts in the Andes, one can well imagine that early coexistence was not very different. The animals profit from food, shelter and warmth, living with the family in the same hut, and are protected from their enemies. Furthermore, the cuy is a crepuscular animal and may have found its perfect environment in the poorly lit dwellings. Man on the other hand may have realized that it was easier to rear the cuy under his supervision and take off adult animals for slaughter whenever needed without going to the trouble of hunting them. It should also not be disregarded that he liked to have this lovable animal for companionship and entertainment. The third stage dates from 1000 B.C. to the time of the Spanish conquest and is characterized by such typical features of domestication as increase in size and greater variation. This was most likely accompanied by increased tameness and improved reproduction.

In 1547, 55 years after the discovery of America by Columbus, Oviedo (according to Cabrera 1953) wrote about an animal in Santo Domingo which he called '*cori*' and which was, in fact, the cuy. Since cavies are not indigenous in the Caribbean, it must be assumed that they were brought to the island by the Spaniards in one of their early voyages from Peru. In Europe, Konrad Gesler described the cuy already in 1554 (Heinemann 1975) and one can expect that the first animals came to Spain soon after the conquest of Peru in 1532. In 1591 the Inca Garcilaso de la Vega, Peruvian historian, mentioned the cuy (then called *coy*) and made reference to the Indians who enjoyed its tasty and nutritious meat (Biblioteca Agropecuaria 1979).

Recent history and present status

In comparison with other domesticated animals very limited selection has been practised in cuys. Only during the last 30 years has some planned breeding been carried out. Leading this activity are the many pet clubs in which hobby breeders are organized, who select mainly for fancy characteristics in their animals. Certain industries and laboratory-animal breeding and multiplying stations are also actively seeking to produce animals which suit their specific research requirements. Finally, countries like Peru, Colombia and Ecuador wish to establish improved breeding stock which exhibit greater vigour, disease resistance, better production traits and improved fertility. The purpose of the undertaking is twofold. Firstly, it is intended to supply sufficient numbers of improved males and females to smallholders and Indian communities of the Andean countries to foster the home production and increase the meat consumption of the rural poor. Secondly, an animal is selected which has appropriate characteristics in all aspects of production ranging from easy handling, feed conversion, reproduction, growth rate, to suitability and profitability for large-scale, intensive rearing.

It is not intended in this contribution to deal at length with breeds and strains and the interested reader is referred to breeder's manuals and handbooks. The *Catalogue of Uniform Strains of Laboratory Animals Maintained in Great Britian* lists strains for Britain and the *Handbook for*

Laboratory Animals for the United States. Turner (1977) describes and illustrates the following cavy breeds which are recognized by the fancy and have official standards:

1. Abyssinian: various colours: rosetted, rough-coated.
2. Agouti: including golden, silver, cinnammon, lemon and orange agouti.
3. Brindle: rare.
4. Crested: with American, English, Himalayan and Agouti varieties.
5. Dutch: various colours; belted.
6. Himalayan: white with black or chestnut points.
7. Peruvian: rosetted, long-haired.
8. Self: black, white, cream, golden, red, chocolate, beige, lilac or saffron; always short-coated.
9. Sheltie: smooth, long-haired.
10. Tortoiseshell: black and red; rare.
11. Tortoiseshell and white: red, black and white.

In Peru cuys are classified according to body conformation and type and quality of hair. Type A is roundish with a short, round face and is more docile than type B which is shy and nervous. Type B has a squared and rectangular body and a long face. In this type the pinna of the ear shows great variation in shape and is sometimes absent. Because of its good temperament type A has better weight gains and feed conversion.

Four hair types are distinguished. Type 1 has short and straight hair which is tight on the body. Nearly always it has a whorl in the front. This is the most common animal and can be of various colours. Type 2 also has short and straight hair but with a number of whorls and rosettes distributed over the body. It is also very common. Type 3 has long and straight hair which can either be close to the body or in rosettes. These animals are less frequent and less fertile because of their hair condition and breeding is difficult. Type 4 has curled hair at birth but on maturing the hair changes to bristle (Biblioteca Agropecuaria 1979, Zaldivar and Chauca 1975).

The genetics of the guinea-pig has been investigated very thoroughly and many research results have been published. The majority of these have been reviewed by Robinson (1975) who lists forty-four mutants out of which eighteen refer·to colour, namely ticked belly agouti, non-agouti,

brown pigment, dark dilution, light dilution, red-eyed dilution, acromelanic albino, diminished, tortoiseshell, yellow, fading yellow, grizzled, pink-eye dilute, roan spotting, piebald spotting, silvering, salmon eye and whitish. The genotypes of the principal colour varieties of guinea-pigs are also reported. Coat characteristics are controlled by six genes which are described as fuzzy, long hair, rough modifier, rough, sticky coat and star.

As a laboratory animal it is important to know about the guinea-pig's immunological and electrophoretic variation and as analytical methods have become more sophisticated in recent years information has become available which has led to a better understanding of the animal's responses to certain treatments. Robinson (1975) describes twelve genes in this field which refer mainly to histocompatibility and immune responses. Also a number of abnormalities have been genetically investigated and genes could be identified which are responsible for defects such as cornea anomaly, dwarfism, congenital palsy, polydactyly, sexual hypogenesis, a form of tremor and waltzing.

To our knowledge no studies have been conducted on the inheritance of skin pigmentation. This trait is of economic importance, because consumer acceptability is greater for unpigmented carcasses. Totally or partially pigmented carcasses give the impression of being rat meat.

Future prospects

Changes in the status quo of cuy breeding are not foreseen. The pet breeders will continue to look for rare genes, but the present population in their hands will not 'explode'. Industry has a demand for cuys for experimentation, but big changes in the existing types of animals are not anticipated. Perhaps cuys for meat production may have a boom for certain groups of people but cuy meat is not readily accepted by all sections of the population and broilers are invading traditional markets. The great advantage of the cuy over poultry is its cheap rearing, both nutrition and management-wise, with low-cost investment in housing. However, the production of cuys for human consumption is restricted to the Andean regions where people are accustomed to its rich but not very tasty meat. In other parts of the world the rabbit will most likely be more acceptable.

There is no need for the conservation of genetic variability at present as both Andean creole breeds and pet breeds are represented by many varieties and no predominant improver breed is threatening to absorb local breeds. On the contrary, the above-mentioned production data stress the necessity for more systematic breeding. Progress could be achieved quite easily as heritabilities of certain production traits are very promising (Jara-Almonte and Castro 1975, Vaccaro *et al.* 1968). Modern breeding methods which are employed, e.g. in poultry-breeding, would guarantee a very rapid improvement if the animal is selected for meat production. In this connection it would be desirable if national effort could be coordinated regionally.

References

Asdell, S. A. (1964) *Patterns of Mammalian Reproduction.* Cornell University Press: Ithaca, New York

Biblioteca Agropecuaria (1979) *Cuy: alimento popular.* Editorial Mercurio: Lima, Peru

Cabrera, A. (1953) Los roedores argentinos de la familia 'Caviidae'. *Publicación Escuela Veterinaria, Universidad de Buenos Aires*, **6**: 1–93

Gade, D. W. (1967) The guinea pig in Andean folk culture. *Geographical Review*, **57**: 213–24

George, W., Weir, B. J. and **Bedford, J.** (1972) Chromosome studies in some members of the family Caviidae (Mammalia: Rodentia). *Journal of Zoology*, **168**: 81–9

Gilmore, R. M. (1950) Fauna and ethnozoology of South America. *Bulletin of the Bureau of American Ethnology. Handbook of South American Indians*, **143** (6): 445–64

Gray, A. P. (1972) *Mammalian Hybrids.* Commonwealth Agricultural Bureaux: Farnham Royal, Bucks, England

Guzman, C. L. (1968) Periodo de engorde en cuyes y el estudio tecnológico de sus carnes. Thesis, Universidad Nacional Agraria: La Molina, Peru

Heinemann, D. (1975) Superfamily cavies. In: *Grzimek's Animal Life Encyclopedia.* Van Norstrand Reinhold: New York

Hückinghaus, F. (1961) Zur Nomenklatur und Abstammung des Hausmeerschweinchens. *Zeitschrift für Säugetierkunde*, **26**: 108–11

Hückinghaus, F. (1962) Vergleichende Untersuchungen Uber die Formenmannigfaltigkeit der Unterfamilie Caviinae Murray 1886 (Ergebnisse der Südamerikaexpedition Herre/Röhrs 1956–1957).

Zeitschrift für wissenschaftliche Zoologie, **166** (1–2): 1–97

Ibsen, H. L. (1950) The guinea pig. In: *The Care and Breeding of Laboratory Animals*, ed. E. J. Farris. Wiley: New York

Jara-Almonte, M. and **Castro Berrios, P.** (1975) Estudios de pesos en cuyes, su interrelación, indices de herencia. *Anales Científicos*, **13** (1–2): 67–72

Paterson, J. S. (1972) The guinea pig. In: *The UFAW Handbook on the Care and Management of Laboratory Animals.* Churchill Livingstone: Edinburgh and London

Robinson, R. (1975) The guinea pig, *Cavia porcellus.* In: *Handbook of Genetics*, ed. R. C. King, Vol. 4: *Mammals.* Plenum Press: New York

Simpson, G. G. (1980) *Splendid Isolation. The curious history of South American mammals.* Yale University Press: New Haven and London

Turner, I. (1977) *Exhibition and Pet Cavies.* Spur Publishing Company: Liss, Hampshire, England

Vaccaro, R., Dillard, E. U. and **Lozano, J.** (1968) Crecimiento del cuy (*Cavia porcellus*) del nacimiento al destete. *Asociación Latino-americana de Producción Animal, Memoria*, **3**: 115–26

Weir, B. J. (1974) Notes on the origin of the domestic guinea pig. *Symposia of the Zoological Society London*, No. 34: 437–46

Wing, E. S. (1977) Animal domestication in the Andes. In: *Origins of Agriculture*, ed. C. A. Reed, pp. 837–57. Mouton Publishers: The Hague and Paris

Zaldivar, M. (1981) El cuy y su producción de carne. In: *Recursos Genéticos Animales en América Latina*, ed. B. Müller-Haye and J. Gelman. Estudio FAO: Producción y Sanidad Animal, No. 22: 129–31

Zaldivar, N. and **Chauca, L.** (1975) *Crianza de Cuyes.* Boletin Técnico No. 81, Ministerio de Alimentacion: Lima, Peru

33

Capybara

E. González Jiménez

Institute of Animal Production,
Central University of Venezuela, Maracay

The first report on this animal, according to Oviedo y Valdés (1959), came from Alvarez-Nuñez in 1541 in Paraguay; thereafter many expeditions reported on this rodent. The first description was by the zoologist Azara in 1802; later Humboldt, in 1820, reported that the Orinoco river Indians used it as meat either fresh or salted. The first drawing of capybara is due to Marcgrave who saw it in the Brazilian rivers in 1648.

The capybara *Hydrochoerus hydrochaeris*, is the largest living rodent. It belongs to the family Hydrochoeridae which diverged from the Caviidae in the Miocene (Mones 1978). Cabrera (1961) divided the species into four subspecies: *H. h. hydrochaeris* in Brazil, Venezuela, the Guianas and Colombia; *H. h. capybara* in northeast Argentina and Paraguay; *H. h. uruguayensis* in Uruguay and eastern Argentina; and *H. h. isthmius* in Panama and the Lake Maracaibo area of Venezuela. The last subspecies is the only one found beyond the Andes.

This animal has many Indian names, but the most common is 'capybara' from the Guarani (Paraguay) dialect. In Venezuela its name is *chigüire*, from the Cumanagotos and Palenque tribes, but the Caribs called it *capigua*, which is similar to the common name in South America. It is also known as *carpincho*.

The American continent did not give rise to many domestic animals but many attempts were made by several Indian tribes; for instance, the Chibchas domesticated deer and the Piaroas of the Orinoco basin, capybaras. Zucchi (1969) reported that Indians ate the meat of chigüires raised close to their houses. Acevedo y Pinilla (1961) reported that many tribes collected young capybaras during the hunting season and kept them until needed for consumption. Nogueira-Neto (1973) has drawn attention to the fact that as early as 1565 Father Anchieta reported on the breeding of the capybara in Brazil, suggesting that they were used as domestic animals by the indigenous population.

In recent years, many attempts have been made towards the domestication of this species. At the Institute of Animal Production in Venezuela, Parra, Esobar and González Jiménez (1978), started a breeding programme from twenty females and five males captured in the Hato El Frio (Apure state). From this time on we have kept them as captive animals and their reproduction has been improved, as shown in Table 33.1. The management is as follows. Groups of five females per male are separated in pens of 120 m², fenced with wire mesh (1.50 m high); 20 per cent of the area is occupied by a pond which this semi-aquatic animal needs and a small roof is required. We give forage such as elephant grass (*Pennisetum purpureum*), and concentrates (18% crude protein) as 30 per cent of the ration. A month before parturition, females are isolated from the group and kept in a 20 m² pen to give birth. Five weeks after the offspring are born, they are weaned and the mother is returned to the breeding pen. Each young is numbered like a piglet and weighed periodically; records are kept for selection. Actually our aim is to have 16 offspring per mother per year, and we are halfway to that goal. However, current efforts are also devoted to exploiting them in their natural non-domesticated state so that their full ecological potential may be harnessed. In Colombia, work aiming at the realization of the potential of capybaras in captivity is in progress, and guidelines for raising them on breeding farms have been provided (Cruz 1974, Fuerbringer 1974).

In natural conditions, even at the slow growth

Table 33.1 Reproductive data on capybara in captivity.

	Mean	S.D.	Number of observations
Birth weight (kg)	1.76	0.20	34
Litter size	3.71	1.34	35
Parturition interval (days)	176.3	3.09	8
Age at sexual maturity (months, females)	15	—	15
Gestation period (days)	150	2.0	15

rate of the young capybaras, the large litter size of this species permits a high animal productivity as shown by Ojasti (1973) with a net productivity of 740 kg/km^2. At present in Venezuela the meat is used salted for consumption during Easter. Catholics are allowed to eat capybara during this period, and this has led to the tradition of consuming this meat during Lent and Easter week.

References

Acevedo y Pinilla, J. M. de (1961). La explotación y comercio del chigüire. Ministerio de Agricultura: Bogotá. Mimeo.

Cabrera, A. (1960) Catálogo de mamíferos de América del Sur. *Revista Museo Argentino Ciencias Naturales Bernardino Rivadavia*, **4** (2): 309–732

Cruz, C. A. (1974) Notas sobre comportamiento del chigüire en confinamiento. *Documento, Primer Seminario Colombo-Venezolano sobre Chigüiros y Babillas*. Inderena, Bogotá. Mimeo.

Fuerbringer, B. J. (1974). *Manual Práctico: El Chigüiro*. Bogotá

González Jiménez, E. (1977) The capybara: an indigenous source of meat in tropical America. *World Animal Review*, No. 21: 24–30

Mones, A. (1978) Filogenia de la familia Hydrochoeridae (Mammalia: Rodentia). *Proceedings of the Second Symposium on Capybara*. Instituto de Producción Animal, Facultad de Agronomía, Universidad Central de Venezuela. Mimeo.

Nogueira-Neto, P. (1973). *A Criação de Animais Indígenas Vertebrados*. Editorial Tecnapis: São Paulo, Brazil

Ojasti J. (1973) *Estudio Biológico del Chigüire o Capibara*. Editorial Sucre: Caracas, Venezuela

Oviedo y Valdés, G. F. (1959). *Historia General y Natural de Indias*. Colección Rivadeneira, Vol. **177** and **121**. Editorial Atlas: Madrid

Parra, R., Escobar A. and **González Jiménez, E.** (1978) El chigüire, su potencial biológico y su cria en confinamiento. *Informe Anual 1978, Instituto de Producción Animal, Facultad de Agronomía, Universidad Central de Venezuela, Maracay*, **1**: 83–94

Zucchi, A. (1969). Nuevos datos sobre la arqueología de los Llanos Occidentales de Venezuela. *Verhandlungen des XXXVIII. Amerikanistenkongresses Stuttgart-München*, **1**: 289–94

34

Chinchilla

Juan Grau

Instituto de Ecologia,
Santiago de Chile

Introduction

The chinchilla is one of the world's most famous furbearing animals. However, very little has been written about it and most of the reliable information available deals with economic aspects.

Nomenclature

Chinchillas are South American Hystricomorph rodents; together with viscachas they form the family Chinchillidae.

Two species are included in the genus *Chinchilla*:
1. *Chinchilla brevicaudata*, earlier *Eryomis chinchilla*, the short-tailed, Peruvian or Bolivian chinchilla.
2. *Chinchilla laniger*, previous called *Mus laniger* or *Eryomis laniger*, the long-tailed or Chilean chinchilla.

General biology

It is likely that the two species were formerly only one, and that the difference in morphology we observe today results from adaptation to the environment. Thus, in accordance with the principle of Bergmann and Allen's rule, *C. brevicaudata*, which lives at altitudes of 3500 m or more and at low temperatures, has a larger body (32 cm long and 600–850 g in weight), smaller ears (3 cm diameter), a shorter tail (10 cm long with 20 vertebrae) and shorter limbs. It also has a longer gestation period (128 days).

In contrast, *C. laniger* living in the coastal mountain range in a rather benign climate, is smaller (25 cm in length and a weight of 400–450 g). Ears are larger (5 cm diameter), limbs longer, and the tail is 14 cm with 23 vertebrae. Pregnancy lasts 110 days.

Anatomically both species have identical genital tracts. Sexual maturity is reached at the age of 6 months. The vagina is permanently sealed with a cementing substance which only softens during the monthly oestrus which lasts 3–4 days. Some of these oestrus periods are incomplete and non-active; only when the time is propitious, i.e. in winter and in midsummer, does coitus lead to conception.

Wild chinchillas appear to be monogamous, but in captivity they can easily adapt to polygamy, with 1 male for 5 or more females.

Pregnancy occurs twice a year. Litter size averages 1.95 in *C. laniger*, but only 1.45 in *C. brevicaudata*. This is the main obstacle to breeding the latter on a large scale.

The two species have identical karyotypes ($2n = 64$), but it is very difficult to obtain hybrids between them. When produced, the hybrid has the *brevicaudata* size and fur. In the experiments we have carried out, the hybrid males were sterile and females fertile. The hybrid females may be crossed with *laniger* or *brevicaudata* males, and have a pregnancy duration intermediate between 100 and 128 days.

The fur fibres of chinchillas are very fine and 25–35 mm long. The roots are dark grey in colour; the ends are blackish or grey, and the middle portion is white. The guard hairs are thicker and longer, giving brightness and bulk to the fur. The fur of chinchillas is of agouti type. It is made up of groups of about 24 fibres emerging from a single follicle. As a defence device, the scared chinchilla sheds locks of hair by the root which block the mouth of a predator.

Geographical distribution and economic importance

Chinchillas have suffered a disastrous persecution

in their native lands, i.e. on both slopes of the Andes. Both species are now in danger of extinction.

Chinchilla brevicaudata, which in the wild state has a much more valuable fur than *C. laniger*, was never successfully bred on a commercial scale, in spite of several attempts since the beginning of the century. However, reproduction in captivity has been achieved, but only in the geographical zone of its original habitat both in Chile and in Argentina, and only at its original altitude does it increase in numbers. Its most important breeding farm was that of Atahualpa, in Conchi Viejo, close to Chuquicamata, in the north of Chile, at an altitude of 3480 m. It was founded in 1931 by Fritz Ferger. At its peak (in 1965) it had 1300 *C. brevicaudata*, but it was closed in 1970 and the stock was distributed to various farms which did not prosper.

The Abra Pampa research station of INTA at 3684 m altitude, in the province of Jujuy, Argentina, is now the world's largest *C. brevicaudata* breeding station, with over 700 animals. It was founded in 1933.

On the other hand, *C. laniger* is more tolerant of lower altitudes. It is now bred in the United States, Canada, Europe (especially Denmark, West Germany, Britain, Spain, Sweden and France) and South Africa, and has come back to South America as a domesticated animal.

Wild species

Chinchilla brevicaudata used to live in the Andes between 7.5 and 30 °S latitude, at over 3000 m altitude, in Peru, Bolivia, Argentina and Chile (See Fig. 341).

Chinchilla laniger lives only in the coastal mountain range of Chile, between 26 and 36 °S.

One hundred years ago chinchillas were so abundant that between 1895 and 1910 Chile exported 2 million pelts to Europe. In 1910 the governments of Argentina, Bolivia, Chile and Peru, agreed to prohibit chinchilla hunting and fur selling but hunting continued in all four countries. In 1929, Chile unilaterally prohibited the hunting of both species, but illegal hunting continued until 1980 after which there was strict control; hunting is now very rare.

It is therefore not surprising that both species are nearly extinct in the wild state.

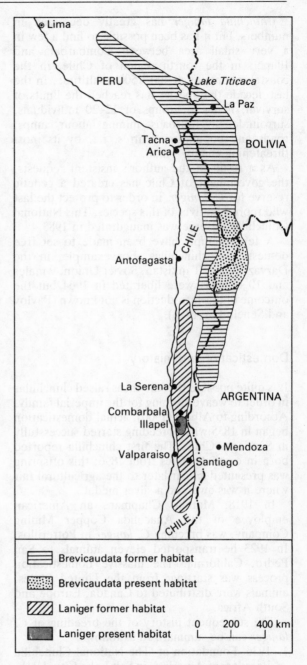

Fig. 34.1 Distribution of wild *Chinchilla spp.*

Legend:
- Brevicaudata former habitat
- Brevicaudata present habitat
- Laniger former habitat
- Laniger present habitat

0   200   400 km

Chinchilla brevicaudata still survives in circumscribed areas in the Andes close to the meeting point of Argentina, Bolivia and Chile, in a desert habitat more than 3500 m high. It lives in small colonies in very inaccessible places.

Chinchilla laniger has greatly decreased in numbers, but it has been possible to find a few in a very small area between Combarbala and Illapel, in the fourth region of Chile, in the coastal mountains at 400–1500 m altitude. In the last decade this species has reached the limits of survival, living in groups of 12–20 individuals, surrounded by railroads, mining labour camps, roads and population centres, i.e. by its most threatening enemy – man.

As a result of the author's insistent requests, the government of Chile has created a genetic reserve for *C. laniger*, in order to protect the last wild representatives of this species. This National Chinchilla Reserve was inaugurated in 1983.

A few attempts have been made to set free domesticated chinchillas. For example, in the Darvaz Range, Tajikistan, Soviet Union, 9 males and 10 females were liberated in 1964 but the outcome of this introduction is not known (Pavlov and Senelnikov 1965).

Domestication and history

It is quite possible that the Incas raised chinchillas in order to weave clothing for the imperial family. According to Albert (1900), actual domestication began in 1855 when breeding started successfully in Santiago, Chile. The first chinchilla reported born in captivity dates from 1896; this offspring was presented a year later to the agricultural fair where it was awarded a silver medal.

In 1918, Mathias Chapman, an American employee of the Anaconda Copper Mining Company, was breeding *C. laniger* in Potrerillos. In 1923 he transported eleven animals to San Pedro, California. The massive domestication process was starting; from the United States, animals were distributed to Canada, Europe and South Africa.

The subsequent history of the breeding of *C. laniger* can be summarized as follows:

1. 1936. Foundation of 'The National Chinchilla Breeders of America', in Salt Lake City, Utah.
2. 1957. The author founded the experimental breeding farm 'Danemanque', still operating, on the Andean slope, 27 km from Santiago, Chile.
3. 1960. Foundation of The Chinchilla Fur Breeders Association of England.
4. 1963. The population of domestic chinchillas

in the United States reached about 500 000 animals, with an average of 130 animals per farm.
5. 1970. Chinchilla industry reached its highest point. After years of selection, the domestic chinchilla reached its largest size and most outstanding fur quality. Some blood of *brevicaudata* was introduced into domestic *laniger* in the United States and Germany and this may have contributed to the larger size. More than ten mutant varieties including, black, beige, lavender and silver, were shown in several exhibitions around the world.
6. 1971 to 1973. Decline in demand for chinchilla furs.
7. 1974. Less critical circumstances for chinchilla ranchers. Thus, chinchilla breeding has recently shown growth and there is a good market for quality furs in the United States and Europe.

The first mutation, Wilson White (dominant), was recorded in 1955 in California. The Black Velvet mutant turned up in the same year in Utah and was developed by Bob Gunning as Gunning Black. Recessive Black was recorded soon after by Ramón Somavia in California. In 1961 Sapphire (recessive) appeared on the Larsen ranch. After that, crossing of mutants produced new colour varieties, e.g. Pink White – a double dominant of Wilson White and Tower Delaney Beige. The Ebony mutation was described in 1978; it is a dark brown with a hint of blue or black and is dominant. There are also three Beige mutants, one dominant and two recessive.

Mutations are not as important in chinchilla as in mink because at the moment it is difficult to find enough pelts of the same shade to make a garment.

Future prospects

Almost 60 years of effective domestication, selection and development of mutant races have brought about a great improvement of chinchilla furs. Further improvements are likely in the future. At the same time efforts must be made to conserve and protect the shattered populations of wild animals which still live in their native mountains. These scanty representatives constitute the genetic pool of the species.

References

Albert, F. (1900) La chinchilla. *Anales de la Universidad de Chile, Memorias Científicas y Literarias*, **107**: 915–34

Cabrera, A. (1960). Acerca de las chinchillas. *Actas y Trabajos del Primer Congreso Sudamericano de Zoología*, **4**: 195–202

Garcia-Mata, R. (1953). La cria de la chinchilla en Argentina. *Revista de Economía, Banco de la Provincia de Córdoba*, **4** (9)

Grau, J. (1963/64). The wild chinchilla today. *National Chinchilla Magazine*, July (1963) Part I: 10–12, Feb. (1964) Part II: 15–18

Grau, J. (1974). *La Chinchilla, su Crianza en Todos los Climas*. Ediciones Científicas Oikos S.R.L.: Buenos Aires and Santiago de Chile

Hansen, E. W., Martiarena. C. A. and **Cabezas. V.** (1972). Origen y evolución del criadero oficial de chinchillas *Chinchilla brevicaudata boliviana*, Brass 1911) en la Sub-estación Experimental Agropecuaria Abra Pampa. IDIA No. 294: 29–64

Hilleman, H. H, Tibbitts, F. D and **Gaynor, A. I.** (1959) *Reproductive Biology in Chinchilla*. National Chinchilla Breeders of America: Middletown, New York

Mann Fischer, G. (1978). Los pequenos mamíferos de Chile. *Gayana. Instituto de Biología, Universidad de Concepción, Chile*, **40**: 256–61

Mohlis, Connie. (1978) *Preliminary Information for the Conservation and Management of Wild Chinchilla (Lanigera) in Chile*. Corporación Nacional Forestal: Santiago, Chile

Ness, N. (1963) The chromosomes of *Chinchilla lanigera*. *Acta Veterinaria Scandinavica*, **4**: 128–35

Pavlov, M. and **Senelnikov, A.** (1965). The introduction of chinchilla into the Darvaz Range, Tajikistan. *IUCN Bulletin* No. 17. Translated from Russian (*Ohota i Ohotnich'e Hozyaistvo*, 1964)

Vidal, O. R., Riva, R. and **Spirito, S.** (1973). Los cromosomas de *Chinchilla brevicaudata*; contribución a la sistemática del género *Chinchilla*. *Physis*, C **32** (84): 141–50

Weir, B. J. (1966) Aspects of reproduction in Chinchilla. *Journal of Reproduction and Fertility*, **12**: 410–11

White, M. C. (1964). Chinchilla research: a bibliography. *National Chinchilla Breeders of America Research Bulletin*, No. 41. Middletown, New York

35
Syrian hamster

Roy Robinson

*St Stephens Road Nursery,
London, England*

Introduction

Among the various hamster species, only one can be held to be domesticated, the golden or Syrian hamster (*Mesocricetus auratus*). It is kept as a popular rodent pet and is also bred extensively in the laboratory for general biological and biomedical studies. A second species, the Chinese hamster (*Cricetulus griseus*), occurs in the laboratory but rarely as a pet.

Reproductive biology

The hamster is remarkably prolific, comparing favourably with the house mouse or rat. Sexual maturity occurs at about 6–8 weeks in both sexes but many instances of precocious maturity in the female have been recorded. The female is polyoestrous, with a regular 4-day oestrous cycle. Litter size varies from 4 to 10, with litters of 12–16 being not uncommon in vigorous strains. Young are weaned at about 21 days and the oestrous cycle recommences within a few days. A primary sex ratio of 150 males to 100 females has been reported but the secondary sex ratio is about 105: 100. A female usually produces 4–5 litters during her reproductive life. Females have *post partum* receptivity of the male but not *post partum* ovulation. Few females have litters beyond 16 months of age but males may be fertile up to 2 years. A breeding pause may be evident during the winter months but can be obviated by conditions of even temperature and constant long-day illumination. Longevity ranges from 1 to 3 years, with few individuals exceeding 2 years.

The various stages of the oestrous cycle have been extensively studied and the principal results have been summarized by Orsini (1961). These observations have practical application in that the period during which the female is receptive to the male can be pin-pointed. This is the late evening of the third day following an early morning copious discharge of a yellowish viscous fluid from the vagina. Males and females can then be placed together with impunity. This system has decided advantages over: 1. having to observe whether the female is receptive (lordotic response), with accompaning risk to the male; or 2. caging pairs or groups together from a young age with constant harassment of the males. The inconvenience of copulation occurring during late evening can be overcome by instituting a reverse-day light regime.

Wild species

The hamsters belong to the family Cricetidae, subfamily Cricetinae, and are indigenous to Eurasia. Many species and other named forms occur in the taxonomic literature (Ellerman 1941) but few of these have been investigated biologically; hence their interrelationships are either unknown or are questionable.

The Syrian hamster is one of a group of medium-sized hamsters inhabiting southeast Europe and Asia Minor. Only a few of the group have been bred in captivity, just sufficient for some species in fact to establish that this is possible, given sufficient attention. The various species are similar in general morphology and coat pattern but differ in details. Their karyotypes differ, probably fundamentally, despite certain gross similarities of individual chromosomes, because hybrids between them are either sterile or very subfertile. Two species have been studied karyologically with the following diploid determinations: Kurdistan or Transcaucasian hamster,

Mesocricetus brandti ($2n = 42$) and Romanian hamster, *M. newtoni* ($2n = 38$). Note that both differ from the diploid count of 44 for *M. auratus*. It is probable that these species have a relatively restricted distribution. The Syrian hamster, for example, probably occurs only in Syria and southern Turkey.

The chromosome numbers for the *Mesocricetus* group of 38–44 are roughly double those for other hamster genera (*Cricetus*, *Cricetulus*) of 20–22 chromosomes. Allopolyploidy or fragmentation has been proposed to explain the difference but this is doubtful since 40 and 42 are modal chromosome numbers for the Muridae. On the contrary, it is more probable that the numbers of 20–22 are unusual and are due to a history of karyological fusion. In fact, the Chinese hamster, *Cricetulus griseus*, is frequently employed in karyological studies on account of the low number ($2n = 22$) and large size of its chromosomes.

Domestication

An adult female and a litter of eight young were unearthed in 1930 from a field in the neighbourhood of Aleppo, Syria. One male and two females thrived in captivity at the Hebrew University, Jerusalem, and the first litter was born in August 1930. Breeding nuclei were subsequently despatched to laboratories in various countries of the world. The hamster has adapted excellently to laboratory regimes and is now firmly established in the repertoire of experimental animals. In 1945 the hamster became recognized as a delightful domestic pet, particularly in Britain, later in the United States and several European countries.

Over the last 35 years, there is some evidence that the animal has become progressively more adapted to domestication (Poiley 1950). There is less cannibalism of young than previously and therefore weaned litter size has increased. A sex difference in pugnacity has always been evident, the females being more aggressive than the males. The young tend to be gregarious but not the adults. It is still not possible to cage post-pubertal females together without persistent and often fatal fighting but it is now possible to house groups of males provided they have been reared together. All young animals are nervous and tend to be jumpy but quickly adjust to regular handling. The use of forceps is not recommended as the animal appears to resent being picked up in this manner and adjustment is delayed or may fail.

The hamster does not require special treatment in the laboratory. It appears to thrive on standard mice or rat pellets, supplemented with grain mixtures (barley, oats, wheat, flaked maize). The animal quickly suffers from dehydration if drinking water is not available. Large mouse cages are ideal. The best results are obtained by controlled mating and by keeping females in separate cages. Colony breeding, in the form of one male and a few females in a large cage, is often more convenient in practice, the females being isolated into separate cages as soon as pregnancy is detected. Pregnancy can be diagnosed by distension of the abdomen or the character of the vaginal discharge as described by Orsini (1961).

A second capture of wild specimens was made in 1970 and taken to an American laboratory. These are being maintained but it is doubtful if they will have any numerical impact upon the general population.

Current and future prospects

The hamster will no doubt maintain its position as a popular pet for children and it has been accepted as a competitive animal at fancy small-stock shows (Robinson 1978). However, the value of the animal resides in its usefulness for research. Biologists are continuously searching for potentially useful new species; few can seriously challenge the established species (mouse, rat, guinea-pig, rabbit) but the hamster comes a close fifth and may even have displaced the guinea-pig. Some idea of the numbers of hamsters bred each year may be obtained from the steady accumulation of known mutants (Robinson 1968, 1975). These would not be discovered if the species were not bred in hundreds of thousands. The inclusion of the animal in the Federation of American Societies for Experimental Biology's handbook on inbred and genetically defined strains of laboratory animals (Altman and Katz 1979) reveals the esteem with which the hamster is held by biologists.

The majority of gene mutants are concerned with coat colour, several with coat texture and hypotrichosis, fewer with anomalies (Robinson 1975, Altman and Katz 1979). Several programmes for establishing inbred strains are in progress and this work will certainly continue, if not expand, as the value of inbred strains for many areas of biological research becomes generally appreciated (Altman and Katz 1979, Festing 1979). Many of the coat-colour mutants have been incorporated into inbred strains, with the twin purposes of preserving the mutants and of characterizing the strains.

The hamster is interesting in possessing two large, flexible and evertible cheek pouches. Furthermore, they appear to be a transplant-priviledged site for many normal and abnormal allografts. The ease with which the pouch can be everted allows investigation of the transplant *in situ*. The flexibility of the walls of the pouch permits the insertion of pellets for pharmacological and chemotherapeutical studies. A review of applications and techniques has been contributed by Handler and Shepro (1968).

Unlike other common laboratory rodents, the hamster can hibernate for short periods of up to 5–7 days, although the period is usually of shorter duration. The notion that the species is a natural hibernator led to the animal being used as a subject for research on hibernation or for low-temperature experiments. Surprisingly, there is considerable variation in the capacity to hibernate among individuals and among stocks. It seems that the species is not an obligate but a facultative hibernator, doing so only in times of prolonged low temperature accompanied by lack of food. In fact, stimuli which cause some individuals to hibernate, cause others to enter a fatal hypothermia while others again remain active. Also, it seems that the condition is not particularly 'deep' because the animal is easily aroused by experimental procedures. Hoffman (1968) has provided a detailed summary of work in this field. There is incidental evidence for a steady fall in spontaneous hibernation in hamster colonies under domestication. This may reflect a genetic change in the propensity to hibernation or improvement in colony management.

Basic information on the gross morphology and general and special physiology may be found in the reviews of Hoffman *et al.* (1968) and Altman and Katz (1979). The latter present an extensive annotation of the many fields of research in which the hamster has featured. Apart from those areas in which the animal is well suited, the animal provides a useful comparison to studies with mice or rats and can help to establish the generality of specific results.

References

Altman, P. L. and **Katz, D. D.** (eds) (1979) *Inbred and Genetically Defined Strains of Laboratory Animals*, Part 2. *Hamster, guinea pig, rabbit and chicken*. Federation of American Societies for Experimental Biology: Bethesda, Maryland

Ellerman, J. R. (1941) *The Families and Genera of Living Rodents*, Vol. 2: *Muridae*. British Museum (Natural History): London

Festing, M. F. W. (1979) *Inbred Strains in Biomedical Research*. Macmillan: London

Handler, A. H. and **Shepro, D.** (1968) Cheek pouch technology. In: Hoffman *et al*. (1968) (op. cit.)

Hoffman, R. A., (1968) Hibernation and effects of low temperature. In: Hoffman *et al*. (1968) (op. cit.)

Hoffman, R. A., Robinson, P. F. and **Magalhaes, H.** (1968) *The Golden Hamster: its biology and use in medical research*. Iowa State University Press: Ames, Iowa

Orsini, M. W. (1961) The external vaginal phenomena characterizing the stages of the estrous cycle, pregnancy, pseudopregnancy, lactation and the anestrous hamster, *Mesocricetus auratus* Waterhouse. *Proceedings of the Animal Care Panel*, **11**: 193–206

Poiley, S. M (1950) Breeding and care of the Syrian hamster. In: *The Care and Breeding of Laboratory Animals*, ed. E. J. Farris, Wiley: London

Robinson R. (1968) Genetics and karylogy. In: Hoffman *et al*. (1968) (op. cit.)

Robinson R. (1975) The golden hamster. In: *Handbook of Genetics*, ed. R. C. King, Vol. 4: *Mammals*. Plenum Press: New York

Robinson, R. (1978) *Colour Inheritance in Small Livestock*. Watmoughs: Bradford

36
Mongolian gerbil

J. D. Turton

Commonwealth Bureau of Animal Breeding and Genetics
Edinburgh, Scotland

General biology

The Mongolian gerbil, *Meriones unguiculatus*, is a cricetid rodent of the subfamily Gerbillinae. The word 'Mongolian' is often omitted, particularly when referring to the species in a non-scientific context, although the shorter form applies to all species of Gerbillinae.

The natural habitat of the species is the desert and semi-desert areas of Mongolia and north-eastern China, where it lives in colonies in galleries or burrows in dry, sandy soil. Its biological and behavioural characteristics have been described by Marston (1976) and can be summarized as follows. In the natural habitat, activity is diurnal and crepuscular, but in captivity gerbils may be active at any time. They rarely vocalize, but frequently exhibit drumming with the hind feet. Animals of both sexes have a prominent mid-ventral sebacious gland close to the umbilicus. This produces a yellowish-brown sebum with a musky odour, which the gerbil uses for scent-marking. Mongolian gerbils have a great capacity for heat regulation, and can tolerate 40 °C for 5 hours without discomfort.

Pelage development and growth in body weight

have been described by McManus and Zurich (1972). Gerbils are hairless at birth, but by 21 days of age the juvenile pelage is complete. A post-juvenile moult occurs between 32 and 38 days of age, and a post-subadult moult between days 58 and 63. The birth weight of pups averages almost 3 g; at sexual maturity, males and females average about 54 and 47 g respectively. Old fat individuals may weigh as much as 100 g.

Gerbils are prone to epileptiform convulsions when stressed. Work on the condition has been reviewed by Vincent *et al* (1979). Such seizures are first observed at about 2 months of age, and 40 per cent of animals may exhibit them in the first 6–10 months of life. Daily provocation of seizures leads to tolerance of the inducing stimulus, and susceptibility is lost within 3–5 days.

Marston (1976) summarized reproductive characteristics as follows. Puberty occurs at 9–12 weeks of age, females being polyoestrous with a 4 to 6-day cycle. Ovulation is spontaneous, and 4–9 ova are shed. The gestation period is normally 24–26 days, but implantation is delayed when females are suckling more than 2 young, and this may result in an interval of up to 42 days between *postpartum* oestrus and parturition. Most females exhibit oestrus within 24 hours of parturition, and a high conception rate can occur from matings at this oestrus. Litter size averages 4.5, and a female may produce up to 10 litters in her lifetime. The young are suckled for 21–28 days. Ovarian cysts are common in ageing females (Norris and Adams 1972).

Information on the genetics of the Mongolian gerbil is sparse. The diploid chromosome number is 44 (Cohen 1970), comprising 16 metacentrics, 14 submetacentrics, 12 acrocentrics and the sex chromosomes. The X-chromosome is the second-largest submetacentric, and the Y-chromosome is a medium-sized metacentric with a secondary constriction in the middle of the long arm.

Albinism occurs, and is inherited as a simple autosomal recessive (Robinson 1973). An autosomal, semidominant spotting mutation has been described (Waring *et al*. 1978). Homozygotes (*Sp/Sp*) are presumed to die shortly before birth, and heterozygotes are mildly anaemic. The latter have white spots on the head, neck and nose, and the feet, tail tip and belly are white.

Susceptibility to epileptiform seizures has been reported as being controlled by at least one variably penetrant dominant allele at an autosomal locus (see review by Vincent *et al*. (1979)). Seizure-resistant and seizure-sensitive strains have been developed by selection for 18 generations.

Domestication

In 1935, twenty pairs of Mongolian gerbils were captured in the region of the Amur river basin of eastern Mongolia and Manchuria (Marston 1976). These animals were taken to Japan, and formed the basis of a closed, random-bred colony. Four breeding pairs were taken to the United States in 1954, and another random-bred colony was established. From there, gerbils were distributed throughout the States, and Continental Europe.

The Mongolian gerbil is an attractive rodent. It produces very little odour, only small amounts of dry faeces, and insufficient urine to damp its bedding. These attributes, together with its behaviour, particularly an innate curiosity, make it a suitable and much sought after children's pet, and large numbers are now kept for this purpose. The species is now used quite extensively in biomedical research. Over the past 5 years, some 600 research papers on the species have been published, most of these on physiological, behavioural and pathological investigations. It is useful as an animal model of human disease, particularly in respect of cerebral infarction, epileptiform seizures, lead nephropathy, and cholesterol absorption and metabolism (Vincent *et al*. 1979).

The annual number of research papers published on the Mongolian gerbil is likely to continue to increase, and the need for more animals for research purposes will generate more work on the breeding, genetics and diseases of the species. Thus, the gerbil is likely to maintain its position as a significant, minor laboratory species and as a household pet.

References

Cohen, M. M. (1970) The somatic karyotype of *Meriones unguiculatus*. A morphological and autoradiographic study. *Journal of Heredity*, **61**: 158–60

Marston, J. H. (1976) The Mongolian gerbil. In: *The UFAW Handbook on the Care and Management of Laboratory Animals*, pp. 263–74. Churchill: Edinburgh, London and New York

McManus, J. J., and **Zurich, W. M.** (1972) Growth, pelage development and maturational molts of the Mongolian gerbil, *Meriones unguiculatus. American Midland Naturalist*, **87** (2): 264–71

Norris, M. L and **Adams, C. E.** (1972) Incidence of cystic ovaries and reproductive performance in the Mongolian gerbil, *Meriones unguiculatus. Laboratory Animals*, **6** (3): 337–42

Robinson, R. (1973) Acromelanic albinism in mammals. *Genetica*, **44** (3): 454–8

Vincent, A. L., Rodrick, G. E., and **Sodeman, W. A. Jr** (1979) The pathology of the Mongolian gerbil (*Meriones unguiculatus*): a review. *Laboratory Animal Science*, **29** (5): 645–51

Waring, A. D., Poole, T. W. and **Perper, T.** (1978) White spotting in the Mongolian gerbil. *Journal of Heredity*, **69** (5): 347–9

37
Muskrat

Gale R. Willner

Department of Fisheries and Wildlife, Utah State University, Logan, U.S.A.

Introduction

The muskrat, *Ondatra zibethicus*, is a member of the order Rodentia, family Cricetidae and subfamily Microtinae, which also includes voles and lemmings. This species is also referred to by the name 'musquash'. There are a total of 16 subspecies of the muskrat in North America. These are morphologically similar but vary with respect to distribution, habitat preference and population status. The common name of the muskrat is derived from the conspicuous odour of secretions from paired perineal musk glands located by the base of the tail. The musk was once used in perfumes and as a lure for baiting traps (Willner *et al.* 1980, Perry 1982).

The muskrat is an important furbearing mammal in North America. The success has been attributed to its wide distribution, abundance, and pelt value. Many aspects of the biology of the muskrat have been investigated, including its ecology, food habits, and reproduction (Willner *et al.* 1980). Giles (1978) and Perry (1982) reviewed techniques for managing the muskrat and its habitat in the wild.

Muskrats in the wild

The muskrat is the largest cricetid. It is chunky in appearance and has a large, blunt head, relatively small eyes, and short, rounded ears. The partially webbed hind feet are broad and fimbriated; the forefeet are much smaller. The tail is nearly as long as the head and body, flattened laterally, scaly and with a sparse fringe of hair (Willner *et al.* 1980).

Pelage colour varies from white and silver through tan, reddish brown and black; it is generally dark brown. The ventral pelage is somewhat lighter than the remainder of the fur. There is no sexual dimorphism in body or cranial measurements. The weight of adults ranges from 700 g to over 1800 g; neonates weigh only about 21 g (Willner *et al.* 1980).

Faecal droppings of the muskrat are small, less than 1.91 cm. Small piles of muskrat droppings are found on partially submerged objects like logs and along runways. The round-tailed muskrat (*Neofiber alleni*) resembles the common muskrat, both in appearance and habits, but it is smaller, has a round tail, and the hind feet are not webbed. Skulls from these two species are similar, except in size. Faecal droppings, food platforms and cone-shaped reed houses may be used to indicate the presence of muskrats, particularly those inhabiting marshes (Willner *et al.* 1980).

The muskrat has a duplex uterus. Reports on the length of the oestrous cycle vary from 2 to 34 days (see Perry 1982 for references). The gestation period varies between 25 and 30 days. The initiation and duration of the breeding season vary with geographical and climatological conditions. Muskrats inhabiting more southerly latitudes of the United States breed throughout the year, with peak reproductive activity occurring in the winter. In more northern latitudes, reproductive activities are confined to the spring and summer months. Peak production of first litters was found to occur in May for muskrats in Wisconsin, southern Quebec, West Germany and northern France. The mean litter size varies from about 4 to 8; 6 or 7 is most common. Northern populations of muskrats produce larger litters than southern populations. The altricial young are born in nests inside dome-shaped houses or burrows in banks (Willner *et al.* 1980).

The diploid number of chromosomes is 54.

Dozier (1948) described variations in pelt colour, ranging from white to black, and their associated genotypes. The common brown colour is a black agouti (*ABC*); the black pelt is a modified agouti pattern. The *A*-gene gives the agouti or wild type coat pattern; the *B*-gene conditions production of the black pigment; and the *C*-gene is for the development of full colour. Albinism is the result of a recessive mutation of the *C*-gene.

Muskrats are widely distributed in North America. They occur in a variety of habitat types such as streams, ponds, and marshes from the Arctic Circle in the Yukon and Northwest Territories to as far south as the Gulf of Mexico, and from the Aleutian Islands east to Labrador and southward along the Atlantic coast to North Carolina. An insular subspecies (*Ondatra zibethicus obscurus*) occurs in Newfoundland. Muskrats are absent from Florida, being replaced by the ecologically equivalent round-tailed muskrat, *Neofiber alleni* (see review by Willner *et al.* 1980).

Muskrat populations, particularly large ones, may cause extensive damage to pond impoundments and ditches by their burrowing activity into banks and dam areas. Muskrats also show preferences for agricultural crops like corn and sugarcane, causing monetary losses to the farmer. 'Eatouts' of marsh vegetation may also be observed in areas where the muskrat is not adequately controlled. Although no specific dollar value is available for the damage caused by muskrats, it is estimated to be in the millions (see reviews by Willner *et al.* 1980, Perry 1982).

Exploitation

Early history

Svihla and Svihla (1931) credited Captain John Smith for being the first to document the muskrat as a commodity in Virginia, in 1621. However, more recent work has shown that the muskrat was used long before that. In Louisiana, muskrat bones were excavated from Indian refuse sites that were carbon dated to 260 B.C (Lowery 1974). In the 1700s, Indians used the fur for clothing (Svihla and Svihla 1931). As a resource, the muskrat has been commercially harvested since 1821 according to records obtained from the Hudson Bay Company in Canada (Elton and Nicholson 1942). Carmichael (1973) presented a history of the muskrat as a fur resource in Manitoba, Canada, between the years 1935 and 1948.

Deems and Pursley (1978) reported the status of the muskrat in the United States and Canada in 1976. The species was used as a food resource in eighteen states and eleven Canadian provinces/territories, and as a fur resource throughout North America except Florida. The annual fur harvest is about 6 million in the United States and 2 million in Canada.

The demand for muskrat, like other furs, depends on fashion. Fur from muskrats was once used for making 'beaver' hats (Lantz 1923). At present, it is used for coats, gloves and trim. The fur industry markets sheared muskrat fur under the name 'hudson seal'. The meat of muskrat is sold as 'marsh rabbit' (not to be confused with *Sylvilagus palustris*). Many recipes are available for preparing and cooking the meat (Dailey 1954).

Fur farming and ranching

The development of fur farms for muskrats, designed to raise them for the purpose of selling pelts, began in the early 1900s in North America. It reached a peak in 1930. The type of farming ranged from breeding muskrats in pens with concrete floors to propagating them in natural conditions with fence enclosures. It was hoped that this type of venture would be as successful as it proved to be for mink. Much of the early literature on muskrat farming emphasized the substantial monetary returns for the minimal investment involved. The Louisiana Department of Conservation (1931) described early farming and ranching operations in the United States. Denmark (1937) discussed the development of fur farms in Canada.

Unlike mink ranching, muskrat fur-farming never developed into a wide-scale practice. Low profit margin was the main reason that many farms failed. As Lantz (1923) noted, the price of the fur does not warrant raising them in captivity. Other reasons why muskrat-farming failed included high losses from handling, fighting, poor sanitation and irregular reproduction (see Perry 1982 for references). Less than 5000 muskrat skins were sold from fur farms in Canada in 1938,

in comparison with the more than 1 million harvested from the wild (Elton and Nicholson 1942).

In Europe muskrat farms were first started in Ireland, Scotland and England and in 1927 there were eighty-eight farms in these three countries. However, some animals escaped and did such damage by burrowing in river banks that in 1932 a government order prohibited the import and keeping of muskrats and the farms were closed. An intense campaign was launched; by 1935 they were under control and by 1939 all the feral muskrats had been exterminated. In France, sixteen farms were set up in 1928 and later; again escapes caused damage and the farms were closed but it has proved impossible to control the feral populations. The same applies to Poland (farms established in 1929 and abolished in 1932), Belgium, Italy, Switzerland and Japan (farms before 1945, first feral animals caught in 1947). There are now no known muskrat farms outside North America.

Introductions and feral populations

The first introduction of muskrat into Europe as a fur resource was made in 1905 when 10 pairs were liberated near Dobřiš on the River Kocaba in Bohemia (Hoffmann 1958). From there they spread over the whole of central Europe. In the east, they went from Poland and Romania into the Soviet Union; in the south, they have entered the north of Bulgaria, have completely settled Yugoslavia and from there are invading Greece and Italy.

In France, the escaped animals spread north into Belgium and reinforced the local escapees. From there, they spread into Luxembourg and the southern Netherlands where they joined the animals coming from Germany. Northern Switzerland was invaded from France.

In Finland muskrats were first introduced in 1922 and were widely stocked over the country. From there they have spread west into Norway and Sweden and east into the Soviet Union.

The present distribution of muskrats in Europe (excluding the Soviet Union) is shown in Fig. 37.1.

In the following countries, the muskrat is now considered a pest and is hunted intensively through the year: France, Belgium, The Netherlands, Luxembourg, Germany (East and West), Austria, Switzerland, Sweden and Norway. In the

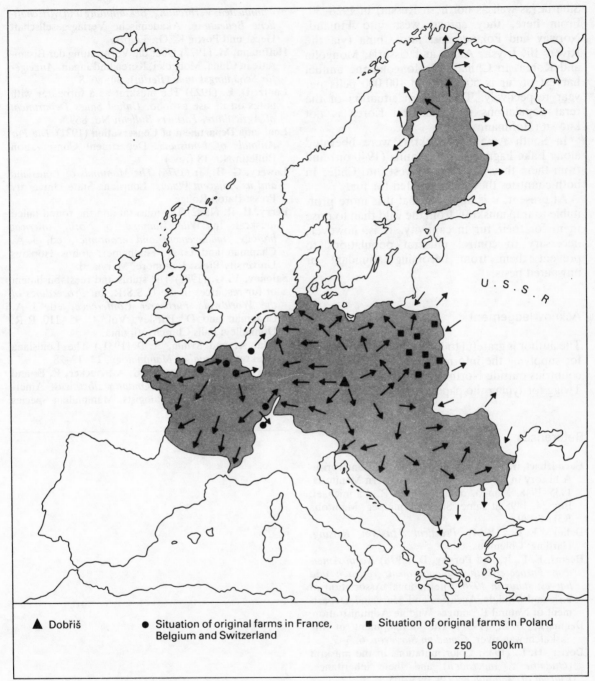

Fig. 37.1 Distribution and spread of the muskrat in Europe (excluding the Soviet Union)

▲ Dobřiš   ● Situation of original farms in France, Belgium and Switzerland   ■ Situation of original farms in Poland

0   250   500km

Soviet Union, it is an important fur resource (Safonov 1981) and about 5 million pelts are cropped annually. It is also cropped on a small scale (50 000–200 000 pelts each) in Bulgaria, Czechoslovakia, Hungary, Poland, Romania, Yugoslavia and Finland. In these countries, the control extends only to river banks and dams (Hoffman, personal communication).

Outside Europe muskrats were introduced into

Siberia (as well as northern Russia) in 1928–30. From here, they spread west into Finland, Norway and Poland, east into China (via the Rivers Ili, Irtysh, Amur and Ussuri), Mongolia and Korea. In China and Mongolia the annual harvest of fur is 500 000 and 100 000 pelts per year respectively. The present situation of the feral populations in Japan and Korea is not known (Hoffmann 1977).

In South America, muskrats were liberated along Lake Fagnano in Argentina (1940 on) and from there they have spread west into Chile. In both countries they are harvested for fur.

At present, it is apparent that it is more profitable to trap muskrats from the wild than to raise them for their fur in captivity. It is, however, necessary to control muskrat populations to prevent them from becoming abundant or unwanted pests.

Acknowledgement

The author is grateful to Max Hoffmann of Berlin for supplying the information about muskrats in countries outside North America and to Kathryn Twigg for typing the paper.

References

Carmichael, R. G. (1973) Innovation and enterprise. A history of fur conservation in northern Manitoba, 1935–1948. Mimeo. obtained from R. Carmichael: Box 24, 1495 St James Street, Winnipeg, Manitoba R3H OW9

Dailey, E. L. (1954) *Practical Muskrat Raising*. Harding: Columbus, Ohio

Deems, E. F. Jr, and **Pursley, D.** (1978) *North American Furbearers: their management, research and harvest status in 1976*. International Association of Fish and Wildlife Agencies and Maryland Department of Natural Resources Wildlife Administration

Denmark, D. E. (1937) The development of the Saskatchewan river. *Canadian Surveyor*, **6**: 3–5

Dozier, H. L. (1948) Color mutations in the muskrat (*Ondatra z. macrodon*) and their inheritance. *Journal of Mammalogy*, **29**: 393–405

Elton, C. and **Nicholson, M.** (1942) Fluctuations in numbers of the muskrat (*Ondatra zibethica*) in Canada. *Journal of Animal Ecology*, **11**: 96–126

Giles, R H. Jr. (1978) *Wildlife Management*. Freeman: San Francisco

Hoffmann, M. (1958) *Die Bisamratte: ihre Lebensge-*

wohnheiten, Verbreitung, Bekämpfung und wirtschaftliche Bedeutung. Akademische Verlagsgesellschaft Geest und Portig K. G.: Leipzig

Hoffmann, M. (1977) Über die Verbreitung der Bisamratte in China, Mongolei, Korea und Japan. *Anzeiger für Schädlingskunde (Berlin)*, **50**: 86–8

Lantz, D. E. (1923) The muskrat as a furbearer with notes on its use as food. *United States Department of Agriculture, Farmers Bulletin* No. 869

Louisiana Department of Conservation (1931) *The Fur Animals of Louisiana*. Department Conservation Bulletin No. 18 (rev.)

Lowery, G. H., Jr (1974) *The Mammals of Louisiana and its Adjacent Waters*. Louisiana State University Press: Baton Rouge

Perry, H. R. (1982) The muskrat and the round-tailed muskrat. In: *Wild Mammals of North America: biology, management and economics*, ed. J. A. Chapman and G. A. Feldhamer. Johns Hopkins University Press: Baltimore, Maryland.

Safonov, V. G. (1981) The status and reestablishment of fur resources in the U.S.S.R. In: *Proceedings of the Worldwide Furbearer Conference*, ed. J. A. Chapman and D. Pursley, Vol. 1: 95–110. R.R. Donnelley: Falls Church, Virginia.

Svihla, A. and **Svihla, R. D** (1931.) The Louisiana muskrat. *Journal of Mammalogy*, **12**: 12–28

Willner, G. W., **Feldhamer, G. A.**, **Zucker, E. E.** and **Chapman, J. A.** (1980) *Ondatra zibethicus*. American Society of Mammalogists, Mammalian Species No. 141

38

House mouse

R. J. Berry

Department of Zoology,
University College London, England

Introduction

Linnaeus wrote of the house mouse that it is 'an animal that needs no description: when found white it is very beautiful, the full bright eye appearing to great advantage amidst the snowy fur. It follows mankind, and inhabits all parts of the world except the Arctic.'

Nomenclature

The Latin word for a mouse is *mus*, which provided a play on words for mediaeval bestiarists who thought that mice arose by spontaneous generation from the soil (*hu-mus*). However the word *mus* is said to be derived from a Sanskrit verb (*mush*), which means 'to steal'. By tautonymy, the species came to be called *Mus musculus*, the little mouse, using the diminutive originally applied to it by Pliny (A.D. 23–79) in his classification of the different kinds of mice, thus distinguishing it from its close relatives the rats (which were also classified by Linnaeus in the genus *Mus*).

Subsequent authors indulged in a riot of naming as mouse-like specimens were collected from almost every part of the world. A major taxonomic simplification was carried out by Schwarz and Schwarz (1943) who combined 133 named forms into 15 subspecies of the original Linnaean species. They suggested that the main differentiation of the species had resulted from commensal forms arising from three basic types; that these commensals had followed the migration of man, and in some places developed distinctive feral forms. The subdivisions of *M. musculus* recognized by Schwarz and Schwarz were *M. m. wagneri* (with a range from the Volga to the Yellow Sea north of 42–44 °N), which has given rise to nine subspecies including *M. m. domesticus*, the ancestor of the laboratory mouse; *M. m. spicilegus* (found in south Russia west of the Volga, Bulgaria, Romania, Hungary and Austria), with one derived subspecies, *M. m. musculus* living sympatrically with it, but also extending into northern Europe including Scandinavia (and thus including Linnaeus' mouse, from his home town of Uppsala); *M. m. manchu* (from southeast Manchuria and the main Japanese islands), with one subspecies; and *M. m. spretus* (from Iberia, southern France and North Africa as far east as Algeria), which has given rise to no commensals.

The Schwarz and Schwarz classification is found in most general accounts of mouse and rodent biology. It reduced interest in the systematics of mouse populations, as well as implying that *M. musculus* is a relatively young species, spreading most importantly from the borders of Iran and Turkestan with the practice of cereal-growing in Neolithic times.

Unfortunately this convenient idea is largely wrong, and work by Marshall and Sage using classical morphological techniques combined with electrophoretic studies of enzymes and proteins has shown that at least seven species of the original *M. musculus* must be recognized (Marshall 1977, and in Morse 1978; Sage, in Morse 1978; Marshall and Sage, in Berry 1981; Thaler, Bonhomme and Britton-Davidian, in Berry 1981).

Most combinations of these seven species can successfully interbreed in the laboratory, but from the point of view of domestication the important species is *M. domesticus*, and the following account concentrates on it. Unless otherwise stated, 'house mouse' should be taken to mean *M. domesticus*. However, introgression may have contributed to the gene pool of the domestic mouse as we know it today.

General biology

Distinctive traits One of the reasons for the enormous success of the house mouse as a commensal is its enormous flexibility, both genetically and phenotypically. The species is a poor competitor with other small mammals in most natural habitats, but has colonized niches as extreme as subantarctic islands, Pacific atolls, high altitudes in the Andes, and food cold-stores. Mice experimentally transferred from a hot to a cold environment adjust physiologically by increasing their metabolic rate and the amount of circulating haemoglobin, and the size of some organs in their offspring changes, most noticeably producing an increase in overall body size and a reduction in tail length, so lessening heat loss (Barnett *et al*. 1975). These phenotypic adaptations are followed by genetical changes which represent the results of natural selection acting on a variable genome under conditions of physiological stress (Berry, in Berry 1981).

House mice are opportunists, with a small number of colonists repeatedly establishing new and often short-lived populations. Because of the amount of inherited variation in the species, this leads to considerable genetical heterogeneity between populations, and it is impossible to determine whether a population is showing adaptation without repeated sampling to monitor any changes which may take place with time. Mice on small islands in temperate latitudes (and in cold-stores) are usually larger than their relatives living under less extreme conditions. This is probably because mice are normally smaller than the physiological optimum, so that they can escape down small holes from predators. This increased size is usually inherited, and persists even when the animals are put into more equable conditions.

House mice are distinguished from *Rattus* species by the possession of much larger first than second and third molars; the third molar is small and sometimes absent. In other respects, mice are very generalized small murid rodents, with mean adult weights ranging from 9 g on a Pacific atoll to 21 g on a subantarctic island. The head plus body length ranges from about 75 mm to almost 100 mm; the head plus body to tail ratio varies from 1.2 to about 0.9. The largest recorded feral mice are from the Faroe Islands of the North Atlantic (they are approximately the same size as standard strains of laboratory mice); the smallest are from Enewetak Atoll in Micronesia.

There is a fair amount of coat colour variation in wild mice. The normal phenotype is dark-bellied agouti. However, local populations, laboratory mice, and 'fancy' animals may range from albino through shades of brown and yellow to black. The commonest variant in nature is partial or complete white belly. The genetics of pigmentation have been extensively studied, and over 40 genes are known which affect the distribution or intensity of colours (Silvers 1979); some of these seem to affect colour only, whilst others influence a range of characteristics. The large and flourishing mouse fancy is principally concerned with producing and showing mouse colour variants: either 'self' (with coats of a single colour – black, fawn, silver, champagne, etc.) or 'marked' – Dutch, tan, variegated, etc.

Reproduction Female mice become sexually mature at around 6 weeks, although later under crowded or cold conditions. Indeed, females born in the autumn in the wild may not come into breeding condition until the following spring. Males apparently attain puberty slightly later than females, but are less affected by fluctuations in the external environment. There is little difference in breeding intensity throughout the year in commensal (including tame) and corn-rick mice; wild-living mice may have a definite breeding season in the more clement season (during the warmth of summer in Britain and northern America but in the damp of winter in South Australia), although those living in a non-fluctuating environment (whether on subantarctic or tropical islands) breed all the year round.

Bronson (1979) has reviewed the environmental factors which control breeding. He concluded that disease, exercise, and light are not critical under natural conditions, but that seven variables may at times be limiting: day–night regulation of diurnal rhythms; total caloric intake; specific nutrients in the diet; temperature variation; agonistic stimuli; specific tactile cues; and priming pheromones. The social environment affects gonadotropin release relatively directly, while pheromones and tactile cues act via sensory reception and brain mediation to control pituitary secretion of luteinizing hormone and prolactin, while dietary and temperature influences affect indirect pathways. Bronson comments that this rather loose type of ambient cueing is unusual, but ideally suited for the colonizing strategy the mouse has evolved. For example, most rodents breed seasonally with their reproduction controlled by photoperiod; this control does not exist in house mice, and dispersing mice are thus able to maximize their rate of increase in a new habitat.

The oestrous cycle varies from 4 to 6 days, oestrus itself occupying less than a day of this period. Fertilization is possible for about 10–12 hours after ovulation; the peak time for ovulation is 03.00 hours; most copulation in laboratory mice takes place between 22.00 and 01.00. Gestation lasts 19–21 days, but implantation (which normally takes place 5 days after fertilization) may be delayed for as long as 2 weeks in suckling females. Parturition usually takes place during the night. It is followed immediately by a *post-partum* oestrus, ovulation occurring 12–18 hours after birth of the young.

Litter size is primarily a consequence of the number of ova shed, which in turn is determined by the size of the mother – the larger the female, the more ova are produced. This correlation disappears in the large island races of mice. Presumably the number of young born in these circumstances is adaptive. If fertilization takes place, virtually all the ova shed are fertilized, but *c.* 10 per cent fail to implant. Another 5–10 per cent of foetuses die between implantation and birth.

Litter size in laboratory mice increases for the first 2–3 litters of any female, and decreases in high parities. In wild-living mice, mean litter size is 4–8, increasing to a maximum in the middle of the breeding season, then decreasing again as young born early in the season begin breeding.

The young weigh approximately 1 g at birth. They are hairless and helpless. The rate of post-natal growth depends on the amount of milk available, and this in turn depends on litter size. In general, mice are close to 10 g when they are weaned at the age of 3 weeks.

Chromosomes House mice typically have 20 pairs of rather similar-sized, acrocentric chromosomes. The X and much shorter Y pair end-to-end at meiosis. The introduction of band-staining techniques has made it possible to identify each chromosome individually.

A few small variants in the amount of centromeric heterochromatin have been found in wild mice and between inbred strains, and a large number of reciprocal translocations and inversions have been induced (and are maintained) in laboratory strains (Miller and Miller, in Morse 1978; Searle and Beechey, in Altman and Katz 1979). However, the most remarkable chromosomal variation is undoubtedly the range of centric fusions (= Robertsonian translocations) found in a variety of areas. Every autosome except no. 19 has now been found to be involved in a naturally occurring fusion. Some fusions are more widespread than others but in general chromosomes seem to associate at random. Although populations with centric fusions would be expected to be fully fertile with normal mice (since they both have the same number of chromosome arms), in practice a great deal of hybrid infertility occurs, and only narrow hybrid zones occur between races (Capanna 1980; Gropp and Winking, in Berry 1981).

Geographical distribution

Wherever man has gone, there has gone also the house mouse – to all intents the species is a weed. In some places it has been unable to establish itself in competition with native small mammals (as over much of tropical Africa) or one of the rat species, but it can fairly be described as cosmopolitan.

A particularly interesting situation exists in Denmark and Germany where *M. musculus* and *M. domesticus* meet and apparently interbreed freely, but retain a narrow hybrid zone (see Thaler, Bonhomme and Britton-Davidian, in Berry 1981).

The economic importance of house mice depends more on their nuisance value as contaminators of stored food and disturbers of lagging, insulation, etc. in buildings, than on the amount of material they consume (Rowe, in Berry 1981). In some parts of the world (most importantly Australia), mice may sometimes reach plague densities and eat significant amounts of food, especially cereals. Their normal diet is catholic and contains a high proportion of insects.

Mice do not seem to be important reservoirs of any human disease. They may be responsible for local outbreaks of mild food poisoning, but they have not been reported to transmit any serious conditions (Blackwell, in Berry 1981).

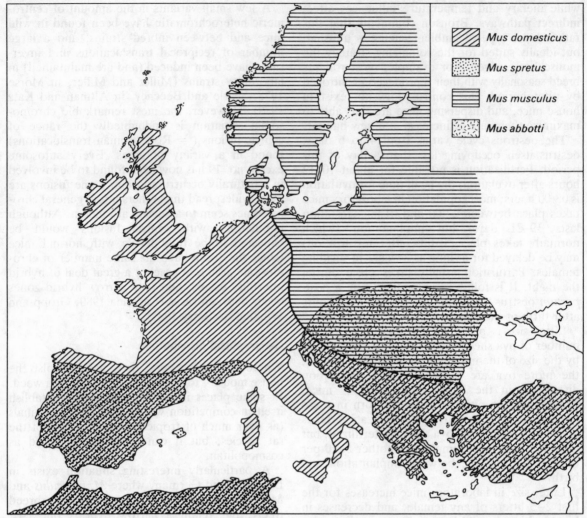

Fig. 38.1 Distribution of the four main *Mus* species in Europe (omitting *M. hortulanus* of the steppes). The range of *M. abbotti* has not been accurately determined and it may not extend as far to the west and north as shown on the map. The species which has spread to most parts of the world as a commensal and domesticated animal is *M. domesticus*. (After Reichstein 1978; Marshall and Stage in Berry 1981; Thaler *et al.* in Berry 1981)

Wild species

As noted above, the taxon traditionally describing the single species *M. musculus* contains at least seven different species (Fig. 38.1).

Mus spretus. This species lives in natural woodland and scrub around the western Mediterranean from southern France to Spain, and from Morocco to Tunisia.

Mus abbotti. This species occurs in natural vegetation and fields at least from Macedonia through Turkey and northwestern Iran to the south coast of the Caspian Sea. Specimens from the northern, more humid portion of this area are brownish sooty on the back, and white or buff with grey bases to the hairs on the belly; those from central and southern Turkey are a paler brown on the back.

Mus hortulanus. This species is characterised by building 'hillocks': a family or small colony piles several kilograms of grain spikes around a cereal stalk, then covers the whole with earth. Its distribution is that of the natural steppe vegetation and the grain fields replacing it in eastern Europe.

Mus musculus. The distribution of *M. musculus* is from Scandinavia, northern Denmark and the River Elbe eastward to at least Yugoslavia and Thessalonica. It lives almost exclusively in manmade habitats, and seems to be completely dependent on buildings during the winter in the north. It apparently does not overlap with the range of any other *Mus*, except where it occupies farm buildings within the range of *M. hortulanus*.

A subspecies, *M. m. wagneri*, lives in northern interior Asia. It has an identical (but smaller) skull than the nominate subspecies, but possesses a pale sandy-brown dorsal pelage, and is pure white on the ventral surface, feet, and underside of the short tail. Its fur is long and silky; even the tail is furry.

Mus domesticus. The European house mouse tends to be confined to buildings, agricultural fields, and vegetation altered by man. It is undoubtedly the ancestor of the 'outbred Swiss albino laboratory mouse' (see pp. 280) which has the same cranial characteristics.

Animals with skulls more like those of *M. musculus* are found in the paler-bellied, brown-backed mice of southern France, Spain and Sicily (hitherto called *brevirostris* or *azoricus*). Similar skulls are seen in emigrants to England, America, and Australia. A more slender skull characterizes all the long-tailed mice from northern Africa eastward through the deserts (*praetextus*) and Iran to Pakistan (*bactrianus*) and up the Himalaya to Kashmir and Nepal (*homourus*).

Mus domesticus apparently adapts to local situations by extreme variations. In northern Africa, there is a geographical mosaic of dark and white-bellied colonies, while in Egypt, Israel, and Syria the pelage is polymorphic for colour. Further east, a desert sandy brown dorsal colour prevails, resembling the gerbils which live in the same area. In North Africa the dark and white-bellied forms often, although not always, segregate in farms and oases respectively. The Himalayan form (*homourus*) is uniformly dark on the back, and white with grey bases ventrally. It is confined to buildings in the winter.

Mus molossinus. Both Asian species (*M. molossinus* and *M. castaneus*) have a skull similar to that of *M. musculus*, but smaller, more rounded, and with a less convex dorsal outline. It is a dark, brownish grey mouse with white feet and a white underside to the short tail. The ventral fur is white with grey bases in Japan and Manchuria, but entirely white in Korea.

Mus castaneus. This species is closely related to *M. molossinus* and interbreeds freely with it in the laboratory. However, the two species are differentiated by allozymes and centromeric heterochromatin, and interbreed nowhere outside the laboratory. It is the Asian equivalent of *M. domesticus* in northwest Europe, with a long tail, dark ventral colour, and a convoluted parietal-squamosal suture. It is almost limited to coastal towns and cities from Bombay to Fukien, and around Taiwan, Indonesia and the Phillipines (although it has become feral in Micronesia). The subspecies of north central India (*tytleri*) has a golden buff dorsum.

These seven species seem to represent biologically distinct groups confused by Schwarz and Schwarz (1943) under the name *M. musculus*. Marshall (1977) has recognized another 14 Asian species which he includes in the genus *Mus*.

A complication within *M. domesticus* is the existence of apparently isolated chromosomal races with considerable infertility with the normal 40-chromosome form (see pp. 275). The most distinct of these is a dark form in Switzerland described in 1840 as *M. (musculus) poschiavinus* on the grounds of its colour, but a century later found to have only 26 chromosomes and little hybridization with normal *M. domesticus* from karyological evidence, albeit indistinguishable on morphological and allozymic grounds. A number of other chromosomal races have subsequently been described from the Rhaetian Alps, the central Apennines, Sicily, Spain, and northern Scotland (Capanna 1980; Gropp and Winking, in Berry 1981), and it is difficult to be certain of their exact systematic position. Because of their lack of differentiation from the normal race, it is probable that they are of fairly recent origin, and it is unknown whether they will persist to become full species.

As described in the next section, there were probably a number of places where mice were taken into domestication, and in view of the fertility of interspecific crosses between members of the *M. musculus* group, it cannot even be assumed that all the genes at present in domesticated stocks originated in *M. domesticus*.

Mice diverged from the voles by the Miocene, and *Mus* speciation probably occurred during the Pliocene. In the Lower Pleistocene, *M. petteri*, a mouse similar in size to *M. musculus* but clearly distinct from it, occurred with hominids at Olduvai. By the middle Pleistocene skeletal remains ascribed to *M. musculus* are recorded from Choukoutien in China, which is also a hominid site; from Binagady in the Caucasus; in travertine deposits near Budapest; on Chios in Greece; and probably at Ubeidiya in Israel. Archaeological sites from prehistoric times have yielded house mice in the Caucasus, Malta, Mallorca, Spain, northern Germany, and England (from an Iron Age settlement at Gussage All Saints in Dorset) (Brothwell, in Berry 1981). Of particular interest are the remains of about seventy-five mice found at the Neolithic site of Çatal Hüyük in Turkey (6500–5650 B.C.), because ribs, scapulae and vertebrae are almost absent. It has been suggested that the remains found there might have been used as an article of clothing, or a ritual offering. By the time of the 17th dynasty of Egypt (1700 B.C.) rodent tunnels (probably of house mice) appear in almost every room at Kahun.

It is a pity that the archaeological evidence is too fragmentary to allow firm statements to be made about the relationships and spread of mice with men in Neolithic times. Notwithstanding, there is little sign of mice in northern Europe before Roman times (Brothwell, in Berry 1981), and it may well be that the present distribution is a very recent result of farming practices and improved communications.

Domestication and early history

Mice have been adored or abominated in many cultures (Keeler 1931). Cats were deified in Egypt around 2800 B.C. and were regarded as embodying all godly virtue; mice probably symbolized evil. At a later date (A.D. 930) a Welsh chief, Hywel Dda, published a standard price-list for cats: one penny for a newborn kitten, tuppence for an inexperienced youngster, but fourpence for a cat once it had caught a mouse. Good mousers were valued: the fine for killing any of the cats in the chief's granary was a ewe sheep and her lamb, or enough corn to cover the dead cat suspended by its tail and with its nose touching the floor. The author of the Lambeth Homilies (1175) knew what he was about in writing, 'When a man will bait his mouse-trap he binds thereupon the treacherous cheese and roasteth it so it will smell sweetly; and through the sweet smell of the cheese, he enticeth many a mouse into the trap.'

However, mice were not evil in all societies. Probably the first written record of the raising and protection of mice by men is in connection with mouse worship in Pontis 1400 years before Christ. Homeric legend mentions Apollo Smintheus (god of mice) in about 1200 B.C. and his worship was still popular at the time of Alexander the Great 900 years later. Mice were worshipped by the Teucrans of Crete who attributed their victory over the Pontians of Asia Minor to their god who caused mice to gnaw the leather straps on the shields of their enemies. Temples were built in which mice were maintained at public expense. According to Aristotle, some at least of these mice were white. (He also stated that mice were generated spontaneously from filth in houses and ships.) The mouse cult spread to other cities and apparently continued as a local form of worship until the Turkish conquest in 1543.

Pliny describes the use of white mice for fortune-telling: 'They are not without certain natural properties with regard to the sympathy between them and the planets in their ascent . . . Soothsayers have observed that it is a sign of prosperity if there be a store of white ones bred.' Probably some of the repute of mice in ancient Greece was due to their alleged medicinal properties. Hippocrates records that he did not use mouse blood as a cure for warts in the same way as his colleagues, because he had a magic stone with lumps on it which had proved an efficient remedy. Galen advocated equal parts of mouse blood, cock's gall, and woman's milk mixed and dried as a cure for cataract. Ever more complicated potions accumulated in the pharmacopoeia during the Dark Ages.

The other main centre of ancient mouse culture was in the Far East. Here mice seem always to have had a high social rating. Albino mice were used as auguries, and government records in China show that thirty albino mice were caught in the wild between A.D. 307 and 1641. In Japan, the mouse of the folk sagas is the symbol and messenger of the god of wealth, Daïkoku. This god is usually represented as standing upon two sacks of rice with a mouse at his feet.

Other mutants than albino have long been known in China. The word for 'spotted mouse' appears in the first Chinese dictionary, written in 1100 B.C. Waltzing mice have been known since at least 80 B.C. There was a mouse fancy established a very long time ago in China and Japan, which clearly valued new and unusual forms. Keeler (1931) records that Yokohama fanciers called a mouse with certain markings a 'Nanking mouse' while fanciers from Shanghai, although it is not far from Nanking, called the same type the 'foreign mouse'. The Japanese fancy mice were derived at least in part from *M. domesticus*.

The Japanese had in their fancy such traits as albinism, non-agouti, chocolate, waltzing, dominant and recessive spotting, and possibly blue dilution, pink-eyed dilution, and lethal yellow. Some of these varieties were brought to Europe by British traders in the mid-19th century, and thence to the United States.

There have been repeated speculations over the history and process of mouse domestication (for references see Smith 1972, Connor 1975), but no agreement about it. Typical characteristics of domesticated mice such as lack of aggression towards introduced mice, failure to freeze when disturbed, handlability (passivity), and restricted agility are found to different extents in different inbred strains, showing the importance of genetical factors. On the other hand, mouse behaviour is very labile and readily adjusts to changed environmental circumstances under conditions where no genetical change or significant inbreeding takes place. Nothing is known about the conditions under which 'fancy mice' were kept when first bred by man. In contrast the antecedents of laboratory mice are fairly well known (see pp. 279–82), but even here it is impossible to say how much unconscious selection for early breeding, fecundity, etc. has played a part in their domestication, and how much their behaviour is a direct phenotypic response to captive conditions (Wallace, in Berry 1981).

The earliest record of the use of mice in scientific research seems to have been in 1664 when Robert Hooke in England used a mouse to study the effects of increased air pressure. He did not state where the mouse came from. Two centuries later a Genevan pharmacist named Coladon bred

large numbers of white and grey mice and obtained segregations in agreement with Mendelian expectation at least 36 years before Mendel published his results on peas. Indeed Sturtevant (1965) has even suggested that Mendel originally worked out his 'laws' of segregation and assortment in mice, but suppressed his mammalian work for fear of antagonizing his ecclesiastical superiors – mediaeval churchmen tended to gloat over the voluptuous and libidinous habits of mice and bred them to observe their wicked actions. One of the earliest demonstrations of genetic segregation in animals after the rediscovery of Mendel's papers was achieved by Cuénot working with albino and agouti mice. He presented his results to the Académie des Sciences in Paris in April 1902.

Recent history

The modern history of mouse genetics can be said to have begun in 1907 when an undergraduate, C. C. Little, began to study the inheritance of coat colour under the supervision of W. E. Castle at the Agriculture School at Harvard. Two years later Little obtained a pair of 'fancy' mice carrying the recessively inherited alleles for dilution (d), brown (b), and non-agouti (a), and inbred their descendants brother to sister, selecting for vigorous animals. This became the first inbred strain (DBA).

Castle did not believe in the value of breeding 'pure strains' of animals, and after army service in the First World War, Little moved to the Carnegie Institute at Cold Spring Harbor where he started the development of a range of inbred strains, largely descended from mice supplied by Miss Abbie Lathrop of Granby, Massachusetts.

Abbie Lathrop was born in Illinois in 1868. Forced by ill-health to give up schoolteaching, she moved to Granby around 1900, failed as a poultry farmer and began raising small animals for sale as pets. She started with a single pair of waltzing mice she got in Granby, and advertized for more animals as orders came in. 'After she had sold 200 or 300 mice Miss Lathrop thought the resources of mouse farming as a business must be very nearly at an end, since the offspring from that number would be enough to supply pets for the entire younger generation, but the orders continued to come in' (*Springfield Sunday Repub-*

lican quoted by Morse 1978). More and more of these orders came from research laboratories, including the Bussey Institute of Agriculture at Harvard. It is likely that Little's original mice came from Miss Lathrop – although it is on record that Bussey staff collaborated with the Boston Cat Club and Mouse Fanciers in their exhibitions (Keeler, in Morse 1978).

Miss Lathrop was initially confused by the increased interest in her mice, 'but the truth came out that the mice are used in laboratories for scientific research as to the causes and treatment of various of the ills that human flesh is heir to. Cancer principally is greatly illuminated by the aid of these little creatures which are subject like people to the ravages of the fearful disease.' This recognition followed contact with Leo Loeb at the University of Pennsylvania about 1908, who had diagnosed as 'cancer' skin lesions which were affecting Miss Lathrop's breeding programme. Between 1910 and her death in 1918, Miss Lathrop published ten papers with Loeb on the incidence of tumours in different lines (see Shimkin 1975). The records of these experiments are now in the Jackson Laboratory at Bar Harbor, Maine, and show that Abbie Lathrop bred mice from the wild, and had produced 12-generation inbred mice by 1915; she must have started inbreeding soon after Little initiated his DBA strain.

From around 1910 until her death, Miss Lathrop's barn and sheds contained more than 11 000 mice, several hundred guinea-pigs, rabbits and rats, and occasional ferrets and canaries. Children were hired at 7 cents an hour to clean the cages.

At Cold Spring Harbor, Little set up a number of brother–sister mated strains. In his line C he mated male 52 to females 57 and 58, all of them obtained from Miss Lathrop. These gave rise to the C57BL and C58 strains. In 1922 he left to become President of the University of Maine and then in 1925 of the University of Michigan at Ann Arbor, taking his mice with him on both occasions. Finally he oversaw the foundation of the Roscoe B. Jackson Memorial Laboratory at Bar Harbor in 1929. In the early years of the Jackson Laboratory work was carried out on the inbred strains A, BALB/c, CBA, C57BL, C57BR, C57L, C58, DBA, N, and I; crosses between the descendants of white-bellied mice caught near Peiping in China in 1926 ('*M. bactrianus*') and

('*M. wagneri*'); together with several mutant stocks.

L. C. Strong joined Little soon after he went to Cold Spring Harbor. It was a time when a paratyphoid epidemic had destroyed most of Little's colony. This obliged Strong (in Morse 1978):

to capture wild mice and start sorting out their hereditary traits through the tedious processes of mate, wait, select, and mate again . . . Out of fear contamination would occur in the blitzed mouse laboratory, we kept the wild mice under the bed in our honeymoon tent. Meanwhile it became obvious that susceptibility and resistance to transplanted tumours were indeed genetically controlled [which] bent my mind to the task of remodelling *M. musculus* . . . Foundations for the development of a better mouse were laid in the summer of 1920. One of Halsey Bagg's albinos, which he claimed were inbred but which proved to be nothing of the sort was chosen as 'great white mother'. Another albino, borrowed from C. C. Little and originating from a commercial colony at Storrs, Connecticut, was sire. The progeny were labelled simply, alphabetically, the A strain . . . [Soon afterwards] an outcross was made between one of these albinos and a survivor from Little's dilute browns. Cancer, although infrequent, was known to occur in both ancestral stocks. This hybrid cross was designed to test the idea that an increase in variability ought to increase the incidence of spontaneous tumours. Continued hybridizing in this stock proved the prediction true. Subsequent mating of a cancer-bearing mouse with a normal one produced the disease in the telltale 3 : 1 Mendelian ratio in the F_2 generation, a result that became the first laboratory proof that cancer is inherited. This historic mating was the start of the so well known C3H high-tumour subline. At the same time, selection towards resistance to cancer set up with this stock produced the C121, the CH1, and the CBA! The latter, selected for longevity, will still outlive any mouse in the laboratory.

There are now many other inbred strains, many of them developed for special purposes from the original few described above (Fig. 38.2). At least three strains have been inbred directly from wild *M. domesticus* by M. E. Wallace (Wallace, and Festing and Lovell, in Berry 1981). A large proportion of the outbred mice used throughout the world derive from the Swiss stock imported into the United States in 1926 from a laboratory in Lausanne, Switzerland. The original population was two males and seven females kept by Dr Clara Lynch in a shoebox in her cabin during the

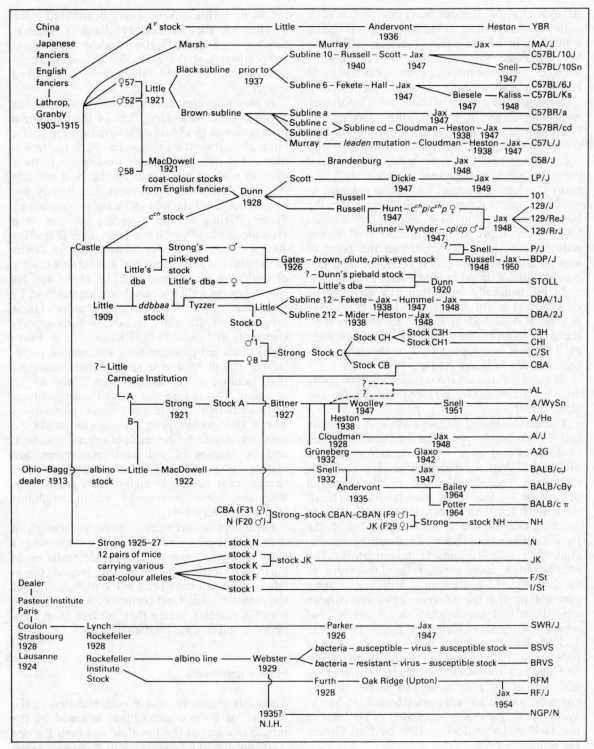

Fig. 38.2 The genealogy of the most commonly used strains of laboratory mice. (After Morse 1978)

Atlantic crossing. Besides being maintained as an outbred colony, these animals have given rise to a number of inbred strains (such as SWR and SJL, together with the strains developed by Webster for resistance and susceptibility to St Louis encephalitis virus and *Salmonella* bacteria).

Although the historical record of laboratory mice is extensive, it is tantalizingly incomplete. For example, we do not know how great was the contribution of wild mice to the Lathrop colony, whether the wild mice were spread throughout or maintained as separate stocks, how many illegitimate matings have occurred, and so on. Nevertheless it can safely be asserted that the bulk of the genes in mice used in laboratories today were drawn from a limited pool in the northeastern United States. From the point of view of laboratory workers it is encouraging that only six alleles found in wild mice at the ninety-eight biochemically monitored loci surveyed by Chapman *et al.* (in Altman and Katz 1979) have not been described in inbred mice. However, there is a much higher proportion of immunological variation not represented in the laboratory (Klein *et al.*, in Berry 1981).

Reviews of the characteristics of laboratory mice are given by Green (1966) and Altman and Katz (1979).

The establishment of the Jackson Laboratory in 1929 was highly significant for the distribution of domesticated mice – or at least for the vast majority of domesticated mice which are used as laboratory animals. Because of the financial depression of the 1930s, the laboratory funds were much less than expected. The staff voted to cut their salaries, live as much as possible off the land, combine households – and sell genetically controlled mice to outside investigators. The income from mice enabled the laboratory to survive, and the continuing distribution of mice speeded up the use of genetically comparable animals by biologists world-wide. A fire in 1947 which destroyed virtually all the mice at the Jackson Laboratory provided an opportunity to redefine the basis of many of the common strains, and allowed the re-characterization of the life histories of mice from inbred strains in the common foundation colonies (Russell, in Morse 1978). C. C. Little was succeeded as Director of the Jackson Laboratory in 1956 by Earl Green. He retired in 1975.

Major laboratories record their interests and results in a *Mouse News Letter* distributed twice a year by the Laboratory Animals Centre, Carshalton, Surrey, England and by the Jackson Laboratory; and the Jackson Laboratory circulates regular lists and bibliographies of inbred strains.

A major development in recent years has been an interest in wild mice. Part of this has come from a desire to obtain new variation for study, but a more important reason has been the recognition that it is scientifically inadequate to treat mice as nothing more than biochemical reacting systems, as laboratory workers find it only too easy to do. Until the mid-1960s the description by Fraser Darling of mice invading his tent on a Hebridean island which had been uninhabited for 40 years tended to be greeted with surprise. Apart from casual observations, the first serious studies of wild mice were those of L. C. Dunn and his colleagues on *t*-alleles, which are transmitted by male heterozygotes to 90 per cent or more of their progeny, but are early lethals as homozygotes (Bennett, in Morse 1978; Klein *et al.*, in Berry 1981). This led to suggestions that mouse populations are divided into small demes, meaning that random gene changes are likely to be common. This in turn was proved to be incorrect by longitudinal studies of island populations where changes in gene frequencies could be demonstrated to be the result of natural selection, and by studies of individual movement and mating groups showing that tight and exclusive demes occur mainly in high-density populations with low turnover (usually within buildings) (Berry, in Berry 1981).

Feral populations seem to thrive particularly in dense grassland on soft soil where burrowing is easy or where loose rocks provide ready-made burrows. Outside buildings the largest populations reported have been in the tussock grass of temperate islands (and corn-ricks which are ecologically similar), where they are free from competition and intense predation.

Future prospects

The early stages of mouse domestication in the temples of Greece and China, followed by the mouse fanciers of the Far East and then Europe were presumably random, with conscious selection for colour, and unconscious selection for breeding traits and perhaps tameness (Berry 1970; Wallace, in Berry 1981; see also Lindzey and Thiessen 1970). During this phase the gene pool in domestication would have been limited, but probably by not a great deal in comparison with the wild: different colonies of Swiss outbred mice founded in 1926 with a total of nine animals remain as heterozygous as island races of mice which have been through a genetical bottleneck during their founding (Rice and O'Brien 1980).

The major differentiation and changes in genetical organization came with the setting up of inbred strains during the 1920s; L. C. Strong, one of the pioneers of using mice for the study of cancer could write of 're-modelling *M. musculus*' (Strong, in Morse 1978). The next and continuing stage has been the use of mutant genes (and chromosomal rearrangements) so that developmental and physiological processes can be studied with an inbuilt experiment and control (Grüneberg 1952; Searle, in Berry 1981). This approach has been refined by the development of 'segregating inbred lines' where only one locus (or, more accurately, one chromosomal segment) are maintained segregating. These lines are formed either by inbreeding stocks containing a mutant gene or, more conveniently, by appropriate management following a mutation in an already inbred strain. They have been particularly important in the study of immunological processes.

The 1960s and 1970s witnessed advances in husbandry techniques, and particularly the establishment of colonies free from specific pathogens in an attempt to reduce non-genetic differences between stocks.

A particularly useful innovation during the same period has been the development of RI (recombinant inbred) lines (see Festing and Lovell, in Berry 1981). These are formed by crossing two inbred strains differing in traits of interest to the experimenter, sib-mating F_1 animals which will be genetically identical with each other but heterozygous at all loci in which the strains differ, and then setting up ten to twenty lines by inbreeding from F_2 mice which will differ because of recombination in the F_1. Comparison between these lines gives information on genetical determinants and linkage (see Searle, in Berry 1981). Some strains and mutants are now kept as deep-frozen embryos rather than conventional breeding stocks (Zeilmaker 1981). Other recent advances in mouse genetics have come from cell fusion and DNA recombinant technology, but these come outside the scope of the present article.

There is a great deal of variation available within inbred strains (Festing and Lovell, in Berry 1981). Notwithstanding, wild mice provide an enormous pool of variants, and considerable effort is now being expended on screening wild mice and incorporating new alleles into laboratory stocks (Chapman, in Morse 1978; Bulfield, in Berry 1981). The domestication of the house mouse is a continuing process.

Acknowledgment

I am grateful to Drs G. B. Corbet, J. T. Marshall, H. C. Morse and A. G. Searle for commenting on this chapter in draft, and to Mr A. J. Lee for drawing the figures.

References

Altman, P. L. and **Katz, D. D.** (eds) (1979) *Inbred and Genetically Defined Strains of Laboratory Animals.* Part I. *Mouse and Rat.* Federation of American Societies for Experimental Biology: Bethesda, Maryland

Barnett, S. A., Munro, K. M. H., Smart, J. L. and **Stoddart, R. C.** (1975) House mice bred for many generations in two environments. *Journal of Zoology*, **177**: 153–69

Berry, R. J. (1970) The natural history of the house mouse. *Field Studies*, **3**: 219–62

Berry, R. J. (ed.) (1981) *The Biology of the House Mouse.* Academic Press: London and New York

Bronson, F. H. (1979) The reproductive ecology of the house mouse. *Quarterly Review of Biology*, **54**: 265–99

Capanna, E. (1980) Chromosomal rearrangement and speciation in progress in *Mus musculus. Folia Zoologica*, **29**: 43–57

Connor, J. L. (1975) Genetic mechanisms controlling the domestication of a wild house mouse population (*Mus musculus*). *Journal of Comparative and Physiological Psychology*, **89**: 118–30

Green, E. L. (ed.) (1966) *Biology of the Laboratory Mouse* (2nd ed). McGraw-Hill: New York and London

Grüneberg, H. (1952) *Genetics of the Mouse* (2nd ed). Nijhoff: The Hague

Keeler, C. E. (1931) *The Laboratory Mouse.* Harvard University Press: Cambridge, Massachusetts

Lindzey, G. and **Thiessen, D. D.** (1970) *Contributions*

to Behaviour – Genetic Analysis. The mouse as a prototype. Appleton-Century-Crofts: New York

Marshall, J. T. (1977) A synopsis of Asian species of *Mus* (Rodentia, Muridae). *Bulletin of the American Museum of Natural History*, **158**: 173–220

Morse, H. C. (ed.) (1978) *Origins of Inbred Mice*. Academic Press: London and New York

Reichstein, H. (1978) *Mus musculus* L. 1758 – Hausmaus. In: *Handbuch der Säugertiere Europas*, Vol. 1, pp. 421–51. Akademische Verlagsgesellschaft: Wiesbaden

Rice, M. C. and O'Brien, S. J. (1980) Genetic variance of laboratory outbred Swiss mice. *Nature*, **283**: 157–61

Schwarz, E. and Schwarz, H. K. (1943) The wild and commensal stocks of the house mouse, *Mus musculus* L. *Journal of Mammalogy*, **24**: 59–72

Shimkin, M. B. (1975) A. E. C. Lathrop (1868–1918): mouse woman of Granby. *Cancer Research*, **35**: 1597–8

Silvers, W. K. (1979) *The Coat Colors of Mice*. Springer-Verlag: New York and Berlin

Smith, R. H. (1972) Wildness and domestication in *Mus musculus*: a behavioural analysis. *Journal of Comparative and Physiological Psychology*, **79**: 22–9

Sturtevant, A. H. (1965) *A History of Genetics*. Harper and Row: New York

Zeilmaker, G. (ed.) (1981) *Proceedings of a Workshop on Embryo Storage and Banking in Laboratory Animals*. Fischer-Verlag: Stuttgart

39
Norway rat

Roy Robinson

St Stephens Road Nursery, London, England

Introduction

The Norway rat is one of the four commonly employed laboratory rodents (the others being guinea-pig, house mouse, and Syrian hamster). Indeed, this is the primary use of the species by man. Small numbers have been bred as pets in England since the middle of the 19th century but the pet rat has never become popular. A considerable quantity of information has been accumulated on the species as a laboratory animal and there are few fields of biological and biomedical research in which it has not featured.

Nomenclature

The Norway rat, *Rattus norvegicus*, was so called because it was once thought, wrongly, to have originated from Norway. It is also called the barn, brown, common, sewer, short-tailed or wharf rat. The black rat (*R. rattus*) is also known as the alexandrine, climbing, grey, long-tailed, roof or ship rat.

The vernacular names for the two species can be confusing. The wild type of both are brown-grey or agouti coloured. There is a form of *rattus* which is almost black due to a dominant gene.

This form is common in Europe and probably gave the species the name of black rat. The Norway rat is also known as the brown rat to distinguish it from the black rat. This is the only significance of the vernacular black/brown distinction.

Reproductive biology

Rats are renowned for their fecundity. Sexual maturity occurs between 50 and 60 days in the majority of animals of both sexes. It may be determined by the opening of the vagina in the female and by the presence of sperm in the semen of males. The female is polyoestrous, with a 4–5 day oestrous cycle. The heat period, which is associated with greatly increased activity, occurs early in the cycle; it begins during the afternoon or evening and lasts for about 12–18 hours. Early mating tends to shorten the period. It may be accurately determined by vaginal smears which contain only cornified cells and an absence of leucocytes. Litter size may vary from 4 to 15 young, with a mean of about 7–8. The number of corpora lutea is regularly in excess of the number of young, indicating a low but constant rate of foetal loss. The life of an ovum is about 12 hours while the life of spermatozoa in the female tract is about 14 hours. Maximum litter size occurs at the third litter and declines after the seventh or eighth.

The gestation period shows little variation at 22 days. There is a *postpartum* oestrus, usually occurring within 12 hours of parturition. If the mother is suckling a large litter, the gestation period may be extended by as much as 7 days. The young are born blind, hairless and pink skinned. The skin darkens with pigment by 2 or 3 days of age; this is followed by a thin covering of erupting hair. The eyes open at about 10–12 days. The young can be weaned at 21 days. If the female has not conceived at the *postpartum* oestrus, the cycles recommence soon after the young are weaned. The secondary sex ratio is approximately 104 males to 100 females. Healthy females may have as many as 10 litters in a lifetime. Litters are born throughout the year with little seasonal fluctuation; if the latter is evident, it can be obviated by a laboratory regime of long-day illumination. The typical life-span is about 2 years, with the females ceasing to breed at about 15–18 months of age.

Karotype

The diploid chromosome number of 42 is within the modal range for murid rodents. In terms of gross morphology, the karotype is undistinguished. There are two large chromosomes, one telocentric and the other subtelocentric, but from then onwards the chromosomes decrease steadily in size, few being easily identifiable individually. The majority of the medium-sized chromosomes are telocentric or subtelocentric while most of the smaller ones are metacentric. The X is a medium sized subtelocentric while Y is a small telocentric.

However, all the chromosomes can be individually identified by means of banding techniques. Several methods of numbering the chromosomes have been proposed, with the inevitable inconsistencies. These have been resolved by the Committee for a Standardised Karotype of *Rattus norvegicus* (1973). The matter of terminology for representing sections of the banded chromosomes has been discussed by Levan (1974) who has devised a system based upon that for human chromosomes. The system would enable individual bands and inter-band regions to be defined in such a manner that eventual finer resolution of the bands can be incorporated.

Wild species

The Norway rat belongs to the rodent family Muridae, subfamily Murinae. The genus *Rattus* consists of a large collection of species and other named forms (Ellerman 1941), many of which, regardless of their taxonomic standing, may be of doubtful biological significance. Missonne (1969) has, in fact, proposed that the genus be subdivided into several genera of equal status. Only a combined taxonomic and biological approach would seem capable of clarifying the situation (cf. Yosida 1973).

The two species which concern us here are *R. norvegicus* (Norway rat) and *R. rattus* (black rat). The latter has never been domesticated. Although the two species are quite distinct, they have an interwoven history and have even been confused in the older literature. They are similar in colour and appearance but they can be readily distinguished by experienced observers. The following is a summary of the chief external differences:

R. rattus	R. norvegicus
Smaller size, slender body	Larger size, stocky body
Muzzle pointed	Muzzle blunt
Ear pinnae large, thin and translucent, with scanty hair	Ear pinnae small, thick and opaque, hairy
Tail slender, often longer than combined length of head and body	Tail stout, never as long as combined length of head and body
Coat soft, with many bristly hairs in adults	Coat soft, with few bristly hairs in adults

The two species have different karyotypes and will not produce viable hybrids. (Hybrid embryos degenerate about mid-term). *Rattus rattus* displays geographical variation of $2n = 38, 40$ and 42 chromosomes, due to progressive evolutionary fusion of two chromosomes (Yosida *et al*. 1971). *Rattus norvegicus* has the uniform karyotype of $2n = 42$ for the whole world. Ellerman (1941) refers to five subspecies or forms of *R. norvegicus* as follows: *R. n. caraco, norvegicus, praestans, primarius* and *socer*. Whether or not these represent valid aboriginal races is unknown. The present-day commensal Norway rat is extremely similar in all regions of the world and individuals will interbreed freely regardless of provenance.

In the Orient, *rattus* probably became a commensal with man as soon as the latter began to establish permanent communities of any size. These would provide a suitable ecological niche in the form of houses, barns and refuse dumps. The species began to spread in company with man, even if it paid a price of a sort – in times of famine even the rats are caught and eaten. There is no sound evidence that the Greeks or Romans were familiar with the rat. It was not until the opening up of trade routes with the Orient that *rattus* became known in western Europe. It is doubtful if the species occurred in appreciable numbers until the 12th century A.D. However, from then onwards it became common, flourishing in most European cities, including Britain, until fairly recent times. The species must bear most of the blame for epidemics of bubonic plague, being the primary host for the pathogen. The *rattus* species reached North America during the 16th century.

Rattus norvegicus reached Europe from the Orient, early in the 18th century probably via Russia. It was first reported in Copenhagen in 1716 and in parts of European Russia and the Balkans by 1727. Rats probably entered England and Ireland in the early 1700s, possibly in ships arriving directly from Russia. By 1750 they had reputedly reached Germany and France and were invading Italy and Spain by the end of the century. The east coast of North America was reached by the late 18th century; presumably the rat was carried there by the early English settlers.

Rattus norvegicus is larger and more pugnacious that *R. rattus* and it soon supplanted the latter in most areas where the two were in contact. In Britain, for example, the only areas in which *rattus* is to be found today are around ports and this may be so only because of replenishment by rats from ships. *Rattus rattus* is still the primary ship rat because of its superior climbing abilities. Upon a world scale, the only inland habitats in which *rattus* has been able to hold its own are woody areas and warmer climates. In Africa, India, southeast Asia and Oceania *norvegicus* is present as an introduced but largely coastal species, being unable to displace the indigenous *rattus* or other species.

Southeast Asia is thought to be the centre of dispersion of murid rodents (Missonne 1969). India, the Malayan peninsula and the larger islands possess many species. *Rattus rattus* has been especially successful in the region, colonizing most of the islands, perhaps inadvertently helped by man, at least in part. The species probably originated in this geographical area. On the other hand, *norvegicus* is evidently a colder-climate species. Although the origin of the species is not known, it seems probable that it evolved in Asia, possibly in central Asia somewhere between the Caspian Sea and Lake Baykal, but possibly as far east as Mongolia or China.

Domestication

The rat was domesticated as as consequence of the once popular 'sport' of rat baiting of the early to middle 19th century. In this sport, a number of rats were released into a large pen and the spectators wagered on how much time would be required by terrier dogs to kill the last rat. Large numbers of rats were required for these spectacles and they were initially procured from the wild. Since wild rats are not easily caught, it is easy to imagine that attempts would be made by fanciers to breed them in captivity. The odd mutant was probably preserved as a curiosity. The albino, black and hooded mutants date from around 1850 in England.

The first report of rats in research was published in France (Philipeaux 1856) but it is not known whether the animals were actually bred in the laboratory or were captive stock. However, breeding experiments were in progress by H. Crampe in Germany prior to 1877 and by H. H. Donaldson by 1893, the latter founding the well-known colony at the Wistar Institute, Philadelphia. The early history of the laboratory rat has been documented by Lindsey (1979).

It is commonly held that domestication involves one of three processes: 1. a form of neoteny by which juvenile characteristics persist into the adult, particularly those which encourage dependency; 2. modification of hormone balance so that the fight or flight instinct is minimized; and 3. reduction of brain size, thereby impairing sensitivity to what might be uncongenial surroundings. The rat has contributed substantially to this concept and study of the species may be commended to those seeking an understanding of the processes involved. That is, for an analysis of the differences between a wild and a tractable domesticated animal.

The wild rat is truly a wild and fearsome creature, whereas the laboratory rat is docile and inquisitive and, given a little petting from an early age, quickly becomes trusting and fearless. The contrast is striking. The most thorough work in this area is that of Richter and associates conducted in the early 1950s. This work could profitably be repeated and extended. Few other species offer similar opportunities for comparative studies. For instance, the house mouse does not appear to exhibit the same degree of tameness as the rat. The rabbit could be useful but the differences between the wild and tame rabbit are largely unknown. Real differences exist between the wild and domestic cat and dog but the study of these species would be beset with practical difficulties.

Profound changes have occurred as a consequence of domestication, modifying reproduction and behaviour as well as the size and functioning of many organs (Richter 1954, Robinson 1965). These changes appear to have been brought about by changes in the activity of the major endocrine glands.

The most obvious changes have taken place in the adrenals, particularly in the cortex, which shows a marked decrease in size. The preputial glands show a similar reduction, as well as the brain, heart, kidneys, liver and spleen, although the reduction has been less, and more variable, for these. The pancreas, thyroid and parathyroids are probably smaller but the variability tends to obscure the situation. On the other hand, the pituitary and thymus show signs of being larger than in the wild rat. There is an increase in the number of Peyer's patches on the intestine.

The gonads and secondary sex structures show relatively little change but they mature earlier and function at a higher level of activity. This probably reflects selection for increased reproductive performance, for the laboratory rat reaches puberty earlier, breeds continuously throughout the year and has larger litters than the wild conspecific. The earlier maturity affects the whole life-cycle, for the laboratory animal rarely lives beyond 3 years, whereas the wild rat may live as long as 4–5 years (Richter 1959). There is a reduction in size but no obvious change in bodily proportions. This is due to earlier cessation of growth, for the wild rat grows at the same rate but for a longer period and becomes the larger animal. Behaviourally, the most obvious change is the reduction in neophobia and fearful behaviour which is so marked in the wild rat; it is present but only weakly expressed in the petted laboratory rat.

History of the laboratory rat

The early history of rat breeding is that of establishment of colonies from which many of the inbred strains and substrains of today are derived. The most well known and oldest stock is the Wistar albino established c 1906 by H. H. Donaldson. This is a random-bred stock, selected for high fecundity and vigour. It is largely a closed colony although small additions of outside animals may have been made in the early stages. Breeding nuclei from this stock were widely distributed throughout the world and most of today's stocks and strains can trace their origins back to it. In particular, H. D. King initiated an inbred strain of Wistar albino in 1909 and this, or a substrain, has survived to the present day as the PA.

Several other inbred Wistar albino strains exist, deriving from the outbred colony at later dates. In fact, the growing appreciation of inbred animals has led many workers to switch from breeding their 'Wistars' as a closed outbred colony to one of inbred lines. Lindsey (1979) concluded, from an analysis of details of 111 inbred strains listed in *Rat News Letter No. 3* (1978), that at least 45 are known to have descended either directly or indirectly from Wistar stock. He commented that the Wistar albino may have contributed more to the gene pool of all modern strains than all other major stocks together.

Other well-known stocks are the Long-Evans, Sprague-Dawley and Osborne-Mendel, all named after the men who initiated the original colonies at the Wistar Institute. All of these commenced their existence as outbred populations and were distributed to many other institutions. The Sprague-Dawley, for example, was subsequently developed as a commercial enterprise, supplying stock to many laboratories. Many substocks have arisen from these colonies and, in turn, provided foundation animals for many of the inbred strains of the present day.

It would seem that the first serious attempt to form a battery of inbred lines for comparative studies was launched by F. D. Bullock and M. R. Curtis in 1921 at Columbia University, New York, progress being consolidated or extended from time to time. In total, eleven lines have survived, many of them with respectable amounts of inbreeding behind them. The best known are the August (several substrains), Copenhagen, Fischer, Marshall and Zimmerman, although these familiar designations are now replaced by the rather faceless (but more easily manipulated) groups of capital letters (such as AUG for August and COP for Copenhagen). Many substrains are being developed for specific purposes. These strains are unique in that the foundation stocks were obtained from dealers and did not possess a direct connection with the Wistar albino.

The most recent development has been the creation of additional inbred strains, almost all of them being produced for specific purposes. Some have doubtless been formed as a result of the vogue for inbred animals but most have come into being as a response to a definite need. This is especially true for the rapidly expanding fields of biochemical genetics and immunogenetics. In the case of the former, the many alleles controlling biochemical differences, as revealed by electrophoretic analysis, produce no visible differences in the phenotype and are best preserved in inbred strains homozygous for alleles at different loci. Once established, such a strain will maintain a constant biochemical profile, needing only an occasional check by subsequent electrophoretic analysis, for possible mutations.

Similarly for immunogenetics, the establishment of inbred strains is useful for maintenance of immunogenic haplotypes. It is also vital in another respect, for the detection of the finer immunogenetic differences at major histocompatibility loci requires a homogeneous genetic background. This is achieved by repeated backcrosses to a standard inbred strain, so that each haplotype is expressed against the same genetic background. The detection of minor histocompatibility or immunological loci is similarly facilitated by the use of highly homogeneous or nearly identical inbred strains. In this manner, arrays of inbred strains can be built up. Once established, these strains should be valuable for areas of research outside the narrow confines of immunogenetics.

Present situation

The rat is one of the most widely-used laboratory animals, for it is relatively robust and eminently adaptable to many experimental procedures. Excellent accounts of techniques, past research and areas for which the rat is particularly suitable may be found in the books by Farris and Griffith (1949), Altman and Katz (1979) and Baker *et al.* (1979). The last two volumes are especially wide-embracing and useful. The topics covered include cancer and cardiovascular research, clinical biochemistry, physiology and techniques, dental research, diseases and aging, gnotobiologic animals, parasitology and toxicology. The chapters by A. L. Kraus on clinical techniques, and by G. B. Briggs and F. W. Oehme on toxicological methods are particularly valuable contributions.

The rat is the favourite experimental animal in the field of mammalian psychology. That is, the study of behaviour in the widest sense, not merely observational but particularly experimental. Some of the earliest work, dating from 1894, used wild rats; later albinos were substituted. Literature surveys have revealed that the rat holds the position of being the mammal most frequently used in ethological research. The standard reference book is that of Munn (1950), who presents a comprehensive overview of studies with the rat. The review encompasses both normal and abnormal behaviour, learning ability (various types of mazes, avoidance and discriminative tests), activity (explorative, spontaneous), temperament (aggression, timidity), preferences of many kinds and more complex studies involving manipulation of several variables simultaneously. However, Munn's account is now very much out-of-date and a comprehensive summary similar to Munn's is urgently needed to draw together the scattered literature. An attempt has been made in this direction to review research for the narrower field of behaviour genetics or psychogenetics (Robinson 1965, 1979).

Genetic research on the rat declined sharply following the death of W. E. Castle in the late 1950s. The house mouse is now the most popular animal for the study of mammalian genetics. However, there has been a reawakening of interest in the rat in recent years. The widespread use of the animal in other fields has meant that a steady trickle of new mutants has been reported. Unfortunately, there has been no concerted effort to conserve these and a number of the older mutants have been lost, whether irretrievably remains to be seen.

Fresh interest in the rat is evidenced by the inauguration of the informal *Rat News Letter* in the spring of 1977; this is devoted to reporting on genetics, notification of new mutants, development of new inbred strains and listing the location of mutants and established strains. Information on *Rat News Letter* can be obtained from the Medical Research Council Laboratory Animals Centre, Woodmansterne Road, Carshalton, Surrey SM5 4EF, England.

Despite the loss of mutants and many inbred strains which could have provided links with the past, several attempts are in hand to exploit those which remain and to build up a battery of inbred strains representative of the current level of genetic research. This entails typing the strains, not only for obvious phenotypic mutants but also for biochemical and immunogenetic variation, as well as more general characteristics, such as behaviour, susceptibility to infectious disease, spontaneous tumours, and longevity. Some 110

entries are contained in the latest list of established strains (Festing 1979b). The more information that becomes available upon inbred strains, the more useful does each one become individually. Festing (1979b) merely summarizes the status and advantages of inbred material over that of random-bred stock but he discusses the subject in more detail in Festing (1979a). Inbred strains can be utilized to produce uniform, exceptionally vigorous, F_1 stock if the situation calls for it. The designation of inbred strains should conform to a standardized nomenclature and Festing's (1979b) listing actively promotes such standardization. The recommended terminology should follow that for the house mouse, which was being revised at the time of Festing's contribution. This revision has now been published (see Lyon 1981). The recommendations hold good for any field which employs inbred material, not merely for genetics as might be supposed.

A rapidly expanding field at the moment is murine immunogenetics and, a little belatedly, the rat is contributing quite usefully (Gunther and Stark 1979). A major histocompatibility complex (*RT1*) has been isolated and is proving to be as complex as that of *H–2* for the house mouse. The major histocompatibility complex (MHC) is a pseudo-allelic complex within which crossing-over is not totally suppressed; it occurs at a low level but sufficiently to establish definable regions of immune activity. These regions are genetic rather than immunological entities, yet they control specific reactions, such as graft acceptance in tissue transplants, mixed leucocyte reaction, or immune response to artificial and natural antigens. These regions form a linear sequence within the complex.

An intriguing difference appears to be emerging between the mouse and rat MHC population genetics in that the number of detectable alleles is much less in the rat than in the mouse. The reason for this is unclear at present. Investigations are in hand to develop inbred reference strains for the various *RT1* alleles, the selected strain being the Lewis, a derivative of the Wistar. Existing stocks and strains are being typed for their *RT1* alleles. Where the latter have been found to be polymorphic, this fact has been determined, and, where appropriate, inbred isogenic substrains have been formed. Since the rat is employed extensively in oncological research, precise information on the MHC genotypes (or haplotypes) can be of considerable importance for the liability rate of specific spontaneous tumours or transplant experiments with tumours.

Future prospects

The future for the rat lies almost exclusively in the field of biomedical research. In fact, there is a new surge of interest in the species at the time of writing, with the publication of a monograph (Baker *et al.* 1979) and a newsletter devoted to disseminating the latest documented information. In some areas of research the rat cannot compete with the house mouse but there is no reason why it should not function as a useful species for assessing the generality of certain research findings. The rat is more expensive to keep, admittedly, but its reproductive performance is almost equal to that of the mouse. The advantage of the animal resides in its larger size and robustness which makes it superior to the mouse and on a par with the guinea-pig. However, the guinea-pig is a notoriously slow breeder. Therefore the rat recommends itself for those areas of research which demand a quick breeding subject which is of moderate size and constitutionally robust.

Although colour varieties of the rat have been accepted by small-livestock fanciers as legitimate exhibition animals, the rat has never been very popular as a pet. The reason appears to be the general odium surrounding the rat, *qua* rat, despite the docile nature of the tame rat and the hardiness of the animal which enables it to withstand rough handling by children.

The recognized colours are albino, black, chocolate and pink-eyed fawn, and hooded pattern in any of the three last-named colours. In the hooded, only the head and a stripe extending along the spine are coloured, the remainder of the body being white. Recently fanciers have recognized a rex coat which differs from the smooth normal by being rough and slightly bristly to the touch. The vibrissae are bent and twisted. This coat type is due to a dominant gene.

References

Altman, P. L. and **Katz, D. D.** (eds.) (1979) *Inbred and Genetically Defined Strains of Laboratory Animals*, Part 1. *Mouse and Rat*. Federation of American Societies for Experimental Biology: Bethesda, Maryland

Baker, D. E. J. (1979) Reproduction and breeding. In: Baker *et al.* (1979) (op. cit.), Vol. I

Baker, H. J., Lindsey, J. R. and **Weisbroth, S. H.** (1979) *The Laboratory Rat*, Vols I and II. Academic Press: London and New York

Committee for a Standardised Karyotype of *Rattus norvegicus* (1973) Standard karyotype of the Norway rat. *Cytogenetics and Cell Genetics*, **12**: 199–205

Ellerman, J. R. (1941) *The Families and Genera of Living Rodents*, Vol. II. British Museum (Natural History): London

Farris, E. J. and **Griffith, J. Q.** (1949) *The Rat in Laboratory Investigation*. Lippincott: New York

Festing, M. F. W. (1979a) *Inbred Strains in Biomedical Research*. Macmillan: London

Festing, M. F. W. (1979b) Inbred strains. In: Baker *et al.* (1979) (op. cit.), Vol. I

Gunther, E. and **Stark, O.** (1979) Overview of the major histocompatibility system of the rat. *Transplantation Proceedings*, **11**: 1550–3

Levan, G. (1974) Nomenclature for G bands in rat chromosomes. *Hereditas*, **77**: 37–52

Lindsey, J. R. (1979) Historical foundations. In: Baker *et al.* (1979) (op. cit.), Vol. I

Lyon, M. F. (1981) Rules for nomenclature of inbred strains. In: *Genetic Strains and Variants in the Laboratory Mouse*, ed. M. C. Green. Fischer: Stuttgart

Missonne, X. (1969) African and Indo-Australian Muridae; evolutionary trends. *Annales du Musée Royal de l'Afrique Centrale: Sciences Zoologiques*, **172**: 1–219

Munn, N. L. (1950) *Handbook of Psychology Research on the Rat*. Houghton Mifflin: Boston, Massachusetts

Philipeaux, J. M. (1856) Note sur l'extirpation des capsules surrénales chez les rats albinos (*Mus rattus*). *Compte Rendu Hebdomadaire des Séances de l'Académie des Sciences*, **43**: 904–6

Richter, C. P. (1954) The effects of domestication and selection of the behavior of the Norway rat. *Journal of the National Cancer Institute*, **15**: 727–38

Richter, C. P. (1959) Rats, man and the welfare state. *American Psychologist*, **14**: 18–28

Robinson, R. (1965) *Genetics of the Norway Rat*. Pergamon Press: Oxford

Robinson, R. (1979) Taxonomy and genetics. In: Baker *et al.* (1979) (op. cit.), Vol. I

Yosida, T. H. (1973) Evolution of karyotypes and differentiation in 13 *Rattus* species. *Chromosoma*, **40**: 285–97

Yosida, T. H., Tschuchiya, K. and **Moriwaki, K.** (1971) Karyotypic differences of black rats collected in various localities of east and southeast Asia and Oceania. *Chromosoma*, **33**: 252–67

40

Giant rat and cane rat

O. O. Tewe, S. S. Ajayi and E. O. Faturoti

University of Ibadan, Nigeria

Giant rat

Wild species

The African giant rat (*Cricetomys gambianus*), otherwise known as the pouched rat, is a burrowing nocturnal rodent. It prefers a cool environment for its burrows. It is commonly found in the forest and savannah regions of Africa south of the Sahara. Two species have been clearly identified. These are: 1. *C. emini* which is found mainly in the deciduous rainforest and forest plantation; and 2. *C. gambianus* whose habitat is confined mainly to derived guinea savannah, edges of forest and around human habitations. Field observations (Ajayi 1975) show that about 33 per cent of their burrows are found inside deserted termite mounds and about 33 per cent underneath roots of trees on abandoned farms in the derived guinea savannah study areas. The burrows are not just simple holes. The entrance is generally by a vertical shaft which leads to a system of galleries and separate chambers, one for the storage of food, one for the deposit of droppings, and another for sleep or, if the inhabitant is a female, for breeding (Rosevear 1969).

These rodents usually lead a solitary life, there being mostly only a single occupant of a burrow except at breeding time when the young are born and are being brought up. They have a high preference for palm fruits while they can also subsist on tubers, grains and vegetables. The breeding activity in the wild appears to be high between the months of August and December. Juveniles have, however, been trapped all the year round.

Breeding in captivity

Attempts at domesticating the giant rat by farmers around Ibadan have largely been limited to trapping the juveniles alive during hunting and rearing these for slaughter when they reach maturity. They are usually kept in wire cages where food from the wild is given to them daily.

The only comprehensive study so far carried out on the domestication of this rodent was initiated by Ajayi in 1972. Wild rats are trapped and kept in rehabilitation cages for 32 days after which they become socially adaptable. They are subsequently transferred into breeding cages which are essentially wooden resting boxes with a rectangular wire-mesh playroom. Each breeding cage holds a pair of rats or a nursing female with its young up to weaning age. Experimental feeding cages are designed to house rats from weaning age to maturity.

The female giant rats attain puberty at about 20–23 weeks. They exhibit an oestrous cycle of about 4 days. The gestation period is about 28 days and young rats are weaned at 21–26 days of age. Sex ratio is 120 males to 100 females. The litter size ranges between 1 and 5 but 4 is most common. So far, the highest number of litters observed was 5 in 9 months. It is thus possible that a female can reproduce 6 times in a year. With an average litter size of 4, a female can therefore be expected to produce about 24 young per year. The reproductive performance of the giant rat is summarized in Table 40.1.

Food preference trials (Tewe and Ajayi 1979) showed that apart from palm fruits, tubers were preferred to grains and vegetables. The most favoured tuber was sweet potato. Nutritional studies (Faturoti 1980) showed that giant rats can tolerate up to 7 per cent crude fibre in their rations. A minimum of 3 per cent oil guarantees optimal growth rate and a level of 22 per cent crude protein is also recommended in their rations.

Table 40.1 Reproductive performance of the African giant rat

Adult weight (g)	692–1220
Female breeding age (wks)	20–24
Oestrous cycle (days)	4–5
Gestation period (days)	28
Weaning age (days)	26
Litter size	1–5
Birth weight (g)	16–28
Weaning weight (g)	66–186
Age at maturity (wks)	24

Chemical analysis of giant rat carcass (Tewe and Ajayi 1979) showed that the meat of this rodent compares favourably with that of domestic animals in its nutrient content. A study carried out by Ajayi (1975) showed that while the domestic rabbit (*Oryctolagus cuniculus*) normally grows bigger than the giant rat, their killing-out percentage is very close, being 50.7 per cent for rabbits and 51.5 per cent for giant rats. The fat content of domesticated giant rats is higher (11.2%) than that of wild ones (5.7%) (Tewe and Ajayi 1979).

A feasibility study carried out at the University of Ibadan indicates that management of the giant rat in an intensive domestication programme is economically viable. Efforts are being made in southern Nigeria to establish large-scale commercial giant-rat rearing enterprises.

Cane rat

Wild species

The cane rat (*Thryonomys swinderianus*) also referred to as the grasscutter, is found in many forest and savannah areas of Africa south of the Sahara. The cane rat lives mostly amongst the coarse and cany grasses, the stalks of which it customarily severs and chews. These grasses occur mostly on damp soil or by the banks of rivers, streams and marshes. It also lives among the lesser savannah grasses where they are dense and afford good cover among tangled herbaceous vegetation. The shelter provided in these areas serves as enough protection; hence, cane rats do not build nests or burrow like the giant rats.

The cane rat is naturally vegetarian, and it is known to cause considerable damage to sugarcane, rice, maize and guinea-corn plots. It also does considerable damage in cultivated crops such as groundnuts, yams, cassava and sweet potatoes. It is a very good swimmer and is nocturnal. For breeding, cane rats excavate shallow pits within their dense cover and line them with grass or leaves. In the wild they reproduce all the year round and they have a litter size ranging from two to five.

Asibey (1979) has carried out some studies on the management of wild populations in Ghana. His studies involved using three techniques for catching the grasscutters so as to find out the relationship and effects of habitat manipulation and continuous exploitation on the numbers and population turnover, using capture–recapture techniques. Methods attempted were cage trapping, pen with cage trapping and using netting with people driving animals into the nets. None of the techniques proved successful as the grasscutters completely avoided these traps. He concluded that the grasscutter is rather conservative and elusive thus making its study in the wild a difficult venture.

Breeding in captivity

The grasscutter has been successfully raised and bred in captivity in Nigeria and in Ghana. As reported by Ajayi and Tewe (1980), inhabitants of the marshy savannah sugar-cane producing area of Nigeria called Bacita have traditionally kept live cane rats in their homes. Asibey (1979) also reported remarkable success with captive stocks. They were easily domesticated. Consequently, their domestication on a large scale is being investigated in Ghana. To this end, interested people are provided with breeding boxes and foundation grasscutter colonies. They are taught to use the same rearing techniques and feed as those used in the laboratory. The animals are bred for home consumption or for cash income. Ajayi and Tewe (1980) have studied their behaviour in captivity. The grasscutter is rather conservative and takes about a month to adjust to confinement. Food preference trials show their high preference for elephant grass (*Pennisetum purpureum*). Sweet potatoes are also highly favoured as compared to cassava, yam and cocoyam. Livestock feed concentrates are not freely consumed. Body weight at maturity is in the range 4.3–6.8 kg with an average killing-out percentage of 63.8.

Basic husbandry techniques involve provision

of large sheds where the rats can move freely and be provided with piles of grass and sugarcane on a daily basis. For satisfactory reproductive activity they should be provided with fairly marshy but tightly fenced areas. There should also be enough plant cover where they can hide.

Available knowledge shows that the domestication of the cane rat should be encouraged whether or not wild populations can be managed through simple habitat manipulation. Indeed data provided in the Wildlife Domestication Unit of Ibadan University show the economic viability of domesticated cane rat colonies.

References

Ajayi, S. S. (1975) Preliminary observations on the biology and domestication of the African giant rat (*Cricetomys gambianus* Waterhouse). *Mammalia*, **39**: 343–64

Ajayi, S. S. and Tewe, O. O. (1980) Food preference and carcass composition of the grass cutter (*Thryonomys swinderianus*) in captivity. *African Journal of Ecology*, **18**: 133–40

Asibey, E. O. A (1979) Some problems encountered in the field study of the grasscutter (*Thryonomys swinderianus*) population in Ghana. In: *Wildlife Management in Savannah Woodland*, ed. S. S. Ajayi and L. B. Halstead, pp. 214–7. Cambridge University Press: Cambridge

Faturoti, E. O. (1980) Studies on dietary crude fibre, fat and protein levels in the African giant rat (*Cricetomys gambianus* Waterhouse), Ph.D. thesis. University of Ibadan

Rosevear, D. R. (1969) *The Rodents of West Africa*. Eyre and Spottiswoode: Margate, Kent, England

Tewe, O. O. and Ajayi, S. S. (1979) Utilisation of some common tropical foodstuffs by the African giant rat (*Cricetomys gambianus* Waterhouse). *African Journal of Ecology*, **17**: 165–73

41
Other furbearers

G. A. Feldhamer

Appalachian Environment Laboratory, Frostburg, Maryland, U.S.A.

and

J. A. Chapman

Department of Fisheries and Wildlife, Utah State University, Logan, U.S.A.

Introduction

This chapter contains information on the distribution, reproduction and potential for domestication of seven diverse species of furbearers. These are beaver (Rodentia: Castoridae); ocelot (Carnivora: Felidae); raccoon (Carnivora: Procyonidae); ermine, sable, marten and wolverine (Carnivora: Mustelidae). As with other furbearers, pelts from these species are used in the manufacture of coats, or as lining and trim on wearing apparel.

Beaver

The Canadian beaver (*Castor canadensis*) is found throughout most of North America from Alaska to eastern Canada, south to northern Florida, Texas, Mexico and California. It was extirpated throughout much of its range during the 19th century, and subsequently reintroduced. The species has been introduced, and is well established in parts of Eurasia (Lahti and Helminen 1974). The European beaver (*C. fiber*) originally occupied most of the forested regions

Table 41.1 Diploid number and characteristics of the chromosomes of the beaver, ocelot, raccoon, ermine, marten and wolverine

Species	2N	Structure	
		Autosomes	**Sex chromosomes**
Castor canadensis	40	metacentric submetacentric subtelocentric	X-submetacentric Y-acrocentric
Castor fiber	48	—	—
Felis pardalis	36	metacentric submetacentric subtelocentric	X-submetacentric Y-minute submetacentric
Procyon lotor	38–42	metacentric submetacentric acrocentric	X-submetacentric Y-submetacentric or subtelocentric
Mustela erminea	44	metacentric submetacentric subtelocentric acrocentric	X-medium-sized submetacentric Y-minute
Martes americana	38	metacentric submetacentric acrocentric	X-submetacentric Y-acrocentric
Gulo gulo	42	metacentric submetacentric acrocentric	X-metacentric Y-acrocentric

of the Palaearctic. Again, following large-scale extirpation, it has been reintroduced extensively throughout the Soviet Union and parts of Europe. The beaver has also been introduced to Tierra del Fuego, Argentina.

The distinction between the two species was documented by Lavrov and Orlov (1973) on the basis of cranial characteristics and chromosome karyotypes (Table 41.1). *Castor fiber* was considered to be a more archaic form. The reproductive characteristics of both species are similar. They are monogamous, and breed once a year, generally between January and March. The gestation period is approximately 107 days and litter size is generally 2–4 young (Wilsson 1971, Jenkins and Busher 1979).

Harvest of beaver pelts in the early 19th century was a crucial factor in the early exploration and colonization of much of western North America. Marten (1892), Jones (1913) and Bailey (1922, 1927) discussed early efforts at beaver-farming in the United States and Canada, including site selection, breeding stock, facility construction, transporting, feeding and marketing. Beaver farming began in Minnesota in the 1920s (Longley and Moyle 1963), but started to decline

in the 1930s. In 1931, 1088 were raised, 538 in 1936, 51 in 1941, and again only 51 animals were pen reared in 1945. In 1963, there were no beaver being reared in pens in Minnesota. Beaver farming has been attempted elsewhere but generally has proved unprofitable, although current attempts are being conducted in Idaho and Utah. Denny (1952) discussed beaver farms in North America and listed eleven states and provinces which had active beaver farms between 1947 and 1948. Table 41.2 is a summary of the number of beaver farms in North America at that time and their general success. Large tracts of land are needed and extreme amounts of forage trees must be supplied if natural food is provided. Fluctuations in prices of fur also mitigate against commercial success. Denny (1952) pointed out 'that the early optimism . . . regarding the success of artificial propagation of beavers has failed to materialize after more than 20 years of effort'.

Since 1934, the farming of beaver in the Voronezh Reserve, in the Soviet Union, has involved basic research in reproduction, genetics, feeding and management , and reintroduction of individuals to the wild. Future research will involve breeding for variation in pelt colouration,

Table 41.2 Location, numbers and general success of beaver farms in North America, 1947–48

State, province, or territory	Number of beaver farms	General success
United States		
Alaska	3	Very unsuccessful
Colorado	None at present	Very unsuccessful
Idaho	None at present	Not successful*
Maine	2	Fair to failure
Michigan	7	Poor, declining number
Minnesota	3	One successful on small scale
Montana	5	Encouraging when natural habitat controlled on marginal land
Oregon	None at present	All resulted in failure
Wisconsin	30	Fair success
Wyoming	392	Wild unconfined beavers – good success; confined beavers – no success
Canada		
Alberta	Not known	Not generally successful
Northwest Territories	4	Two on profitable basis, two just started in 1948

* Currently (1980) being raised in Idaho and Utah (Walter Howard, personal communication)
From Denny (1952)

improved fur quality and reproduction, and increased resistance to disease (Semenov and Lavrov 1981). Results could eventually be applied towards commercial production.

Ocelot

The ocelot (*Felis pardalis*) occurs only in America in various habitat types from Arizona, New Mexico and southern Texas, south to Ecuador, Paraguay and northern Argentina. The species does not occur, however, in the high-elevation areas of Bolivia and southern Peru (Guggisberg 1975). The reproductive capacity of the ocelot is fairly low; up to 3 litters may be produced yearly, with only 1–2 young per litter.

Spotted cats such as the ocelot and the closely related margay (*F. wiedii*) have been used extensively in the fur trade. However, their current status as federally endangered species will undoubtedly minimize future trade. Ocelots are also popular as 'exotic' pets. An estimated 75–90 per cent of those trapped never survive to become pets, however (Petersen 1971).

Captive breeding programmes for the ocelot by

individuals, organizations such as the Long Island Ocelot Club, and several zoos in the United States have met with limited success. Although breeding and parturition occur, females may kill or abandon their young. Captive breeding of ocelots is improved by simulation of their natural habitat, lack of disturbance or artificial insemination. Petersen (1971, 1974) proposed guidelines for the importation, trapping and ownership of exotic felids. He also suggested that regional centres be established to further development of captive breeding programmes for endangered felids.

Raccoon

The raccoon (*Procyon lotor*) occurs in southern Canada from British Columbia to Nova Scotia, south throughout the United States except for portions of the Rocky mountains, and into Mexico and Central America. The species was introduced into the Soviet Union in 1936, Germany, France and other European countries, and is well established (Corbet 1978).

Raccoons generally breed from January to

March. The gestation period is about 64 days; mean litter sizes range from 2 to 5. The species may be polymorphic for chromosome number (Table 41.1).

Early attempts at farming of raccoons proved unsuccessful. In the United States and the Soviet Union, captive breeding programmes currently are not on a commercial basis. Some experimental farming of raccoons is being conducted in Finland (Valtonen, personal communication to I. L. Mason).

The Mustelids

The ermine (*Mustela erminea*) is also called the short-tailed weasel or stoat. Its geographic range includes tundra, boreal and deciduous forest zones of the Palaearctic and Nearctic regions. In the former region, they occur south to the Caucasus, Manchuria and northern Japan (Corbet 1978). The southern limit of their range in North America is about 39 °N latitude from California in the west to Maryland in the east. The marten (*Martes americana*) occurs only in North America. It is found throughout Alaska and northern Canada, south to the Great Lakes in the east, and to mid-California, Utah and Colorado in the west. The sable (*Martes zibellina*) was originally found throughout the northern coniferous forests from Scandinavia to eastern Siberia, south to North Korea and northern Japan. It was extirpated throughout most of the Soviet Union by the beginning of the 20th century. The sable has since been reintroduced to most of its former range. The marten and sable are very closely related species. With the European pine marten (*M. martes*) and the Japanese marten (*M. melampus*), they form an allopatric group that Hagmeier (1961) suggested may be conspecific. Darker pelts of marten have been sold as 'sable' or 'American sable' in the past. The wolverine (*Gulo gulo*) is found in the Palaearctic from the northern tundra of Norway to eastern Siberia, south to about 58 °N latitude in Europe, and the Ussuri region in Asia. It also inhabits the tundra and boreal forest regions of North America, in limited numbers. Wolverines are very rare in the contiguous United States (see Wilson 1982). North American wolverines were once considered a separate species, *G. luscus*; however, the two forms are now regarded as conspecific (Kurtén and Rausch 1959).

Like most mustelids, ermine, marten, sable and wolverine exhibit delayed implantation. Thus, the period from conception to parturition is about 270 days, while the embryo is actually implanted for only 21–28 days (30–40 days in wolverine). All four species are polygamous and breeding occurs in the early summer. Litter sizes are generally 1–6 young; 2–3 kits are most common. One litter is produced per year.

Martens have been raised in captivity in North America for many years by government agencies to study breeding and reproduction, feeding and management (see Markley and Bassett 1942). They have also been raised in zoos, and by private individuals on a commercial basis. Various aspects of commercial holding facilities, foods and mating techniques have been documented (see for example, Yerbury 1947, Orsborn 1953). These relatively small-scale commercial efforts continue. However, as with fur trapping, financial success or failure is often dependent upon yearly fluctuations in fur prices. Marten are fairly easy to work with in captivity, probably much more so than ermine or wolverine. As a result, although ermine and wolverine have been studied in captivity, commercial domestication of these species has not been attempted.

Captive maintenance of the sable has been conducted in the Soviet Union for nearly 50 years. The objectives are similar to those described for the beaver; propagation of animals for reintroduction to the wild, and research dealing with enhanced breeding potential, genetics and disease resistance. Similar research has been done on captive wolverine. Presently, captive maintenance is entirely on an experimental, not an economic basis. However, propagation on a commercial scale may be feasible in the future (V. L. Lavrov, personal communication).

Domestication of beavers, ocelot. ermine, marten, sable or wolverine is not being attempted elsewhere in Asia or in Africa at the present time.

References

Bailey, V. (1922) Beaver habits, beaver control and possibilities in beaver farming. *U.S. Department of Agriculture Bulletin* No. 1078

Bailey, V. (1927) Beaver habits and experiments in

beaver culture. *U.S. Department of Agriculture Technical Bulletin,* **21**: 1–40

Corbet, G. B. (1978) *The Mammals of the Palaearctic Region: a taxonomic review.* Cornell University Press: Ithaca, New York

Denny, R. N. (1952) A summary of North American beaver management 1946–1948. *Colorado Game and Fish Department Current Report,* **28**: 1–58

Guggisberg, C. A. W. (1975) *Wild Cats of the World.* Taplinger: New York

Hagmeier, E. M. (1961) Variations and relationships in North American marten. *Canadian Field-Naturalist,* **75**: 122–38

Jenkins, S. H. and Busher, P. E. (1979) *Castor canadensis. Mammalian Species,* **120**: 1–8

Jones, J. W. (1913) *Fur-farming in Canada.* The Mortimer Company: Ottawa

Kurtén. B. and Rausch, R. (1959) Biometric comparisons between North American and European mammals. *Acta Arctica,* **11**: 1–44

Lahti, S. and Helminen, M. (1974) The beaver *Castor fiber* (L.) and *Castor canadensis* (Kuhl) in Finland. *Acta Theriologica,* **19**: 177–89

Lavrov, L. S. and Orlov, V. N. [Karyotypes and taxonomy of modern beavers (*Castor*, Castoridae, Mammalia)]. *Zoologicheskii Zhurnal,* **52**: 734–42. (In Russian with English summary)

Longley, W. H. and Moyle, J. B. (1963) The beaver in Minnesota. *Minnesota Department of Conservation Technical Bulletin,* **6**: 1–87

Markley, M. H. and Bassett, C. F. (1942) Habits of captive marten. *American Midland Naturalist,* **28**: 604–16

Marten, H. T. (1892) *Castorologia or the History and Tradition of the Canadian Beaver.* Drysdale: Montreal

Orsborn, E. V. (1953) More on marten raising. *Fur Trade Journal of Canada,* **31**: 14, 34–35

Petersen, M. K. (1971) Some alternative approaches for perpetuation of endangered wild felids. In: *Proceedings of a Symposium on the Native Cats of North America, their Status and Management,* eds. S. E. Jorgensen and L. D. Mech, pp. 133–7. U.S. Department of the Interior, Fish and Wildlife Service: Twin Cities, Minnesota

Petersen, M. K. (1974) Preservation of wild felids by captive breeding. In: *The World's Cats,* Vol. II: *Biology, Behavior and Management of Reproduction,* ed. R. L. Eaton, pp. 148–56. World Wildlife Safari: Winston, Oregon

Semenov, V. A. and Lavrov, V. L. (1981) The role of the Voronezh State Reserve in re-establishing the beaver in the U.S.S.R. Paper presented at the World-wide Furbearer Conference, Frostburg, Maryland, U.S.A.

Wilson, D. (1982) The wolverine. In: *Wild Mammals of North America,* ed. J. A. Chapman and G. A. Feldhamer, pp. 644–52. Johns Hopkins University Press: Baltimore

Wilsson, L. (1971) Observations and experiments on the ethology of the European beaver (*Castor fiber* L.). *Viltnevy,* **8**: 115–266

Yerbury, H. (1947) Raising marten in captivity. *Fur Trade Journal of Canada,* **25**: 14, 30–2

42

Domestic fowl

R. D. Crawford

*Department of Animal and Poultry Science,
University of Saskatchewan, Canada*

Introduction

Domestic fowl or chickens are known and kept throughout the world. They are the most widely utilized of all poultry species.

They were domesticated from one or more species of southeast Asian junglefowl. Those accepting evidence for descent exclusively from red junglefowl refer to the domestic form as *Gallus gallus*, and those believing in a polyphyletic origin refer to it as *Gallus domesticus*.

There are many English language words for the various age and sex categories. The usual terms are 'chick' (newly-hatched), 'cockerel' (juvenile male), 'pullet' (juvenile female), 'cock' or 'rooster' (adult male), and 'hen' (adult female). There are many other specialized terms, some of which refer to intended food use, and there are many colloquial variations. The usual general terms are 'domestic fowl' and 'chicken'; there are regional preferences for these terms and in some areas the two are used interchangeably. In some countries, the word 'poultry' refers exclusively to chickens and in others it refers to all domestic avian species.

Domestic fowl are so well known and ubiqui-tous that a description of morphological features is unnecessary. Delacour (1965) stated that the genus *Gallus* is isolated from all others, and its relationship to other Phasianidae is neither very close nor very striking.

Egg production of wild species is variable (Delacour 1965). Under research-farm conditions, red junglefowl produced 62.5 eggs in a year compared with 181 eggs from White Leghorns (Hutt 1949). Wild birds probably do not begin reproducing until they are 1 year of age; under controlled environment, industrial chickens begin laying by 5 months of age. A reasonable expectation from industrial egg-production chickens is 250 eggs in a year with an average weight of 62 g per egg. Hatchability from breeder flocks should exceed 90 per cent. Reproductive performance of industrial meat chickens is not as good. Breeder flocks of these specialized lines are expected to produce about 150 eggs per female in a breeding season of 10 months; hatchability of eggs is about 85 per cent. The incubation period for both wild and domestic fowl is 21 days.

The diploid chromosome number of 78 is generally accepted. Considerable progress has been made in constructing a linkage map, including assignment of linkage groups to specific chromosomes and chromosome arms (Somes 1978a).

The 1981 FAO *Production Yearbook* (FAO 1982) provides data indicating a total of 6.5 thousand million chickens in the world, the equivalent of 1.4 birds per caput. Actual total is probably much higher since vast numbers of broilers marketed within the year would not have been included. All countries of the world have chickens in large numbers. They are one of the least expensive and most efficient sources of animal protein in the care of mankind. Probably no other domesticated animal species has such universal importance as a food source.

Wild species

The standard reference source on junglefowl, ancestors of domestic fowl, is Delacour (1965). Four species are known, one of them being divided into five subspecies. They occupy the warmer parts of southern Asia. Figure 42.1 illustrates normal ranges.

The red junglefowl (*Gallus gallus*) most closely

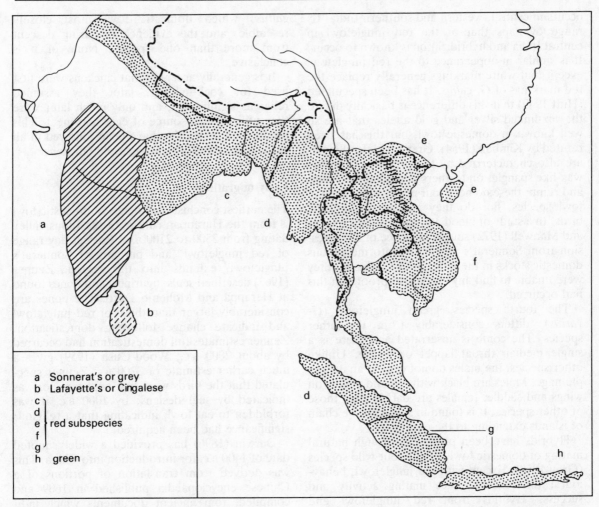

Fig. 42.1 Distribution of junglefowl in southern Asia. (Modified from Delacour 1965)

resembles domestic fowl. Its phenotype is assumed by geneticists to be the wild-type reference. Plumage colour and pattern are the same as found in Brown Leghorns and other domestic stocks having the black-breasted-red plumage phenotype. Delacour (1965) feels certain that all domestic fowl are descended from it and that the other three species made no contribution. This species is the most widespread, extending from northwestern India to southern China and Hainan, and south to Sumatra and Java. It has been introduced or has become feral on many Pacific islands.

Five subspecies of red junglefowl are listed by Delacour (1965), distinguished by slight differ-

ences in size and colouring; they intergrade where ranges overlap. The five are identified as Cochin-Chinese, Burmese, Tonkinese, Indian, and Javan. The Indian subspecies is said to be the tamest and most easily domesticated. The Javan form occurs with the green junglefowl (*G. varius*) in Java but the two have different habitats and do not interbreed naturally. The Javan subspecies is the only one to have been studied cytologically; the chromosome number is 78 and the karyotype appears to be the same as that of domestic fowl (Ohno *et al.* 1964).

LaFayette's or Cingalese junglefowl (*G. lafayettei*) is found only in Sri Lanka. It is similar to the red junglefowl, except that males have red plumage on the breast, and the voice is quite different.

Sonnerat's or grey junglefowl (*G. sonnerati*)

occurs in central, western and southern India. Its range overlaps that of the red junglefowl in central India and hybridization is known to occur. It is similar in appearance to the red junglefowl except that white markings generally replace the red markings of *G. gallus*. It has been speculated (Hutt 1949) that this difference is basically due to the sex-linked silver and gold alleles that are so well known in domestic fowl, but this has been refuted by Kimball (1954). Grey junglefowl males are also characterized by the presence of horny wax-like spangles on feathers of the hackle, wings and rump; these do not occur in any other jungle-fowl species, nor do they appear in domestic birds. In a study of blood and egg proteins, Baker and Manwell (1972) sought evidence of introgression from Sonnerat's junglefowl into indigenous domestic stocks in the same part of India but they were unable to find any conclusive proof that this had occurred.

The fourth species, green junglefowl (*G. varius*), differs considerably from the other species. The comb is unserrated and there is a single median throat lappet or wattle. Unlike other species, the males do not exhibit an eclipse plumage. Males are black with red and yellow on wings and saddle; females are darker than those of other species. It is found in Java and the chain of islands extending to the east of it.

Hybrids have been produced through natural mating of domestic fowl with all four wild species (Gray 1958). Except with red junglefowl, behavioural differences reduce mating activity and success. Hybrids from red junglefowl and domestic birds are fully viable and they can reproduce normally. Hybrids produced from domestic fowl and the other three wild species are viable; F_1 males generally have better reproductive ability than F_1 females, perhaps indicating longer evolutionary separation of parental stocks. Delacour (1965) records that green junglefowl males are frequently raised in captivity in their native islands to be crossed with domestic females; the hybrid male so produced is prized for its loud long-drawn monosyllabic crow.

Domestication and early history

Hutt (1949) reviewed the evidence favouring descent exclusively from the red junglefowl (*G. gallus*) which domestic fowl most closely resemble, and the evidence favouring descent from more than one species. None of it is conclusive.

It is generally accepted that chickens were first used for cock-fighting, later they assumed religious significance, and only much later were they utilized as a source of eggs and meat. This progression appears to have been followed in all early civilizations which possessed chickens.

Early migrations

The earliest conclusive evidence of domestic fowl is from the Harappan culture of the Indus valley dating from 2500 to 2100 B.C.. Present-day range of red junglefowl, and perhaps of Sonnerat's junglefowl, extends into this region. Zeuner (1963) described seals, figurines, and bones found at Harappa and Mohenjo Daro. The bones are considerably larger than those of red junglefowl and indicate change following domestication. Zeuner estimates full domestication had occurred by about 2000 B.C.; Wood-Gush (1959) gives a much earlier estimate of 3200 B.C. Zeuner speculated that the birds were first kept for sport, as indicated by seal designs; by 1000 B.C. it was forbidden to eat fowl, indicating that a religious significance had been acquired.

Darwin (1896) has provided a widely quoted date of 1400 B.C. for introduction into China. This was derived from translation of portions of a Chinese encyclopaedia published in 1609 and compiled from ancient documents which indicated that chickens were regarded as creatures of the west and were introduced about 1400 B.C. Brown (1929) credits transport to Buddhist priests journeying from India to China about 1500 B.C. A brief review of recent Chinese archaeological literature (Ho 1977) indicates much earlier association of man and fowl in China. Remains have been found at northern and western sites which are believed to predate the Indus valley culture. It is not clear from Ho's review whether the bones are from wild junglefowl or whether they are from domesticated birds. They probably indicate at least that the red junglefowl had a much greater range than it does now (Fig. 42.1). Ho states that combined evidence shows that domestication of the chicken in China antedated that of the Indus valley civilization by centuries.

If Chinese domestication of chickens was the first, presumably that stock did not persist or it remained isolated.

Coltherd (1966) has published a critical review and interpretation of the confusing and conflicting literature pertaining to early presence of domestic fowl in Egypt. He interprets the chick hieroglyph which is found in the earliest inscriptions to represent a quail or a guinea-fowl and not the domestic fowl. The first fairly definite representation of a chicken occurs among graffiti on blocks from a temple of the Middle Kingdom near Thebes. It is dated at about 1840 B.C. About this time trade routes linking Egypt, Mesopotamia and India had been established and there was a flourishing commerce in exports and imports. Political chaos occurred after the fall of the Middle Kingdom, disrupting normal trade, and chickens may have disappeared. During the reign of Thuthmose III (1479–47 B.C.) Egyptian trade was re-established and expanded. The royal annals of this period include the phrases 'four birds of this country; they bear every day', which have been interpreted from their context as referring to chickens from an area east of the Tigris. This evidence is supported by several archaeological finds. One is a bird's head depicted in a scene on the walls of the tomb of Rekhmara, vizier of Thuthmose III, at Thebes. The scene depicts tribute from Keftiu, which is believed to be modern Crete. A line drawing of the bird's head appears in Carter (1971). To a geneticist, relative proportions of the head furnishings, and the tri-lobed rounded comb indicate a pea-combed heterozygote; facial features are those of south-east Asian chickens as is the pea comb. A second piece of archaeological evidence is an ostracon depicting a rooster found in association with the tomb of Tutankhamun and probably contemporary with his burial in 1338 B.C. It is redrawn in Zeuner (1963) and in Carter (1971). The comb is too large for a junglefowl, which the bird otherwise resembles. And thirdly, a silver bowl from the reign of Seti II (1200–1194 B.C.) depicts a cockerel similar to that on the ostracon of a century earlier. The next five centuries were times of steady political decline and recession in trade. No traces of domestic fowl have been found in this period. From 664 to 525 B.C. the record again becomes rich, perhaps because of close ties with Greece. Thereafter, under Persian influence the bird became very well known. It is possible, and consistent with developments elsewhere, that chickens did not become a part of domestic livestock in Egypt until this period of Greek and Persian influence. Earlier presence may represent curiosities or exotic trade items only.

The 1st millennium B.C. was a period of rapid expansion in keeping of domestic fowl. From the Indus valley civilization, it must have reached Iran at an early date; Brown (1929) gives a date of about 550 B.C. but it probably was much earlier. Chickens are depicted on Assyrian coins of the 8th century B.C. In Zoroastrian literature of the period 2000–700 B.C., the cock was known as the herald of the dawn (Wood-Gush 1959), and Zeuner (1963) states that it played an important part as guardian of good against evil.

From Iran, chickens were carried westward to Mesopotamia and Asia Minor (Zeuner 1963). The Medes may have brought them to the Euphrates–Tigris basin, leading to the Greek reference to chickens as the Median bird. Chickens were common in the Mediterranean and Asia Minor by the 6th century B.C. although there are scattered evidences of their presence in Greece and Italy in the two preceding centuries. The Egyptian records indicate occasional presence very much earlier. At first, the Greeks used chickens only for sport, but the Romans adopted them as a food source almost immediately and developed husbandry and production technology to a high level (see Lind 1963. Wood-Gush 1959, Zeuner 1963).

Taran (1975) has summarized new evidence concerning the presence of domestic fowl in ancient Judea. Earlier interpretation had been that these birds did not reach the area until the 2nd century B.C. (Wood-Gush 1959) which was consistent with the lack of mention of them or their eggs in any part of the Old Testament. Taran describes a series of archaeological finds consisting of potsherds and seals clearly depicting fowl of the fighting kind which date back to the 7th century B.C. Presumably the birds were not used as a food source during that time. The contemporary Egyptian record for this period is also rich in depicting domestic fowl (Coltherd 1966).

According to Brown (1929), chickens were distributed to Europe by two routes. The first route was via Iran and the Mediterranean and has been well documented. The second proposed by Brown was via China and Russia. The Russian and Balkan chickens seen by him during his travels in the early 20th century were distinctly Asiatic in type; local tradition was that the birds had been introduced by the Mongols about 500 years earlier, which is reasonably compatible with Brown's estimated date of the 13th century A.D. for introduction to Russia. Presumably from there they entered central Europe. Peters, as reviewed by Wood-Gush (1959), also suggested two routes of entry to Europe, both from Iran; one was via the Aegean Sea and the other via the Black Sea reaching central Europe through southern Russia. Carter (1971) has presented a persuasive argument on linguistic grounds about the spread of domestic fowl. He traces gradual change of the Sanskrit word *kukuta* from India across Asia and through central and northern Europe until the word becomes 'chicken' and 'cock' in England; similarly he traces the Sanskrit word *pil* which evolves into *gallus* and 'pullet'. If correct, his evidence favours distribution according to Peters' proposal.

Zeuner (1963) discounts reports of very early presence of domestic fowl in central Europe. He says that the Celts had chickens in the 1st century B.C. or maybe a little earlier, but since they had close contacts with Italy, the birds probably came from there rather than from the southeast. In Switzerland, bones of domestic fowl have been found at an archaeological site dating from 58 B.C. or earlier.

Julius Caesar specifically mentioned that domestic fowl were present in Britain before the Roman conquest, but Zeuner (1963) feels that the birds had not been there for long. Caesar mentioned that the Britons would not eat them. The earliest pre-Roman site where they have been found is dated at A.D. 10–42. Local tradition in Cornwall is that chickens had been introduced first by the Phoenicians, and a similar tradition persists in Scotland (Carter 1971). Several Roman archaeological sites include remains of chickens and it is certain that they had brought them. Similarly, Zeuner (1963) credits the Romans with popularizing chickens elsewhere in Roman Europe, both within their sphere of dominance and beyond its borders.

Information concerning introduction of domestic fowl to Africa, except for Egypt, is scanty. Sauer (1969) indicates that they were present at the time of first white contacts. Among the stocks were chickens with black feathers, meat and bones, described from Mozambique in 1635. Similar black birds were known in India then but not in Europe. Carter (1971) points out that the words for domestic fowl in both East and West Africa have an Indian root. Trade between India and the east coast of Africa was well developed at an early date so that introduction of domestic fowl from India is probable.

Available information concerning the origin and introduction of domestic fowl in southeast Asia and the Pacific islands has been reviewed by Carter (1971). Names for chickens in China and in the Philippines are similar and can be traced to the Zhou dynasty, 1122–249 B.C.; they bear little relationship to the Sanskrit words that spread elsewhere in Asia and to Europe. Japanese words are totally different, as are their chickens, perhaps indicating an origin distinct from those of China. Evidence based on both morphology and linguistics suggests introduction to the Pacific islands from both India and China.

Pre-Columbian chickens in America

The controversy concerning pre-Columbian presence of chickens in the Americas has occupied agricultural historians for many years. The subject has been reviewed by Wood-Gush (1959) and by Sauer (1969). It has been very extensively analysed by Carter (1971) who assembled biological, cultural, and linguistic evidence, and related these to contemporary situations in Asia and Europe. The evidence is cumulative and strongly suggests introduction of chickens from across the Pacific to South America many centuries before European conquest. Conclusive evidence would be the finding of skeletal remains in pre-Columbian archaeological sites, but thus far none has been found.

Part of Carter's (1971) analysis consists of estimating rate of spread of domestic fowl. Using the historical record for Europe and Asia, he estimated a rate of 1.5–3 km per year in diffusion of chickens from the Indus valley through Europe. The data for South America, assuming Spanish introduction, indicate a much more rapid spread, roughly calculated at 80–160 km per year. The first authentic landing of Europeans in South America was by Cabral on the Brazilian coast in A.D. 1500, and perhaps chickens were introduced then. The Spanish certainly introduced their own stocks later. During the next 40 years, Spanish and Portugese exploration was extensive. Reports from each new area explored – southern Brazil, the Orinoco, Peru, the Amazon, northern Argentina, and the Paraguay river – indicate the abundant presence of chickens. Carter feels that this rate of spread following a first introduction in A.D. 1500 is much too rapid to be possible. He makes the suggestion instead that introduction and spread may have paralleled and been contemporary with the spread of maize from Mexico into South America, believed to have begun about 1500 B.C. The rate of diffusion if contemporary with maize would be similar to that calculated for spread of chickens in Asia and Europe.

Linguistic evidence is that the names used are non-Spanish, except in areas where it is known that chickens were first introduced by the Spanish. Because of great variation in names, Carter suggests plural introductions or prolonged prehistory or both. He notes that three names, one in South America, one in Central America, and one in Mexico, seem directly related to Asiatic names for the chicken.

In tropical forested South America, chickens and eggs were not used as a food, but feathers were used in decoration. This was in keeping with Asiatic practice, but not with 16th century European custom where chickens served primarily for food. The birds were widely used for cockfighting. In areas under Spanish influence – Cuba, Puerto Rico, and the Rio Grande – no spurs were used as in Spain. In Brazil, birds were fought with blunt natural spurs as in Japanese and Polynesian custom. In all of the rest of Central and South America, Asiatic slashers, birds of Asiatic type, and Asiatic names were used.

The zoological evidence is of most interest in the context of the present review. Descriptions of the American stocks indicate presence of several well-known single-gene mutations. Among those most probably present were pea comb, silky, frizzle, rumplessness, crest, naked neck, fibromelanosis (black skin), blue egg, and ear-tuft. Authors agree that most of these traits are Asiatic and not European. It has been argued that, if the traits are Asiatic, their presence in the Americas must have been pre-Columbian. But the Spanish were actively trading in southeast Asia and Africa concurrently with their exploration of the Americas and presumably had access to both European

and Asiatic stocks. A reappraisal of mutant origins is perhaps warranted.

Pea comb is characteristic of modern breeds known to have originated in southeast Asia. It is not frequent among the traditional stocks of Europe or the Mediterranean but it has occurred there.

Silky, frizzle, and crest are all ancient mutants. Earliest reports of them refer to Asia where they are currently very prevalent. Genetic rumplessness has also been known for many centuries; its geographic origin has not been recorded. All of these mutants are likely to have been Asiatic in origin. But all were known in the Mediterranean area by at least A.D. 1600 since all of them are illustrated in Aldrovandi (Lind 1963). The Spanish could have obtained them there. However, it may be significant that other characteristics well known in Europe at that time – polydactyly, rose and duplex comb, white eggshell – are not mentioned as existing among the early American chicken stocks.

The naked-neck condition is mentioned only once in Carter (1971) as being present in Brazil under the name of naked-neck Madagascars. Carter's authority assumed them to be Portugese imports, but Carter considers them to be pre-1500 imports. Early literature does not mention this mutant. But it must be of considerable antiquity since it currently is distributed world-wide among indigenous stocks in tropical areas where it seems to have a selective advantage through superior heat tolerance.

Fibromelanosis (black skin) is another mutant for which there is no record in Europe at the time of Spanish conquest of the Americas. The condition was well known in Asia and Africa by 1635 (Sauer 1969) and is thought to be of Asian origin. There is no clear indication that it was present among South American chickens at the time of first Spanish or Portugese contact. Darwin (1896) refers to its presence in Paraguay about 1800, and Carter (1971) makes reference to it several times.

Blue-egg and ear-tuft mutations are peculiar to American chickens. Both were first drawn to the attention of the scientific community by Castello (1924), genetics of the blue-egg trait was described by Punnett (1933), and inheritance of ear-tuft was not deciphered until very recently (Somes 1978b). If they were present in American chickens of A.D. 1500, it is puzzling why no mention was made of them in early records since both traits are strikingly unusual and unlike anything known before. It is also peculiar that Darwin, whose prime interest was variation, did not come upon them during his extensive travels within South America during the voyage of the Beagle. It is probable that both mutations occurred within South America, but there is no evidence to indicate whether mutation was pre- or post-Columbian.

Overall, zoological evidence favouring pre-Columbian introduction of chickens to the Americas is weak. The other kinds of evidence reviewed by Carter (1971) present a stronger case. Until additional circumstantial and historical evidence accumulates, or until there is discovery of skeletal remains from a known pre-Columbian site, the controversy over presence of chickens in the Americas before A.D. 1500 will remain unresolved.

Early breeds and morphological variations

Many well-known single-gene mutations had accumulated by the time that first documents were written. Some of these had been incorporated into local breeds, perhaps through artificial selection and also perhaps through geographic isolation. Although poorly documented, it is probable that local types or breeds had evolved wherever chickens were kept, partly through selection but most likely through the genetic effects of isolation.

Aldrovandi (Lind 1963) and Wood-Gush (1959) have reviewed the writings of Columella, Aristotle, and Pliny concerning poultry husbandry and breeding in the Graeco–Roman world of the 1st century A.D. The Greeks bred their birds primarily for fighting and the Romans bred them for food. Several distinct breeds were recognized then – Rhodian, Median, Tanagran, Chalcidian, and Adria – and it was accepted that the best birds were obtained through crossbreeding. Presence of most of the variations in shape and colour known to us now is indicated in the very early literature, with the possible exception of sex-linked barring which has had extensive utilization in recent years.

Aldrovandi (Lind 1963) described a number of distinct stocks known in A.D. 1600 and illustrated some of them. He included Bantam, Paduan, Turkish and Persian. His illustrations clearly indicate presence of rose and buttercup combs, polydactyly crest, rumplessness, dwarfism, feathered shanks, silky and frizzle.

An inventory of broad groupings of types of chickens known in Great Britain late in the 19th century was given by Darwin (1896). He believed that most of the chief kinds had been imported to Britain but that many subbreeds were still unknown there. His list comprised thirteen stocks, briefly noted as follows. He considered Game (fighting) breeds to be not much different from the wild type. The gigantic Malay breed was well known; it had upright carriage with very long head, neck and legs, and a drooping tail; feather coat was tightly held to the body. This breed persists relatively unchanged to the present. Darwin listed as a separate group the Chinese chickens called Cochin or Shanghai; these were very large, with feathered legs and profuse soft downy plumage, phlegmatic disposition and yellowish skin. The Dorking was described as a large five-toed bird, with a compact square body. The Spanish breed was described as the stock now known as White-faced Black Spanish, and included as subbreeds the Blue Andalusian and a Dutch variety. Hamburghs were well known, there being a number of colour varieties. All crested chickens were combined as a Crested or Polish breed; a number of subbreeds not necessarily related to the main one were listed – Sultan, Ptarmigan, Ghoondook, Crève-coeur, Horned, Houdan, and Guelderland; some of these are still currently popular among hobbyists. Bantams were listed separately; Darwin stated that they were ancient and originally from Japan; he characterized them by small size alone and indicated that there were several subbreeds including Cochin, Game and Sebright. Rumpless fowl were listed separately as were Silkies and Frizzles. The Silkies had fibromelanosis (black skin) as did a normally feathered breed which he called Sooty. Darwin also listed Creepers or Jumpers which are clearly recognizable as characterized by the lethal creeper gene.

Darwin's inventory includes virtually all of the types and mutants known from earlier literature. It is a useful list since it was a starting point for the sudden booming interest in selective breeding of chickens which began in the mid-19th century. It is curious that the list barely includes the layer and broiler stocks which now totally dominate the industrial poultry business of the world.

Recent history and present status

Since the mid-19th century, domestic fowl have undergone a rate of evolution exceeding that which occurred in all of the centuries following domestication. Despite predictions of genetic plateaux being reached, production performance continues to change at a rapid rate. But change has been accompanied by a pronounced reduction in variability, causing concerns about future developments.

Pure breed development and the 'hen craze'

There is a vast contemporary popular literature documenting the increasing interest being shown in chickens beginning about 1850. This was coincident with rapid industrial and agricultural development of North America and Great Britain especially, but was witnessed in other parts of the world, too. Birds which had been barnyard scavengers began to assume much greater importance, and large sums of money were spent on their acquisition, care and breeding. Skinner (1974) has documented this phenomenon in the United States.

The impetus seems to have been exotic stocks brought to the west from China. These birds were much bulkier than any seen before, they had very profuse soft feathering, and left the impression of having great size. They were variously called Cochins, Shanghais, and Chinas. Prominent personalities became involved in their breeding, which served to popularize them. They were rapidly developed into several distinct breeds – Cochins, Langshans, Brahmas – which attained immense popularity in the 'hen craze' of the late 19th century. Concurrently attention was given to local stocks, selecting them for uniformity of appearance so that they would achieve breed or variety status. And adventurous breeders began purposely to cross existing breeds to come up with something new through recombination of major characteristics.

According to Skinner (1974) these developments led to exhibitions of stock, which inevitably led to competitive showing. Disagreements among exhibitors led to the writing of breed descriptions and to the adoption of breed standards. Each development led to increased interest by the public so that attention to poultry breeds and breeding reached unprecedented heights.

Brown (1929) has provided an illustrated description of the breeds and varieties which were developed during this era, and of the old breeds which were made more uniform. His material pertains especially to the British experience. There were similar volumes written in other countries. Breeds had become grouped according to area of origin – American, Asiatic, British, French, Italian, etc. – and they were divided into varieties characterized by a particular colour or colour pattern or comb type. Several new breeds were developed in Great Britain, notably Sussex and Orpington. In the United States, many new breeds evolved, including Plymouth Rock, Wyandotte, Rhode Island Red, New Hampshire, and Jersey Giant. Canada contributed only the ill-fated Chantecler.

The interest in poultry was almost exclusively in achieving perfection of show specimens. Little attention was paid to productivity in eggs or meat. Unfortunately, the movement leaders became entrenched in their own affairs and were unwilling to recognize a public demand for greater productivity (Skinner 1974); as a result, breeders concerned about productive and reproductive performance gradually drifted away from those who were concerned only with show points. Eventually the show breeders found themselves isolated and regarded as a hobbyist group, while the commercial breeders went on to establish a world-wide food production industry. Interest in breeding of exhibition stocks has resurged during the past decade as a hobby, incidentally serving as an important repository of rare genetic material. Standards for exhibition birds continue to be revised and reprinted at intervals; those best known are the American Standard of Perfection, the American Bantam Standard, and British Poultry Standards.

The breeds which currently dominate world production of eggs and meat were developed during this period. Leghorns arrived in the United States between 1828 and 1831 (Skinner 1974) and many importations followed. They were imported to England at a later date but had become popular by 1876 (Brown 1929). They represented the indigenous stock of Tuscany and took their name from the port city of Leghorn (Livorno). Colouring was not uniform at first, but fanciers soon had isolated a large number of varieties. Only the Single Comb White Leghorn remains in commercial use as the exclusive layer of white-shelled eggs. Brown-shelled eggs currently are derived from crosses involving several minor breeds, all of which were developed after 1850 as dual-purpose (eggs and meat) stocks. Barred Plymouth Rock, Rhode Island Red, and New Hampshire were all developed in the United States. Australorp (Australian Orpington) was a selection from the older English breed. Present-day broilers are based heavily on a cross of Cornish with White Plymouth Rock. The Cornish was developed in England from Asiatic fighting stock; according to Brown (1929) the white variety is unrelated to the original Dark Cornish. White Plymouth Rocks were derived as sports of the original breed in the United States. Nearly all of the vast numbers of breeds and varieties developed by the fanciers or derived from ancient sources have been discarded by the industry in favour of these few.

Breeding and selection for productivity

The Americans have been world leaders in poultry genetics and selection, their progress and output far exceeding that of others. Productivity improvement in other countries has generally followed the American example. Unlike the situation in plant breeding where development of new varieties has been the realm of public institutions, poultry breeding has traditionally been the responsibility of private individuals with technical guidance from the public servants.

Warren (1974) pointed out that two completely independent events at the beginning of the 20th century permitted poultry breeding to progress from an art to a science. The first event was the rediscovery of Mendelian principles of heredity. The second was the development of a reliable and workable trapnest. In recent years, individual caging has been used for breeder housing and the trapnest has fallen out of use.

Both private breeders and research workers at first attempted to improve productivity by assuming control of all characteristics by a few gene pairs. This proved unsatisfactory and they soon began to use progeny testing as a means of identifying superior breeding individuals. It was found that sex-linked characteristics could be used to provide auto-sexing, or distinction between males and females at hatching, based on differences in morphology or colouring. This technique has immense importance now in both broiler

and layer industries, to permit sex-separate rearing of broilers, and to permit immediate discard of males in laying stocks. Various comparative measurement procedures were introduced to aid in selection of breeding stock. Egg-laying contests became very popular, but because of poor control over pre-test and test environments, results from these contests had only limited value and they were gradually discontinued. Official schemes to encourage recording of production characteristics on the breeders' home premises were also developed, usually under the name of record of performance (R.O.P.); these served a useful purpose in choosing breeding stock on the basis of recorded production rather than on visual appearance only, but they have now ceased functioning in nearly all countries.

Between 1930 and 1950 the concepts of heritability, genetic correlation, selection differentials, and prediction equations came into general use by advanced breeders and by research workers. Progeny testing was largely discarded as being too slow a procedure and it was replaced with collateral relatives selection. This period was also the time when breeders began to exploit the effects of crossbreeding in commerical production (Warren 1974). It was generally found that hatchability, growth rate, and viability were improved through crossing, and that egg production was frequently increased. Breed crosses were used very extensively, particularly in the developing meat industry. Although strain crosses did not have the heterotic effect of breed crosses, it became necessary to use them in breeding stocks for commerical white-egg production; as with Holstein-Friesians in dairy cattle, the difficulty lay in the fact that there was no other breed equivalent in capability to the White Leghorn, and strains within the breed had to be used.

Further refinements in breeding and selection procedures have been introduced since 1950. Geneticists addressed the question of genotype × environment interaction, which had immense implications for breeders who intended global expansion of their stock sales. Hull and Gowe (1962) developed a hypothesis indicating that large and important interactions would be found only when environmental effects were very large and genetic differences were wide. Since commercial stocks are now very similar, and since the environment provided for industrial poultry throughout the world is very uniform, breeders have been able to carry out selection work at one location and confidently sell their stocks for production throughout the world; had interactions been shown to be of importance, then the present-day franchise system of stock distribution could not have evolved. Lengthy consideration was given to utilizing immunogenetics in improvement of stocks; it was held that genes for antigenic factors might be used as markers to identify superior production genotypes (Gilmour 1969). But the promises appeared not to materialize and immunogenetics receives little attention now in commercial breeding operations (Warren 1974). Random sample tests came into vogue as a means of comparative testing, fully replacing the old R.O.P. programmes. In these tests, random samples of a breeder's commercial stock are compared with those from others throughout an entire life-cycle and under rigidly controlled environmental conditions. Acceptance of random-sample testing by the poultry industry clearly marked the transition in poultry breeding from the individual bird being the unit to the flock or strain or crossbred combination being the unit; this fundamental philosophy has been of immense importance in facilitating the massive industrial exploitation of poultry, an exploitation which had not yet occurred with other species of domesticated animals. As a further refinement in evaluating progress, geneticists developed the use of random-bred controls as a selection yardstick (Gowe *et al*. 1959) and these controls have become part of the random-sample testing procedure.

Since 1950, the poultry industry has narrowed its choice to three distinct kinds of stock. In all three, the final commercial product is a multiple cross. White Leghorns emerged as the exclusive breed in use for production of white-shelled eggs. Crosses of various breeds – Rhode Island Red, New Hampshire, Barred Plymouth Rock, and Australorp particularly – were intensely selected for egg production to serve the special markets for brown-shelled eggs that exist in some parts of the world. And crosses basically involving White Cornish and White Plymouth Rock became the choice for broiler meat production.

There has been severe attrition in the number of individuals and companies involved in breeding (Warren 1974).Very large breeding companies emerged, supplanting smaller organizations because of the superior quality of their products,

superior business procedures, and immense financial backing. Many of the smaller organizations became distributors or agents for the larger ones. Most companies became absorbed into multinational corporations, giving them the resources to expand stock sales globally under a franchise system. The end result has been that world-wide sales of breeding stock for both layers and broilers rests in the hands of a very few multinational corporations who utilize only a tiny fraction of the genetic diversity that has accumulated over past centuries.

Present status

It is convenient to classify present-day domestic fowl in three very broad categories which relate to their ownership and utilization. Industrial chickens are those bred by the gigantic multinational corporations for use world-wide in mass production of white-shelled and brown-shelled eggs and of chicken broilers. Middle-level stocks are those kept in moderate-sized flocks under reasonably good management conditions; they comprise local breeds under traditional farm conditions; the vast array of hobbyist or exhibition breeds can be included in this category. The third category is that of native or indigenous stocks, kept under conditions of minimal care and having very low productivity; feral populations can be included in this group.

Industrial chickens   At the present time there are eleven multinational corporations which breed and distribute stocks for industrial egg production. The corporations are called primary breeders. Each offers white-shelled egg layers, and nearly all of them also offer brown-shelled egg layers. These eleven provide nearly all of the industrial breeding stock for the world. The primary breeder conducts the breeding and selection work, based on four purebred grandparent lines (A, B, C, and D). These are mated to produce a two-way cross male parent (AB) and a different two-way cross female parent (CD). Parent stocks are distributed throughout the world where they are used as multiplier flocks by local franchised hatcheries. The parents (AB and CD) are mated to produce a four-way cross commercial pullet (ABCD) which is used as the industrial egg-layer; males are usually discarded at hatching. The layer bears the name of the primary breeder. Production performance of the birds is superb. Under suitable industrial conditions, expectation from white-shelled egg stocks is 250–270 eggs per bird in a production period of 12 months; egg size is very uniform, averaging about 62 g; body size at sexual maturity is about 1250 g and increases to about 1700 g by the end of production. The birds are selected for a productive life in intensive cage housing. To reduce body maintenance and housing costs, size of the birds is being steadily reduced by the primary breeders; some utilize a sex-linked dwarfing gene to achieve size reduction. Performance of brown-shelled egg layers has improved to the point that it is nearly equivalent to that of the Leghorns; but feed efficiency is still somewhat poorer and body size is still greater than desired.

It is estimated that there are currently also eleven primary breeders of chicken broilers, owned by multinational corporations, supplying breeding stock for the broiler industry of the world. Breeding and distribution of broilers involves the same four-way crossing procedure as utilized in layer production; the primary breeder owns the four grandparent lines, the multiplier-distributor holds the two crossbred parents, and the broiler grower has the final four-way cross. The paternal grandparents evolved from White Cornish and maternal grandparents are based on White Plymouth Rocks. Selection on the male side is primarily for growth rate, and on the female side for reproduction. The commercial meat birds are marketed at various ages and weights for particular markets. In North America, the largest demand is for birds at a live weight of 1900 g. Males reach this weight at about 44 days of age, and females at 49 days, with feed conversion rates of 1.9 and 2.0 respectively. The sexes are usually reared separately. Performance improvement has been incredibly rapid; it is estimated that currently the age to reach the 1900 g market weight is being reduced by 1 day per year.

Middle-level stocks The traditional poultry industry of developed countries utilized the stocks called middle-level. These included the many breeds and varieties developed in the late 19th century as well as those surviving from antiquity. In most instances they were bred for production of both eggs and meat. With the advent of industrial chicken production, it has been usual to find that utilization of middle-level stocks declined as consumers began to rely on the inexpensive, uniform, and high-quality products so readily available from merchants. In some countries, middle-level stocks have virtually disappeared from commercial production. In a recent survey of genetic resources in Canada (Crawford, unpublished), it was found that only eleven strains of middle-level chickens remained in the entire country; 35 years earlier, all of Canada had been supplied with eggs and meat from these stocks. A similar situation prevails or is approaching in other countries.

Middle-level chickens for the most part now survive only among hobbyists, who keep them for interest and for competitive exhibition. A vast array of breeds and varieties, bantams and normal-sized birds, can still be found in many countries. Surveying of Canadian poultry genetic resources (Crawford, unpublished) indicated that nearly all of them were present in Canada although in very small numbers. Unfortunately the Canadian hobbyists are multipliers and exhibitors; they are not breeders. They rely on a few dealers for their stocks, rather than selecting their own, so that genetic base for these breeds is probably very small and rapidly decreasing, leading to concerns about safe preservation of the genetic variability which they carry. A similar situation has been found to prevail in the United States (Skinner 1974).

Indigenous and feral stocks Native or indigenous chickens are considered to be those which have persisted in the village economies of long-settled countries. Most of them lead a precarious existence as scavengers, receiving almost no human care. Even so, they have served as a continuing food supply, as a recreation resource, and in some instances have retained importance in religious and cultural rites. Where a middle-level industry has developed, and especially where industrial production has become dominant, the indigenous stocks have rapidly disappeared. Indigenous chickens have undergone some evaluation and utilization in locally improved poultry husbandry, particularly in Arab countries and in India. But elsewhere, almost nothing is known about local native chickens and their attributes.

Current knowledge concerning feral populations of domestic fowl is scanty. Darwin (1896) mentioned occurrence of many such populations, particularly on oceanic islands, but there has been no current survey or inventory to determine whether any of these persists. The only feral stock which has been studied recently is that of Northwest Island in the Australian Great Barrier Reef (McBride *et al.* 1969); the birds are of Japanese origin and may have been living independently of man since 1899.

Future prospects

Concern has frequently been expressed that plateaux in selection response have been reached or will soon be reached in performance of industrial chicken stocks. However, performance continues to improve, especially in broiler stocks, and primary breeders are optimistic that selection progress in egg numbers and quality, and in meat yield, will continue to be made.

Further change in meat-bird performance should be carefully monitored to avoid past mistakes. There is an apparent trend by primary breeders to move broiler parent stocks from an environment of floor pens and natural mating to one of cage housing where artificial insemination will be used. This change could permit further increase in growth rate and body weight since constraints imposed by normal mating behaviour and normal locomotion would be removed. There is a distinct danger that such developments might result in loss of reproductive fitness such as has occurred in commercial turkeys. In that species, artificial breeding is required and leg strength of market birds has become a matter of serious concern.

Concern should also be expressed about the near-monopoly which exists in breeding of industrial stocks. There are eleven primary breeders of egg production chickens and eleven of chicken broilers which provide multiplier stock to most of the industrial sector of the world. Some countries are not served by them; Australia and France have remained insular. But most of the world relies on a few multinational corporations for their eggs and broiler meat. This situation is dangerously close to one of monotypic populations which can be vulnerable to genetic disaster as discussed with reference to turkeys (see Chapter 47, p. 325).

In an exceedingly comprehensive review, Sheldon (1980) has reviewed the current state of poultry genetics as reflected by research activity and acquisition of knowledge. He concluded that the present rate of acquiring new knowledge in poultry genetics is not sufficient to allow productive application of the new genetic engineering technology which will be developing in the next decade. He urged that research and research funding continue at an accelerated pace.

Conservation of genetic variability

Since genetic needs for future change are unknown, and since there has already been tremendous uncontrolled loss of genetic material through stocks which have been discarded, conservation of germ plasm has become a matter of serious concern. It is held by some that the need for preservation of genetic variability is greater in poultry, especially in chickens, than it is in any other form of domestic animal because of the way in which the poultry industries have evolved.

Conservation has received international consideration. It was the topic of the fourth FAO expert consultation on animal genetic resources (FAO 1973), and was again considered in a summary conference (FAO/UNEP 1981). Throughout, FAO member nations have been urged to become involved in cataloguing, evaluating, and preserving their resources.

Cataloguing has been in progress in some countries, and there has been some action in establishing conservation flocks. A catalogue of poultry held at Canadian research and teaching institutions has been compiled annually (Crawford 1981), and a survey has also been conducted in Canada to estimate commercial holdings of industrial and middle-level stocks (Crawford, unpublished). An inventory of chicken stocks in the United States was first published in 1972 and it has had several revisions (Somes 1978c); each subsequent edition includes listings from additional countries. In Australia, an expert panel on conservation of poultry genetic resources issued a report (Standing Committee on Agriculture 1979) which urged preparation of a register of poultry genetic stocks including commercial, institutional and fancier holdings; a beginning has been made in compiling that register.

Only a few centres are currently maintaining live collections. One of the largest is at Parafield Poultry Research Centre in Australia and there is another in Romania. A third collection is kept at the University of Saskatchewan in Canada. Although there are hobbyist groups whose main interest is preservation of rare breeds, e.g. the Society for Preservation of Poultry Antiquities in the United States, and the Rare Breeds (Poultry) Society in Great Britain, Crawford's unpublished study indicates that long-term preservation must become the responsibility of public agencies if it is to be secure; a similar conclusion was reflected in the study by the Australian panel.

When field technology for long-term preservation of sperm cells and ova becomes available, the task will be made much easier. Meanwhile, erosion of genetic resources in domestic fowl is proceeding and effective action for conservation cannot be delayed much longer.

References

Baker, C. M. A. and **C. Manwell** (1972) Molecular genetics of avian proteins. XI. Egg proteins of *Gallus gallus*, *G. sonnerati* and hybrids. *Animal Blood Groups and Biochemical Genetics*, **3**: 101–107

Brown, E. (1929) *Poultry Breeding and Production*. Benn: London

Carter, G. F. (1971) Pre-Columbian chickens in America. In: *Man Across The Sea – Problems of pre-Columbian contacts*, ed. C. L. Riley, J. C. Kelley, C. W. Pennington, and R. L. Rands, Ch. 9. University of Texas Press: Austin, Texas

Castello, S. (1924) The *Gallus inauris* and the hen which lays blue eggs. *Proceedings of Second World's Poultry Congress, Barcelona*, pp. 113–18

Coltherd, J. B. (1966) The domestic fowl in ancient Egypt. *Ibis*, **108**: 217–23

Crawford, R. D. (1981) *Catalogue of Poultry Stocks Held at Research and Teaching Institutions in Canada* (14th ed). University of Saskatchewan: Saskatoon.

Darwin, C. (1896) *The Variation of Animals and Plants Under Domestication* (2nd ed). Appleton: New York

Delacour, J. (1965) *The Pheasants of the World*. Country Life: London

FAO (1973) *Report of the Fourth FAO Expert Consultation on Animal Genetic Resources* (*Poultry Breeding*). Food and Agriculture Organization of the United Nations: Rome

FAO (1982) *FAO Production Yearbook 1981*, Vol. 35. FAO: Rome

FAO/UNEP (1981) *Animal Genetic Resources Conservation and Management*. Report of the FAO/UNEP Technical Consultation held in Rome, 2–6 June 1980. FAO: Rome

Gilmour, D. G. (1969) Blood group research in chickens. *Agricultural Science Review*, **7**: 13–22

Gowe, R. S., Robertson, A. and **Latter, B. D. H.** (1959) Environment and poultry breeding problems. 5. The design of poultry control strains. *Poultry Science*, **38**: 462–71

Gray, A. P. (1958) *Bird Hybrids. A check-list with bibliography*. Commonwealth Agricultural Bureaux: Farnham Royal, Bucks, England

Ho, Ping-Ti (1977) The indigenous origins of Chinese agriculture. In: *Origins of Agriculture*, ed. C. A. Reed. Mouton: The Hague and Paris

Hull, P. and **Gowe, R. S.** (1962) The importance of interactions detected between genotype and environmental factors for characters of economic significance in poultry. *Genetics*, **47**: 143–59

Hutt, F. B. (1949) *Genetics of the Fowl*. McGraw-Hill: New York

Kimball, E. (1954) Genetics of junglefowl plumage patterns. *Proceedings of Tenth World's Poultry Congress*, Edinburgh, **1**: 2–4

Lind, L. R., (1963) (trans.) *Aldrovandi on Chickens. The ornithology of Ulisse Aldrovandi (1600) Volume II, Book XIV*. University of Oklahoma Press: Norman

McBride, G., Parer, I. P. and **Foenander, F.** (1969) The social organization and behaviour of the feral domestic fowl. *Animal Behaviour Monographs*, **2**: 127–81

Ohno, S., Stenius, C., Christian, L. C., Becak, W., and **Becak, M. L.** (1964) Chromosomal uniformity in the avian subclass *Carinatae*. *Chromosoma*, **15**: 280–8

Punnett, R. C. (1933) Genetic studies in poultry. IX. The blue egg. *Journal of Genetics*, **27**: 465–70

Sauer, C. O. (1969) *Agricultural Origins and Dispersals. The domestication of animals and foodstuffs* (2nd ed). M.I.T. Press: Cambridge, Massachusetts

Sheldon, B. L. (1980) Perspectives for poultry genetics in the age of molecular biology. *World's Poultry Science Journal*, **36**: 143–73

Skinner, J. L. (1974) Breeds. In: *American Poultry History 1823–1973*, ed. J. L. Skinner, Ch. 1. American Printing and Publishing: Madison, Wisconsin

Somes, R. G. Jr (1978a) New linkage groups and revised chromosome map of the domestic fowl. *Journal of Heredity*, **69**: 401–403

Somes, R. G. Jr (1978b) Ear-tufts: a skin structure mutation of the Araucana fowl. *Journal of Heredity*, **69**: 91–6

Somes, R. G. Jr (1978c) *Registry of Poultry Genetic Stocks*. Bulletin 446, Storrs Agricultural Experiment Station. University of Connecticut: Storrs

Standing Committee on Agriculture (1979) *The Conservation of Genetic Material*. Report of the Expert Panel appointed by the Animal Production Committee of Standing Committee on Agriculture: Canberra

Taran, M. (1975) Early records of the domestic fowl in ancient Judea. *Ibis*, **117**: 109–10

Warren, D. C. (1974) Breeding. In: *American Poultry History 1823–1973*, ed. J. L. Skinner, Ch. 8. American Printing and Publishing: Madison, Wisconsin.

Wood-Gush, D. G. M. (1959) A history of the domestic chicken from antiquity to the 19th century. *Poultry Science*, **38**: 321–6

Zeuner, F. E. (1963) *A History of Domesticated Animals*. Hutchinson: London

43
Pheasants and partridges

T. H. Blank

Formerly Research Director, Game Conservancy, Hampshire, England.

Pheasants

Pheasants are large polygamous game birds, the males having very distinctively coloured plumage. They have the ability to fly swiftly for relatively short distances and under favourable conditions show a large annual rate of increase which can sustain fairly heavy cropping. Although when reared in captivity their food conversion rate is lower than that of poultry, pheasant flesh is of good quality. Because of these characteristics the pheasant, originally prized for its ornamental and culinary qualities, is nowadays a highly esteemed quarry species.

Wild species

Delacour (1977) recognizes forty-nine species of pheasants in sixteen different genera. They may be divided into two groups – the ornamental pheasants with forty-seven species and the game or true pheasants with only two species – the common pheasant, *Phasianus colchicus*, and the Japanese green pheasant, *P. versicolor*. The common pheasant has been separated into thirty-

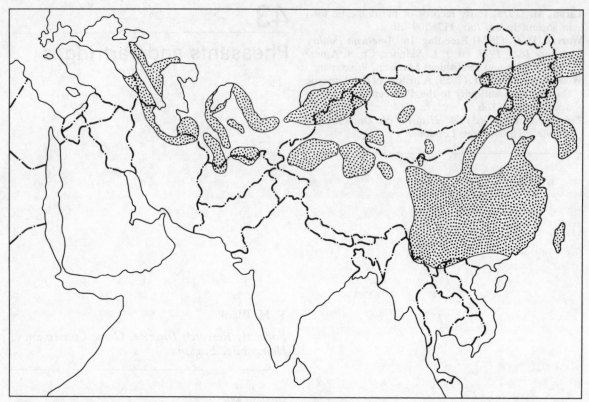

Fig. 43.1 Natural distribution of the pheasant (*Phasianus colchicus*). (From Delacour 1977)

one subspecies throughout its extensive range and three subspecies of the Japanese green pheasant have been recognized in Japan. All these subspecies interbreed freely in captivity and in the wild where ranges overlap.

While the many subspecies show variations in breeding biology it is generally accepted that in the wild most pheasants are polygamous. Pheasants never rear more than a single brood in a season but this may involve laying several unsuccessful clutches. Average size of first clutch is 10–15 eggs and incubation, carried out by the hen, covers a period of 25 days. Care of the chicks is the exclusive responsibility of the hen.

The pheasant is a south Palaearctic and northeast Oriental species which occurs naturally in a variety of habitats in a discontinuous belt of varying width across Asia from the Black Sea to the Pacific Ocean (Fig. 43.1). Essentially it is a bird of the "edge" – i.e. where two different types of habitat meet – showing a preference for open fields and cropland well interspersed with nesting and escape cover in the form of small areas of wood, scrub, marshland or hedgerow.

Thirty-one subspecies of the common pheasant (*P. colchicus*) have been identified, mainly on plumage differences of the males. In the eastern part of their range several subspecies have a more or less well-developed white ring around their necks while in the west – where the first European introductions came from – the white neck ring is absent. In Europe, however, the original introductions of the western black-neck subspecies, *P.c. colchicus*, have interbred with subsequent introductions and the present-day wild pheasant with its very mixed origins is indistinguishable from the hybrids reared in captivity.

In Roman times the pheasant *P. c. colchicus* was introduced into western Europe and in more recent years, particularly during the 18th and 19th centuries, several different subspecies were brought in and contributed to the typical wild European pheasant of very mixed ancestry. Before the year 1800 pheasants from England were introduced to the eastern United States, but it was nearly a century later that the first successful releases occurred in Oregon 1881 and New Jersey 1889, the latter with mixed European stock. More recently successful pheasant introductions have been made in New Zealand and captive breeding stocks established in Australia and South Africa.

In its native countries pheasants were never sufficiently numerous to constitute a food source of any economic importance. In European countries where the pheasant has been introduced, however, the relatively large number of birds maintained by intensive rearing programmes (in Great Britain alone it is estimated that 8 million pheasants are reared annually) supports a recreation industry worth many millions of pounds in addition to contributing to the country's food supplies. The annual legal harvest of pheasants in the United States has been estimated at between 16 and 18 million in good years (approximately 20 000 tons of food) and in Great Britain may be as high as 8 million!

Domestication and history

Legend attributes the introduction of the pheasant into Europe to Jason and the Argonauts about 1300 B.C. Pliny the Elder (A.D. 23–79) records the same original home for the pheasant (the Rion valley in Georgia, once known as the Phasis valley in Colchis – and hence the scientific name) while Palladius (A.D. 350) quoting from earlier writings, details methods for rearing and fattening pheasants. Throughout the Middle Ages, as well as being reared for food, the pheasant was firmly established as a quarry species in many European countries – to be hunted only by members of the land-owning classes.

Although the advent in the 17th century of the muzzle-loading hand-gun and the use of small shot in place of single bullets made it possible to shoot birds in flight and popularized game-bird shooting as a recreational sport, it was not until the middle of the 19th century that the breech-loading gun, with its facilities for rapid reloading, led to the demand for artificially high concentrations of pheasants. Coinciding with the onset of the Industrial Revolution in western Europe and the resulting accumulation of wealth, the high

costs of producing such large numbers of pheasants were readily accepted. In a relatively short space of time the pheasant rearer's prime objective showed a fundamental change from producing food for the table to providing a living target at which guns could be aimed. The cost of producing the end-product, the bird on the table (incidental though it was to the sport provided), had increased enormously since, on the average, only one in three of the released pheasants would be recovered by shooting.

The relatively small-scale pheasant-rearing programmes that had been carried out up to the mid-19th century (primarily to produce meat for the tables of the rich) did not apparently affect the characteristics of the stock. "Tameness" may have been bred into it but this was to some extent unintentionally counteracted by the periodic trapping of wild stock from the surrounding countryside or its importation from abroad.

From the middle of the 19th century onwards the scale of pheasant rearing in western Europe, and particularly in England, rapidly increased. Not only was it practised on more and more private estates but the late 19th and early 20th centuries also saw the establishment of many game farms devoted to the propagation of pheasants. Since the primary objective of pheasant rearing was now the provision of targets which most closely resembled the wild bird – in size, appearance, flight habits and above all in ability to survive in the wild – no selective breeding away from the wild stock was attempted.

At first the emphasis was laid on the production of eggs for sale, since incubation was then carried out by broody hens. Many game farmers attempted to maintain fairly pure stocks of at least three subspecies of pheasant – usually *P. c. colchicus*, *P. c. torquatus*, and *P. c. mongolicus* – because some customers showed preferences (often based on erroneous assumptions) for certain varieties.

The advent of the Second World War halted pheasant rearing throughout most of Europe; but with the post-war cessation of food rationing, and the adoption of many of the new poultry production techniques, the scale of pheasant rearing rapidly increased. Incubation methods used for poultry, with modifications to accommodate the somewhat different conditions required during the hatching stages, led to the game farmer's concentration on the production of day-old chicks and the decline of egg sales. Pelleted foods, developed for poultry rearing, were well suited to rearing pheasants and the mass production methods now being employed.

Brooder lamps and brooder houses, accommodating units of 250 or even 500 pheasant chicks, now take the place of the broody and coop units on the open rearing field of 40 years ago. Indeed, it is only by employing such methods that the high costs of pheasant poult production have been contained.

Today the vast majority of pheasants are reared for restocking – to provide a shootable surplus of birds that are indistinguishable from the wild ones. To offset the artificiality of the initial period of intensive rearing, 7- to 8-week-old poults are given an acclimatizing period of several weeks in the specially prepared habitat of a release pen.

In recent years studies have been made by the Wildlife Service of the U.S.A. of the precise environmental requirements of the different subspecies of pheasants in their native Asian habitats with a view to filling any similar but untenanted ecological niches in the United States. Another, and perhaps more common approach, has been to release hybrid stock from Europe which provides a large gene pool in which natural selection can work.

Future prospects

While pheasants have been reared in captivity for at least 2000 years their domestication is still in its infancy. This is partly because no selective breeding was practised while food production was the primary objective of pheasant rearing up to the middle of the 19th century. From then on, when the production of pheasants for sporting purposes was almost the sole object of rearing, it became an intrinsic part of rearing policy that the qualities of the wild bird should be retained. Wild stock was reintroduced into game-farm stock at frequent intervals. Pheasant restocking is an expensive business and every pheasant shot will cost the sportman four or five times its carcass value. Therefore pheasant rearing for sporting purposes will remain the prerogative of the more affluent societies; and even there the ethics of rearing birds for such purposes are coming under increasing criticism.

Pheasant production directly for the table is still at the experimental stage. Although it has been shown that with artificial lighting regimes egg-laying can be induced throughout the year and selective breeding can produce 'jumbo' strains, it is in countries without competition from 'wild' shot pheasants that most interest in the project is being shown.

Partridges

Although partridges – grey *Perdix perdix*, red-legged *Alectoris rufa*, rock *A. graeca* and chukhar *A. chukhar* – have long been kept in captivity, rearing has until recently been carried out only on a very small scale. Now, as in the case of pheasants, partridge rearing in relatively large numbers has been developed primarily for sporting purposes – either in attempts to introduce partridge species to suitable ecological niches in new countries or to provide a shootable surplus in countries where annual partridge production from wild stocks is insufficient to meet the demands of sportsmen. In either case, the objective has been to produce birds which retain the qualities of the wild ones and no conscious attempts have been made at domestication.

In the United States, chukhar – mainly from varieties introduced from India – have been bred in fairly large numbers and released in suitable habitats in several states where they become established – particularly in Nevada, California, Washington and Idaho – and rearing for the augmenting of wild stocks has become the prerogative of private enterprise.

In Europe rearing and releasing grey partridges in the post war period enjoyed a temporary boom, but the relatively low returns from shooting, and the failure to increase wild breeding stocks by this method, have led to its decline. Red-legged partridges on the other hand, which have been reared and released in increasing numbers in recent years, have provided a higher recovery rate by shooting (approximating to the 30–40% yielded by reared pheasants) although increases in wild breeding stocks have rarely been achieved.

In addition to the pure red-legged partridge a hybrid of doubtful origin (red-legged × rock partridge or red-legged × chukhar partridge) is being reared in some European countries, not only for supplementing wild stocks but also for producing young birds for the table. Although its ability to breed successfully in the wild is somewhat suspect, this shortcoming is offset by its advantages as a game-farm bird. Good egg production (60–70 eggs) has been achieved in relatively small wire-floored pens by pairing hybrid hens (of rock or chukhar crossed with red-legged partridge origin) with cock red-legged partridges. Selection for high egg production and amenability to close penning is still in its infancy, however, and the success of this development will depend on the maintenance of the relatively high price commanded by the carcases of the young (12-week-old) hybrid partridge.

References

Bachant, J. P. (1971) History of ring-necked pheasants in Ohio. *Ohio Game Monographs No. 4.* Ohio Department of Natural Resources: Ashley

Blank, T. H. (1967) The recovery of reared pheasants on an English pheasant shoot. *Transactions of the 8th International Congress of Game Biologists,* pp. 361–5. State Game Research Institute: Helsinki

Delacour, J. (1977) *The Pheasants of the World* (2nd ed). Saiga Publishing: Hindhead, Surrey, England

Westerskov, K. (1963) Evaluation of pheasant liberations in New Zealand based on a 12-year banding study. *Wildlife Publication No. 71.* New Zealand Department of Internal Affairs: Wellington

44
Peafowl

Iain Grahame

Daw's Hall Wildfowl Farm, Suffolk, England

Wild peafowl

This chapter will be confined to the Asian peafowl of the genus *Pavo*, since the Congo peafowl, *Afropavo congensis*, the only true pheasant endemic to Africa which was 'discovered' as recently as 1936, is not a true peafowl and is not domesticated. The Asian peafowl are *Pavo cristatus* and *P. muticus*, more commonly known as Indian peafowl and green peafowl, respectively.

There are several differences between the two species. Green peafowl are longer in the leg, neck and crest, and both sexes have bright, metallic colouring. Linnaeus' appelation *muticus* (or 'mute') is misleading, but the male's call is nevertheless less loud and strident than that of *cristatus*. Although there is no overlap in the wild between *cristatus* and *muticus*, hybridization between the species frequently occurs in captivity, the offspring of such matings being entirely fertile.

Peafowl are readily distinguished from other Galliformes, particularly by the brilliant colouring, erect crests and long trains of the males. These trains, sometimes mistakenly called tails, in fact comprise the upper tail coverts (numbering over

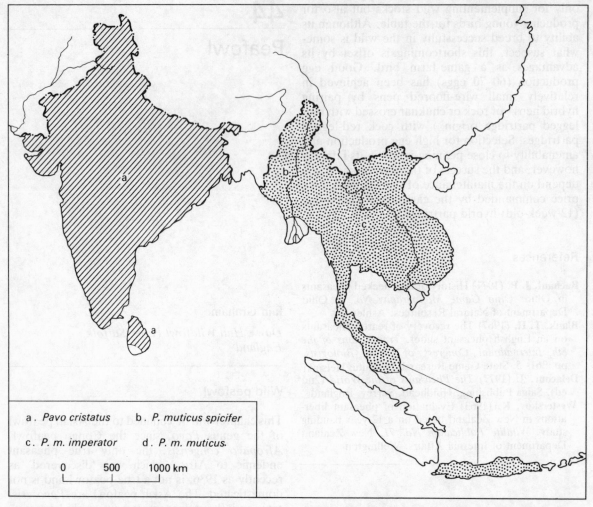

a . *Pavo cristatus*   b . *P. muticus spicifer*

c . *P. m. imperator*   d . *P. m. muticus.*

0   500   1000 km

Fig. 44.1 Distribution of wild peafowl (Delacour, 1951)

200) which lie on top of twenty graduated rectrices. The tail coverts, with their spectacular ocelli, form the central theme of the peacock's display. They are spread fan-shaped above the bird's back while he pirouettes in front of the female, calling and rattling his long quills. Although the train is not fully developed until peacocks are 3 years old, immature birds will attempt to display in their first year. Peahens also have crests, but can be distinguished by their shorter legs, much shorter tail coverts and, in *cristatus*, the lack of irridescent plumage.

Chromosome counts have been performed for five wild pheasants plus the domestic fowl (all in the family Phasianidae), for the turkey (Meleagridae) and guinea-fowl (Numididae). In all of these the diploid number $2n$ is $78 \pm$ (i.e. with a margin for error); in *P. cristatus* $2n = 66 \pm$, i.e. 33 pairs. The number has not been determined for *P. muticus*. Comparisons of macrochromosomes indicate that, while *P. cristatus* is distinct from other Phasianidae so far investigated, *Pavo* shows closer resemblance to *Numida* than to *Coturnix*, *Gallus* or *Phasianus*.

Peafowl are most commonly found between sea level and 1000 m. Polygamous by nature, they usually occur in large or small flocks in open country and in scattered woodland, often close to cultivation. They are omnivorous, feeding mainly on fruit, seeds, insects and small vertebrates. Females nest on the ground and clutches consists of 4–8 eggs. The incubation period is 28 days.

As will be seen from Fig. 44.1, *P. cristatus* is confined to the Indian subcontinent, while three distinct subspecies of *P. muticus* occur in areas eastwards. There is no evidence that it ever occurred on the isolated land masses of Sumatra and Borneo.

Domestication and early history

It is hardly surprising that a bird of such gorgeous hues and proud gait should have had a long association with man. Peacocks in fact have been famous in art, in legend, in literature and in religion for over 3000 years. In Greek mythology the peacock was the emblem of the goddess Hera. In China and Japan they have been regarded as symbols of beauty and power since at least the 12th century, when Chinese trade missions first reached the east coast of the Malay peninsula. Both Malayan and Javan peoples were then Hindu, with scattered Buddhist monasteries.

To Buddhist and Hindu peafowl are either sacred, or (depending on sect) of very special significance. They are regarded as bodily representations of the mythical bird Fung, which symbolizes the feminine aspect of the harmonious combination Ying and Yang. The Buddha himself is often depicted riding on a displaying peacock. Just as the bird readily devours young snakes and other pests, so does the Buddha astride his avian steed dispense new hope while destroying evil, unwholesome thoughts. Train feathers have long adorned both Hindu and Buddhist shrines.

Phoenicians transported them to Egypt and to Asia Minor, and there are references to their existence in ancient Assyria and Babylonia. The earliest biblical reference to peafowl is 1 Kings 10: 22, 'For the King [Solomon] had at sea a navy of Tharshish with the navy of Hiram: once in three years came the navy of Tharshish, bringing gold, and silver, ivory and apes, and peacocks.' Although some controversy still exists as to the locality visited by these ships, historians currently favour either southern India, Ceylon or Malacca. If it was indeed the latter, then *muticus* would have been the species collected.

Many importations of peafowl to Europe are known to have been carried out by the armies of Alexander the Great. Greeks, then Romans, kept and reared these birds for the banqueting table, where they were served surrounded by their own plumage.

Early Christians regarded the peacock as a symbol of the resurrection of Christ, and the bird was often painted on the walls of catacombs and incorporated in the mosaics of early churches.

From Rome they were taken to France, England and other European countries, where they continued to be a gourmet's delight until gradually supplanted by turkeys from Mexico in the 15th century. Culinary habits, however, take time to die. Writing in the mid-18th century, Oliver Goldsmith recounts how they were still being served on special occasions, when, for example, 'our citizens resolve to be splendid'.

From the same writer we learn that 'peacocks were brought from the East Indies; and we are assured that they are still found in vast flocks, in a wild state, in the islands of Java and Ceylon'. From this it would seem clear that importations were of both *cristatus* and *muticus*.

Old birds are tough and disappointing to eat, and the taste is very dependent on the birds' diet. Young birds fed on corn or mustard are apparently delicious, while those feeding on certain fruits, notably figs, have a bitter flesh.

Recent history and present status

Their spectacular plumage has long made Indian peafowl popular in captivity. They are currently represented in almost every zoo in the world, and appear to be equally common in parks, gardens and backyard aviaries.

Green peafowl, by contrast, are comparatively scarce in captivity. Not only are they susceptible to cold, but males are often pugnacious both to human beings and to their own kin.

The most recent survey of captive Galliformes, conducted by The World Pheasant Association in 1979, produced the following figures:
Indian peafowl – 4696
Green peafowl –315

Such censuses are often misleading. As census coordinator I can estimate the likely degree of inaccuracy. Statistics for Indian peafowl, scattered as they are across almost the entire world and kept by potentates and paupers, are inevitably underestimated. Captive stocks of these must certainly be over 6000 and probably in excess of 10 000.

Figures shown for green peafowl will be more accurate, for it is far easier to learn the whereabouts of birds requiring such specialized management. After allowing for some degree of error, their numbers are unlikely to be in excess of 500, of which 50 per cent are kept in the United States. Judging from the ignorance displayed on census forms, many of those who keep green peafowl in collections do not know to which subspecies they belong. Due to their less gaudy plumage, very few of the Burmese form (*P. m. spicifer*) are kept in captivity, but there is a very real danger that aviculturists will unwittingly create hybrids between the two more spectacular races, *P. m. muticus* and *P. m. imperator*.

Although degeneration is a common symptom of captivity, there is no evidence that domesticated peafowl have suffered in this way. Under proper conditions of management fertility is extremely high. The only noticeable effect of centuries of domestication is a slightly heavier bird that is a little lower on the legs (Delacour 1951). Several spectacular mutations of *P. cristatus* are known, noticeably an albinistic form (White peafowl) and one described by Sclater in 1866 as *P. nigripennis* (Black-shouldered peafowl). Apart from these and less stable mutations like the 'Pied', there is a fine hybrid form known as Spalding's peafowl, best obtained by mating a male *P. nigripennis* with a female *P. muticus*.

Nowadays most Hindus regard peafowl as sacred birds, so in many areas *P. cristatus* receives full protection. Large flocks frequently gather round temples where they are fed.

Future prospects

Indian peafowl have shown over at least 3000 years how successfully they have adapted to domestication. They are long-lived, hardy, and only the females are liable to fall to predators if not enclosed during the nesting season. Despite what is probably an equally long association with man, future prospects for the survival of green peafowl are far less encouraging.

Helped in no small way by protection resulting from superstition or religious scruple, the status of *P. cristatus* in the wild is very healthy. Sadly, the same cannot be said for *P. muticus*. According to the latest revision of the *Red Data Book* (ICBP 1979), *P.m. muticus* is now restricted to two reserves in west and east Java, where their total population is estimated at less

than 250. *Pavo m. imperator* has become greatly reduced in Thailand, and information is lacking on its present status in Yunnan and Burma. *Pavo m. spicifer* is likewise known to have suffered a serious decline throughout its range. All races are consequently now regarded as endangered, the principle causes of their reduction being illicit hunting and habitat destruction in the interests of agriculture. Much must now depend on establishing viable breeding stocks in captivity, particularly in the Western world.

References

Bergmann, J. (1980) *The Peafowl of the World*. Saiga Publishing: Hindhead, Surrey, England

Delacour, J. (1951) *The Pheasants of the World*. Country Life: London

Goldsmith, O. (1774) *An History of the Earth, and Animated Nature*. Nourse: London

Grahame, I. (1979) WPA census of Galliformes. *World Pheasant Association Journal* **4**: 72–5

ICBP (1979) *Endangered Birds of the World: the ICBP Red Data Book*. International Council for Bird Preservation and Smithsonian Institution Press: Washington. (IUCN *Red Data Book* Vol II *Aves*)

Ray-Chaudhuri, R. (1973) Cytotaxonomy and chromosome evolution in birds. In: *Cytotaxonomy and Vertebrate Evolution*, ed. A. B. Chiarelli and E. Capanna, pp. 425–83. Academic Press: London

45

Japanese quail

Noboru Wakasugi

Laboratory of Animal Genetics,
Nagoya University, Japan

Introduction

The Japanese quail belongs to the family Phasianidae. Its scientific name is *Coturnix coturnix japonica*. In the United States it is sometimes called 'coturnix quail' to distinguish it from the bobwhite (*Colinus virginiana*) and California quail (*Lophortyx californicus*).

Newly hatched chicks of domestic Japanese quail weigh 5–8 g. They show rapid growth and attain sexual maturity around 42 days of age. Adult males weigh 90–130 g and adult females 110–160 g. Egg production rate for the first year is about 80 per cent under controlled lighting conditions. Reproductive activity of the Japanese quail is more strongly influenced by the change in photoperiod than that of the domestic fowl. The eggs are mottled and resemble those of wild birds. They weigh 7–11 g (7–8% of the body weight of the hen), indicating high rate of metabolic rate for egg formation. The incubation period averages 17–18 days with a range of 16.5–20 days.

Chromosome number is $2n = 78$; there are 6 pairs of large chromosomes, 6 pairs of medium-size chromosomes and 27 pairs of microchromo-

somes. Sex chromosomes are of large size; the Z-chromosome is metacentric and the W-chromosome is acrocentric, contrasting with that of the domestic fowl which is submetacentric (Benirschke and Hsu 1971).

The number of domestic quail in Japan is about 6 million. Commercial quail production is carried out, as in the chicken industry, in specialized units such as hatcheries, raising farms, processing farms for eggs and meat. Farms of various sizes are distributed through the centre and south of Japan. Eggs are packed in thin plastic cases and sold at food stores. For longer storage they are canned or packed in vinyl bags after boiling and removal of the shells. They are used also as ingredients for other foods such as mayonnaise and cakes. The meat is valued as a gourmet food not only in Japan but also in other countries such as France, Italy and the United States.

Wild species

The distribution of wild Japanese quail covers Japan, Korea, east China, Mongolia, Siberia and Sakhalin, i.e. the area between 100 ° and 150 °E and 17 ° and 55 °N (Fig. 45.1). The distribution of related subspecies is as follows.

European quail *C. c. coturnix*, breeds in the wide area ranging from the coast of the White Sea to North Africa and from the British Isles to the west of Lake Baykal and migrates to tropical Africa, Arabia and the south of India in winter. *Coturnix c. conturbans* inhabits the Azores. *Coturnix c. confisa* inhabits Madeira and the Canary Islands. *Coturnix c. inopinata* inhabits the Cape Verde Islands. *Coturnix c. africana* lives in tropical Africa, the Comoro Islands, Madagascar and Mauritius. *Coturnix c. ussuriensis* is considered to be the same as *C. c. japonica* and some researchers consider *C. c. erlangeri* and *C. c. africana* to be the same (Kawahara 1978, Kiyosu 1979).

Surveys of wild quail were made in Japan and Korea from 1925 to 1933. The total number of quail tagged and released was 9786, of which 758 were recovered. It was confirmed that they usually breed in the north of Japan in spring and summer and migrate to the south in autumn. Their migration distance was estimated to be 400–1000 km. A few quail released at Kyushu in the south of Japan were recaptured in Korea.

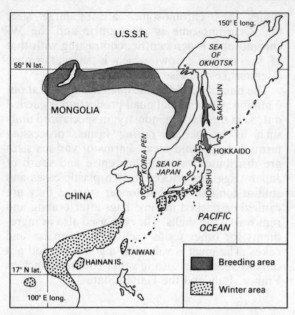

Fig. 45.1 Distribution of wild Japanese quail. (From Kawahara 1978)

Some quail breed in the mountains of their winter habitat without migrating to the north. They like a temperate climate and the northern limit of their winter habitat is around 38 °N. This coincides with the isotherm for 0 °C monthly average temperature. They usually nest in grassland and lay 7–14 eggs. Nesting is performed by the female only. The male courts other females when his partner begins nesting. In this survey the tagged quail were rarely recaptured later than the fourth year after releasing. Therefore, their life-span is estimated to be 3–4 years. The number of wild quail caught in Japan per year from 1924 to 1933 varied between 200 000 and 500 000. Their capture is now prohibited.

A recent study on wild Japanese quail has revealed that their body weight and shank length are smaller by about 20 per cent than those of the domestic quail. The other characteristics of wild quail in captivity are late sexual maturity, high incidence of non-layers and low egg-production rate. The cross between the wild and domestic Japanese quail produced fertile F_1 hybrids. Repeated backcrosses to either wild quail or domestic quail were successful (Kawahara 1973, 1978). These facts suggest that there are no differences in the karyotype between the wild and domestic Japanese quail. The colony of wild quail

maintained in the laboratory became tame after the sixth generation. However, their body weight and reproductive performance were still inferior to those of the domestic quail at the eighth generation, although their performance improved, possibly due to unconscious selection under the breeding conditions in the laboratory (Kawahara 1976).

Domestication and history

The Japanese quail was originally valued as a song bird. Tradition says that the enjoyment of its rhythmical call originated about 600 years ago (Yamashina 1961). However, it is uncertain whether the ancestors of these song quail were domesticated in Japan or introduced from China or Korea. A quail book which is considered to have been published in 1648–51 described the craze of the song quail (cited by Ohmori 1918). The raising of song quail was most popular among the Samurai (warrior) caste in the feudal age. It continued among common people after the Meiji renovation in 1868 and there was a craze also in the modern age (Oda 1917, Ohmori 1918). There seems to have been a number of variations in the quail call and it is inferred that the lines or closed colonies showing a peculiar call were bred and that quail song contests were frequently held.

The present domestic Japanese quail was bred from the song quail by enthusiastic breeders around 1910. In those days, eggs were hatched in petroleum incubators, or under broody hens (Japanese bantam breeds of the domestic fowl were usually employed). They vigorously engaged in the improvement of the quail for higher egg-production rate and established good layers (Oda 1917, Ohmori 1918). Such improved quail were propagated commercially mainly in Aichi prefecture in the centre of Japan. The first prosperity of the quail industry occurred in the 1930s and by 1941 there were about 2 million domestic quail in the country.

The domestic quail were almost exterminated in Japan during the Second World War. The commercial quail was restored from the few that survived the war. However, the original song quail was unfortunately swept away. The commercial quail rapidly recovered and the number of quail in the country again reached the total of about 2 million in the 1960s (Yamashina

1961). At present the number fluctuates around 6 million.

Japanese quail were introduced to the United States several times in the 1930s to 1950s as game birds. This was unsuccessful. Another attempt was made to use them, this time as laboratory animals in the late 1950s (Padgett and Ivey 1959). Around that time, they were also introduced into Europe and the Middle and Near East where they have been used for both commercial purposes and scientific research. Attempts to use them as laboratory animals have also been made in Japan. Their advantages are small size, short generation interval and high egg-production rate. Other characteristics are high metabolic activity, strong sensitivity to light and resistance to disease. However, the fact that they are very sensitive to inbreeding depression is a serious disadvantage (Wakasugi and Kondo 1973). Many mutant genes have been discovered, e.g. morphological mutants, mainly plumage colour (Truax and Johnson 1979), and biochemical mutants (Kimura 1978). Immunological studies have also started (Katoh and Wakasugi 1980). Efforts have been made to establish various genetic stocks with mutant genes or particular characteristics and a registry of these interesting stocks has been compiled (Somes 1981).

Future prospects

As already mentioned, there are three kinds of Japanese quail, e.g. wild quail, commercial quail and laboratory quail. Interaction between commercial and laboratory quail breeding would stimulate their mutual development. Mutants found at the commercial level would be useful for laboratory work and at the same time genetic variety would be conserved. Conversely, introduction of stocks with sex-linked genes to the commercial quail breeder would be useful for auto-sexing of newly hatched chicks. Such a breeding method may lead to the differentiation of breeding stocks and production stocks as practised in the chicken industry. This will also be a convenient method to make use of heterosis. It would be interesting to observe how the three kinds of Japanese quail diverge from each other with the passage of time.

References

Benirschke, K. and Hsu, T. C. (1971) *Chromosome Atlas: Fish, Amphibians, Reptiles and Birds*, Vol. 1., Folio Av-5. Springer-Verlag: Berlin, Heidelberg and New York

Katoh, H. and Wakasugi, N. (1980) Studies on the blood groups in the Japanese quail: detection of three antigens and their inheritance. *Developmental and Comparative Immunology*, 4: 99–110

Kawahara, T. (1973) Comparative study of quantitative traits between wild and domestic Japanese quail (*Coturnix coturnix japonica*). *Experimental Animals*, 22 (Supplement): 138–50

Kawahara, T. (1976) [History and useful characters of experimental strains of Japanese quail (*Coturnix coturnix japonica*)]. *Experimental Animals*, 25: 351–4. (In Japanese)

Kawahara, T. (1978) [Body traits and behaviour in wild Japanese quail *Coturnix coturnix japonica*]. *Tori (Bulletin of the Ornithological Society of Japan)*, 27: 105–12. (In Japanese with English summary)

Kimura, M. (1978) [Protein polymorphism in poultry]. *Japanese Poultry Science*, 15: 43–54. (In Japanese)

Kiyosu, Y. (1979) [*The Birds of Japan*], Vol. 11. Kodansha: Tokyo. (In Japanese)

Oda, K. (1917) [*Fifteen Years' Experiment for Japanese quail Breeding*]. Oda-Chorui-Kenkysho: Tokyo. (In Japanese)

Ohmori, S. (1918) [Ten Years' Japanese Quail Breeding]. Tobundo-Shoten: Tokyo. (In Japanese)

Padgett, C. S. and Ivey, W. D. (1959) *Coturnix* quail as a laboratory research animal. *Science*, 129: 267–8

Somes, R. G. Jr. (1981) *International Registry of Poultry Genetic Stocks*. Storrs Agricultural Experiment Station: University of Connecticut, Storrs, Connecticut

Truax, R. E. and Johnson, W. A. (1979) Genetics of plumage colour mutants in Japanese quail. *Poultry Science*, 58: 1–9

Wakasugi, N. and Kondo, K. (1973) Breeding methods for maintenance of mutant genes and establishment of strains in the Japanese quail. *Experimental Animals*, 22 (Supplement): 151–9

Yamashina, Y. (1961) Quail breeding in Japan. *The Journal of the Bombay Natural History Society*, 58: 216–22

46
Guinea-fowl

P. Mongin and M. Plouzeau

*I.N.R.A. Station de Recherches Avicoles,
Nouzilly, France*

General biology

The domestic guinea-fowl, *Numida meleagris*, belongs to the family Numididae. Both its English and its Latin name point to its African origin. For other names see below.

Guinea-fowls originated in Africa. There are five genera in the family Numididae and *N. meleagris* is one of the four species of *Numida*. Only two species have been domesticated: *N. meleagris* is native to West Africa but it can also be found in Morocco, in the Cape Verde Islands and in southwest Arabia where it was probably introduced; *N. ptilorhynca* is native to East Africa. The latter is distinguished by the blue colour of the wattles and the presence of a brush of hair at the base of the beak. All the species are found in the wild state and are hunted but flocks of free-ranging *N. meleagris* and *N. ptilorhynca* are maintained in Madagascar and in Réunion.

Guinea-fowl are a gallinaceous species and possess the characteristics of alectropodes, that is to say, a sternum with very developed posterior notches and a thumb which is raised with respect to the other digits.

The Numididae have sometimes been classed as a subfamily of the Phasianidae due to their close resemblance; their immunological characteristics are especially similar (Mainardi 1959). The genus *Numida* is distinguishable from the other Numididae by the cephalic bump (hence the name helmeted guinea fowl) and the blue or red wattles.

The sexes are very similar and can in fact be distinguished only by some details of the barbules. Wild guinea-fowl live in groups and are monogamous by nature but, under intensive rearing, good reproductive performance can be obtained using 1 male for 3 or 4 females. Guinea-fowl subjected to normal seasonal variations in light and temperature lay their first egg at 18 weeks of age. Free-range birds can produce up to 100 eggs in their first laying season extending from April to September in the northern hemisphere. In the wild, the guinea-fowl lays about 15 eggs in a laying period lasting 2 months. Held in controlled environment buildings and using artificial insemination, sexual maturity can be postponed to as late as 32 weeks of age (in order to avoid the production of eggs too small to be incubated), and it is possible to obtain around 170 eggs in 43 consecutive weeks from which around 110 chicks will hatch. During the second year of lay one may expect a further 20–30 offspring.

The number of chromosomes in the guinea-fowl has not been clearly established; $2n$ lies between 74 and 80 according to Piccini and Stella (1970) and Denisova and Khrustalev (1974).

Numida meleagris produces fertile hybrids when crossed with *N. mitrata* and sterile hybrids (predominantly male) when crossed with *Acryllium vulturinum* (vulturine guinea-fowl), *Gallus gallus* and *G. domesticus*, and *Pavo cristatus* (peafowl). Hybrids have also been reported with *Meleagris gallopavo* (turkey) and *Phasianus colchicus* (pheasant) (Gray 1958).

The greatest number of studies on hybrids have been carried out in France (Petitjean 1969) and the Soviet Union (Gromov 1970) on the domestic fowl × guinea-fowl cross. These studies showed that the hybrids had a short life expectancy; the few females produced died within 48 days of hatching. Their conformation was intermediate between those of the parents. Their diploid chromosome number was 78 but they were sterile. The incubation period of eggs from guinea-fowl fertilized by domestic cock is 23.6 days against 26–27 days for eggs fertilized by the male guinea-fowl. Domestic fowl × guinea-fowl, turkey × guinea-fowl and peacock × guinea-fowl crosses were first reported by Cornevin (1895).

Nowadays, the guinea-fowl is distributed world-wide. In some European countries (France, Italy, the Soviet Union, Hungary) it is raised intensively in the same way as the chicken for egg and meat production. Europe dominates world production of guinea-fowls, 80 per cent of which are produced in France. In the United States, studies are being made with a view to establishing industrial production of this species. In Japan and Australia, planned production of guinea-fowl has just got under way (Smetana 1974).

Elsewhere its introduction has resulted either in a return to the wild state as a game bird (e.g. Jamaica, Central America and Malaysia), or in its establishment as a semi-domesticated species on family smallholdings. Guinea-fowls introduced to the West Indies in the 16th century and allowed to return to the wild state show feather colour variations which are nowadays maintained by selection, although these birds were probably not domesticated at the time of their introduction, despite the claim of Buffon (1749–88).

Domestication and early history

The relative importance and use of the guinea-fowl have varied throughout history and from civilization to civilization (Cauchard 1971). It appears as a sacred animal on Egyptian bas-reliefs. Its traces can be found in 5th century B.C. Greece where it was given the name *melanargis* (black-white). This name became changed to *meleagris* under the influence of mythology. Ovid recounts that Meleager's sisters wept unceasingly after his death until Artemis changed them into guinea-hens but their feathers were stained ever-more by their tears. It has been recorded that the guinea-fowl was offered as a sacrifice to Athena, but it was also a well-known if expensive dish. This was also true in Rome where, since the founding of the empire, its African origin was known: Varro called it 'the African chicken', Pliny 'the chicken of Numidia' and Columella identified two varieties, one with red and the other with blue wattles.

Guinea-fowl were known throughout the Roman empire but, for reasons which remain unclear, they seem to have disappeared completely in the Middle Ages with the possible exception of England, where 13th-century records refer to it as *avis africana*, and France where it was called *perdrix de Terre-Neuve* (Saint-Loup 1895). The name *poule de Guinée* appears to have been used first in 1555 by Belon (cited by Saint-Loup). This was the period when the Portuguese returning from their African explorations reintroduced guinea-fowl to Europe, the West Indies, Central America, India and Malaysia. They gave it the name *pintada* or painted chicken and this changed to *pintade* in French while the name 'guinea-fowl' stayed in English and *gallina de Guinea* in Spanish. Italians call it *faraone*, another reference to its African origin. In all cases the species referred to is *Numida meleagris*.

Recent history and present status

Guinea-fowl outside their native African surroundings have adapted perfectly to farms where they are reared to the present day. They live in free-ranging flocks keeping a degree of independence and nearly always laying far from the homestead. For about the last 20 years, European poultry farmers and breeders wishing to diversify their products have become interested in this species which lays well and is sought after for the quality of its meat which resembles that of game birds. Unfortunately the habits of the guinea-fowl are such that traditional techniques used in breeding are not suitable. To overcome this, Petitjean in 1963 perfected an artificial insemination technique requiring breeding females to be housed in individual cages. It turned out that the animal is well suited to this type of husbandry, both egg production and fertility showing improvement.

Nowadays the following characters are being selected for: body weight, carcass yield, laying intensity, fertility, hatchability and egg weight. New strains for egg and for meat production are being formed. In France and Italy industrial production is mainly directed towards meat production while the Soviet Union has concentrated on the production of eggs for human consumption.

The wild *Numida meleagris* has a dark grey feather with white spots (*perlé*). The spots are usually, but not invariably, large and round. In the domestic guinea-fowl the wild colour type is commonest but several others have been developed: violet, lilac, buff (*chamois*), white; all with or without spotting.

As for the genetics of plumage colour, according to Colonna Cesari (cited by Lamblard 1970), spotting (*perlé*) is due to a dominant gene *P*. The other colours are due to recessive genes which possibly act by diluting the black pigment of the grey variety to a greater (buff and isabelle) or less (lilac) extent. The gene for lilac is *l*, for buff (*chamois*) *c* and for isabelle (sex-linked) *is*. The grey colour (spotted or not) occurs when an individual carries at least one dominant allele at each of the three loci: lilac, buff and isabelle. The grey phenotype can therefore cover a variety of heterozygotic genotypes. The grey non-spotted variety is called 'violet'. The white guinea-fowl is apparently a true albino but the pale buff is sometimes wrongly called 'white'.

Although the feathers are not marketed, the colour of the plumage is nonetheless a character taken into account in selection programmes with the view to producing on the one hand homogeneous flocks and on the other hand autosexing varieties (Kuit 1974, Veitsman 1968). Recent studies have shown that differences in feather colour correspond to small differences in body weight or laying ability. Breeders may exploit this fact (Auxilia 1970, Voroshilova 1974).

The proportion of different colours varies between countries. Compared to the grey-spotted type there are more white guinea-fowl in the Soviet Union than in France and more lilacs in Italy than in the Soviet Union or France.

Future prospects

Breeders have been working on guinea-fowl only since the 1950s. It is surprising that this animal which has been farmed since early times in Africa has retained all the characteristics of its wild counterpart (feather morphology, social behaviour) when subjected to modern intensive rearing methods employing battery cages and artificial insemination.

The guinea-fowl is a species with potential. Its meat yield is higher than that of the chicken; its eggs have a similar composition and while they may be smaller, the shell is much stronger. The guinea-fowl is hardy and adapts easily to all types of climate. The new strains created in Europe show excellent production characteristics although they cost more than traditional poultry. The guinea-fowl has retained its fundamental characteristics. It has been observed that the domesticated guinea-fowl will, after a period of adaptation, reproduce with its wild conspecific and this is encouraging for other countries possessing either the domesticated or the wild species which they wish to exploit.

References

Auxilia, M. T. (1970) Attitudine delle faraone di razza grigia e paonata alla produzione delle uova. *Annali dell' Istituto Sperimentale per la Zootecnia*, **3**: 107–24

Buffon, G. L. (1749–1788) *Histoire Naturelle des Oiseaux*, Vol. 2. Imprimerie Royale: Paris

Cauchard, J. C. (1971) *La pintade*. Editions Henri Peladan: Uzes

Cornevin, C. (1895) *Traité de Zootechnie Spéciale: les oiseaux de Basse-cour*. Baillière: Paris

Denisova, R. A. and **Khrustalev, S. A.** (1974) [The normal karyotype of the guinea-fowl.] *Tsitologiya i Genetika*, **8**: 535–6, 569. (In Russian)

Gray, A. P. (1958) *Bird Hybrids*. Commonwealth Agricultural Bureaux: Farnham Royal, Bucks, England

Gromov, A. (1970) [Interspecies hybrids]. *Ptitsevodstvo*, **20** (3): 15. (In Russian)

Kuit, A. R. (1974) From France, 'guinea capital', perhaps a new industry here? *Broiler industry*, **37** (3): 32–6

Lamblard, J. M. (1970) La pintade dans le Gard. *Revue de l'Elevage*, édition spéciale: La production moderne des viandes de canard, dindon, oie, pintade, pigeon et caille, 35–40

Mainardi, D. (1959) Immunological distances among gallinaceous birds. *Nature*, **184**: 913–14

Petitjean, M. J. (1969) De quelques applications de l'insémination artificielle en aviculture. Essais d'hybridation interspécifique coq × pintade et pintade mâle × poule. *Revue de l'Elevage*, **24** (5): 123–31

Piccini, E. and **Stella, M.** (1970) Some avian karyograms. *Caryologia*, **23**: 189–202

Saint-Loup, R. (1895) *Les Oiseaux de Basse-cour*. Baillière: Paris

Smetana, P. (1974) Some aspects of guinea fowl production. *Proceedings 1974 Australasian Poultry Science Convention, Hobart*, 302–306

Veitsman, L. N. (1968) [Guinea-fowl in U.S.S.R.: the

biological basis of guinea-fowl rearing]. Thesis of the Academy of Sciences, Novosibirsk. (In Russian)

Voroshilova, S. (1974) [Growth and development of guinea-fowl broilers]. *Ptitsevodstvo*, **24** (7): 20. (In Russian)

47
Turkey

R. D. Crawford

Department of Animal and Poultry Science, University of Saskatchewan, Canada

Introduction

Nomenclature

Domesticated turkeys were derived from the native wild turkey of North America. Those taken to Europe at the time of the Spanish conquest were descendants of the Mexican turkey, *Meleagris gallopavo gallopavo*. It is quite possible that other subspecies from southwestern North America contributed to the domesticated stocks of that time. Beginning in the 16th century, turkeys were brought from Europe to settlements of eastern North America where there was frequent interbreeding with the wild Eastern subspecies, *M. g. silvestris*. According to Schorger (1966), Linnaeus utilised the binomial *M. gallopavo* for the domesticated bird in his tenth edition (1758) of the *Systema*. That binomial continues to be accepted since there is no doubt about the species of origin, and since at least two subspecies of it contributed to the modern domesticated turkey. Nevertheless, the binomial which was chosen is unfortunate since it creates confusion with other poultry species similarly named – guinea-fowl, peafowl, and chickens – to which the turkey is only distantly related. When first introduced to Europe, the turkey was frequently thought to be a guinea-fowl, and hence was termed *meleagris*; by others it was thought to be a cross between a rooster and a peafowl, and hence the term *gallopavo* (Schorger 1966).

Several possible derivations of the popular name are recorded in the literature. Perhaps the most plausible is that when the birds were first introduced to Europe anything foreign was said to be from Turkey, and this word eventually became associated with the species (Schorger 1966).

The English language words for describing turkeys are 'tom' or 'stag' or 'gobbler' (adult male), 'turkey hen' (adult female), and 'poult' (youngster). Other terms for certain age and sex categories are known but they are not in common use.

General biology

Turkeys are the largest domesticated gallinaceous bird and they have all the general characteristics of that Order. They also have several traits which distinguish them from others.

They are perhaps best known to the general public by the display of the male. It is similar to that of the peacock, and undoubtedly was the main cause of confusion between the two species when turkeys were first introduced to Europe. During the strut it is the wing and tail feathers which are most conspicuously displayed but nearly all feather tracts are involved. It is essentially a mating display, containing elements of aggressive behaviour.

In the mature male, a thick mass of fibrous tissue develops over the breast and crop; Audubon called it the breast sponge (Schorger 1966). Prior to the mating season in wild turkeys, it becomes filled with fat and serves as an energy reserve. The breast sponge is present in domesticated toms and results in some wastage of edible meat.

In both sexes, there is a fleshy protuberance at the dorsal base of the beak. It is popularly called a snood, and in technical literature it is called the frontal caruncle. It remains rudimentary in females, but in males during both aggressive and sexual excitement it can become rapidly expanded and lengthened to dangle over the side of the beak; retraction is also rapid. It is usually removed from industrial stocks to avoid fighting injury and disease entry; the operation is called 'desnooding'.

The head and neck of both sexes are relatively bare of feathers, and caruncles form on the lower neck, more noticeably in males. In the mature male, skin colour can change rapidly to blue, red, and white as part of behavioural displays. The literature seeking to explain the mechanism for colour change has been reviewed by Schorger (1966); the mechanism is not yet fully understood.

The turkey tom has a long hair-like appendage on the breast, commonly called a beard. It is usually absent in females. The beard is composed of primitive contour feathers which first appear at sexual maturity and are never moulted (Schorger 1966). It may have a role in mating behaviour, but its function, if any, is not clearly understood. The autosomal recessive white mutation which is used to provide white feathers in all industrial stocks does not act on the beard; it remains black.

Reproductive rate of wild turkeys is relatively low (Schorger 1966). They do not become sexually mature until the second year. A clutch normally contains about 15 eggs, and if the first nest is destroyed a second clutch may be laid. Nesting success is only about 35 per cent. Incubation time is 28 days.

Domesticated turkeys reproduce at a much higher rate. Under appropriate conditions they reach sexual maturity by about 30 weeks of age. There are no very good data available for unimproved stocks, but it is generally accepted that reproductive fitness has declined with selection for extremes of breast and thigh muscling. A reasonable expectation from turkey broiler breeder hens is 95 eggs yielding 65 poults in a period of 6 months. Hens are kept for only one breeding season.

Parthenogenesis occurs in certain domesticated stocks and it has received extensive study (Olsen 1974). The products (so-called 'parthenogens') are diploid, having the normal 80 chromosomes. Diploidy in unfertilized ova is restored by suppression of the second polar body. Selection for parthenogenetic development has been successful; 22.4 per cent of unfertilized eggs incubated have given rise to embryos and 3.2 per cent of the eggs hatched. All the 'parthenogens' are males. They can reach sexual maturity and sire normal offspring.

The relationship of turkeys to other gallinaceous birds is distant. Gray (1958) has reviewed the literature pertaining to hybridization. Fertilization, usually in artificial insemination, can be achieved with domestic fowl, peafowl, pheasant and guinea-fowl. Hybrids with domestic fowl die during incubation. Those with peafowl and guinea-fowl are said to reach adulthood.

Numbers and distribution

The 1981 FAO *Production Yearbook* (FAO 1982) provides data indicating a total of about 124 million turkeys in the world. These data can serve only as a guide; some reporting countries do not distinguish among poultry species and their numbers are included with data for chickens; numbers for other countries are estimated and may have considerable error.

Turkeys are a major poultry species in the Americas, Europe, and the Soviet Union. In Africa they are important only in South Africa. In Asia and Oceania they have only minor importance. Israel has the largest consumption of turkey meat per caput, followed by Canada and the United States.

Wild turkeys

According to Aldrich (1967), all wild and domesticated turkeys belong to one highly variable species, *M. gallopavo*. The only near relative is the ocellated turkey, *Agriocharis ocellata*, of tropical Mexico and Central America. Gray (1958) indicated that hybrids can be produced in reciprocal matings of the two species; the hybrid males are fertile; one hybrid female has been studied and she produced only soft-shelled eggs. Since ranges of the two species are far apart, and since the ocellated turkey was never domesticated, introgression is not likely to have occurred.

Six subspecies of wild turkeys are recognized (Aldrich 1967). There is no evidence that they occupied territory beyond that in which they have been found in historical times. Figure 47.1 from Aldrich (1967) indicates present and historical

ranges of these subspecies. It is assumed that they evolved from pheasant-like ancestors either in North America or in Asia. The species is known to have existed during the Pleistocene, between 15 000 and 50 000 years ago, in eastern and southwestern United States.

The six subspecies are the Eastern, Florida, Rio Grande, Merriam's, Gould's, and Mexican (Aldrich 1967). Populations in moist deciduous forest regions of the eastern United States are more deeply pigmented than those of the drier sections of the southwest; northern populations, and those living at high altitudes are larger and have shorter legs, feet and bills than those of southern areas and lower altitudes. Where ranges adjoin, distinction between subspecies is not clearly marked. The Eastern subspecies has brown tips to the tail feathers and upper tail coverts, and a bronze tone to the body feathers. Those of the southwest of the continent have very pale tips on their tail feathers and upper tail coverts so that they appear to be white-rumped; their body plumage has a greenish tone so that they appear blacker than the Eastern form.

There have been no planned studies of hybridization among the subspecies and with domesticated turkeys. It is not likely that there are any serious barriers to hybridization since faulty reproduction would have been noticed by poultrymen and recorded in the literature. No studies of chromosome complement in wild turkeys have been reported.

Wild turkeys are a popular game bird in North America. Natural populations have persisted in some localities and introductions outside traditional ranges have sometimes been successful. Birds are propagated at game farms for release, and hobbyists raise them too. Care has not always been taken to keep subspecies distinct during captive propagation, and reintroductions have not always been with subspecies originally native to the area. Hence it might be difficult to find pure populations of each subspecies in the wild. Schorger (1966) observed that it is doubtful if many wild birds east of the Mississippi are free of admixture with the domesticated form. Aldrich (1967) believes that the Mexican subspecies, ancestor of the first domesticated turkeys sent to Europe, may now be an endangered bird since its known distribution in southwestern Mexico has been greatly reduced.

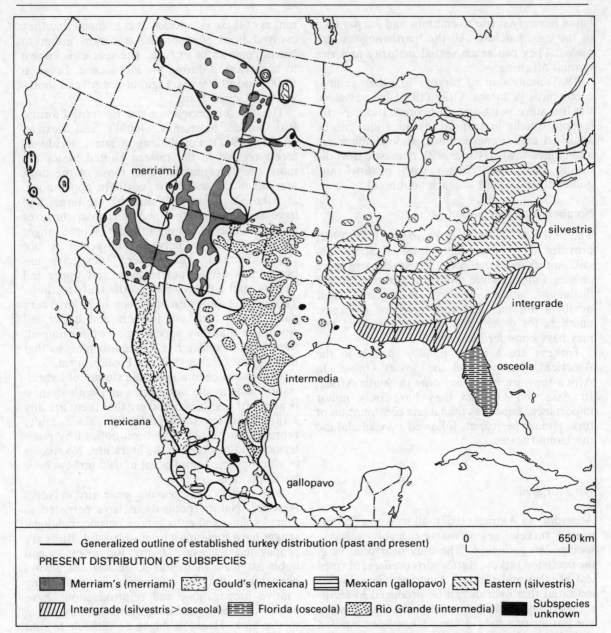

Fig. 47.1 Distribution of the wild turkey in North America. (From Aldrich 1967)

Domestication and early history

Domestication in Mexico and Central America

Schorger (1966) has published an extensive historical review concerning the wild turkey and its domestication.

Very little is known about first domestication of turkeys in Mexico. Schorger mentions only that bones of the domestic bird from the Palo Blanco period, 200 B.C.–A.D. 700, have been found in the valley of Tehuacan which is in the present-day Mexican state of Puebla. By the 16th century, it is known that they were being raised throughout central and southern Mexico, and in Guatemala, Honduras, Nicaragua, Costa Rica, and some of the Caribbean islands. Since the range of wild turkeys did not extend south of the Rio Balsas valley near present-day Mexico City (Fig. 47.1), domesticated birds kept to the south must have been acquired through trade. All turkey bones found in Yucatan archaeological sites have been of the ocellated species, and since domesticated turkeys were far less numerous there than in the Mexican domain, Schorger suggests that they were not being raised in Yucatan until shortly before the arrival of the Spaniards.

Frequent reference is made to turkeys in Spanish literature of the 16th century. It is not always clear whether reference is to wild or to domesticated turkeys, and there is apparent confusion with other native species such as curassows and chachalacas. Columbus may have been the first European to see turkeys when he landed on the coast of Honduras on 14 August 1502; he was given food by the natives which included *gallinas de la tierra*. There may have been an earlier discovery. Schorger records that Vicente Yáñez Pinzón was at the Gulf of Paria in Venezuela in 1500 and was there given birds for consumption and some to transport to Spain; the birds were probably turkeys. Similarly, Pedro Alonso Niño is recorded as discovering the turkey on the coast of Cumana, Tierra Firme, in 1499 and of taking it to Europe in 1500. Schorger draws attention to recently discovered documents from 1511 and 1512 which clearly refer to transport of turkeys to Spain. The first, dated 24 October 1511 and signed by the Bishop of Valencia, was an order for each ship from the Islands and Tierra Firme to bring to Seville ten turkeys, half males and half females, for breeding. The second, dated 30 September 1512 from the king, refers to two turkeys which had arrived in Spain on a ship from Hispaniola. Date of first discovery and of importation to Spain may remain uncertain. But by 1530, turkeys were apparently firmly established on farms in Spain.

Cortés began the exploration of Mexico in 1519 where he found domesticated turkeys in abundance. For instance, Montezuma required large numbers to support his household and his menagerie; it is recorded that a thousand were needed daily. Turkeys were the most conspicuous and cheapest source of meat at the time. Spanish livestock were not yet prevalent. In a manuscript completed about 1570, Sahagún noted colour variants among the domesticated birds; the wording of Schorger's translation is important to

geneticists since it indicates the presence of self-colours rather than colour variations within a plumage: '. . . (the birds) . . . were of various colors; some white, others red, others black, and others brown'.

All authors indicate that the turkeys taken to Europe by the Spaniards were domesticated from the wild Mexican subspecies. However, there is a strong possibility that other subspecies contributed also. Schorger's map (1966: 6) of locations where turkeys were raised in the 16th century includes large areas of the range of Gould's and Rio Grande turkeys, in addition to that of the Mexican subspecies (Fig. 47.l). He says that although races of wild turkeys with white rumps can develop as much wildness as the Eastern form, they appear to be more easily domesticated. He provides reference to a domestication of Rio Grande turkeys and to easy taming of Gould's.

Domesticated turkeys in southwestern United States

Schorger (1966) has reviewed the extensive literature concerning domestication of Merriam's turkey by the agricultural Pueblo people of the southwestern United States. Because of the great geographic distance between this culture and that of central Mexico, Aldrich (1967) doubts that domesticated birds or the idea of domestication were exchanged between the two. Hence the Pueblo domestication must have been a separate one. There is no evidence that their birds have contributed to modern domesticated turkeys. It would be of interest to learn whether descendants of the Pueblo people still possess turkeys related to their ancient stock.

Archaeological findings indicate that turkeys were being kept in captivity by Pueblo people about A.D. 500–700 and that the birds were probably domesticated. There is no doubt about domestication by A.D. 700–900. Most evidence indicates that the birds were confined in corrals within dwellings rather than being allowed to roam freely. Remains of these corrals have been described as containing vast quantities of turkey manure, eggshell fragments, and turkey bones. The Pueblo people made extensive use of turkey feathers for ornament and in weaving. Spurs were used by them and by other native peoples as arrowheads, and feathers were used in arrows as well. Bones were used in a variety of tools. But, at least initially, turkeys as a food source were of minor importance.

Pinkley (1965) has provided an anecdote concerning wild Merriam's turkeys becoming habituated to man and developing into a serious nuisance around settlements in present-day national parks. It is suggested that the same may have occurred around pueblos. Confinement of birds might have been necessary to avoid damage to crops, resulting in eventual domestication.

Introduction to Europe

Following Spanish discovery and introduction to Spain, turkeys were distributed throughout Europe with extreme rapidity. Schorger (1966) has reviewed the literature on first introductions. A pair of turkeys was sent from Hispaniola to Rome soon after 1520; the female was white. Turkeys were present in France by 1538 and were well known by 1552. The first concrete date for their presence in England is 1541; although there are many references from earlier dates, these could refer to other species. It is generally stated that turkeys were present in Norway and Denmark by the middle of the 16th century. They are listed in an inventory of the king's farm in Sweden in 1556, and were being raised on private farms by 1600. They were not known in Germany before 1530 but were used as food at a wedding banquet in 1560. Since the reproductive rate of turkeys is not especially high, and since birds of that time were not sexually mature until their second year, probably because of inadequate husbandry, an explanation must be found for the speed with which they became a common kind of poultry. Although historians have not addressed this question, a search of the records might well indicate that there were massive shipments of live birds from the Americas in the years following first discovery of the domesticated species, which would rapidly increase the European population.

Pre-Columbian turkeys in Europe

There is also the possibility that turkeys were already present in Europe and Asia before the Spanish conquest of the Americas. Schorger (1966) has reviewed the controversy concerning discovery of turkeys depicted on the frieze of a mural in Schleswig cathedral, supposedly painted in 1280. It has been suggested that Norse

voyagers may have brought turkeys to northern Europe from North America. The mural was restored in 1890 or 1891 and it is now believed that the restoring artist added the turkeys; they were not part of the original 1280 mural.

Bökönyi and Jánossy (1958) have reported the finding of signet rings in Hungarian graves of the 10th to the 13th centuries A.D. which bear images of birds. Two of these they interpret to be turkeys. The one depicted in Fig. 21 of their original paper bears a resemblance to a turkey, but there are stylized lines above the head which could indicate a crest as in peafowl. Their Fig. 22 is very stylized and not very turkey-like but there is an apparent appendage from the end of the beak which could represent a snood. They also report the finding of a tarsometatarsus in excavation of a 14th century stratum of the Royal Castle of Buda, and provide a diagram of it; they believe it is a turkey bone. Schorger (1966) has had the bone examined and concluded that it is from a peafowl.

In a discussion of evidence for the presence of pre-Columbian chickens in America, Carter (1971) mentions intended publication of evidence for introduction of the turkey to Europe prior to the Spanish conquest. In a personal communication (1975) he indicated that his evidence was linguistic. If chickens reached the Americas from southeast Asia before the 16th century by crossing the Pacific, it might have been possible for turkeys to be transported the other way across the Pacific to reach Asia and later Europe before the Spaniards had brought them there. However, there is still no firm evidence that this actually occurred.

Domesticated turkeys in eastern North America

When Europeans began colonizing the east coast of North America they brought turkeys with them. The stock was descended from that which had been taken to Europe from Mexico in the 16th century. Most reports of the time indicate that these birds were much smaller than the local Eastern subspecies of wild turkey; they were much blacker in appearance and had white tips to the tail feathers. Turkeys were taken to Jamestown in Virginia in 1607 or soon thereafter; they were brought to Massachusetts in 1629; by 1698 they were plentiful in Pennsylvania and New Jersey; the French at Quebec city had turkeys as early as 1647 (Schorger 1966).

It is generally agreed that the Eastern wild turkey had not been domesticated by the native people, but it was a common practice to raise captured poults (Schorger 1966). Attempts were later made to domesticate the subspecies but most of these ended in failure because of the extreme wildness of the birds.

It is also generally agreed that there was extensive crossing of domesticated and wild stocks. Crosses entered the wild population, and there was frequent crossing into farm flocks, from both captive wild turkeys and from those running free. The crosses were much larger and more vigorous than the domesticated parent, indicating heterosis.

The hybrid between domesticated Mexican and wild Eastern turkeys had the wild-type plumage pattern of both, but acquired the bronze tones of the Eastern subspecies' plumage. It was much larger and much more vigorous than the Mexican stock. Under the name of American Bronze it was exported to Europe, particularly to Great Britain where it became very popular. It also became the stock used commercially throughout North America.

Recent history and present status

Small (1974) has provided a detailed account of the development of the turkey industry in the United States. Development has followed a similar pathway in other countries. Turkey has great importance as a food in the United States, mostly because of its association with the Thanksgiving harvest festival. According to folklore, wild turkey was served as the main entrée during the three-day Thanksgiving feast of the Pilgrims and Indians in November of 1621. Thanksgiving has become a major American festival, celebrated on the fourth Thursday in November, and turkey is the traditional food. A similar feast is held in Canada on the second Monday of October, also with turkey as the featured food, but the occasion has much less social importance there than in the United States.

Until the end of the Second World War, turkeys were kept in traditional ways. Reproduction was seasonal, using both natural and artificial incubation. Poults were reared outdoors, usually wandering freely. They were fattened for market in late autumn or early winter. A number of colour varieties were recognized. But all stocks were much alike – larger and meatier than the wild form, but otherwise unchanged.

Since the mid-1940s, a highly efficient turkey industry has evolved. Birds are raised under industrial confinement conditions throughout the year. Two main market products are derived – a small turkey, called a broiler in North America, for home consumption, and a heavy turkey for institutional use and further processing into convenience foods. Breeding is in the hands of a very few multinational primary breeders whose stock is distributed to industrial growers throughout the world. Most market birds are a three-way cross, propagated by artificial insemination. Nearly all of them are white-feathered. Growth rate and feed efficiency are high, making the turkey one of the cheapest sources of meat in most countries. Carcasses are eviscerated and may be sold fresh or frozen, whole or in portions. The events – social, political, economic, and genetic – leading to this efficient food-production system have been reviewed by Small (1974).

Intense selection for weight and conformation have caused major changes in the domesticated turkey. In only a few decades, they have resulted in more rapid evolution than that which had occurred previously in the several centuries following domestication.

Development of Broad-Breasted Bronze

Probably the most important genetic event was the development of turkeys with greatly hypertrophied breast and thigh muscles. According to Davey (1974), the first improvement should be credited to Mr Jesse Throssel who had emigrated from England to Canada in 1926. In England, he had been breeding turkeys for meat quality, hatchability and early maturity. He imported one tom and two hens of this stock in 1927 and within a year had obtained 75 progeny from them. Small (1974) stated that some of these birds were exhibited in the United States where they created great excitement. They were very large and had exceptionally wide breasts; toms reached weights of 18 kg and hens of 13 kg at 9 months of age. Toms were bought from Mr Throssel for use in American breeding flocks, producing a hybrid progeny superior to either parent. Within a very few years, virtually all North American breeding stocks were changed from the narrow-breasted wild-type to the new broad-breasted kind, either through mass selection within a population or through crossing with the English–Canadian stock. Heavy muscling has been retained and increased in commercial turkeys, vastly improving their meat yield. Considerable publicity has attended recent efforts of breeders and research workers who are attempting now to produce a 45-kg turkey.

Development of the Beltsville Small White

About the same time as commercial turkeys were being selected for increased muscling, the United States Department of Agriculture began developing a new breed to meet consumer demand for a relatively small turkey having good meat yield (Marsden 1967). The objectives were white plumage, small market size, early market age, heavy muscling, and good reproductive ability without having to use artificial insemination. First matings were made in 1935. Stock was released to other institutions in 1941 and to private breeders in 1945. It was recognized as a new standard variety in 1951.

The new variety had a very broad genetic base. Hatching eggs had been obtained from one source of Narragansetts, three of Bronze, six of White Hollands, one of White Austrian turkeys from Scotland, and from three sources of wild turkeys. Marsden (1967) has provided details of the mating plan used in combining these stocks, and of the selection procedures utilized in developing the new variety.

Beltsville Small Whites became very popular in the United States and throughout the world. At the peak of their popularity in 1954, Marsden (1967) estimated that 19 million were being raised in the United States.

According to Small (1974), it is unlikely that any purebred Beltsville Small Whites remain in existence. However, they made a tremendous contribution in developing the small strains used to produce the broiler turkey of the modern industry, both by contributing genetic material and by providing a model for selective breeding.

Decline of reproductive ability

Although poorly documented in the research literature, it is now generally recognized that there is a negative correlation between changes in the growth rate and body weight complex and reproductive fitness. This applies to turkeys where increases in size and muscling have been accompanied by a pronounced decline in reproductive abilities.

Fertility from natural mating has been worst affected. As conformation, particularly of males, has changed through selection, coordination has been affected and males have become less able to complete copulation successfully. In the normal mating behaviour sequence (Hale *et al.* 1969), receptivity of the female terminates when the vagina has been everted; if not inseminated, she will remain infertile for a variable number of days until she again becomes receptive to the male. As a result, fertility from natural mating in commercial breeding flocks has declined so drastically that it has become necessary to use artificial insemination. The practice of artificial insemination is now universal in the breeding of industrial stocks, and natural mating ability is no longer included in selection programmes for sire lines.

Concurrently, there has been a decline in fertilizing ability of semen, in numbers of eggs per female, and in hatchability of eggs. The extent of these declines has been difficult to measure since 'unimproved' stocks have not been maintained as controls of populations under selection for improved meat production performance. These problems have been somewhat resolved by adopting a crossbreeding programme in which sire lines are selected primarily for growth traits and dam lines are selected primarily for reproductive ability. The commercial sires are usually a pure strain and there is routine individual selection of toms for semen quality prior to the breeding season. Commercial dams are usually a two-strain cross having superior egg numbers and hatchability of eggs augmented by hybrid vigour.

Reproductive performance is improving, but it is still rather poor. A reasonable expectation from turkey broiler breeding stocks is 95 eggs and 65 poults per hen in a production period of 6 months' duration. From heavy turkey stocks, 85 eggs yielding 55 poults per hen should be obtained in a breeding season of 5 months.

Utilization of single-gene effects

Single genes with major effects on morphology or colour are used extensively in commercial breeding of chickens. More use of such genes will probably be made in turkey breeding in the future.

Commercial turkeys are now almost universally white-plumaged, due to action of the autosomal recessive white gene. White turkeys were known in early Mexico and the same mutant may have been involved.

Although genetic capability for autosexing using colour genes exists (Somes 1977), it has not yet been utilized by commercial breeders. The autosexing technique currently has immense importance in chicken breeding.

Similarly, turkey breeders have not made use of genetically determined white and yellow skin, a trait of considerable importance in broiler chicken breeding. Most turkeys, including the wild-type, are white-skinned. A yellow-skinned condition apparently exists in a French breed called Noir de Sologne but its genetics have not yet been studied.

Unimproved stocks

All industrial stocks are white-feathered, broad-breasted, and are reproduced using artificial insemination. Some stocks remain in use in developing countries, and as farm-flock stocks in developed countries, retaining natural mating ability, coloured feathers, and a more normal body conformation. A recent survey of Canada's poultry genetic resources (Crawford, unpublished) revealed that only one such stock remains available in Canada. A French breeding company (ADETEF 1980) has developed a strain of traditional farm turkeys which is popular on the French market and which has been exported to a number of developing countries including Mexico, the original home of the domesticated turkey. Undoubtedly, others exist as indigenous stocks elsewhere but little is known about them.

In countries where poultry hobbyism is popular, some fancy stocks of turkeys are bred. Most of these retain wild-type conformation and size, and are distinguished by a variety of plumage colours and patterns.

Future prospects

There is cause for concern about future use of turkeys as a food product, particularly with regard to loss and exhaustion of genetic variability.

Only a few multinational breeding companies supply the world market with industrial stocks. One North American company is immensely big and appears to have no serious competition. There is a strong possibility that a monotypic population may develop which could be vulnerable to genetic disaster. This topic has received intensive study in crop plants as a result of the severe maize blight which occurred in the United States in 1970 (NAC 1972), and it is considered by plant breeders to be a problem of major proportions. The topic has not yet become of concern to animal breeders, but it should be considered particularly with regard to industrial poultry breeding where monotypic populations are developing.

Concern should also be expressed about turkeys reaching the limits of biological fitness as they are selected for extremes of body size and muscling. Reproductive fitness is seriously impaired now and artificial insemination is required. It is likely that breeding stocks will soon be housed in cages (if enough genetic variability remains to fit them to that environment), where leg strength will have less importance than it does now in a floor-pen situation. The changed environment may permit selection for even greater weights and muscling which could cause further decline in reproductive fitness. Unimproved stocks are so vastly different in growth performance from current industrial stocks that breeders have little interest in reverting to their use.

The need for conservation of genetic variability is perhaps more critical in this species than it is even for domestic fowl, and is far more urgent than for most domesticated mammalian species. A partial list of stocks in Canada and the United States has been compiled (Somes 1977), but elsewhere there has been little activity and the presumed existence of indigenous stocks in developing countries, especially in Central and South America, has had no documentation.

References

ADETEF (1980) La société Bétina. *Bulletin de l'Élevage Français*, No. 14: 13–15

Aldrich, J. W. (1967) Historical background, and Taxonomy, distribution and present status. In: *The*

Wild Turkey and Its Management, ed. O. H. Hewitt, Ch. 1–2. The Wildlife Society: Washington, D. C.

Bökönyi, S. and **Jánossy, D.** (1958) Data about the occurrence of the turkey in Europe before the time of Columbus. *Aquila*, **65**: 265–9

Carter, G. F. (1971) Pre-Columbian chickens in America. In: *Man Across The Sea – problems of pre-Columbian contacts*, ed. C. L. Riley, J. C. Kelley, C. W. Pennington and R. L. Rands, Ch. 9. University of Texas Press: Austin, Texas

Davey, A. D. (1974) Canada. In: *American Poultry History 1823–1973*, ed. J. L. Skinner, Ch. 17. American Printing and Publishing: Madison, Wisconsin

FAO (1982) *FAO Production Yearbook* 1981, Vol. 35. FAO: Rome

Gray, A. P. (1958) *Bird Hybrids. A check-list with bibliography*. Commonwealth Agricultural Bureaux: Farnham Royal, Bucks, England

Hale, E. B., Schleidt, W. M. and **Schein, M. W.** (1969) The behaviour of turkeys. In: *The Behaviour of Domestic Animals*, ed. E. S. E. Hafez, (2nd edn), Ch. 17. Ballière, Tindall and Cassell: London

Marsden, S. J. (1967) The Beltsville Small White turkey. *World's Poultry Science Journal*, **23**: 32–42

NAC (1972) *Genetic Vulnerability of Major Crops*. National Academy of Sciences: Washington, D.C.

Olsen, M. W. (1974) Frequency and cytological aspects of diploid parthenogenesis in turkey eggs. *Theoretical and Applied Genetics*, **44**: 216–21

Pinkley, J. M. (1965) The pueblos and the turkey: who domesticated whom? *American Antiquity*, **31** (2) Part 2: 70–2

Schorger, A. W. (1966) *The Wild Turkey: its history and domestication*. University of Oklahoma Press: Norman, Oklahoma

Small, M. C. (1974) Turkeys. In: *American Poultry History 1823–1973*, ed. J. L. Skinner, Ch. 12. American Printing and Publishing: Madison, Wisconsin

Somes, R. G. Jr (1977) *Registry of Poultry Genetic Stocks*. Bulletin 446, Storrs Agricultural Experiment Station: University of Connecticut, Storrs, Connecticut

48

Common duck

G. A. Clayton

Formerly Department of Genetics, University of Edinburgh, Scotland

Introduction

Delacour (1959, 1964) considers that the domestic duck derives from the green-headed mallard, *Anas platyrhynchos platyrhynchos*, widely distributed throughout the northern hemisphere. The mallards are included among the dabbling ducks or surface feeders in the tribe Anatini in the subfamily Anatinae of the Anatidae the enormous family which includes most waterfowl.

General biology

Domestic ducks are much more variable in colour, size and gait than the wild mallard. Farmyard ducks generally appear coarser and less tightly feathered than their wild relatives and in most instances have lost the ability to fly. Most domestic ducks are much larger than the wild type. The majority of breeds retain a characteristic horizontal posture and its associated waddling walk, but some have evolved a stance which is practically erect which permits them to walk and run rapidly on land with apparently greater ease, hence the name 'runner'. The anatomical changes associated with this altered gait still seem to be the subject of conjecture as are the selection pressures which have brought it about (Clayton 1972).

In the wild, mallard tend to be monogamous and to form pair bonds during the breeding season. After the duck has laid her clutch of 10–12 eggs and begins to brood she is abandoned by the drake. The eggs take 28 days to incubate and the brooding and rearing of the young is performed solely by the female. Under domestic conditions much of this behaviour is lost and duck breeders do not need to keep more than 1 drake per 4–6 females to maintain a high level of fertility in the eggs produced.

Duck are indeterminate layers. This means that if eggs are removed from the nest or even if a young brood is destroyed, the duck has the ability to lay more to replenish the clutch or raise another brood. It follows that clutch size of itself does not indicate the total number of eggs that a duck is capable of producing. Many indeterminate layers in the wild are capable of much higher egg production than might be assumed from a knowledge of the naturally occurring clutch size, and Delacour (1964) gives a number of examples in ducks.

Ducks, like other birds, respond sexually to changes in day length, their gonads maturing in the spring and regressing in the autumn. Using artificial light ducks may be brought into lay at any season of the year. By maintaining or extending day length, egg-laying, which would otherwise cease, is prolonged and may be greatly increased. In the wild the mallard starts breeding approximately 12 months after hatching but well-grown domestic ducks may be brought into full production by 6 months of age if exposure to light is regulated appropriately.

Like all birds the duck has a great many chromosomes, most of them small. Bloom (1969) gives the modal diploid ($2n$) number as 78 and Mott *et al.* (1968) have established the usual avian sex-determining mechanism, i.e. Z/Z (males) and Z/W (females). The latter authors have demonstrated differences in chromosome morphology between the common duck and the muscovy from a study of hybrids between the two species.

Geographical distribution

The domestic duck is distributed throughout the world. It has its greatest economic importance in southeast Asia as a source of meat and eggs. Under peasant conditions, particularly if there is much water about, the duck appears to be more disease-resistant, to forage more successfully and to produce more eggs than the chicken. About 75 per cent of all domestic ducks are to be found in south and east Asia, the principal countries being Vietnam, Indonesia, Thailand, China, Bangladesh, the Philippines and Burma. Ducks would and do flourish in the temperate zones of the world but are of comparatively minor economic importance compared with the domestic chicken.

Wild species

Delacour (1956) lists thirty-eight species in the genus *Anas*. Grey (1958) records forty-eight interspecific or intergeneric crosses involving the mallard, of which six produce fertile offspring, a further eight offspring of doubtful fertility and the remaining thirty-four produce many instances of viable but sterile hybrids. In view of the ease of domestication of the mallard and some related species and their ability to hybridize it would be surprising if some of the latter had not made a genetic contribution to the domestic stock.

Despite a considerable range in size, colour and appearance all breeds of domestic duck interbreed freely. The ease with which mallard, hatched from eggs taken from the wild, may be reared in captivity is well known. There can be little doubt that these birds have been domesticated on many occasions and that gene exchange between the wild and domestic ducks has occurred and is still occurring.

Domestication and early history

The history of the domestic duck is obscure but, like the chicken, the range of types emanating from the Far East points to southeast Asia as a major centre of domestication. Zeuner (1963) summarizes the limited amount of information available and its very paucity suggests that the duck was of minor economic importance in Europe and the Middle East in early historical and prehistoric times. Although the Egyptians, Greeks and Romans fattened wild duck in captivity the earliest reference to domestic ducks in Europe that Delacour (1964) can find is as recent as the 12th century A.D. Delacour's suggestion that ducks were not domesticated in Europe before the Middle Ages is given strong support by Harper (1972) who undertook a survey of archival and classical literature searching for evidence on this question.

According to Yeh (1980) ducks have been domesticated in China for at least 3000 years. Pottery models of ducks and geese dating to about 2500 B.C. have been found in China suggesting the possibility of an even longer history of domestication (Watson 1969), but more archaeological evidence is badly needed.

Unlike the chicken, whose selection and dispersion by man in earlier times owes a great deal to its aggressive nature and spectacular manner of fighting, the placid duck is an unsuitable vehicle for man's bloodthirsty and gaming instincts. From the outset the role of the humble duck must have been to supply eggs and meat, a function it is admirably fitted to perform in the hot wet environment of the paddy fields, rivers and canals of southeast Asia.

The domestic duck differs in at least one important respect from its progenitor, the wild mallard. In almost all cases it has lost the ability to fly. The wild mallard is a powerful flier capable of instant flight from water or land; it can achieve high speeds and can cover immense distances. It is an interesting comment on the power of selection that an attribute such as flight, vital to survival in the wild, should be lost so readily. As Darwin and others have noted, domesticated mallard increase appreciably in size in a few generations and soon lose the ability to fly. Body size is a trait which responds readily and rapidly to selection and, in the environment of the farmyard, larger ducks would have an advantage in competition for food and are probably more prolific. Such birds would leave more descendants than smaller ones resulting in a genetic trend towards increasing size, unless or until countervailing pressures stopped the process. Increased size hampering flying ability would be strongly selected against in the wild but the farmyard, offering protection against predators and a regular supply of food, eliminates flying as essential to survival.

Most early poultry books written in English include passing references to ducks and, without exception, refer to their great prolificacy. Where

it is mentioned, a breed called the Indian Runner is singled out for its exceptionally high fecundity. The Indian Runner, so-called because of its considerable running ability, gave rise to strains called Penguins, a term which aptly describes the upright stance characteristic of the Runner but developed by fanciers to an exaggerated degree in the case of the former. The Indian Runner was imported to England in the 19th century where it made a major contribution to egg-laying breeds, notably the Khaki Campbell. The latter owes its name to a Mrs Campbell who in the closing years of the 19th century kept a small flock of about 12–15 ducks. She managed to acquire a single Indian Runner female that had reputedly laid a large number of eggs and crossed it with a Rouen, a large meat-type breed. At some early stage Mrs Campbell introduced a wild mallard into the cross and in a few generations extracted the intermediate-sized dark-fawn khaki-coloured duck which now bears her name. The Khaki Campbell soon acquired a reputation for prolificacy confirmed by Jansen, a Dutch breeder who, in 1923 in a flock of recently imported Khaki Campbells, recorded an average of 240 eggs per duck in a laying year (Clayton 1972). The point of this digression is to establish the fact that ducks were capable of extremely high levels of egg production before the application of modern selection and breeding techniques, a fact which is amply confirmed by reports published by the Food and Agriculture Organization of the United Nations and by various institutes of which that by Kingston *et al.* (1979) is a good example. The latter report 245 eggs per duck housed from a flock of 305 in 360 days and they provide circumstantial evidence that this figure is conservative. The point to note is that these performances were recorded on native ducks, the Alabio duck of Indonesia in the case of the example given, and their qualities are entirely attributable to their history in the hands of the peasants of southeast Asia and owe nothing to Western genetics and technology.

The techniques, practised to this day in southeast Asia, for the artificial incubation of eggs are remarkably simple and effective and may have developed well before the last 2000 years over which the Chinese are known to have practised the art. If artificial means were used to incubate and rear progeny the duck would be freed from the constraint, otherwise indispensable to the survival of the species, to brood and raise the next generation. Under such circumstances there would be strong, if unconscious, selection pressure in favour of those birds which laid the most eggs and this would certainly involve selection against the brooding tendency which causes laying to cease. It is noticeable in most domestic ducks, particularly the egg-laying strains, that the instinct to brood has virtually disappeared and that these birds, if not confined, will lay eggs wherever they happen to be. That this trait is commented upon in poultry books at the beginning of the 20th century indicates that loss of the nesting and brooding instinct is not a recent phenomenon.

Droving has been a part of traditional duck husbandry in the Far East for centuries if not millennia. The ducks, foraging in paddy fields as they go, are driven slowly along by their owner on their way to market, a journey which might cover several hundred miles and take 6 months to complete. This form of husbandry seems the most likely explanation for the considerable walking and running ability of the Indian Runner, and the native ducks of the Far East which resemble it. This mobility is conferred by a more upright posture which enables the animal to walk rather than waddle. Ducks leading this nomadic existence and capable of laying eggs as they travelled could have contributed their genes to the next generation and this might be part of the explanation for the carelessness these animals display in their egg-laying habits.

Natural selection over the course of evolution has endowed the ducks with the ability to produce more eggs when circumstances warrant, a property shared by many ground-nesting birds, which man has been able to turn to his great advantage. By freeing the duck from the constraints of nesting or brooding and from predation, starvation or seasonal changes in day length, man has allowed ducks to express a variability previously forbidden by natural selection. If we add to this the husbandry methods of the peasants of the Orient we can account for much of the variety and prolificacy of our modern breeds.

Recent history and present status

Fully intensive production of ducks has only been achieved in the last 20 years. Ducks have been produced in large numbers under semi-intensive conditions in many countries, Holland, England (Norfolk) and the United States (Long Island) being good examples. Fully intensive production can only be achieved in a completely controlled environment wherein all phases of the life-cycle may be regulated at will. In most climates this can only be accomplished in specially designed buildings where factors such as temperature, ventilation and day length are controllable. The older or semi-intensive methods of production have used outside facilities for some or all phases of reproduction and rearing and this is relatively costly in terms of labour, food wastage, weather hazards, predation and the impossibility of precise regulation of reproduction in the absence of light control. The peculiarities of the duck have made it much more difficult to adapt to fully intensive systems than, for instance, the chicken, but recent technical developments have now made it possible to produce ducks to prearranged schedules on an industrial scale. One such enterprise in England hatches 6 million–7 million ducklings annually. Technical advances of this nature are important in ensuring a future for the duck industry.

Most of the developments towards intensive production have taken place in Europe and North America and have been directed towards the production of meat or table ducks. Interest in ducks as egg producers has waned in the Western countries in the face of fierce competition from the chicken-egg industry. Until now the duck has not been exposed to the breeding and selection techniques practised over the last 50 years on an unprecedented scale on the chicken. Had this been the case there is little doubt that some hyper-efficient egg-laying strains of ducks would have been evolved. Egg-laying strains of ducks capable of averaging 300 eggs in a laying year currently exist and may be capable of further improvement in efficiency. It is possible that egg-layers of this kind may have a useful contribution to make in those parts of the world where chickens encounter serious disease problems.

The duck as a meat producer has been the subject of intense development, particularly in England, over the last 20 years. The most noticeable change has been the improvement in growth rate reflected in live weights of 3.5 kg, sexes combined, attainable in 7 weeks in the case of some of the larger strains. Twenty years ago the time required would have been 9 weeks and the time saved substantially reduces production costs. Differences between the sexes in body size are of considerable commercial interest because of the effect on size variation of carcasses produced. At about 8 weeks of age the female is about 6–7 per cent smaller than the male (this ratio does not change much at later ages) in the case of the domestic duck but in the muscovy the difference is much greater, approximately 35–40 per cent. This sexual dimorphism is one of the striking differences between these two species of duck.

The domestic duck exhibits many variants in shape, size and colour, which fact suggests a long history of domestication. Many of these varieties have been given the status of breeds, a somewhat ambiguous term referring mainly to morphological characteristics which a set of individuals have in common. Delacour (1959) lists seventeen separate breeds but he appears to have confined himself to Europe and North America. The countries of Asia all have their own names for local races as well qualified for breed status as are the Pekins, Aylesburys, Rouens, Khaki Campbells, etc., of Europe.

Domestic ducks range in adult body size from the diminutive Call weighing less than 1 kg to the largest meat strains of Pekin and Aylesbury origin weighing as much as 6 kg. It should be noted that the ancestral form is unlikely to exceed 1.5 kg and almost all the domestic varieties are larger. It is only the small varieties of domesticated duck like the Call and East India, another small but dark-coloured breed, which fly well. These small breeds are kept for ornamental or decoy purposes and have been selected for small size by breeders.

Breeds of domestic ducks exist in a wide range of colours. Solid white ducks of which the Call, Pekin and Aylesbury are examples are homozygous for recessive white. If any of these breeds is crossed with a coloured breed the first-generation progeny are coloured, but the latter if mated *inter se* produce progeny approximately 25 per cent of which are white. By backcrossing these white segregants to the coloured grandparental breed and repeating the process a number of times, a strain may be produced virtually identical with the original coloured strain but solid white in colour. This process may be the explanation for white varieties of well-known coloured breeds, e.g. the White Campbell, a derivative of the famous Khaki Campbell.

Future prospects

Duck eggs have not made a major contribution to human diet except in some countries of south-east Asia. Although the duck is very prolific it has a large appetite and, because food is the major element in production cost, is unlikely to become more efficient than specialized egg-laying strains of hen as far as cost per unit weight of egg is concerned. The duck cannot be kept permanently on wire floors because its feet are unable to stand the strain, whereas the chicken manages to survive and produce in wire-floored battery cages which lend themselves to large-scale factory production. Factors such as these make it unlikely that the duck will replace the hen as a source of eggs for human consumption except possibly in hot wet areas with poor hygienic conditions where the duck's superior viability gives it an advantage.

In considering the relative advantages of ducks and chickens as egg producers it should be borne in mind that very little effort has been made to improve the efficiency of egg production in ducks in comparison with the immense amount of work that has been done on the chicken. There is no doubt that modern methods of breeding and selection could effect considerable improvements, if only by reducing body size, thus lowering food maintenance requirement but preserving fecundity of some of the egg-laying breeds like the Khaki Campbell.

In most parts of the world the duck is, and will continue to be, raised as a source of meat. As a meat animal the duck has two major advantages. The first is that it may be killed as young as 6–7 weeks and yet preserve its palatability. The second is its phenomenal growth rate which means that acceptable market weights can be attained relatively rapidly.

A valuable by-product of the duck is feathers and, more importantly, down. Down is by far the best material to provide insulation for bedding and clothing and it has yet to be surpassed by anything synthesized by the chemical industry. In a world of increasing energy costs duck down is likely to remain a valuable commodity and continue to provide a welcome source of additional revenue to duck producers. Duck down has been and still is regarded as a by-product of duck processing but should the commodity increase in value breeders might be well advised to increase the down yield of their ducks by selection. Nothing is known about the heritability of down yield in ducks but if it is in any way analogous with wool it should be possible to effect considerable increases in yield.

A serious disability from which the table duck suffers is the high fat content of the carcass which may amount to as much as 30–35 per cent of the eviscerated weight. This is unacceptably high except in countries such as China where fat is desired. There is no doubt that selection methods could be devised which would reduce fat and improve lean meat yield and as these improvements are effected the table duck is likely to improve its competitive position and increase its share of the market.

At present duck meat comprises a negligible proportion of total meat consumption in most countries but technical advances permitting large-scale completely intensive production, and genetic improvements which are being effected, ensure a future for ducks in the agricultural economies of the world. In the Far East duck production seems likely to continue along traditional lines until producers are no longer able to meet the food requirements of the burgeoning cities when factory systems will come into their own.

References

Bloom, S. E. (1969) A current list of chromosome numbers and variations for species of the avian subclass Carinati. *Journal of Heredity*, **60**: 217–20

Clayton, G. A. (1972) Effects of selection on reproduction in avian species. *Journal of Reproduction and Fertility* (Supplement 15): 1–21

Delacour, J. (1956, 1959, 1964) *The Waterfowl of the World*. Vols 2, 3 and 4. Country Life: London

Gray, A. P. (1958) *Bird Hybrids. A check-list with bibliography*. Commonwealth Agricultural Bureaux: Farnham Royal, Bucks, England.

Harper, J. (1972) The tardy domestication of the duck. *Agricultural History*, **46** (3): 385–9

Kingston, D. J., Kosasih, D. and **Ardi, I.** (1979) The rearing of Alabio ducklings and management of the laying duck flocks in the swamps of South Kalimantan. Report No. 9. Centre for Animal Research and Development: Ciawi, Bogor, Indonesia

Mott, C. L., Lockhart, L. H. and **Rigdon, R. H.** (1968) Chromosomes of the sterile hybrid duck. *Cytogenetics*, **7**: 403–12

Watson, W. (1969) Early animal domestication in China, In: *The Domestication and Exploitation of Plants and Animals*, ed. P. J. Ucko, and G. W. Dimbleby. Duckworth: London

Yeh, Hsiangkui (1980) [Fossil ducks and Pekin ducks]. *Fossil*, **1**. (In Chinese)

Zeuner, F. E. (1963) *A History of Domesticated Animals*. Hutchinson: London

49

Muscovy duck

G. A. Clayton

Formerly Department of Genetics,
University of Edinburgh,
Scotland

Introduction

Nomenclature

The muscovy duck is native to Latin America where it is known as the 'criollo' (i.e. native) duck. It is one of the greater wood ducks, genus *Cairina*, included in the tribe Cairinini, the perching ducks, belonging to the subfamily Anatinae a branch of the Anatidae.

The muscovy has had many names both scientific and popular. Aldrovandi (1603), one of the greatest of the early European naturalists, included the muscovy with most of the other species of duck, which it resembles so closely, in the genus *Anas*. To the muscovy he gave the specific designation *cairina* in the mistaken belief that this bird originated in or about Cairo in Africa. Linnaeus in 1766 rechristened the muscovy *Anas moschata* in deference no doubt to the name 'musk' or 'muskovia' by which the bird was known to many Europeans. It remained for Fleming in 1822 to confer separate generic status upon the muscovy and designate it *Cairina moschata* by which name it is known to this day.

The reason for the name 'muscovy' is unknown. There is no evidence that Moscow or Russia has had any part in the history of this duck. The bird has no odour which might suggest musk or musky. Other speculations include a corruption of Mexico whence the Spaniards might have obtained specimens or that the name may come from the Muisca Indians of central Colombia. That the name is of long standing is attested by Ligon (1657) who encountered what he describes as 'muscovia ducks' when he visited Barbados some 10 years earlier. What adds greatly to the confusion is the plethora of names by which the muscovy was known. Popular names included Barbary duck, Guinea duck, Cairo duck, *canard de Barbarie*, *Bisamente* (musk duck), *Stummente* (mute duck), *Flugente* (flying duck), *La-Plata-Ente*, *Rothautente* (redskin duck), *Gansente* (goose duck), and various equivalents of musk, e.g. *musque*, *moschus*, *muscaat*, *muskus*. In Indonesia the muscovy is known as the Manila or *Eentok* duck. The Oxford English Dictionary quotes Goldsmith's *Natural History* 1776, VI, 130: 'The Muscovy Duck, or more properly speaking, the Musk Duck, so called from a supposedly musky smell . . . a native of Africa.' It is evident that many Europeans considered the muscovy to be of African origin which makes the suggestion that the name is a corruption of Mexico or Muisca somewhat improbable.

General biology

Warzenente (warty duck), yet another German name for the much-labelled muscovy, aptly describes the feature which distinguishes it most sharply from the common domestic duck. In the muscovy there is an area of bare skin from the bill to just above and behind the eye, unlike the common duck which has a completely covered head. Much of this bare skin is covered in what superficially resemble warty outgrowths or caruncles which, in the case of the male, culminate in a large red fleshy knob at the base of the bill between the nostrils. This feature becomes much more conspicuous during the breeding season.

The Muscovy differs from the common duck in that the male is very much larger than the female and appears to be considerably more aggressive. In the breeding season muscovy males fight one another with great violence using their strong heavily clawed feet and wings. The muscovy appears to be polygamous, unlike the common duck which, at least in the wild or semi-natural state, forms pair bonds. Mating appears to take the form of an assault with the female seemingly making desperate but seldom successful efforts to escape. Mating can occur successfully on dry land or in water.

The muscovy has a much slower growth rate than the common table duck. At 8 weeks of age the latter, especially the females, will have grown much faster and will in fact be approaching mature body weight. The muscovy requires some 4–6 weeks longer to attain the same live weight but the males, maturing at 5–6 kg, will eventually exceed all but the largest strains of table duck. The reason is simple. The table duck with its enormous appetite stores the surplus energy from its excessive intake as fat. The muscovy has a much more restrained appetite and consequently is much less fat. Although the muscovy takes longer to reach a given market weight the carcass is superior because it contains more lean meat and less fat.

Under domestication the muscovy behaves, and may be treated, much as other ducks. It is extremely hardy; access to swimming water is not essential and it may be reared intensively. As with so many other birds, under conditions of expert husbandry it demonstrates a surprising prolificacy. Two 20-week laying cycles separated by a 10-week moult and rest period should achieve production of 120–150 hatching eggs per duck housed. Considering that the muscovy has never been exposed to sophisticated breeding programmes this is a remarkable advance on the normal clutch size of 12–14 eggs that it lays in the wild. The muscovy may be brought to sexual maturity by 30 weeks of age and egg size, which increases with age, ranges from 65 to 85 g. The incubation period for the muscovy, 35 days, is 1 week longer than for the common duck and hatchabilities of up to 75 per cent of eggs set have been achieved. On a flock basis a ratio of 1 male to every 3–4 females is sufficient to maintain good fertility.

The chromosomes of the muscovy, like all avian species, are small in size and large in number. Bloom (1969) quotes a modal diploid number of 80. Mott *et al.* (1968) were unable to determine the exact diploid number but produced strong circumstantial evidence that the standard sex-determining mechanism in birds (Z/Z males and Z/W females) applies. They also found morphological differences between the first and fourth pairs of chromosomes of the muscovy and the common duck.

Relatively little is known about the genetics of the muscovy. Under domestication it undoubtedly has increased in size and shows a good deal of colour variation ranging from almost pure black to pure white. Ten loci, one of which is sex-linked, are known to be polymorphic for genes affecting plumage colour. Hollander (1970) may be consulted for references.

Geographical distribution

The muscovy is to be found almost anywhere where domestic poultry can be kept. Its hardiness probably gives it an advantage in extreme tropical conditions, especially if standards of management leave room for improvement. It is found in Europe generally existing in small flocks in farmyards and village ponds although some large flocks are known in France. It is common in southeast Asia, especially in Indonesia and Taiwan, and in the countries of Middle and South America. It is widespread in Africa and is often encountered in villages, especially in West Africa.

Wild species

The wild muscovy is an elegant creature; its long black feathers have a metallic green and purple irridescence with contrasting wing coverts of pure white. The domesticated form in comparison is an ungainly overgrown coarser version lacking much of the beauty of the plumage and displaying considerable variation in plumage colour.

The wild muscovy is only to be found in Central and South America. It is found from Mazatlan in Mexico southwards to the coast of Peru in the west and to the River Plate estuary in Argentina in the east (Delacour 1959). It is a bird of the tropics and its preferred habitats are tree-lined rivers and forested swamps. Unusually for a duck, the wild muscovy roosts in trees where it builds its nests in forks or holes often high above the ground. Clutch size in the wild ranges from 8 to 20 eggs and the incubation period is 35 days. Two related species, *Cairina hartlaubi* and *C. scutulata*, occupy similar ecological zones in West Africa and southeast Asia respectively (Delacour 1959). All three species are very similar in their habits.

Gray (1958) does not report any attempts at interspecific hybridization although *C. moschata* has been crossed with many other genera of the Anatidae. The most common cross, muscovy × domestic duck, produces a vigorous hybrid which itself is sterile as appears to be the case with other intergeneric hybrids.

Domestication and early history

Historical, osteoarchaeological or other evidence is lacking concerning the antiquity of domestication. When the Spaniards arrived in the New World at the end of the 15th century they found the Indians in possession of a domesticated duck, almost certainly the muscovy, which the Spaniards called *pato casero* (house duck) (Sauer 1952). The domesticated muscovy is called *pato criollo* to distinguish it from *pato real* the wild type in the Argentine. The terms *creole*, *criollo*, *creollo* mean 'of native (indigenous) origin', although it is sometimes taken to imply Spanish descent. It is assumed that the muscovy was kept by the Indians as a source of food but whether it served other purposes, religious, sacrificial, fighting or ornamental, is unknown. Delacour (1964) tells us that the Spanish conquistadors found muscovy ducks domesticated on the northern coast of Colombia and also in Peru and that this was the only domestic bird of the South American Indians, who reared them in large numbers. Gilmore (1950) states that the centre of domestication of the muscovy is unknown but that the evidence points towards the area of high cultures in the central Andean region, probably Peru, where representations of the muscovy duck on pre-Columbian pottery are common.

Bosman, who visited the Guinea coast of West Africa in 1699, quoted by Jeffreys (1956) says:

Next are the Ducks, which have been few years known on this coast. I cannot tell from what country they were brought; but they have no manner of affinity with those of Europe, nor indeed are they much like them; being one half larger, and of another colour, commonly white, or black, white and brown mixt. The drakes have a large red knob on their bills, almost like the turkies, only it does not hang so loose, but firmer, and is very like a cherry.

Undoubtedly a muscovy from the description but unknown to Bosman at least in his native Holland. The significance attached by Jeffreys (1956) to Bosman's ignorance is vitiated by evidence of the presence of the muscovy in Italy at least a century before, provided by Aldrovandi (1603). Belon (1555) who illustrates what is almost certainly a muscovy which he describes as 'La grosse Cane de la Guinée' supplies further evidence of the presence of this bird in Europe. Jeffreys' case for a pre-Colombian introduction of the muscovy to Africa from South America gains no support from his assumption, now shown to be false, that the bird was unknown in Europe long after it had been established in Africa.

If the wild muscovy resembles other members of the duck family it will domesticate rapidly and with ease. If this is the case it is reasonable to expect repeated instances of domestication and recruitment from the wild. Delacour (1959, 1964) writes that the ready domestication of muscovies is proved by the fact that pure wild stock bred in captivity become heavier and develop large face caruncles after two to three generations and that wild muscovies do well on ponds and lakes.

It is not known how long the domesticated muscovy has been separated from its wild progenitor. It might be as little as 600–700 years which, in the case of the muscovy, would correspond to generations, but our lamentable ignorance of Inca history makes it impossible to put an upper date on this figure. A feature of domestication is the proliferation of types which ensue, with those species with the longest history of domestication showing the greatest range of variation. On this criterion the domesticated muscovy is a comparative latecomer to the fold differing only from the wild type in being larger and showing greater variation in plumage colour. There is no reason to suppose that it is not fully compatible genetically with the wild type, but there does not appear to be any direct evidence on this question.

It is surprising that Aldrovandi, who was born in 1522 only 29 years after Columbus' epic voyage, should have been able to state quite catogorically in Chapter 28 of his *Ornithologiae* that the muscovy duck, which he has drawn and describes with considerable accuracy, came from Cairo. The extensive contact between Spain and North Africa in the early 16th century might be the explanation for the distribution of the muscovy in Africa although the rate of dispersion in North Africa seems to be unexpectedly high. The success of the muscovy in Africa may have been due, in part at least, to its hardiness and tropical adaptation enabling it to fill the niche which might otherwise have been occupied by the domestic duck which is less tolerant of hot dry conditions. There does not appear to be any hard evidence whatsoever establishing the presence of the muscovy in any part of the Old World before the 16th century. Despite the objections of Jeffreys (1956) the hypothesis that the muscovy was dispersed about the world by the Spaniards and Portuguese in the 16th century is still the most tenable, and would not conflict with the earliest known reference to the muscovy in Taiwan which appeared in the *Taiwan Fu Chi* (*Annals of Taiwan*) of 1693 (Hwei-Huang 1973). A relatively recent introduction into the Far East is also suggested by the Chinese designation of the muscovy as 'foreign duck' as distinct from the common duck for which the term 'Pekin duck' is almost a synonym and which the Chinese have reared for at least 3000 years.

Jeffreys' point (1956) about the names that Europeans gave the muscovy suggesting African origins should not be overrated. *Meleagris gallopavo*, unquestionably a Spanish introduction of the early 16th century, was and still is called a turkey because Europeans, long unaware of its history, tended to attribute eastern or African origins to the novel, bizarre or exotic. The muscovy itself was known to some as a turkey duck.

Recent history and present status

In southeast Asia, especially in Indonesia, the muscovy has assumed a special importance as an effective 'incubator' where it is used to hatch eggs of the common duck, *Anas platyrhynchos*. The latter has tended to lose its brooding instincts in the course of its long history of domestication and, if obliged to hatch its own eggs, would suffer greatly reduced production. The muscovy duck may be persuaded to remain broody for as long as 20 weeks, and, if allowed a rest period of 8–12 weeks, will resume brooding again and again. The muscovy can continue as an efficient and productive 'incubator' for 5 or more years, and this is an important reason for its presence and popularity in countries such as Indonesia.

In Europe it is in France that the muscovy has its greatest concentration. To the French, probably the most demanding nation on earth concerning the quality of their food, the muscovy, known to them as *canard de Barbarie*, finds favour as a source of meat. The hybrid from the cross muscovy by the common duck, *Anas platyrhynchos*, known in France as *mulard*, is raised for its liver and for meat. In the rest of Europe the muscovy is as yet of minor economic significance though the situation may change as more people become aware of its superior carcass attributes.

The *mulard* has not received the attention it deserves except perhaps in Taiwan where for centuries large numbers have been produced. The muscovy seems to be used almost invariably as the male parent of the hybrid and presumably this is because of the greater, hence cheaper, egg production of the common duck. Unfortunately there appears to be little or no information available on the reciprocal cross. Mating can be accomplished naturally but in Taiwan the common duck, known locally as *Tsaiya*, is so much smaller than the muscovy drake that human assistance is often provided. Hwei-Huang (1973) reports that muscovy males require training to copulate under these circumstances and that fertility of eggs produced is 40–80 per cent. He also describes an artificial insemination technique which seems to work well and offers a number of advantages over natural mating including a tenfold reduction in the number of drakes required and improved fertility.

The *mulard* appears to have a number of advantages over either parent. The large size difference between the sexes characteristic of the muscovy is not reflected in the hybrid which is closer to the common duck in this respect. The *mulard* is considerably larger than the native Taiwan duck – one of the reasons for its popularity for the table – but whether it would display comparable heterosis if the maternal breed was one of the large duck strains remains to be established. In terms of growth rate, food conversion and carcass quality the *mulard* seems to be intermediate between the parental species but this could amount to a considerable economic advantage.

Future prospects

Among the domesticated animals the muscovy must be one of the least developed. Dispersed around the globe it continues to exist in small numbers in backyards and villages much like the domestic hen in previous centuries. It has not yet been exposed to modern breeding and selection methods and whatever potential it may have for meat or egg production remains to be exploited. The reason is, presumably, that the muscovy, a comparative latecomer in the Old World, had to compete with the domestic duck, an established species with natural advantages in fecundity and growth rate. The very large difference in size between the sexes characteristic of the muscovy would prove to be a disadvantage in most circumstances particularly with large-scale producers to whom uniformity of product is important.

Despite a slower growth rate the muscovy is an efficient food converter and produces a carcass with a high lean-meat content. There seems to be no reason why the muscovy should not be improved for growth rate, a highly heritable trait, and if this were done it may yet come to make an important contribution to the world's poultry meat supply, either in its own right, or by producing the hybrid with the common duck which combines many of the advantages of both parents.

Acknowledgements

I am deeply grateful to Dr H. M. Hine of Edinburgh University for translating Chapter 28 of Aldrovandi's *Ornithologiae*. I wish to thank Dr R. A. Dankin of Jesus College, Cambridge, and F. M. Lancaster of the Harper Adams Agricultural College, Shropshire for useful references and suggestions and J. C. Powell of Cherry Valley Farms Ltd, Rothwell, Lincolnshire, England for information on intensive production.

References

Aldrovandi (1603) De Anate Cairina. *Ornithologiae*, hoc est de Avibus Historiae. Lib. XIX, Chap. 28. For bibliographical details consult Lind, 1963, Aldrovandi on Chickens, University of Oklahoma Press, Library of Congress Catalogue Card. No. 63-8989

Belon, P. (1555) *L'Histoire de la Nature des Oyseaux*, Ch. 19. Paris

Bloom, S. E. (1969) A current list of chromosome numbers and variations for species of the avian subclass Carinati. *Journal of Heredity*, **60**: 217–20

Delacour, J. (1959, 1964) *The Waterfowl of the World*. Vols 3 and 4. Country Life: London

Gilmore, R. M. (1950) Fauna and ethnozoology of South America. *Smithsonian Institution Bureau of American Ethnology, Bulletin* No. 143 (6): 345–464

Gray, A. P. (1958) *Bird Hybrids. A check-list with bibliography*. Commonwealth Agricultural Bureaux: Farnham Royal, Bucks, England

Hollander, W. F. (1970) Sexlinked chocolate coloration in the Muscovy duck. *Poultry Science*, **49**: 594–6

Hwei-Huang, H. (1973) The duck industry of Taiwan. *Animal Industry Series No. 8*. Chinese American Joint Commission on Rural Reconstruction: Taipei, Taiwan

Jeffreys, M. D. W. (1956) The Muscovy duck. *The Nigerian Field*, **21** (3): 108–11

Ligon, R. (1657) *A True and Exact History of the Island of Barbados*. Printed for Humphrey Moseley at the Princes Armes in St. Paules Churchyard, London

Mott, C. L., Lochart, L. H. and **Rigdon, R. H.** (1968) Chromosomes of the sterile hybrid duck. *Cytogenetics*, **7**: 403–12

Sauer, C. O. (1952) *Agricultural Origins and Dispersals*. American Geographical Society: New York

50
Goose

R. D. Crawford

*Department of Animal and Poultry Science,
University of Saskatchewan, Canada*

Introduction

Geese have been kept as domesticated birds for centuries. They have always been a minority species and even yet have not received the commercial or industrial exploitation accorded to other poultry such as chickens, ducks and turkeys.

Nomenclature

They are assumed to have been derived from the wild genus *Anser*. Those typical of Europe were domesticated from the greylag, *A. anser*, and those of southeast Asia from the swan goose, *A. cygnoides*. Between these two geographical areas, indigenous geese probably represent hybrids of the two original stocks. In North America, domestication of the Canada goose, *Branta canadensis*, was started during the early years of settlement, but when fully-domesticated *Anser* stocks became available, the Canada goose was abandoned as a domestic form and remains now only as a wild bird or as an ornamental. *Branta* and *Anser* can hybridize but the offspring are sterile so that introgression has not occurred.

The usual English language words for describing geese are 'gander' (adult male), 'goose' (adult female), and 'gosling' (youngster). No terms are in common usage to describe other age-sex categories. The general term for the bird is 'goose', and the plural is 'geese'.

General biology

The goose is a semi-aquatic bird, quite capable of living and reproducing without access to swimming water. If water is available, it will spend part of the day swimming and resting on it. But most of its feeding activity takes place on land. The goose is a grazing bird, able to utilize large quantities of forage in its diet. According to Cowan (1980), plant cell walls are ruptured in the gizzard, allowing digestion of cellular contents; there is relatively little digestion of fibrous plant material. Perhaps because it evolved in the Arctic where spring and summer day lengths are very long, the goose characteristically feeds for prolonged periods, even in darkness. It has no crop, but there is an enlargement at the end of the gullet proximal to the gizzard which serves as a temporary food-storage organ.

Growth rate is very rapid. High-temperature brooding is required until 3 weeks of age only, and birds of most breeds are ready for marketing by 15 weeks of age or earlier.

Reproductive rate is low. Although they have limited reproductive capacity in the first year, geese are not fully mature until 2 years of age. It is generally believed that reproduction is best in the second year and that it remains good until the fifth year. Egg numbers per year seldom exceed 40, nearly all of these being laid during the period of increasing day length. It is not unusual for a goose to undergo a short cycle of egg production in the autumn when day length is decreasing. Pairing, usually for the entire reproductive life, is characteristic of wild geese but domestic birds can be mated successfully in large flocks where they form polygamous groupings. Artificial insemination is feasible but not widely used. Broodiness has been retained by domestic geese and natural incubation and brooding are still exploited by some goose raisers.

Life-span can be exceedingly long. Ives (1951) cites anecdotes of individual domestic birds living for more than 100 years. In commercial breeding, they are kept until at least 5 years of age, which is quite different from the situation with other poultry species which are discarded after one breeding cycle.

Numbers and distribution

There appear to be no recorded statistics on numbers of domesticated geese in the world. The FAO *Production Yearbook* does not provide a separate listing for them. Saleyev (1975) reported that geese were second in popularity to chickens in the Soviet Union. They have been traditionally popular throughout Europe and they probably continue to be kept in large numbers there. The proceedings of the Society for the Advancement of Breeding Researches in Asia and Oceania in 1979 (SABRAO) workshop on animal genetic resources in Asia and Oceania makes only occasional reference to geese. In Canada, about 180 000 were slaughtered in registered stations in 1979 as compared with 50 000 in 1975, reflecting a growing interest in production of specialty foods; these figures do not include birds raised for home consumption so that total census figures may be double those shown.

In general, it can be said that geese are frequent only in northern latitudes. Few are kept in hot climates, and there only the Chinese breed adapts well.

Geese are kept mainly to produce meat for roasting. In Europe, especially in France and in Hungary, there is some force-feeding of geese to yield fatty livers for use in pâté de foie gras. Feathers are harvested for clothing insulation; both down feathers and outer feathers are used and are collected at time of slaughter; plucking of live geese has declined or disappeared as a custom in most countries, but it is still practised in Hungary (Clayton, personal communication). In North America and perhaps elsewhere, geese are used to a limited extent for weeding of crops such as strawberries, asparagus, and cotton; the small and active Chinese breed is preferred for this purpose. Use of geese as guardians continues as, for instance, among villagers in the Andes. The sport of goose fighting flourished in early Russia (Ives 1951) but there is no reference to the sport persisting into the present.

Wild species

Early writers, including Darwin (1896), agree that domesticated geese of Europe are derived from the greylag. Delacour (1954) states that the western greylag (*Anser anser anser*) was the ancestral form, and that the eastern greylag (*A. a. rubirostris*) of Asia was never domesticated. The western subspecies breeds in northern and eastern Europe and into the Soviet Union where it intergrades with the eastern subspecies; in winter it migrates as far as North Africa and Iraq.

Early writers also agree that domesticated geese of the Orient were derived from the swan goose (*A. cygnoides*). Evidence for this origin seems to have been based mainly on plumage colour and pattern which are identical in swan geese, Brown Chinese and African breeds. These three are also similar in having more upright carriage than European stocks, and in having a black bill. The Chinese and African breeds have a very large knob at the base of the upper beak and this protuberance is faintly evident in the wild species.

Cytological evidence of origin would be desirable, but little attention has been paid to such studies with geese. Two reports on chromosome numbers and morphology using modern techniques have been published (Bhatnagar 1968, Hammar 1966). Further studies are in progress in the avian cytogenetics laboratory at the University of Minnesota (Shoffner, personal communication). Bhatnagar (1968) studied White Chinese, and Hammar (1966) studied domesticates of both *A. anser* and *A. cygnoides*; both authors found the diploid number of chromosomes to be about 80, and this is the number now accepted for breeds of both species (Bloom 1969).

Unfortunately, greylag geese have not been studied cytologically, and it can only be assumed that the karyotypes of *A. anser* and of European domesticated stocks are identical. Karyotypes of Chinese geese and their presumed ancestor, *A. cygnoides*, are identical (Shoffner, personal communication). Tentative evidence has been found that the Emden (European), and presumably the greylag, differ from Chinese and swan geese in the morphology of the fifth chromosome (Shoffner, personal communication); the fifth chromosome of Emdens is submetacentric and that of Chinese and swan geese is almost metacentric. Gray (1958) cites many references to crosses between Chinese and greylag, and between Chinese and European breeds. Hybridization can take place in both directions.

Hybrids are fertile among themselves and in backcrosses, but hatchability in backcrosses is somewhat reduced, which might be expected if parental chromosomes differ.

Delacour (1954) classified true geese as a single genus, *Anser*, with nine species, and a large number of subspecies. Many of these can hybridize and the hybrids are often fertile (Gray 1958). Shoffner (personal communication) has prepared karyotypes of five of these species and has found major differences among them in morphology of the second chromosome. He found that the karyotype of the bar-headed goose (*A. indicus*) appears to be identical with that of the swan goose and suggests that the possibility of domesticated birds having been derived in part from bar-headed geese should not be overlooked.

Delacour (1954) classified the brant (or brent) goose as a single genus, *Branta*, with five species and a very large number of subspecies. Shoffner's unpublished work indicates that this genus differs from *Anser* mainly in the configuration of the fifth chromosome. Gray's checklist (1958) shows that many hybrids have been observed between the two genera but that usually the hybrids are sterile. This situation is analogous to that of domestic and muscovy ducks which can hybridize but yield a sterile 'mule'.

Domestication and early history

Zeuner's (1963) account of domestication and early history of geese remains definitive, and there is very little new information to be added. He states that geese were doubtless kept by man from the Neolithic period onwards, i.e. from about 3000 B.C., and that domestication probably first occurred in southeastern Europe. The greylag breeds readily in captivity and hence it is probable that there have been many separate domestications which have continued to the present.

According to Zeuner (1963), geese are well documented from the Old Kingdom of Egypt (2686–1991 B.C.), but there is no good evidence that the birds were domesticated at that time. He interprets them to be greylags and white-fronted geese (*A. albifrons*). If white-fronted geese were domesticated by the Egyptians, they did not persist as a domesticated form. By the time of the Egyptian New Kingdom (1552–1151 B.C.), the greylag was fully domesticated, as it was in Europe.

Zeuner (1963) has described the geese of ancient Greek and Roman civilizations. Greylags had been domesticated in Greece and they are mentioned in Homer's *Odyssey*. The Romans made extensive use of them. They used both meat and eggs, and practised force-feeding to produce fatty livers. Birds were plucked twice yearly to obtain down feathers for cushions and upholstery. The first mention of using quills for writing was made in the 5th century A.D. The fat from geese was used for a variety of medicinal purposes. There is a well-known story of an attempted invasion of Rome by the Gauls in 390 B.C. which was averted by alarm calls of the sacred geese in the Temple of Juno.

Zeuner (1963) makes only brief mention of the Chinese goose, indicating that nothing is known about its early history in the Orient. Delacour (1954) states that it was raised for many centuries in China, beginning probably more than 3000 years ago and that it spread from there to India, Africa, and eventually to Europe.

Recent history and present status

Development and utilization of domesticated geese in the 19th and 20th centuries have been documented by Johnson and Brown (1913), Ives (1951), Merritt (1974) and others.

Darwin (1896) commented that 'although the domestic goose certainly differs somewhat from any known wild species, yet the amount of variation which it has undergone, as compared with that of most domesticated animals, is singularly small'. Nearly a century later, the same comment is valid. Breeds described in 19th-century writings have changed very little. Only a few new ones have been developed, none of them much different from the old breeds.

The dominant breeds of the Western world – Emden, Toulouse, Chinese, and African – were already established before agricultural writings became abundant in the mid-19th century. As these breeds became more prevalent, local mongrel stocks were gradually supplanted.

The Emden, from the German city of the same name, along with derivative breeds, has become the predominant meat goose in the west. A recent survey of poultry genetic resources in Canada

(Crawford, unpublished) indicated that Emdens are the only stock now being bred commercially.

The Toulouse, an indigenous stock of southern France, has been widely used as a meat producer. But where commercial production prevails, white-feathered birds are preferred and the grey Toulouse has declined in numbers. Toulouse and derivative breeds are the ones used for fatty liver production.

The two Chinese varieties, White and Brown, are well documented as having originated in the Orient. They have changed very little since first brought to the west.

Origin of the African breed remains in doubt. Recorded information indicates that it reached the west via the United States early in the 19th century (Johnson and Brown 1913; Ives 1951). At that time, the New England coast was an important shipping and whaling centre, and arrival of livestock from distant foreign ports was not unusual. Descriptions of the geese now called Africans were remarkably similar, even though they were variously attributed to origins in Zanzibar, Hindustan, Hong Kong, etc., and were identified under a great variety of names. Whatever their origin, it is likely that they were a well-developed and widespread type, and they have remained relatively unchanged to the present. Because of rather poor reproductive rate, Africans are not now being utilized commercially.

A relatively new breed, the Pilgrim, has had some popularity and emotional appeal in North America. It is a medium-weight breed in which males are white, and females are grey; sex of goslings can be identified by down colour. Auto-sexing of this kind was well known to early writers (Ives 1951). Folklore indicates that this stock may have been brought to Massachusetts from The Netherlands by the Pilgrim fathers in 1620 (op. cit.).

Very little information is available concerning present-day stocks in the Soviet Union. Pimenov and Smirnov (1974) and Pimenov (1976) have described a large number of breeds not known in the west, some of which are being held as gene conservation flocks. Saleyev (1975) has provided a detailed account of procedures being used in the Soviet Union for large-scale production of broiler goslings, and has briefly described some of the genetic stocks being used. Crossbreeding is extensively used, with egg-laying breeds such as Kuban, Rhine, and Gorky as maternal parents and heavy breeds such as Kholmogory, Vladimir, and Vishtines as sires.

Future prospects

In reviewing the North American goose industry, Merritt (1974) observed that the breakthrough in production efficiency that characterized the chicken and turkey industries has not yet occurred for geese, and that no widely copied production system of producing market geese at the optimum stage of growth has evolved. A continuation of traditional production practices is not likely to lead to a well-structured industry. Until a suitable production system is defined, the goose industry in North America and elsewhere will probably remain small.

There are indications that North American producers are attempting to develop a suitable production system and that an efficient industry may be developing. In Canada, the numbers of geese processed in registered facilities have quadrupled within a few years. In the United States, producers are trying to establish roast goose as a holiday food for St Valentine's Day, thereby increasing demand and consumption. A production system for broiler goslings is already at an advanced stage in the Soviet Union (Saleyev 1975).

If industrial production is established, it is inevitable that unsuitable stocks will rapidly be discarded. There has already been erosion of goose genetic resources in countries where other poultry species have been industrialized. In Canada, the only purebred stock still available commercially is the Emden and other traditional breeds are held now only by hobbyists. In the Soviet Union (Pimenov 1976) there is a State collection of breeds. Other countries should similarly preserve their resources against future needs.

References

Bhatnagar, M. K. (1968) Mitotic chromosomes of White Chinese geese. *Journal of Heredity*, **59**: 191–5

Bloom, S. E. (1969) A current list of chromosome numbers and variations for species of the avian subclass Carinatae. *Journal of Heredity*, **60**: 217–20

Cowan, P. J. (1980) The goose: an efficient converter

of grass? A review. *World's Poultry Science Journal*, **36**: 112–16

Darwin, C. (1896) *The Variation of Animals and Plants Under Domestication*, Vol. 1 (2nd ed). Appleton: New York

Delacour, J. (1954) *The Waterfowl of the World*. Vol. 1. Country Life: London

Gray, A. P. (1958) *Bird Hybrids. A check-list with bibliography*. Commonwealth Agricultural Bureaux: Farnham Royal, Bucks, England

Hammar, B. (1966) The karyotypes of nine birds. *Hereditas*, **55**: 367–85

Ives, P. P. (1951) *Domestic Geese and Ducks*. Orange Judd: New York

Johnson, W. G. and **Brown, G. O.** (1913) *The Poultry Book*. Musson: Toronto

Merritt, E. S. (1974) Geese. In: *American Poultry History 1823–1973*, ed. J. L. Skinner, Ch. 14. American Printing and Publishing: Madison, Wisconsin

Pimenov, B. (1976) [A farm specializing in collecting goose breeds]. *Ptitsevodstvo*, No. 4: 32–34. (Abstract in *Animal Breeding Abstracts*, **45** (9): 523, No. 5641)

Pimenov, B. and **Smirnov, B.** (1974) [Goose breeds and their potential]. *Ptitsevodstvo*, No. 5: 30–3. (Abstract in *Animal Breeding Abstracts*, **43** (2): 57, No. 534).

Saleyev, P. (1975) Ways of increasing goose meat production in the U.S.S.R. *World's Poultry Science Journal*, **31**: 276–87

Zeuner, F. E. (1963) *A History of Domesticated Animals*. Hutchinson: London

51
Swan

M. A. Ogilvie
The Wildfowl Trust, Slimbridge, England

Wild swans

There are six species of swan of which only the mute swan, *Cygnus olor*, has been domesticated. Hybrids have been reported occasionally between *C. olor* and other swans, and with geese, but mainly in captivity and they are normally infertile (Gray 1958).

Cygnus and *olor* come from Latin words for 'swan'. The species was termed 'mute' by contrast with the very vocal whooper swan (*C. cygnus*) and Bewick's swan (*C. columbianus*) which also occur in Britain. In fact the mute swan makes a wide range of snorts and hisses.

Together with geese, swans form one tribe, Anserini, within the family Anatidae. They are separated from other tribes on the basis of their large size, similar plumage in both sexes, lack of patterning in the downy young, and reticulated, not scutellated, tarsal surfaces (Johnsgard 1968). The mute swan is distinguishable from other all-white swans by its curved neck, pointed tail, and orange bill with a basal black knob.

Mute swans do not normally breed until they are 3–4 years old, though some birds, especially females, may do so at the age of 2 years. The pair bond usually lasts for life, though divorce does occur (Minton 1968). Laying begins in late March, the eggs being laid at intervals of 1–2 days to reach a clutch size of 5–9. The eggs are incubated only by the female, for about 36 days. Hatching is synchronous, and the young are precocial and nidifugous. They are self-feeding, but brooded and defended by their parents. They fledge in about 130 days. Most young are driven off the breeding territory by the adult male during the autumn, though others may accompany their parents to a wintering area, remaining there when the adults return to the breeding territory the following spring (Cramp and Simmons 1977).

The chromosome number is $2n = 80$ (male) and $2n = 79$ (female) (Yamashina 1952).

The main range of the mute swan is in northwest Europe where there are probably around 135 000 (Atkinson-Willes 1981). Separate populations occur in the southern Soviet Union, Mongolia and China. None of these is thought to number more than a few thousand (Ogilvie 1972). The mute swan has been introduced to North America, Australia, New Zealand, and South Africa, where it has become established, though numbers remain small (op. cit.)

Domestication

There is no evidence that mute swans have ever been fully domesticated. Instead of taking them completely into captivity, man has protected them and then harvested the young for food. In a few places, the cygnets were separated from their parents at hatching and reared in pens until required. Mostly, though, the parents reared the cygnets, which were then caught up before they could fly. This method, which could be called semi-domestication, had obvious advantages in saving labour and food, but probably produced fewer young overall.

It is probable that the mute swan has been kept in captivity and reared for food in many countries, but only in Britain was this taken to the lengths of recording the ownership of almost every swan in the country. This was done largely through an elaborate system of bill marks. A full and detailed history of ownership has been recorded by Ticehurst (1957) from which some of the following information is taken.

The first records of swan ownership date back at least to the 12th century A.D.. The species probably occurred in the wild in Britain before that and was not introduced as has been suggested. The swan was a royal bird and the Crown granted subjects the privilege of owning and keeping swans, under the conditions that the birds were marked with their owner's bill mark, and were kept pinioned. Ticehurst believed that by the 15th century virtually every mute swan in the country was owned in this way. It seems more probable, however, that in, for example, the extensive fens of East Anglia, wild swans continued to live through this period.

The main reasons for keeping swans have been summarized by Ticehurst as: 1. as a mark of distinction, especially if the birds were a royal gift; 2. in order to use them as important presents, to show esteem or gain favour; and 3. for profit and/or food. The mute swan was highly thought of for its meat and was, before the introduction of the turkey, an important dish at banquets and festivals.

There are now only a handful of places where swans are looked after. The general public feed swans, particularly on lakes and rivers in towns, and these birds become hand-tame. Only at Abbotsbury, Dorset, are young still reared in pens, though subsequently all released, while broods are still rounded up in the traditional way for marking on the Thames and in Norfolk.

No change is foreseen in this situation.

References

Atkinson-Willes, G. L. (1981). The numerical distribution and the conservation requirements of swans in northwest Europe. *Proceedings of the International Waterfowl Reseach Bureau Symposium*, Sapporo, Japan Feb. 1980. International Waterfowl Research Bureau: Slimbridge, Glos.

Cramp, S. and Simmons, K. E. L. (1977) *The Birds of the Western Palearctic*, Vol. 1. Oxford University Press: London

Gray, A. P. (1958) *Bird Hybrids*. Commonwealth Agricultural Bureaux: Farnham Royal, Bucks, England

Johnsgard, P. A. (1968) *Waterfowl*. University of Nebraska Press: Lincoln, Nebraska

Minton, C. D. T. (1968) Pairing and breeding of mute swans. *Wildfowl*, **19**: 41–60

Ogilvie, M. A. (1972) In: *The Swans*, by P. Scott and the Wildfowl Trust, pp. 29–56. Michael Joseph: London

Ticehurst, N. F. (1957) *The Mute Swan in England.* Cleaver-Hume: London

Yamashina, Y. (1952) Classification of the Anatidae based on the cyto-genetics. *Papers from the coordinating committee for research in genetics*, **III**: 1–34

52
Pigeons

R. O. Hawes

Department of Animal and Veterinary Sciences, University of Maine at Orono, U.S.A.

Introduction

Nomenclature

The world over, the 'pigeon' is *Columba* livia; *columba* is the Latin for dove and *livia* means leaden or blue-grey in colour (cf. livid). The word 'pigeon' is of French origin, originally *pijon*, from the Latin *pipio*, a young chirping bird. All domestic breeds of pigeons and all wild and feral populations are classified as '*C. livia*. The wild ancestor is considered to be the rock pigeon or rock dove of Europe and Asia. Goodwin (1967) treats the feral pigeon separately from the wild bird. This seems appropriate considering the long and continuing infusion of genetic material from the domesticated species.

Domestic forms of the rock pigeon occur in every country except at the polar caps. Feral birds probably occur in every sizeable town in the world with the possible exception of a few small islands.

The only other species of Columbidae which is considered 'domesticated' is *Streptopelia risoria*, the Barbary dove, a form of the African collared dove (*S. roseogrisea*). It is technically incorrect to give the domestic form species status. However, since the term *risoria* is one of long standing, Goodwin (1967) feels that the two populations should be treated separately.

The word 'dove' is of Anglo-Saxon origin from *dyfan* meaning to dip or *diefan* – to dive. The Barbary dove is almost exclusively a captive species. Feral flocks do exist in several Florida (United States) cities and in the city of Los Angeles, California. Other names for this bird are 'ring dove', 'domestic collared dove', 'ringed turtle dove', 'laughing dove', and 'ringneck dove' (the United States name).

The names 'pigeon' and 'dove' are synonymous when applied to the family *Columbidae*. There is no scientific distinction between these two terms and there is no one species for which the word 'dove', taken alone, seems to be absolutely proper.

General biology

The pigeon's body is adapted for a life in the air and on land. The wings are constructed to give great speed and manoeuverability in flight while the legs and feet are such that the bird can walk or perch with ease. Pigeons are generally arboreal, nesting and roosting in trees. Some species prefer cliffs, caves, buildings or the ground.

Food of wild species consists of seeds, fruits and herbage together with some invertebrate animals. Feral pigeons consume a wide array of material including bread, popcorn and peanuts (in parks), meat scraps, spilled grains of many kinds at mills, wharves, railway stations and grain elevators. Pigeons drink by inserting their beaks into water and sucking up a continuous draught of liquid. This is almost unique among birds although Goodwin (1967) reports that certain arid-land species drink in this manner since they are often in danger when visiting water.

The pair bond in domestic varieties and in ferals will last until death or removal of mate. Goodwin (1967) doubts that pairing for life occurs in wild species that migrate or wander after the breeding season.

Both sexes take equal part in nest-building, incubation and care of young. Typically in wild, feral and domestic pigeons 2 eggs are laid. Clutch size in domestic and feral Barbary doves is also two. Incubation time for typical pigeons is 16–17 days; for the Barbary dove it is 13–15 days. For the first 4–5 days of life, the young are fed 'crop milk', a substance unique to the Columbidae, and composed of epithelial cells from the crop lining. The causative hormone for the production of crop milk is prolactin and the pigeons' crop-sac response to this hormone continues to be used as an assay method (Farner and King 1973, Levi 1957).

Sexual maturity in domestic pigeons, as measured by age at first egg, is approximately 150 days. Variation exists depending on season of hatch. Life-span in domestic birds can be as long as 15 years. Reproductive success declines after 5 years but egg-laying and semen production can continue up to 11 years (Levi 1957).

The chromosome number for the pigeon is at least $2n = 80$ and for the Barbary dove $2n = 76$ (Stock *et al.* 1974). Variations exist in chromosome counts for many avian species due to the large number of microchromosomes.

Wild species

There is some agreement that pigeons had their origins in Gondwanaland and reached Africa, South America and Australia before continental break-up, or at least while the sea crossings remained relatively short. The genus *Columba* is represented in both the Old and New Worlds but authorities differ in their interpretation of this situation. Some (Mayr 1946, 1964; see Murton and Westwood 1977) feel that North America was colonized by dispersal from Eurasia while Murton and Westwood feel that the similarity of North American and European *Columba* is due to convergence.

The relationship between man and pigeon started toward the end of the last Ice Age when the bird's distribution was limited to a relatively narrow band of moderate climate in the northern hemisphere. Some would consider this bird to be a self-domesticating species which seeks out human fields and settlements. Pigeons thrived as agriculture thrived but modern intense farming methods are not as conducive to their success as were the older systems.

The possibility that more than one Eurasian species contributed to the many domestic pigeon breeds cannot be dismissed with complete finality. However, since the masterly presentation by Darwin (1868) in favour of monophyletic origin, no sound evidence has been produced to discount his theory.

The only zoologist who seriously attempted to promote the polyphyletic theory was Alessandro Ghigi (1908, 1922; see Cole 1930). Ghigi was impressed with the similarity of the snow pigeon (*C. leuconota*) and the domestic Gazzi Modena breed – but crosses between the two proved that their patterns were not genetically homologous. There are traits in certain wild species which resemble traits in domestic breeds; *viz.* the feather pattern of the speckled pigeon (*C. guinea*) which resembles checker (*C/C*), the lack of wing bars in the wood pigeon (*C. palumbus*) which resembles barless (*c/c*), and the extended black of the black wood pigeon (*C. janthina*) which resembles spread (*S/S*). However, it is more probable that these and the many other modifications found in domestic birds have arisen as a result of spontaneous mutation under domestication.

Goodwin (1967) describes the wild form of *C. livia* as about the size of the average feral pigeon but more compact in shape. The wild birds tend to be broader across the shoulders and more muscular, to have a shorter tail, a more slender beak and a smaller cere than ferals or domestics. The blue-grey colour of the rock pigeon with its two black wing-bars and black tail-bar should be familiar to all. Whitman (1919) argued that the blue checker colour pattern (*C/C*) was the ancestral type rather than blue-bar, but it is difficult to confirm this.

The present remaining wild populations inhabit the Faeroes, Shetland and Orkney Islands, the Hebrides, Ireland and Scotland, countries bordering the Mediterranean Sea, eastern Europe and western Asia, Arabia, India and Ceylon, Transcaspia and Turkestan, and Africa north of the equator (Fig. 52.1). In some instances it is difficult to distinguish accurately between wild and feral populations.

Goodwin (1967) considers the blue checker colour-pattern (*C/C*) to be the most common in ferals. This is the pattern of the Old English dovecote pigeon. However, in many areas the wild type pattern (*c⁺/c⁺*) is probably maintained by infusion of Racing Homer genes from birds that fail to return home. Extensive studies on feral populations in regard to reproduction, food habits, colour patterns, etc. have been carried out by Murton *et al.* (1972).

In the Barbary dove there is little, if any,

Fig. 52.1 Distribution of present wild populations of the rock pigeon. (Redrawn from Goodwin 1967)

difference between captive and feral forms.

Hybridization between *C. livia* and other *Columba* species generally results in low egg fertility, high squab mortality, infertile F₁ females and F₁ males of low fertility (Gray 1958, Hollander and Miller 1981). In these crosses most workers used domestic birds to represent *C. livia*. Levi (1965) states that the wild pigeon is easily domesticated and crosses readily with domestic ones, but gives no data for this conclusion. When Whitman (1919) crossed a 'domestic' rock pigeon with a domestic breed (African Owl) the F₁ were fertile in both sexes.

Crosses between *S. risoria* and several other *Streptopelia* species give similar results to those of interspecific *Columba* crosses (Gray 1958). Hybridization between the pigeon and the Barbary dove is not easy but has been accomplished. Female hybrids rarely survive; males survive but exhibit low fertility when backcrossed to doves (Cole and Hollander 1950: see Hollander and Miller 1981).

Domestication and early history

There is evidence from ancient Iraq that man was attempting the domestication of wild rock pigeons at least as long ago as 4500 B.C. Dove figurines have been found at Arpachiyah and Assyrian reliefs show them frequently (Zeuner 1963). By the 5th century B.C. the Greeks had begun to show domesticated pigeons on their statues and vase drawings. A large number of renderings show people holding pigeons that are clearly not resisting (Levi 1957). Based on an analysis of Sanskrit, Old Persian and Latin words for pigeon, Kligerman (1978) places the time of domestication at between 1500 B.C. and 600 B.C. To capture pigeons for religious purposes or for consumption is a far step from domestication. Domestication infers the hand of the human in governing the feeding and breeding of a captive species. Thus even the keeping of pigeons in dovecotes which is referred to in Old Testament writings (Isaiah 60 : 8) does not equal 'domestication'. Dovecote culture continues to be practised in the east, especially in Egypt, still with a minimum of human intervention. The step from capture to domestication is not easily identified and may well have lasted hundreds of years. Zeuner (1963) suggests that the Persians brought the domestic pigeon to Europe.

The pigeon made its first impression on man, not in a culinary sense, but as a symbol of love and fertility. The gentle cooing, the courtship, and the attachment of mates were qualities not lost on early man. The pigeon became the symbol of the Asiatic goddess of love, Astarte (Kligerman 1978: 51–66). It was used by the Hebrews for sacrifices. Those who were too poor to sacrifice a sheep or goat could offer instead turtle-doves or young pigeons (Leviticus 1 : 14–17; 14 : 22, 30). The Christian religion transferred the symbol to heavenly love and the dove became emblematic of the Holy Spirit. A symbolic white dove still appears at many occasions where love and tranquility are the themes such as funerals, Christmas, St Valentine's Day, and Easter; it is also used as an emblem of peace on postage stamps and flags.

A form of pigeon husbandry called dovecote culture began in the east in ancient times and brought pigeons close to man in a more lasting relationship. Mud or stone towers were constructed which provided pigeons with a nesting place and allowed humans to gather the young for culinary purposes. Pliny the Elder confirms domestication in Roman times by describing several distinct breeds of pigeon. While the Romans respected the religious aspects of the pigeon they also enjoyed consuming them and large prices were paid for breeding stock (Darwin 1868).

Pigeons became popular in England in the mid-13th century, being introduced from France. Large dovecotes were added to manorial buildings and some manors produced thousands of pigeons annually. The right of having a pigeon-house was confined to the lords who could punish in court anyone who attempted the same privilege. The lord's pigeons became a chief grievance of the peasants who saw their seed devoured by these pests. In 1651 it was reckoned that there existed 26 000 pigeon-houses in England alone (Kligerman 1978). The dovecote pigeon became so well established that it was considered by early writers as a separate species (*C. affinis*). There was no biological reason for this distinction and the name has since been disregarded.

The unique homing ability of the pigeon was recognized and used in Egypt in Roman times. Messages were tied to the bird's leg or to its neck. The Roman emperor, Nero (A.D. 54–68), used them to send the results of sports to his friends and relations (Zeuner 1963). Pigeon racing was well recognized in Palestine about A.D. 200–220. It is only in the past century that pigeon racing as a sport and pastime has been developed in Europe and the United States. Extensive use was made of homing pigeons in both world wars and in the Korean conflict. They were credited with saving many lives and were often decorated for their efforts.

Recent history and present status

The development of the domestic pigeon has come about entirely under the hand of man. As Darwin pointed out in 1868, there are no wild species at present to account for the diversity found in domestic birds. Levi's (1965) ambitious *Encyclopedia of Pigeon Breeds* must be consulted to appreciate the many combinations of pattern, colour and other morphological traits. Levi states that there is no record of domestic pigeon breeds native to Germany, France, Great Britain or North America. The breeds of these areas can be traced to combinations of breeds imported from the original area of domestication: Persia, India or Asia Minor. Only recently Hollander and Petitt (1980) visited Turkey, which has figured prominently in the origin of many breeds, and observed pigeons which are probably the living ancestors of the highly developed Oriental Frills.

Darwin stated that in the mid-19th century there were probably over 150 kinds of pigeons that bred true and were separately named. Most had been developed before 1600. Probably no records exist, if in fact there ever were records, to tell us how these early breeds were produced. Hollander and Petitt (1980) state that even at present there are no journals or organized pigeon societies in Turkey, an important country in breed development. Pigeons arrived in the New World with early French, English and Dutch settlers. In the United States there is little recorded information relative to pigeons prior to 1850, although birds arrived in Canada as early as 1606 and were being raised 'commercially' in Virginia in 1712 (Levi 1957). The breeds that seem to have a common source are grouped as follows by Levi (1957):

1. Persian Wattled pigeons – probably developed from birds imported into western Europe from the Barbary states. Such breeds as the Scandaroon, French Carrier, English Carrier, Barb, Racing Homer, Show Homer and Exhibition Homer were developed from this source.
2. Persian Tumblers – it is thought that the original flying Tumblers came from Persia and from this group came such breeds as Flying Tumblers, Exhibition Tumblers, Nuns, Helmets, Rollers, Tipplers and Parlour Tumblers.
3. Italian pigeons – birds bred in Roman times also presumably came from the east. These birds were bred for size and became known as the Roman Banquet pigeon. The Runt, Spanish, Mondain, Maltese, Hungarian, Florentine, King, Strasser and German Lark are descendants of the Roman birds. Possibly the Pouters also belong in this group.
4. Indian pigeons – the Fantail group and the Indian Mookee were developed in India.

5. Oriental pigeons – this group consists of the Oriental Frills and the Turbits, probably from Turkey, and the Owls from North Africa.
6. German Toys – thought to have been developed from the German field pigeon which may have been of eastern origin but became feral in Europe. Present breeds classified as Toys are the Gimpel, Lark, Spot, Shield, Starling and Swallow.
7. Jacobins – probably from India; they came to the west via Cyprus.

For further information on individual breeds, consult Levi (1957 or 1965).

The Racing Homers constitute a large and important segment of the pigeon fancy. Individual birds may bring prices as high as $20 000 each. Pigeon racing is the national sport of Belgium and it has been calculated that this country has over 125 000 racing pigeons. The International Pigeon Fliers Federation, with headquarters in Europe, has twenty-four member countries. Extensive research has been done on homing ability and yet the complete story of how pigeons navigate remains a mystery (Keeton 1974).

There has also been selection for exaggeration of certain behaviour traits. Performance flying (tumbling, rolling, highflying), shaking (oscillation of the neck), pouting (inflating the crop) and trumpeting (protracting the last note of the display coo) are characteristics valued in certain breeds. These traits, as well as several single gene mutations which have arisen since domestication, were recently reviewed by Hollander and Miller (1981).

The production of squab is a rather specialized business and very labour-intensive. There are, however, farms in the United States with over 2000 pairs of breeders. Breeds such as the Carneau, Giant Homer, Utility King, Mondain and Pioneer have been developed for meat production. Growers expect 12–14 squabs per pair per year. Young birds are marketed at a dressed weight of 330–380 g at about 4 weeks of age. Dovecote culture is still an important enterprise in Egypt and other Middle Eastern countries.

The Barbary dove is kept either for ornamental purposes or for its interesting call. The American Dove Association recognizes twenty-three 'varieties' of ringneck dove based on various combinations of plumage colours and patterns with or without the silky plumage mutation.

Future prospects

Except in the case of a few large squab plants there has been little organized breeding similar to the sophisticated selection systems employed in chickens. The selection for exhibition, racing and performance flying rests in the hands of dedicated amateurs. It is doubtful that this situation will change in the near future.

The pigeon has traits that make it poorly adapted to intensive agriculture: 1. monogamous matings; 2. altricial young; and 3. the space requirement for flight. Thus it is likely that it will remain a specialty food item.

As with other domestic species there is concern regarding the decline and loss of germ plasm; societies have been organized to preserve certain rare breeds. However, unless the breed in danger has some unique trait not found in other birds it is questionable how much time and effort should be committed to its survival merely to maintain a 'breed'. Colours, patterns, and various morphological traits can always be recombined from existing birds and, given a little time, almost any breed can be recreated.

The future of the feral pigeon looks bright and the popularity of domestic pigeons will remain high world-wide.

Acknowledgement

Appreciation is extended to Dr Willard F. Hollander for suggestions in the writing of the manuscript.

References

Cole, L. J. (1939) The origin of the domestic pigeon. *Proceedings of the Seventh World's Poultry Congress and Exposition*, Cleveland, Ohio, pp. 462–66

Darwin, C. (1868) *The Variation of Animals and Plants under Domestication* (2nd edn). Orange Judd: New York

Farner, D. S. and King, J. R. (1973) *Avian Biology*. Academic Press: London and New York

Goodwin, D. (1967) *Pigeons and Doves of the World*. British Museum Publication No. 663. British Book Centre: London

Gray, A. P. (1958) *Bird Hybrids. A check-list with bibliography*. Commonwealth Agricultural Bureaux: Farnham Royal, Bucks, England

Hollander, W. F. and Miller, W. J. (1981) Hereditary variants of behavior and vision in the pigeon. *Iowa State Journal of Research*, **55** (4): 323–31

Hollander, W. F. and Petitt, R. M. (1980) We saw the original Frills in Turkey. *American Pigeon Journal*, **69**(6): 16–17

Keeton, W. T. (1974) The mystery of pigeon homing. *Scientific American*, **231** (Dec.): 96–107

Kligerman, J. (1978) *A Fancy for Pigeons*. Hawthorn Books: New York

Levi, W. M. (1957) *The Pigeon* (2nd edn, rev.). Levi Publishing: Sumter, South Carolina

Levi, W. M. (1965) *Encyclopedia of Pigeon Breeds*. T. F. H. Publications: Jersey City, New Jersey

Murton, R. K. and Westwood, N. J. (1977) *Avian Breeding Cycles*. Clarendon Press: Oxford

Murton, R. K., Thearle, R. P. J. and Thompson, J. (1972) Ecological studies of the feral pigeon (*Columba livia*). I. Population, breeding biology and methods of control. II. Flock behaviour and social organisation, *Ecology*, **9**: 835–89

Stock, A. D., Arrghi, F. E. and Stefos, K. (1974) Chromosome homology in birds: banding patterns of the chromosomes of the domestic chicken, ringnecked dove, and domestic pigeon. *Cytogenetics and Cell Genetics*, **13** (5): 410–18

Whitman, C. O. (1919) *Orthogenetic Evolution in Pigeons*. Posthumous works of Charles Otis Whitman, ed. Oscar Riddle. The Carnegie Institution of Washington: Washington, D. C.

Zeuner, F. E. (1963) *A History of Domesticated Animals*. Hutchinson: London

53

Canary, bengalese and zebra finch

Malcolm Ellis
Wadebridge, England

Canary

The canary is kept as a cage bird in many countries; indeed, after the budgerigar, it is the most popular cage bird. It was formerly used extensively down mines to detect the presence of poisonous gases.

It is descended from the wild canary, *Serinus canarius* (formerly regarded as a race of the serin, *S. serinus*, and named *S. s. canariensis*) of the family Fringillidae, subfamily Carduelinae.

When the male (cock) canary is in full breeding condition he sings lustily; when the female (hen) is in full breeding condition, she calls to the singing male and searches out and carries nesting material. The clutch usually numbers 4–5 eggs. As each egg is laid, a common practice in captivity is to remove it from the nest and replace it with a dummy egg until the fourth day, when the proper eggs are replaced. These are incubated by the female for 13–14 days. The dummy eggs help ensure that all the eggs hatch at approximately the same time and the young develop equally. The young leave the nest after 16–20 days and 10 days or so later are independent. Usually there is a second clutch.

The canary has the diploid chromosome number $2n = 80$ (Hsu and Benirschke 1975).

The wild canary lives on the Canary Islands, Madeira and Azores. It is said to be established as a feral population on Sand Island, Midway, in the Hawaiian chain (Peterson 1940). The male canary is coloured mainly yellow and greyish with blackish streaking. The female is slightly duller. Compared to most domesticated canaries, the wild canary is smaller and less brightly coloured but shows fairly well-defined sexual dimorphism.

The Fringillidae can be characterized as primarily arboreal finches. The bill is very hard and more-or-less conical in shape and adapted for extracting or shelling seeds.

Restall (1975) records that on Gran Canaria fanciers cross a male wild canary with a domesticated female. Most accounts of hybrids do not make it clear whether they refer to wild or domesticated birds. Rutgers (1977) lists several species said to have hybridized with the canary; probably he is referring to the domestic canary. Some hybrids which he lists are questionable. Restall is among those authors who state that attempts have failed to produce hybrids with the chaffinch, *Fringilla coelebs*, and brambling, *F. montifringilla*; Restall also discounts hybridization with the saffron finch, *Sicalis flaveola*. Hybrids with the yellow bunting, *Emberiza citrinella*, pileated finch, *Coryphospingus pileatus*, and red-crested finch *C. cucullatus* are also questionable.

It is a common practice to pair domesticated canaries with European Carduelinae species. The hybrids may be known as mules because they are usually sterile. Hybrids have proved possible with other Carduelinae, among them African serins, American purple finch, *Carpodacus purpureus*, black-headed or hooded siskin, *Carduelis notatus* (*Spinus ictericus*) and red or hooded or red hooded siskin, *C.* (formerly *Spinus*) *cucullatus*. Hybrids with this last species have proved fertile and have been the source of new varieties of canary (see p. 358). The male red hooded siskin has red, black and white plumage. By comparison the female is quite dull. This species is slightly smaller than the wild canary. It is native to Colombia, Venezuela, Trinidad, Monos and the Gasparee Islands.

According to Rogers (1975) the black-headed or hooded siskin and black-headed canary or alario finch, *Serinus* (formerly *Alario*) *alario*, also give fertile hybrids with the canary.

Zeuner (1963) considered it quite conceivable that the Guanche, the prehistoric inhabitants of the Canary Island, caught and kept the canary prior to the Spanish conquest in the 15th century. It is usually considered that it was during the first half of the 16th century that the canary was introduced as a cage bird into Europe; Zeuner stated that birds even reached America and India. The earliest known published reference to this bird was by Gesner in 1555. At first the Spaniards retained a monopoly on canary breeding, perhaps for nearly a hundred years, by exporting only male birds. According to an account by Giovanni Pietro Oliza, published in Rome in 1622, this monopoly came to an end when a Spanish ship bound for Leghorn (Livorno) was wrecked and the canaries which it was carrying escaped or were released and reached the Italian island of Elba and became established there. In the 17th and early 18th centuries Italy became the centre of canary breeding. In 1709, Hervieux, a Frenchman, wrote a book about the canary, which contained descriptions of twenty-nine 'varieties', many of which were not truly separate varieties but merely plumage variations. During the 19th century some small towns in the Harz mountains in Germany became the most famous centres of canary breeding. Many of the present varieties were developed in the British Isles. The most important change during recent times has been brought about by the developments from fertile hybrids between the red-hooded siskin and the canary.

There exists a wide variety of types and colours, differing in many ways from the wild canary and red hooded siskin. Among them there is considerable diversity in size, shape, colour and markings. The most diverse are the Belgian and Scotch which have an extraordinary hump-backed appearance; the frilled which have long feathers curling outwards; and the Gloster corona, Lancashire coppy and crested Norwich which possess a circular crest.

Among the oldest of present varieties are the frills, lizard and Norwich plainhead. Others which have been long established are the Belgian; Border (once known as the Cumberland); Gloster (consort and corona); Lancashire (coppy and plainhead); Scotch (also known as the 'bird o'

circle' and originally, the Glasgow don); and Yorkshire. During the 1940s the Fife was developed in Scotland.

Each of the varieties named above got its present name from the region, in many instances in the British Isles, where it was developed and/or because of its highly characteristic plumage. Usually they are known as type-breeds and each should conform to an accepted standard of appearance. One, the London variety, disappeared about 1934. With the long-established roller and the comparatively recent malinois from Belgium and Spanish timbrado (classified in 1962), looks are secondary to the song. These and others bred specially for their singing may be known as song canaries.

During recent years a number of colour mutations have developed which are collectively known as 'new colour' canaries. Another comparatively recent development is 'red factor' canaries, those developed from hybrids between the canary and male red hooded siskin. This hybrid is called the copper hybrid. Because new colour and red factor canaries are comparatively new, the names for many have yet to be standardized. There is much interest in producing differently coloured canaries, especially a truly red, also a black canary.

Bengalese

The bengalese or bengalese finch is known only as a domesticated bird. It is kept in many countries. The origin of the name is unknown. The bengalese is of Asian origin and it is possible that rightly or wrongly it was associated with Bengal. Bengal is within the range of the ancestral species, though not of the race from which the bengalese is considered to have descended. It has also been called the bengalee or pied mannikin. Especially in North America, it may be known as the society finch.

Bengalese nest exceptionally readily. The sexes can accurately be determined only by behaviour, the courtship of the male, particularly his song, being the most reliable guide. Four to eight eggs are laid and incubation lasts 12–14 days. The young leave the nest after about 21 days and 14 days or so later are independent.

The bengalese is considered to have descended

from *Lonchura striata* and, although long domesticated, it may still be listed as *L. striata*, dom.; *L. domestica*; or with the old generic names *Munia* or *Uroloncha*.

The striated, white-rumped, white-backed or sharp-tailed finch, mannikin or munia, *Lonchura striata*, belongs to the family Estrildidae, subfamily Amadinae. The Estrildidae are widespread in warmer parts of the Old World. Most are small to very small; the Amadinae can be characterized as small, stockily built, stout-billed species. The Estrildidae feed mostly on seeds. The nest is an untidy bundle with a short horizontal entrance. Usually there are 4–6 white eggs. Both sexes share nest building, incubation and care of the young. The nestlings are remarkable for the decoration on their palate and tongue. The young mature quickly and may breed when only a few months old.

Lonchura striata ranges from India and Sri Lanka, southwards to Sumatra and eastwards to China and Taiwan. It has a number of races; these tend to be divided into two subspecies groups, called striated finches and sharp-tailed finches. The latter is called the *L. s. acuticauda* group by Eisner (1957).

Eisner reached the conclusion that the bengalese is a domesticated form of sharp-tailed finch developed in Japan from stock probably originating from southeast China or Taiwan. The most comprehensive account of the species available to Eisner, listed one race from China, *L. s. squamicollis*. This is not included in a later account (Howard and Moore 1980), which lists *L. s. swinhoei* from southern China and Taiwan. Eisner mentions that Taiwanese specimens are extremely similar to those from Nepal. They are of the race *L. s. acuticauda*, the distribution of which includes Bengal.

Suggestions have been made that the bengalese is the result of hybrids between two or more species of mannikins and including the silverbill *L. malabarica* (formerly *Euodice malabarica*). In some instances confusion has occurred because the striated and sharp-tailed finches have been treated as separate species. Eisner concluded that there is no reason to suppose that it is a hybrid with any other bird.

Taka-Tsukasa (1922) quoted an old Japanese book about cage birds, that the bengalese comes from China. He stated that it was imported into Japan about 200 years earlier (early 18th century). It is not made clear whether this was the wild ancestor or domesticated bengalese. Taka-Tsukasa states that during the first 50 years it seems that the white and cinnamon (fawn) varieties did not appear in Japan.

There are comparatively few varieties of bengalese. The self chocolate and chocolate-and-white closely resemble *L. striata*. The crested variety, because of the crest, is least like it. The white variety is considered prone to poor sight and blindness. The self chocolate, self fawn and crested are comparatively recent. Dilutes are another recent development. Frilled and silver varieties are being established.

The bengalese is commonly kept partly or wholly for its remarkable capability as a foster parent. Most frequently it is used to hatch the eggs and rear the young of Erythurinae.

Zebra finch

After the budgerigar and canary, the zebra finch is probably the next most common cage bird. It seems to be named on account of the black-and-white bars on the rump and tail. In Australia it may be known as the chestnut-eared finch, because of the male's bright ear-coverts. The domesticated zebra finch almost certainly descended entirely from Australian stock, namely *Poephila guttata castanotis*. This name, as well as *Taeniopygia guttata castanotis*, *T. castanotis* and *Poephila castanotis*, although inappropriate, is often applied to the domesticated bird.

The zebra finch, if allowed to, will often nest all year round. Unless the supply of nesting material is limited there is a tendency for the nest and eggs to be covered by a second nest and eggs and perhaps even a third or more. The clutch usually consists of 4–5 eggs but may number 2–7 or more. Incubation lasts 12–16 days. After 20–21 days the young leave the nest and 7–14 days later are independent.

The zebra finch belongs to the family Estrildidae, subfamily Erythurinae, which is predominately Australasian. *Poephila g. castanotis* is by far the most common member of its family in Australia. Several others are popular cage and aviary birds and it is likely that some will be domesticated.

Immelmann (1965) lists twelve to fourteen species (dependent upon classification), plus the bengalese, which have hybridized with the zebra finch. Several are of the subfamily Erythurinae, others Amadinae. Immelmann does not state whether the hybrids were fertile. *Poephila g. guttata* lives on some of the Lesser Sunda Islands. *Poephila g. castanotis* occurs virtually all over Australia. Domestication was comparatively recent, since when there has been no marked change in distribution. Possibly the zebra finch was first introduced into British about 1840, as a pretty little cage bird.

The wild type may be known as the grey or normal. A number of varieties have been developed; these include: fawn (cinnamon, isabel); cream; silver; white; chestnut-flanked white (marked white, masked white, marmosette); pied (variegated); penguin (white-throated, white-bellied, silver-wing, blue-wing).

The domesticated varieties differ from the wild zebra finch principally by their colour patterns. One variety is very similar to the wild type but the beak is pale orange or yellow instead of red. The most exceptional change is the development of a crested variety. The wild type and most varieties show well-defined sexual dimorphism.

In 1958 the Zebra Finch Society (Great Britain) announced that the wild type and its colour mutations would no longer be considered foreign birds. It had been established a considerable time: it is thought that the white appeared first in an aviary in Australia in 1921; the fawn probably in Australia about 1936; while about the same period the pied was bred in Denmark.

References

Eisner, E. (1957) The bengalese finch. *Avicultural Magazine*, **63**: 101–8

Howard, R. and **Moore, A.** (1980) *A Complete Checklist to the Birds of the World*. Oxford University Press: London

Hsu, T. C. and **Benirschke, K.** (1975) *Chromosome Atlas*, Vol. 3. *Fish Amphibians, Reptiles and Birds*. Springer-Verlag: New York

Immelmann, K. (1965) *Australian Finches*. Angus and Robertson: London

Peterson, R. T. (1940) *A Field Guide to Western Birds*. Houghton Mifflin: Boston, Massachusetts

Restall, R. L. (1975) *Finches and Other Seed-eating Birds*. Faber and Faber: London

Rogers, C. H. (1975) *Encyclopedia of Cage and Aviary Birds*. Pelham Books: London

Rutgers, A. (1977) *The Handbook of Foreign Birds*, Vol. 1, (4th edn rev.). Blandford Press: Poole, England.

Taka-Tsukasa, N. (1922) Aviculture in Japan. *Aviculture Magazine* (3rd ser.), **13**: 15–22

Zeuner, F. E. (1963) *A History of Domesticated Animals*. Hutchinson: London

54

Budgerigar

Christopher Lever

Winkfield, Berkshire, England

Introduction

Melopsittacus undulatus was placed in the family Psittacidae, but in Australia it is now assigned to a new family, Platycercidae, the broad-tailed parrots (Condon 1975). Australian and English vernacular names include 'betcherrygah,' 'betherrygah,' 'boodgereegah', 'canary parrot', 'flightbird', 'grass parakeet', 'lovebird', 'scallop parrot', 'shell parrot' 'singing parakeet', 'undulated grass parakeet' (used by John Gould in 1840), 'warbling grass parakeet' and 'zebra parrot'. Aboriginal dialect names include *atithirra* (north-central Northern Territory), *cinpara* or *cinparu* (west-central Queensland), *garrmbil* (northern Queensland), *kilkilkari* (northwest South Australia), and *kolyabiddy* and *padda-moora* (both west Western Australia).

Young cocks may produce spermatozoa within 60 days of leaving the nest. This rapid sexual development, a physiological adaptation to an arid environment, enables young birds to reproduce quickly in optimum conditions (Forshaw 1980). The average clutch contains 4–6, occasionally 8, eggs which are normally laid on alternate days. Incubation, by the hen only, lasts for 18 days, and commences with the first egg.

In the wild there are two main breeding seasons: August to January in the south and June to September in the north. Breeding can, however, take place at any season, and may be stimulated by heavy rainfall preceded by a period of abundant food in conjunction with optimum temperatures and photoperiod (Wyndham 1978). Budgerigars may have several broods. They have 8 pairs of large and 6 pairs of small chromosomes (Rogers 1935).

Budgerigars are bred in domestication principally in Europe, North America, South Africa, Australia and Japan.

Wild species

Budgerigars are endemic to Australia, where they are widely distributed throughout the arid inland regions; they occur on several offshore islands but not in Tasmania (see Fig. 54.1). According to Immelmann (1972) budgerigars are Australia's most abundant parrot.

Budgerigars are gregarious and nomadic, their movements and enormous fluctuations in numbers being largely governed by the availability of water and seeding grasses. Drought is the budgerigar's worst enemy, and despite its ability to survive considerable periods without water, prolonged drought causes many deaths (Wyndham 1978, Forshaw 1980). In times of drought budgerigars may disperse to coastal regions.

The budgerigar forms a monotypic genus believed by Forshaw (1980) to be transitional between the grass parakeets *Neophema* (of which there are seven species found mostly in the southern half of Australia) and the monotypic ground parrot (*Pezoporus wallicus*), which is confined to parts of the coast of southwestern and southeastern Australia and to Tasmania.

Wild budgerigars are light green in colour (the yellow and blue mutants which have been observed are escapees from captivity) and are smaller than domesticated varieties.

Budgerigars are alleged to have hybridized with the African lovebirds *Agapornis* and with *Neophema*, but no viable offspring have ever been recorded.

Budgerigars were at one time roasted and eaten by Aboriginals. In the Aboriginal Eora (Port Jackson) dialect *budgeri* = 'good' (to eat), while *gar* and *kar* occur as elements in the names for

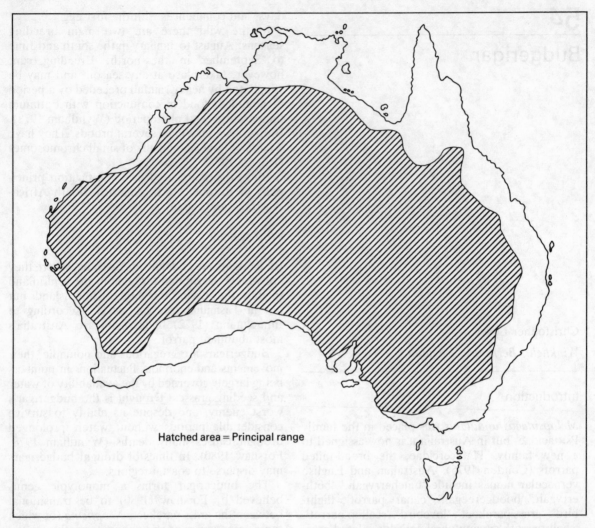

Fig. 54.1 Distribution of budgerigars in Australia. (From Royal Australian Ornithologists' Union Atlas of Australian Birds, 1983)

various parrots in the dialects of other tribes in eastern New South Wales.

Domestication

The first live budgerigars in Europe were two brought back to England in 1840 by John Gould from Australia, where in that year (or before) they had first been bred in captivity by his brother-in-law, Charles Coxen. They were introduced to France in about 1845, to Belgium and Holland in about 1850, to Germany in 1855, and to the United States and South Africa in about 1909. Belgium and Holland were the earliest countries to breed budgerigars in domestication on a large scale, followed by France (especially at Toulouse from the early 1880s), Germany (where they were first bred in 1855 in Berlin by the Gräfin von Schwerin), Britain, North America and Australia. Yellow mutants were first developed in Belgium in 1871–72; blue mutants first appeared in Holland in 1878.

Recent history

Genetics of colour

The genetics of colour in the budgerigar have been

extensively studied, especially by Duncker (1928, 1933), Steiner (1932), Crewe and Lamy (1935), Rogers (1935), and Taylor and Warner (1961).

Analysis of the various colour varieties can be expressed in terms of factors affecting: (1) the intensity of melanic pigment; (2) the ground colour (yellow or white) of the plumage; and (3) the tone of blue or green in the various combinations of (1) and (2).

1. According to Steiner, the intensity of the melanic pigment is governed by three allelomorphic factors, R_n, R_g and R_p. R_n produces normal intensity of black pigment as in the wild form; R_g produces 'grey-wing' varieties, and R_p 'pallid' types. When the factor of the yellow lipochrome pigment, L, is present, the 'pallid' form is yellow; when it is absent this form is white. R_n is dominant to both R_g and R_p, but the R_gR_p bird is intermediate in shade between the pure 'grey-wing' R_gR_g and the 'pallid' R_pR_p forms.
2. When the yellow lipochrome pigment is present its interaction with the different intensities of melanic pigment results in various shades of green; when it is absent a corresponding series of blues appears. L is completely dominant to l.
3. The various tones of green and blue depend on a pair of factors, S–s, which affect the structure of the feathers. Where S is present this becomes modified and blue is less vivid. As a result the normal light green of the wild form becomes olive green, while blue changes to a dullish mauve. For this pair of factors there exists a heterozygous form, Ss birds in the green series being dark green, while in the blue series they become 'cobalts'.

The S–s pair, with the L–l pair, show the genetical phenomenon of linkage, these two pairs occurring in the same chromosome.

From the evolutionary viewpoint the budgerigar is of special interest because all the various colour forms depend on five mutational changes which have occurred since domestication in a species which has never successfully hybridized with any other. Charles Darwin based his theory of evolution through natural selection largely on the effects of artificial selection on wild forms brought into domestication; some of his examples of domesticated varieties, however, may have originated in hybridization, whereas in the case of the budgerigar all the different varieties originate from a single wild form without adulteration through hybridization.

Feral populations

Feral colonies of budgerigars have frequently become established, most recently on Tresco in the Isles of Scilly (Lever 1977), but they became extinct in the hard winter of 1978/79. The only known feral population at present has existed for at least 20 years in urban and suburban districts on the western and eastern coasts of south and central Florida, U.S.A., where flocks with an increasing preponderance of wild-type light green birds now number up to 8000 individuals.

Other domesticated parrots

Two other Psittaciformes have long histories of domestication; these are the African lovebirds (*Agapornis*) and the cockatiel (*Nymphicus hollandicus*).

Lovebirds – of which there are nine species, in the family Psittacidae – occur in tropical Africa and Madagascar; they comprise a group of closely related and almost completely allopatric forms, most of which hybridize freely in captivity. They have been known in domestication since about 1860, but not until the 1970s, when many colour mutations became available, was breeding in captivity firmly established (Low 1980).

The distribution of the cockatiel, which forms a monotypic genus and subfamily in the family Cacatuidae, closely coincides with that of the budgerigar. It has been kept in domestication since before 1850, but it was not until 1950 (in the United States) that the first colour mutation occurred; this and others became freely available in North America and Europe in the late 1960s and early 1970s, since when breeding in captivity has become extremely popular. The breeding of cockatiel mutations is still in its infancy, and the development of new ones can be confidently anticipated (Low 1980).

References

Condon, H. T. (1975) *Checklist of the Birds of Australia*, Part 1: *Non-Passerines*. Royal Australian Ornithologists Union: Melbourne

Crewe, F. A. E., and **Lamy, R.** (1935) *The Genetics of the Budgerigar*. Authors: Edinburgh

Duncker, H. (1928) Die Vererbung der Farben bei Wellensittichen. *Vögel ferner Lander*, **1**

Duncker, H. (1933) *Budgerigar Matings and Colour Expectations*. Budgerigar Society: Chaddesden, England

Forshaw, J. M. (1980) *Australian Parrots*. Lansdowne: Melbourne

Immelmann, K. (1972) *Die Australischen Plattschweif-sittiche*. Ziemsen: Wittenburg

Lever, C. (1977) *The Naturalised Animals of the British Isles*. Hutchinson: London

Low, R. (1980) *Parrots. Their care and breeding*. Blandford Press: Poole, England

Rogers, C. H. (1935) *Budgerigars and How to Breed Them*. Iliffe: London

Steiner, H. (1932) Vererbungsstudien am Wellensittich *Melopsittacus undulatus* (Shaw). *Archiv der Julius Klaus-Stiftung für Vererbungsforschung, Sozialanthropologie und Rassenhygiene*, **7**(1/2): 37–202

Taylor, T. G., and **Warner, C.** (1961) *Genetics for Budgerigar Breeders*. Iliffe: London

Wyndham, E. (1978) Ecology of the budgerygah *Melopsittacus undulatus* (Shaw) (Psittaciformes: Platycercidae). Ph. D. thesis, University of New England: Armidale, New South Wales

55
Ostrich

W. R. Siegfried

FitzPatrick Institute of African Ornithology, University of Cape Town, South Africa

Introduction

The ostrich, *Struthio camelus*, is the sole species of the family Struthionidae (order Struthioniformes), and is the largest living bird. Ostriches can only have evolved from flying ancestors, but the phylogenetic relationships between ostriches and other avian species are still uncertain.

Fully-grown males attain between 100 and 130 kg and stand about 2.2 m when erect; females are slightly smaller. Males attain their black and white adult plumage when about 2 years old, but are not necessarily fully mature sexually. Females and immatures are much duller, having greyish brown plumage. The young wear a spiky, black-tipped, buffy plumage until they are about 4 months old.

Ostriches have a variety of vocalizations, but normally are silent. During the breeding season, territorial males utter a deep booming call. Timing of breeding is very variable according to habitat, being irregular and opportunist in arid areas; elsewhere breeding is more regular and mainly in the dry season. Breeding behaviour also varies, and includes monogamy and polygamy.

Although several females may lay eggs in the same or several different nests, only one (the major) female incubates the eggs at each nest. The clutch of a major female includes 5–11 eggs, and incubation proper begins about 16 days after laying the first egg (eggs are laid on alternate days). Captive birds can be induced to lay up to 90 eggs by continual removal. The male incubates at night, and the female during the daytime. Major hens can recognize their own eggs, and eject from the nest those of other females if more than 20–25 eggs are deposited in the same nest. An egg weighs, on average, 1.5 kg. Incubation requires about 45 days. Broods from many nests form crèches accompanied and guarded by one or several adults, not necessarily the parents of the chicks, for up to 9 months.

Ostriches are endemic to Africa, occurring chiefly south of the Sahara; a remnant population occurs in Morocco. Formerly more widespread, ostriches were exterminated in Arabia in the 1960s. The extinct Arabian population has been recognized as a distinct race *S. c. syriacus*. There are four extant races: *S. c. camelus* (Morocco, Mauretania, Ethiopia, Sudan and Uganda); *S. c. molybdophanes* (Ethiopia, Somalia and Kenya); *S. c. massaicus* (Kenya and Tanzania); and *S. c. australis* (Zimbabwe, Botswana, South Africa and Namibia). Fertile hybrids have been produced from crosses between *S. c. australis* and *S. c. camelus*, and *S. c. molybdophanes* and *S. c. massaicus*. *Struthio c. australis* now occurs as a pure wild population only in the northern parts of its original range, because of adulteration by feral birds of hybridized ancestry involving *S. c. camelus* which was introduced into South Africa.

The total domesticated population of ostriches, centred in the Oudtshoorn district of South Africa, embraces some 100 000 birds. This is the only region in the world where domesticated ostriches are farmed intensively on a large scale. Currently, the industry is worth approximately $U.S. 20 million annually, 40 per cent of which is accounted for by sale of feathers.

Domestication

From very early antiquity, ostriches have provided man with food, clothing, utensils and adornment. Moreover, the ostrich features frequently in prehistoric rock art and folklore, and in the scriptures and earliest written records of man. Apart from being hunted for their flesh and plumes, ostriches were kept in captivity, tamed and perhaps semi-domesticated by the early Egyptians, Greeks and Romans. Captive animals were fattened for the table, and Egyptian and Roman ladies of noble birth rode ostriches on ceremonial occasions. Subsequently, a few unsuccessful attempts were made in Europe to rear ostriches for their meat, including one quaint venture in England in 1680. However, undoubtedly man's long-standing fascination with the ostrich feather as an article of adornment was the foundation for the bird's domestication. Unlike the feathers of other birds, the barbs of the ostrich feather are equally long on both sides of the central shaft. This, incidentally, is why the ostrich feather was adopted as a symbol of justice and truth in ancient Egypt.

The wearing of ostrich plumes, as high-fashion attire, was promoted seriously by Marie Antoinette in France during the second half of the 18th century. At about this time, many South African farmers were already keeping a few tame ostriches whose feathers were cut every 2–3 years for sale to the public. However, until the 19th century, all ostrich feathers supplied to Europe were obtained from wild birds hunted and killed in North Africa and Arabia. The first records of feathers of 'wild' ostriches being exported from South Africa date from 1838.

The ever-growing increase in popularity of ostrich feathers in the fashion world of the first half of the 19th century, and the concomitant decline of ostriches in the wild, prompted the Acclimatization Society of Paris in 1859 to offer premiums for the successful domestication of ostriches in Algeria and Senegal, and for breeding the birds in Europe. The only really successful response to this offer was achieved in Algeria. However, at about the same time independent attempts at domestication were taking place in South Africa. In addition, ostrich farming subsequently was tried out in Asia, Australia, and in both North and South America. These trials generally ceased with the sudden collapse of the feather trade in 1914. Subsequently, ostrich farming has been successful in South Africa alone.

Recent history

South Africa's virtual monopoly in ostrich farming began in the Cape Province in the early 1860s. In 1865 there were 80 birds in captivity at the Cape, and by 1904 there were 357 000 birds in captivity there, many having been hatched by artificial incubation.

Between 1885 and 1905, several importations of birds belonging to the North African race were made, to improve the quality of feathers produced by Cape birds. By 1910, selective breeding had resulted in a number of well-established Cape ostrich strains, such as Riempie, Evans and Barber.

A lucrative trade in ostrich wing feathers persisted for about 50 years, until it crashed disastrously with the onset of the First World War. Approximately 1 million ostriches were being farmed for their feathers in the boom years soon after the turn of the century, and the value of annual exports of feathers alone was approximately $U. S. 8 million. After the slump in 1914, ostrich farming remained depressed until after the Second World War when it revived, mainly through a diversification of consumer products, and grew steadily to its present scope. There is no part of the bird that is not put to some use, and the vagaries of high fashion no longer solely determine the success of ostrich farming. Indeed, while trade in prime, high-fashion wing feathers is still somewhat in the doldrums, there is a ready and stable market for indifferent body feathers – formerly regarded as relatively inferior – used in the manufacture of utilitarian articles such as dusters.

Future prospects

Although the ostrich industry has achieved economic stability, the management of birds on many farms is still relatively primitive. More particularly, there is considerable scope for improvements in the artificial incubation of eggs and the selective propagation of breeding stock. In part, these deficiencies reflect the fact that, in spite of its extraordinary interest, the ostrich has received little attention from scientists, most scientific studies of the bird in the wild and in captivity having been carried out in the last decade or so. This new scientific interest could foreshadow future changes in the ostrich as an animal useful to man, as well as in the industry itself. At present, the industry is based more on the semi-controlled breeding of semi-tame birds than on completely controlled propagation of truly domesticated stock.

Reference

Brown, L. H., Newman, K. B. and Urban, E. K. (1981) *The Birds of Africa*, Vol. 1. Academic Press: London and New York

Osterhoff, D. R. (1979) Ostrich farming in South Africa. *World Review of Animal Production*, **15** (2): 19–30.

Siegfried, W. R. and Brooke, R. K. (1983) *The Ostrich and Ostrich Farming: a bibliography*. FitzPatrick Institute, University of Cape Town: Cape Town

Smit, D. J. V. Z. (1963) *Ostrich Farming in the Little Karoo*. Bulletin No. 358, Department of Agricultural Technical Services: Pretoria, South Africa

56

Cormorant

Zhang Zhong-ge

Beijing Agricultural University, China

The Chinese train cormorants to catch fish, but they are not bred in captivity. The usual species is the common cormorant, *Phalocrocorax corbo*, belonging to the family Phalocrocoracidae of the order Pelecaniformes. Popular names in China are 'water crow', 'fishing eagle' and 'black hare'.

Wild cormorant

The distribution of the wild bird extends from eastern Asia through central and southern Asia to southern and western Europe.

The cormorant has a narrow body up to 80 cm long and it weighs about 1.87 kg. Its plumage is black with a bronze sheen. On cheeks and throat are yellow patches of bare skin. The pharynx and oesophagus can dilate fivefold so that it can swallow large fish without difficulty. It has a long hooked bill which clamps a fish like pliers.

Cormorants inhabit rivers, creeks, lakes and the sea coast. They do not avoid people. They are excellent swimmers and are often seen swimming in small flocks. They can dive to a depth of 1–3 m and occasionally to 10 m for 30–40 seconds or even up to 70 seconds.

During the mating season cormorants flock together and build nests in low-lying riverside or reed marshes. In north China cormorants lay eggs, about 5 in one brood, in March, April or May. The eggs hatch in approximately 28 days. The male and female share in building the nest, incubating the eggs and in feeding the young. They have a peculiar way of feeding – the parent bird opens its mouth wide enough for the young bird to put its bill inside the mouth and reach the semi-digested fish down the long oesophagus.

Exploitation

China has the earliest record of taming cormorants for the purpose of catching fish. As early as 221 B.C. (in the ancient Chinese book *Er Ya*) and again during A.D. 25–230 (in *Yi Wu Zhi – Local Records of Foreign Matter* by Yang Fu) it is stated that cormorants can dive deep under water to catch fish and people living in gorges fed the birds for this purpose. In the Tang dynasty the great poet Du Fu wrote:

Every family has 'black hare' and for
Every meal they eat fish

These lines were written about A.D. 759–768 when the poet lived in Sichuan. It can be inferred that China began to tame cormorants earlier than that time. Japan is among the countries with early records of taming cormorants. It is recorded in the ancient Chinese book *A History of Japan* from *Records of Sui Dynasty* (A.D. 590–617) that 'Japan has fertile soil and more water than land. The Japanese tie small rings round the necks of cormorants and make them dive into rivers to catch fish, with a daily catch of over one hundred.' (Li Fan, Qian Yanwen, *et al.* 1979). Li Shizhen (1518–93), the great pharmacologist and biologist in the Ming dynasty, describes in his *Compendium of Materia Medica* the thriving industry of cormorant keeping in his time. He writes, 'Fishermen in the south generally keep scores of cormorants for each boat, which are made to catch fish.' This testifies to the large scale of cormorant keeping in ancient China.

Training of a cormorant begins several days after the bird is caught. First its wings are clipped to prevent it flying away. Before it is driven to the water, a long string is tied to the foot of the cormorant with the other end fastened to a stick on the bank. A grass ring is tied round its neck to prevent it swallowing its prey. When the trainer finds that it has caught a fish, he makes a peculiar whistling sound, which recalls the cormorant; he then feeds it with small fish. After eating, the cormorant is driven back into the water again. The training proceeds in this manner for a month and then continues from a fishing boat. The cormorants under training are first made to perch on the gunwales and then are driven into the water to catch fish. This stage of training lasts for another month. It is only when the cormorants become obedient to the fisherman's orders that the training is completed. Then the string is removed from the foot, but the ring (usually made of nutgrass flatsedge) remains round its neck to prevent it from swallowing big catches. Young cormorants learn to catch fish by imitating their elders and need no special training by men in the actual catching.

Cormorants generally have a life-span of 13–15 years or, in a very few cases, of 20 or more years. Cormorants of 2–10 years old are in their prime of life for catching fish. After that age their fishing capacity declines. A cormorant, therefore, has a working life of about 10 years.

References

Li Fan, **Qian Yanwen, et al.** (1979) [*A History of Living Beings*, Book 5 (Origin of Plant Cultivation, Origin of Animal Domestication, Functions of Microbes)]. Kexue (Science) Press. (In Chinese)

Li Shizhen (1957) [*Compendium of Materia Medica.*] People's Publishing House. (In Chinese)

57
Crocodile, turtle and snake

A. de Vos

formerly Wildlife Officer,
FAO, Rome

Crocodiles

Wild species

Crocodilians are classified into three families – the true crocodiles (Crocodylidae) with thirteen species, the alligators and caimans (Alligatoridae) with seven species and the Gavialidae with a single representative, the gharial (*Gavialis gangeticus*).

The true crocodiles comprise three genera: *Crocodylus*, *Osteolamus* and *Tomistoma*, found throughout the tropical regions of the world. The alligators are found primarily in the Americas and include the genera *Alligator*, *Caiman*, *Melanosuchus* and *Paleosuchus*. The gharial now survives only in India and Nepal.

Crocodiles have holding, not cutting teeth. All species are strictly carnivorous and most will eat a variety of animal material live or dead, but predominantly fish. The gharial eats fish throughout its life. Large crocodiles are capable of pulling down extremely big mammals.

Crocodiles are able to regulate their body temperature only to a limited degree by behavioural activities such as sunbathing, gaping and seeking shade or water. The metabolic rate can be controlled voluntarily over a wide range. This allows a crocodile to be dormant, reduce its metabolic rate and thus utilize its food and oxygen supply very slowly.

Initiation of breeding depends on the species and the age and size of the animals concerned; in the larger species breeding starts between 8 and 15 years and in the smaller species between 4 and 6 years of age.

Wild crocodiles are usually territorial during the mating season. All species appear to lay only one clutch of eggs each year. The mean clutch size is, therefore, the annual level of egg production. Clutch size increases with increase in female size. The number of eggs varies with the species, and for big crocodiles ranges from 30 to 70 eggs. Hatching occurs after about 80 days.

The eggs are either buried 20–50 cm deep in the earth or in a nest mound of rotting vegetation, depending on the species. Warm conditions are critical to hatching success. Both juveniles and adults grow best at temperatures of 27–31°C.

The twenty-one living species of crocodilians all have a greater or lesser commercial value. Among these, members of the genus *Crocodylus* are particularly in demand, including the Nile crocodile, (*C. niloticus*) the estuarine crocodile (*C. porosus*) and the freshwater crocodile of the island of New Guinea (*C. novaeguineae*). The American alligator (*Alligator mississippiensis*) is also much valued. Hundreds of thousands of individuals have been killed for the skin trade, but the meat of crocodilians has also been eaten by certain indigenous people, and by-products such as the penis and glands have also commercial value.

Breeding and rearing in captivity

Although haphazard attempts at breeding crocodiles in captivity on farms date back to the 1930s, it was not until the 1950s that more serious efforts were made. This was precipitated by a general decrease in the wild populations of almost all crocodile species, and a rapid increase in the value of their skins and other products.

Successful breeding of Nile crocodiles took place initially in 1970 in Zimbabwe. Until then eggs were collected and hatched at three government rearing-stations. This was followed by breeding in capitivity of the American alligator,

the freshwater crocodile of Papua New Guinea, the estuarine crocodile, the marsh crocodile and the caiman. None of these breeding operations are large-scale, with the exception of that of Mr Yangprapakorn near Bangkok, Thailand, who has successfully reared thousands of crocodiles of different species.

Success in rearing young crocodiles has been demonstrated in several parts of the world using a variety of pen designs, stock densities and staple feeds. Young crocodiles require scrupulous attention to basic essentials – including hygiene.

In the early 1970s a network of village rearing pens was developed in Papua New Guinea. Some large, technically more advanced rearing stations have been included in this system to provide an outlet for small, live crocodiles that are surplus to the village rearing capacity but too small for skin production. The collection of small crocodiles rather than eggs has been encouraged because the chances of success at the village level are greatest if a start is made with young crocodiles. So far, captive breeding has not been encouraged because that requires facilities and resources not easily acquired by the majority of the village people.

Future prospects

Apart from the American alligator and the New Guinea crocodile, crocodilians are threatened with extinction because of habitat destruction and over-utilization. It is for this reason that nations ratifying the Convention of International Trade in Endangered Species (CITES) have agreed to operate far-reaching trade restrictions on all crocodile species listed on the endangered species schedules of the Convention. The present approach is, however, not to enforce these restrictions on farm-raised stock. This should be an incentive to further development of crocodile farms, particularly of *C. niloticus* and *C. porosus*. Breeding and rearing of alligators is now in progress in the southern parts of the United States, and in Papua New Guinea. At the Madras Crocodile Bank successful breeding has been obtained with the mugger (*C. palustris*). Morelet's crocodiles (*C. moreletii*) have produced three successful clutches in the Atlanta Zoo.

There is no doubt that within the tropics crocodiles can be bred and reared in captivity on a commercial basis. The economics of the business will depend very largely on local labour and food costs. The use of crocodiles as an economic resource need not involve the extermination of wild populations.

Turtles

Marine turtles so far have not been bred and reared in captivity on a commercial basis, although the green turtle (*Chelonia mydas*) has been raised from eggs or hatchlings in confinement. The young are given special feed and protection. Places where this activity is carried on include Réunion, the Cayman Islands, Surinam, Australia (Torres Strait islands) and Indonesia. Experimental breeding of marine turtles has been started in the Cayman Islands.

Small tortoise species are bred and raised for the pet trade in several countries, including Japan and the United States.

Snakes

Although snakes are being kept for the production of venom and skins on a commercial basis, these are generally wild-caught specimens. Therefore, so far, there is no questions of true domestication of snakes.

References

Bolton, M. (1980) Crocodile management in Papua New Guinea. *World Animal Review*, No. 34: 15–23
De Vos, A. (1979) A manual on crocodile management. Mimeo., FO: MISC/79/30 (draft). FAO, Rome
Yangprapakorn, U., McNeely, J. A. and **Cronin, E. W.** (1971) In: *Crocodiles*, IUCN Publications (new ser.) Supplementary Paper No. 32, pp. 98–101

58
Edible frogs

D. D. Culley

School of Forestry and Wildlife Management, Louisiana State University, Baton Rouge, U.S.A.

Introduction

Consumption of edible frogs could probably be traced back to cave man. These excellent-tasting animals have commanded an intense interest by man down through the centuries. The American bullfrog (*Rana catesbeiana*) is so highly esteemed that it has been introduced into Mexico, South America, several Asian countries and Pacific islands, and Europe. However, there exist over twenty other large species of anurans equal to the American bullfrog in quality. Not one species has been intensively cultured on a commercial scale, or extensively cultured on a serious commercial scale (confining a wild population of frogs to fenced-in ponds and allowing them to reproduce and grow under natural conditions).

The international human food market is currently supplied with wild-caught frogs. Just how many species contribute to this pool is unknown, but the dominant species include *Rana catesbeiana, R. tigrina, R. hexadactyla, R. esculenta*, and possibly *R. ridibunda*. There are several other species eaten locally in many countries. The species shown in Table 58.1 are reported to be consumed. Most of the informa-

Table 58.1 Large edible frogs of the world

Continent and data sources	Species	S–V length*	Culture techniques	Notes
Africa (Wager 1965, Cochran 1967, Stewart 1967, Zahl 1967, Lammotte and Perrett 1968, Gewalt 1977)	*Conraua goliath*	30 cm and above	Not developed	Largest edible frog (weight 4 kg or more). Eaten locally and rarely
	Conraua robusta	→ 14 cm	Not developed	May be eaten locally
	Pyxicephalus adspersus	→ 22 cm	Not developed	Aggressive and inflicts painful bite. Small legs. Possibly eaten
	Rana fuscigula	8 cm and above	Not developed	May be eaten locally
	Rana vertebralis	→ 15 cm	Not developed	Largest *Rana* in South Africa. May be eaten locally
Asia (Taylor 1966, Cochran 1967, Das *et al* 1975)	*Glyphoglossus molossus* (narrow-mouthed toad)	8 cm (may reach 10 cm)	Not developed, but common in rice paddies where it is extracted from holes	Eaten locally (Thailand)
	Rana acanthi	→ 10 cm	Not developed	Probably eaten
	Rana blythi	→ 27 cm	Not developed	Eaten locally
	Rana hexadactyla	→ 13 cm	Some extensive culture reported. Intensive culture under study (India)	Processed legs widely exported
	Rana magna	→ 13 cm	Not developed	Eaten locally (Philippines)
	Rana moodiei	→ 13 cm	Some extensive culture reported	Probably recognized now as *R. cancrivora*[†]. Eaten locally (Philippines)
	Rana tigrina (Indian bullfrog)	→ 25 cm	Intensive culture not developed but under study (India)	Processed legs widely exported
	Rana vittigera	8 cm (may reach 10 cm)	Not developed	Probably recognized now as in the *R. limnocharis*[†] complex. Eaten locally (Philippines)
Europe (Cochran 1967, Arnold and Burton 1978)	*Rana dalmatina* (agile frog) *Rana esculenta* (edible frog) *Rana lessonae* (pool frog) *Rana ridibunda* (marsh frog) *Rana temporaria* (common frog)	10–15 cm	Intensive culture techniques poorly developed but some extensive culture reported	All consumed to varying degrees. *R. ridibunda* is usually the largest

Table 58.1 (continued)

Continent and data sources	Species	S–V length*	Culture techniques	Notes
North America (Conant 1958, Cochran 1967, Nace *et al.* 1974, Priddy and Culley 1971)	*Rana aurora*	→ 12 cm	Not developed	Rarely eaten
	Rana catesbeiana (bullfrog)	→ 20 cm	Intensive techniques reasonably well developed (U.S.A.) Extensive culture widely reported	Introduced into several European, South American and Asian countries. Widely consumed
	Rana grylio	12 cm (may reach 16 cm)	Not developed	Only eaten locally
	Rana hecksheri (river frog)	→ 15 cm	Not developed	Only eaten locally
	Rana pipiens complex	9 cm (may reach 16 cm)	Intensive techniques reasonably well developed (U.S.A.).	Consumption declining
South America (Mertens 1960, Cochran 1967, Duellman 1978)	*Battachophrynus microphthalmus*	→ 20 cm	Not developed but some work in progress	Totally aquatic, cold-water resident. Leg size reduced. Heavily eaten locally
	Caudiverbera caudiverbera (water frog)	→ 20 cm	Not developed	Eaten Previously classified as *Calyptocephalella gayi*
	Ceratophrys cornuta (horned frog)	→ 15 cm	Not developed	Reduced legs. Consumption doubtful
	Ceratophrys dorsata (horned frog)	→ 13 cm	Not developed	Reduced legs. Consumption doubtful
	Leptodactylus fallax (mountain chicken)	→ 15 cm	Not developed	Eaten
	Leptodactylus pentadactylus (bullfrog)	→ 22 cm	Extensive culture techniques under study	Eaten
	Rana dunni	→ 12 cm	Not developed	Eaten locally
	Rana palmipes	→ 13 cm (possibly more)	Not developed	Eaten
	Telmatobius culeus (Lake Titicaca frog)	→ 20 cm	Not developed	Reduced legs. Totally aquatic, cold-water resident. Eaten by some ethnic groups

*S–V length: Distance from snout (nose) to vent (cloacal opening)
[†]Taxonomic classification provided by Dr Robert F. Inger, Field Museum of Natural History, Chicago, Illinois

tion on consumption was obtained by the author over several years of correspondence. Frogs with a maximum body length of less than 10 cm are not included in the list but these smaller frogs are consumed.

Culture systems

Extensive culture has a long history (Chamberlain 1897, Stoutamire 1932, Louisiana Department of Conservation 1938) and is still the dominant interest. These systems normally involve construction of fenced ponds to prevent the escape of the frogs. Flowing water is frequently provided to aid in sanitation and maintain water of satisfactory quality. In some cases, more elaborate concrete raceways are constructed and floating feeding trays provided to confine live food. However, the numbers produced to an edible size probably will not exceed 400 per ha per year which is hardly sufficient for a commercial operation.

In contrast, under an intensive culture system where the frogs are confined within buildings and where living food (fish and crayfish) is produced in ponds and transferred to the frogs, literally thousands of frogs can be produced with food coming from a 1-ha pond (see next section).

Thus, based on current knowledge, frogs must be cultured like many domesticated animals today, confined where they can be tended daily, provided with ample food, treated for diseases, maintained in sanitary conditions, under temperature and light control, and bred when needed. The techniques developed in the author's laboratory and tested on a commercial scale (Priddy and Culley 1971, Culley *et al.* 1978) with *Rana catesbeiana*, and in the Amphibian Facility at the University of Michigan (Nace *et al.* 1974, Nace and Rosen 1978) with *R. pipiens* (leopard frog) are the only two systems demonstrating that thousands of edible frogs can be produced on a continuous basis. The outdoor systems of Kawamura (Laboratory for Amphibian Biology University of Hiroshima), a patented system (Stearns 1939), and others visited by the author, could be developed commercially, but in all it was necessary to confine the frogs in order to secure total control.

The keys to domestication

Controlled breeding

Controlled breeding of edible frogs is confined to a few species under laboratory-type conditions; primarily *R. catesbeiana* and the *R. pipiens* complex (Easley *et al.* 1979, 1981, Culley *et al.* 1981, Nace 1968). Other species, cultured extensively, breed on a seasonal cycle as dictated by the natural environmental conditions.

It is reported that under natural conditions periods of 3–5 years are required for frogs to grow to an edible size. But when wild edible frogs are confined, maintained at appropriate temperatures and provided with ample food, their growth rates are rapid (Culley and Gravois 1970) and a few individuals reach 450 g in 4 months (Culley *et al.* 1978). *Rana catesbeiana* can be cultured to a marketable size (for research) in 3–5 months post-metamorphosis and for human consumption in 5–8 months. The tadpole stage requires only 8–10 weeks.

We have bred large 4-month-old bullfrogs (450 g) successfully. The first generation offspring grew about 30 per cent faster than wild stock held in our system (Culley, unpublished) and the selection programme is still ongoing.

Food supply

Frogs seem to require live food; this is a major deterrent to intensive culture. However, in tropical climates, production of certain species of fish in the Poecilidae family has exceeded 10 000 kg per ha per year (Shang and Baldwin 1980). An acceptable frog of edible size would weigh about 200 g. The production of *R. catesbeiana* to this size, for example, would require about 600 g of food. Thus, 16 667 bullfrogs could theoretically be produced on living food from a 1-ha pond.

The current wholesale price for frozen, packaged, wild-frog legs sold in the United States seldom exceeds $13 per kg. It would take about eight cultured bullfrogs to yield 1 kg of processed legs, bringing in only $1.62 per frog, and cost about $2 to produce the frog (Culley and Baldwin 1981). Thus with today's prices and current status of culture the United States market would not support commercial frog culture for processed legs. In other countries where higher prices are paid for legs, labour costs are lower, and smaller legs accepted, it might be economically feasible to culture the bullfrog, particularly if selection for rapid growth can be achieved, the front legs and back are processed with the hind legs, and the skin is used for leather.

References

Arnold, E. N. and **Burton, J. A.** (1978) *A Field Guide to the Reptiles and Amphibians of Britain and Europe.* Collins: London

Chamberlain, F. M. (1897) Notes on the edible frogs of the United States and their artificial propagation.

In: *United States Bureau of Fisheries Report for 1897*, pp. 249–61

Cochran, D. M. (1967) *Living Amphibians of the World*. Doubleday: New York

Conant, R. (1958) *A Field Guide to the Reptiles and Amphibians of the United States and Canada*. Houghton Mifflin: Boston, Massachusetts

Culley, D. D. and Baldwin, W. J. (1981) The feasibility of mass culture of the bullfrog in Hawaii. Unpublished paper available from Fisheries Section, Room 249 Agricultural Center, Louisiana State University: Baton Rouge, Louisiana 70803

Culley, D. D. and Gravois, C. T. (1970) Frog culture, a new look at an old problem. *American Fish Farmer*, 1 (5): 5–9

Culley, D. D., Horseman, N. D., Amborski, R. L. and Meyers, S. P. (1978) Current status of amphibian culture with emphasis on nutrition, diseases and reproduction of the bullfrog, *Rana catesbeiana*. In: J. W. Avault (ed.) *Proceedings World Mariculture Society*, 9: 653–70

Culley, D. D., Othman, A. M. and Easley, K. A. (1981) Predicting ovulatable bullfrogs (*Rana catesbeiana*) by morphometry. Unpublished paper available from Fisheries Section, Room 249 Agricultural Center, Louisiana State University: Baton Rouge, Louisiana 70803

Das, C. R., Rao, P. L. N., Mohanty, S. N. and Panigrahi, V. (1975) A note on some interesting observations on the breeding behaviour of *Rana hexadactyla* Lesson in captivity. *Journal of the Inland Fisheries Society of India*, 7: 120–2

Duellman, W. E. (1978) *The Biology of an Equatorial Herpetofauna in Amazonian Ecuador*. University of Kansas Press: Lawrence

Easley, K. A., Culley, D. D., Horseman, N. D. and Penkala, J. E. (1979) Environmental influences on hormonally induced spermiation of the bullfrog, *Rana catesbeiana*. *Journal Experimental Zoology*. 207: 407–16

Easley, K. A., Culley, D. D. and Othman, A. M. (1981) Reproductive control of the female bullfrog, *Rana catesbeiana*. Unpublished paper available from Fisheries Section, Room 249 Agricultural Center, Louisiana State University: Baton Rouge, Louisiana 70803

Gewalt, V. W. (1977) Einige bemerkungen über fang, transport, und haltung des goliathfrosches (*Conraua goliath* Boulenger). *Jena Zoological Garten* N. F., 47: 161–92

Lamotte, M. and Perrett, J. L. (1968) Revision du genre *Conraua* Nieden. *Bulletin Institut fondamental d'Afrique noire*, 30 ser. A. (No. 4): 1603–44

Louisiana Department of Conservation (1938) *Frog Industry in Louisiana*, Bulletin No. 26. Department of Conservation, Division of Fisheries: New Orleans, Louisiana

Mertens, R. (1960) *The World of Amphibians and Reptiles*. Harrap: London

Nace, G. W. (1968) The amphibian facility at the University of Michigan. *BioScience*. 18: 767–75

Nace, G. W., Culley, D. D., Emmons, M. B., Gibbs, E. L., Hutchison, V. H. and McKinnell, R. G. (1974) *Amphibians, Guidelines for the Breeding, Care, and Management of Laboratory Animals*. National Academy of Sciences: Washington, D.C.

Nace, G. W. and Rosen, J. K. (1978) Sources of amphibians for research. In: *Cloning: Nuclear Transplantation in Amphibia*, ed. R. G. McKinnel, pp. 251–75. University of Minnesota Press: Minneapolis

Priddy, J. M. and Culley, D. D. (1971) The frog culture industry, past and present. *Proceedings Annual Conference Southeastern Association of Game and Fish Commissioners*, 25: 597–601

Shang, U. C. and Baldwin, W. J. (1980) Economic aspects of pond culture of topminnows as an alternative baitfish. In: *Proceedings World Mariculture Society*, ed. J. W. Avault 11: 592–5

Stearns, J. E. (1939) Tanks simplify bullfrog culture. *Illustrated Mechanix*. 1939 (6): 51

Stewart, M. M. (1967) *Amphibians of Malawi*. State University of New York Press: New York

Stoutamire, R. (1932) *Bullfrog Farming and Frogging in Florida*. Bulletin No. 56. Florida Department of Agriculture: Tallahassee, Florida

Taylor, E. H. (1966) *Amphibians and Turtles of the Philippine Islands*. Asher: Amsterdam (translated; original published 1921)

Wager, V. A. (1965) *The Frogs of South Africa*. Purnell: Johannesburg

Zahl, P. A. (1967) In quest of the world's largest frog. *Journal National Geographic Society*, 132 (6): 146–52

59

Common carp

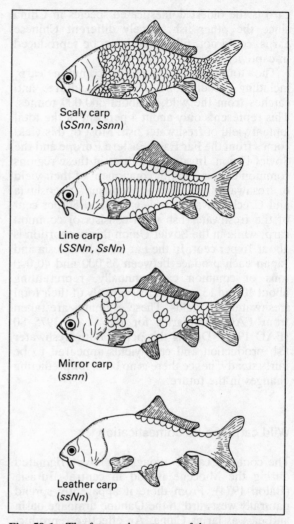

Scaly carp
(SSnn, Ssnn)

Line carp
(SSNn, SsNn)

Mirror carp
(ssnn)

Leather carp
(ssNn)

Fig. 59.1 The four scale patterns of the common carp.

G. W. Wohlfarth

Fish and Aquaculture Research Station, Dor, Israel

Introduction

The common carp *Cyprinus carpio* is the only species in the genus *Cyprinus*. It has been suggested that this species should be divided into two subspecies, i.e. the European carp, *C. carpio carpio* and the eastern or Asian carp, *C. c. haematopterus* (Balon 1974). A sub-classification into varieties has been suggested to describe the four different scale patterns of the common carp, i.e. scaly (wild type), mirror, line and leather (Alikhuni 1966), (see Fig. 59.1). These sub-classifications are not justified since spawning between the groups is accomplished with the same ease as within each group, and the resulting progeny are fertile. The major variation of scale pattern is a result of Mendelian segregation, involving two unlinked loci, with two alleles at each locus (Probst 1949, 1950). A number of intergeneric crosses have been attempted between the common carp and other cyprinids, the most successful being with *Carassius carassius* (Alikhuni 1966). Progeny from these intergeneric crosses showed low fertility, often almost total sterility.

The diploid chromosome number of the common carp was originally estimated as $2n =$ 104, though later investigations suggest that $2n =$ 100 (Ojima and Hitotsumachi 1967). The carp's chromosomes are short and rather uniform in length and there is no cytological evidence for a heteromorphic pair of sex chromosomes.

Fish farming has been practised in China for about 4000 years and in Europe for several hundred years (Mann 1961). The common carp was the first fish species to be domesticated, due principally to its ease of spawning in captivity. It is the major fish species in European aquaculture, but a relatively minor component in Chinese aquaculture systems (Tang 1970). The common carp is the oldest domesticated species in China since the other fish, largely different Chinese carps, could not, till recent years, be reproduced in captivity.

The total annual yield of common carp, including production from aquaculture and catches from the wild, is about 300 000 tonnes. This represents only about 4 per cent of the total annual yield of freshwater fish. Most of this yield comes from the Far East, Eastern Europe and the Soviet Union. In some countries in these regions common carp is a major component of their yield of freshwater fish. Thus in Romania, Yugoslavia and Czechoslovakia between 40 and 80 per cent of the freshwater fish yield consists of common carp, while in the Soviet Union this proportion is about 16 per cent. In the Far East, Indonesia and Japan each produce between 35 000 and 40 000 tons of common carp annually, representing about 10 and 15 per cent respectively of their total freshwater fish yields. These estimates are taken from FAO summaries for the years 1975–80 (FAO 1981). During this period both freshwater fish production and carp yields appeared to be fairly steady; hence there is no basis for predicting changes in the future.

Wild carp and its domestication

The common carp is thought to have originated during the Miocene period in central Eurasia (Balon 1974). From there it apparently spread naturally westward to the Danube drainage basin and east as far as China. An effective reproductive barrier appears to have been formed by this geographic isolation, resulting in two races of the common carp. The divergence between these races resulting from their geographic isolation increased in response to their separate domestication in these two areas.

The natural distribution of the common carp in both Europe and Asia appears to have taken place only during the last post-glacial period. This is deduced from the fact that carp did not penetrate naturally to America via the Behring land bridge or to the British Isles. Carp were transported from the Danube area to Italy by the Romans and, like marine fish, stocked into holding ponds (*piscinae*) to ensure a supply of fresh table fish (Balon 1974). This cannot be regarded as the beginning of the domestication of

carp in Europe since there is no record of the Romans using fish spawned in captivity for farming.

The first true domestication took place between the 7th and 13th centuries in monastery ponds. Presumably these were originally used to ensure a supply of fresh fish for days on which the consumption of meat was prohibited by religious regulations (Balon 1974). This led to fish farming and the domestication of carp.

Differences between the Chinese and European races of the common carp

The common carp was domesticated separately in Europe and China. Aquaculture techniques practised in these regions differ widely. Genetic adaptation in response to these different techniques resulted in the differences between the European and Chinese cultivated carp (Moav *et al*. 1975, Wohlfarth *et al*. 1975a).

Since the 19th century methods of cultivating carp in Europe include the use of spawning ponds for fry production, separate raising of each year-class, growing the fish at low stocking rates during their second and third years and the use of supplemental feed (Mann 1961). Harvesting is carried out at the end of summer, by draining the ponds and removing all the fish. Brood stock is selected by choosing the largest and highest-backed fish from this harvest. Most carp grown in Europe possess the (mutant) mirror scale pattern.

In China carp have been grown for centuries in subsistence polyculture ponds. The number of carp per unit area is high and not controlled, since common carp spawn in these ponds. The other fish, mainly various species of Chinese carps are stocked, since they do not reproduce in captivity. Little if any supplemental feed is used, the fish subsisting largely on 'natural food' produced in the ponds. This natural production is stimulated by heavy and frequent manuring. Other nutrient inputs consist of various waste products and refuse. The ponds are drained irregularly, if at all. Harvesting is carried out by frequent seining and selective cropping of the larger individuals, the smaller ones being returned to the ponds. Brood stocks, consisting of carp left in the ponds, are continuously augmented by fish from the wild. These differences in husbandry have brought about differences between the two carp races in the following traits.

Fecundity. The Chinese carp is by far the more fecund, hence its name big belly carp or fat belly carp (Alikhuni 1966), the bellies of both sexes being swollen by their large gonads. The very harsh conditions under which the Chinese carp reproduced, compared to European spawning ponds, must have resulted in very high egg and fry mortalities. The response of the Chinese carp was to produce larger numbers of gametes. In the Chinese carp the gonado-somatic index is higher, and onset of sexual maturity earlier than in the European carp.

Viability. The Chinese carp is much more viable than the European.

Seine escapabality. The Chinese practice of harvesting by seining tends to remove the more easily caught individuals, leaving the 'escapees' to reproduce. This is equivalent to selection for seine escapability. European carp, which were not subjected to this selective pressure, are much more easily seined than Chinese carp (Wohlfarth *et al*. 1975b).

Growth rate. Growth of Chinese carp immediately after hatching is faster than that of European (Hulata *et al*. 1976). This is seen as a response to the harsh conditions in the all-purpose Chinese fish-pond, carp fry being more sensitive than older fish. In later growth, however, the European carp is superior. Selective harvesting in Chinese ponds is equivalent to selection for slower growth, whereas in Europe the larger fish are consciously selected.

Body shape. The European carp is a high-backed fish, as a result of conscious selection for this trait, whereas the Chinese carp has a much more elongated body.

It may be noted that in most of the traits in which the races differ, the Chinese carp are more similar to wild carp than the European. The Chinese carp should be regarded as a semi-domesticated animal, since wild fish are frequently added to breeding stocks.

The effects of domestication on the European carp

At least two comparative studies have been carried out between the European wild and domesticated carp. Wild carp were taken from the Danube (Rudzinski 1961) and from Lake Valence in Hungary (Steffens 1964), spawned in captivity and their progeny compared with domesticated European carp. It is to be doubted whether these wild carp are identical with wild carp before domestication. Some fish always escape from ponds into the surrounding drainage basin, leading to some introgression of the domesticated carp's hereditary material into wild carp populations.

Most of the differences found between wild and domesticated carp (Table 59.1) are explicable as a result of domestication. The faster growth and higher back of the domesticated carp are a direct result of artificial selection for these traits. The later sexual maturity of the domesticated carp may be due to the partial relaxation of natural selection for early sexual maturity. It may also be due to a correlated response to artificial selection for faster growth, which may be accompanied by delayed sexual maturity. The tameness of the domesticated carp is possibly due to relaxation, under domestication, of natural selection for wild-type behavioural patterns. The lower number of erythrocytes and the lower concentration of sugar in the blood of the domesticated carp may be a result of lower requirements for bursts of rapid activity in the domestic environment.

The effects of domestication on the Japanese common carp and a comparison between Japanese and European domesticated strains

Wild Japanese carp, domesticated strains of the Japanese carp and domesticated strains of the European carp were found to differ in a number of traits. The progenitors of the wild Japanese carp were collected from Lake Kasumiga; the domesticated Japanese carp were of the Yamato and Asagi strains, while the domesticated European carp, both scaly and mirror, had been introduced to Japan in 1968.

The European domestic carp grew faster than the Japanese domestic carp, with the Japanese wild carp showing the slowest growth. These growth differences were more pronounced when supplemental feed was applied than in a fertilized pond, hinting at a genotype × environment interaction. The same ranking between the three groups was found for body length, number of gill rakers on the first arch, ratio of body height to body length and ratio of intestinal length to body

Table 59.1 Comparison between wild and domesticated European carp (From Steffens 1964)

Traits			Comments
Skeletal traits	No. of vertebrae No. of gill rakers No. of pharyngeal teeth No. of intermuscular bones		No difference between
Proportional size of:	Heart Kidneys Pancreas Swim bladder Fillet		Wild and domesticated
Disease resistance	To infectious dropsy		
Blood contents	No. of erythrocytes Conc. of haemoglobin Conc. of blood sugar		Higher in wild
Body shape	Length: height Length: width Length: head length		Higher in wild
	Proportional head size Proportional gut length Proportional mouth opening		Higher in domesticated
Composition of flesh	% dry matter % fat % protein		Higher in wild when no supplemental food presented
Growth	Weight, length		Faster in domesticated
Sexual maturity	Age (years)	females males	4 in domesticated, 3 in wild 3 in domesticated, 2 in wild
	Proportional gonad weight at 3 years	females males	Higher in wild No difference
Air bladder	Relation of anterior: posterior chamber		Higher in domesticated
Behaviour	Ease of handling, seining		Domesticated tamer

length (Suzuki and Yamaguchi 1980). These groups also differ in behavioural characteristics. In angling tests the European mirror carp was by far the most catchable on artificial bait and the Japanese wild carp the least catchable. The two domestic Japanese strains, though differing between themselves, showed intermediate catchabilities. No differences in catchability between the groups were found when earthworms were used as bait. Suzuki and Yamaguchi (1980) conclude that these differences are a result of domestication and that the European carp has been more intensely domesticated than the Japanese.

Current breeding methods

Selective breeding has not been carried out in China with any of the fish cultured. Presumably this is due to the inability, till recent years, of reproducing any of the Chinese carps, while the common carp often constitutes only a minor component of the fish stocked. The traditional European method of selection was recently tested in a series of controlled experiments. No response to selection for faster growth was obtained, i.e. progeny of the larger individuals did not grow faster than progeny of random samples (Moav and Wohlfarth 1976). Other recent findings show that growth rate, fecundity and disease resistance are reduced and the incidence of skeletal deformations increased by inbreeding (Moav and Wohlfarth 1973, Hines *et al.* 1974). Heterosis for growth rate has been demonstrated repeatedly and is applied in commercial fish farming in Israel and Hungary (Wohlfarth *et al.* 1965, Bakos 1979).

Other breeding plans include the use of wild carp strains. A strain of carp able to grow at low temperatures was produced by crossing a cold-resistant wild carp from the River Amur (on the border between Siberia and Manchuria) with a domesticated European strain (Kirpichnikov 1972). After a complex system of intercrosses and back crosses, the 'Ropsha' carp was isolated; it apparently grows faster than either of its progenitors in northeastern Soviet Union and is more resistant to infectious dropsy and to an airbladder disease. Another study was undertaken to improve manure utilization of carp, by crossing a Chinese strain to a domesticated European strain. The resulting crossbred showed a faster growth rate than its parents, or other crossbreds, when manure was the major nutrient input (authors' unpublished data).

Selective breeding of a different type led to the production of Japanese fancy carps. These consists of blue, grey, red, orange, yellow and white fish as well as multi-coloured individuals (Axelrod 1973). These fish were first produced in the 19th century in an isolated mountainous area of Japan (Niigato Prefecture), by isolating rare coloured variants from production ponds. The heredity of the different colours and colour combinations is not known. Some are claimed to have been 'stabilized', but in many cases, the frequency of individuals with the desired colour pattern is less than 1 per cent. Originally these fancy carp were of purely Japanese origin, but later some hybridization with European carp was practised, resulting in fish with mirror and line-scale patterns.

Future prospects

Presently available breeding knowledge is not fully utilized in practical aquaculture. Methods of selective breeding of common carp are applied in Israel, Hungary, the Soviet Union and possibly China. The slow growth rate of European carp resulting from inbreeding (Moav *et al.* 1964, 1975), could be improved by crossbreeding between different strains. In the Far East, substitution of crossbreds between the big belly carp and European carp for purebred big belly carp has been shown to result in a considerable increase in growth rate but a decreased tolerance to anoxia (Sin 1982). Similarly use of the cold-resistant Ropsha carp should make possible carp culture in areas in which low temperatures is the major limiting factor.

References

Alikhuni, K. H. (1966) Synopsis of biological data on common carp, *Cyprinus carpio* L. 1958 (Asia and the Far East). *FAO Fisheries Synopsis*, **31**: 1. FAO: Rome

Axelrod, H. R. (1973) *Koi of the World, Japanese Colored Carp*. T.F.H. Publications: Hong Kong

Bakos, J. (1979) Crossbreeding Hungarian races of common carp to develop more productive hybrids. In: *Advances in Aquaculture*, ed. T. V. R. Pillay and

W. Dill. Fishing News Books: Farnham, Surrey, England

Balon, E. K. (1974) Domestication of the carp, *Cyprinus carpio* L. *Royal Ontario Museum Life Sciences Miscellaneous Publication*

FAO (1981) *Yearbook of Fisheries Statistics*, Vol. 50. Catches and landings 1980. FAO: Rome

Hines, R. S., Wohlfarth, G. W., Moav, R. and Hulata, G. (1974) Genetic differences in susceptibility to two diseases among strains of the common carp. *Aquaculture*, 3: 187–97

Hulata, G., Moav, R. and Wohlfarth, G. (1976) The effects of maternal age, relative hatching time and density of stocking on growth rate of fry in the European and Chinese races of the common carp. *Journal of Fish Biology*, 9: 499–513

Kirpichnikov, V. S. (1972) Methods and effectiveness of breeding the Ropshian carp. Communication 1. Purposes of breeding, initial forms and systems of crosses. *Genetika*, 8: 65–72

Mann, H. (1961) Fish cultivation in Europe. In: *Fish as Food*, Vol. 1, ed. G. Bergstrom. Academic Press: London and New York

Moav, R., Hulata, G. and Wohlfarth, G. (1975) Genetic differences between the Chinese and European races of the common carp: I. Analysis of genotype-environment interaction for growth rate. *Heredity*, 34: 323–40

Moav, R. and Wohlfarth G. (1973) Carp breeding in Israel. In: *Agricultural Genetics*, ed. R. Moav. Wiley: New York

Moav, R. and Wohlfarth, G. (1976) Two way selection for growth rate in the common carp (*Cyprinus carpio* L.). *Genetics*, 82: 83–101

Moav, R., Wohlfarth, G. and Lahman, M. (1964) Genetic improvement of carp: VI. Growth rate of carp imported from Holland relative to Israeli carp and some cross-bred progeny. *Bamidgeh*, 16: 142–9

Ojima, Y. and Hitotsumachi, S. (1967) Cytogenetic studies on lower vertebrates: IV. A note on the chromosomes of the carp (*Cyprinus carpio*) in comparison with those of the funa and the gold fish (*Carassius auratus*). *Japanese Journal of Genetics*, 42: 163–7

Probst, E. (1949) Vererbungsversuchen beim Karpfen. *Allgemeine Fischerei Zeitung*, 74: 436–43

Probst, E. (1950) Der Todesfaktor bei der Vererbung des Schuppenkleides des Karpfens. *Allgemeine Fischerei Zeitung*, 75: 339–70

Rudzinski, E. (1961) Vergleichende Untersuchungen über den Wildkarpfen der Donau und den Teichkarpfen. *Zeitschrift für Fischerei und deren Hilfswissenschaften*, 10 N.F.: 105–35

Steffens, W. (1964) Vergleichende anatomisch-physiologische Untersuchungen an Wild- und Teichkarpfen. *Zeitschrift für Fischerei und deren Hilfswissenschaften*, 12 N.F.: 725–800

Sin, A. W. (1982) Stock improvement of the common carp in Hong Kong through hybridization with the introduced Israeli race 'Dor-70'. *Aquaculture*, 29: 299–304

Suzuki, R. and Yamaguchi, M. (1980) Meristic and morphometric characters of five races of *Cyprinus carpio*. *Japanese Journal of Ichthyology*, 27: 199–206

Tang, Y. (1970) Evaluation of balance between fishes and available fish foods in multispecies fish culture ponds in Taiwan. *Transactions American Fisheries Society*, 99: 708–18

Wohlfarth, G., Lahman, M., Moav, R. and Ankorion, Y. (1965) Activities of the carp breeders union in 1964. *Bamidgeh*, 17: 9–15

Wohlfarth, G., Moav, R. and Hulata, G. (1975a) Genetic differences between the Chinese and European races of the common carp: II. Multicharacter variation – a response to the diverse methods of fish cultivation in Europe and China. *Heredity*, 34: 341–50

Wohlfarth, G., Moav, R., Hulata, G. and Beiles, A. (1975b) Genetic variation in seine escapability of the common carp. *Aquaculture*, 5: 375–87

60

Goldfish

Zhang Zhong-ge
Beijing Agricultural University, China

Introduction

The goldfish derives from China and all goldfish today are descendants of fish from that country. The Chinese for (wild) goldfish is *ji-yu*; the domestic variety is called *jin ji-yu* (golden ji fish) which is commonly abbreviated to *jin-yu*. In Japanese this becomes *kingyo*.

Its scientific name is *Carassius auratus auratus* and it is a member of the family Cyprinidae of the order Cypriniformes. There are several theories concerning its closest relatives which are discussed below.

Goldfish have a small head and wide lips. The belly is round and swollen and the body covered with mucus. The tail fin is two-lobed but in fancy breeds it may be trifurcate or quatrefoil and in these it hangs down loosely. In extreme cases the tail fin may be as much as one-and-a-half times the length of the body. The body is entirely covered with scales which are typically golden but there are many colour varieties. The gold colour only develops after 50 days of age. The varied colours and forms are the result of artificial selection and represent a work of art from China.

Goldfish grow fastest at 30 °C. Adult length seldom exceeds 30 cm and weight 900 g. Goldfish are very resistant to heat and disease and can survive in poorly oxygenated waters for a long time.

Wild goldfish are eaten but the domestic variety was used only in medicine, e.g. as a cure for dysentery.

Wild species

Wild goldfish exist only in eastern Asia (chiefly China). Their natural colour is: back – olive green or brown; sides – golden or silver; underside – silver. Their natural habitat is weedy rivers and lakes. The body is long and laterally compressed.

In China the wild goldfish is thought to be a variety of the crucian carp, *Carassius carassius* (Chen Zhen 1959). In Europe and the Soviet Union the two species are regarded as distinct (Syvorov 1948, Wheeler 1969). The crucian carp is a wild species found naturally from eastern England across Europe and Siberia as far as the Lena river. A closely related species, the gibel or Prussian carp, also occurs over much of this range from the southern Baltic States eastwards to the Pacific Ocean and is found in the Amur river. For many years the two were confused (for they look very similar) but the second was given the name *Carassius gibelio*; occasionally the two were treated as subspecies so that the second became *Carassius carassius gibelio*. More recent studies have shown that the gibel is more closely related to the goldfish than to the crucian carp and it is treated now as a goldfish subspecies (*C. auratus gibelio*). Modern opinion of the three forms is therefore as follows.

1. *Carassius a. auratus*, the goldfish, distributed in China and north Vietnam in the wild form and widely kept as the domesticated goldfish.
2. *Carassius a. gibelio*, the gibel carp, distributed from eastern Europe to the Pacific coast of Siberia.
3. *Carassius carassius*, the crucian carp, distributed from England to central Siberia and farmed as a secondary crop in fish-farms in inland Europe.

The members of the genus *Carassius* differ from the carp, *Cyprinus carpio*, in lacking barbels on the lips, and internally by having only a single row of pharyngeal (throat) teeth (the carp has three rows); the teeth also differ in shape.

The crucian carp differs from the *C. auratus* subspecies in less obvious characters, notably in having a finely serrated rear edge to the spine in the front of the dorsal and anal fins, between six and eight branched rays in the anal fin, and twenty-six to thirty-one gill rakers on the first gill arch; *C. auratus* has coarsely serrated spines in the fins, five or six branched rays in the anal fin, and thirty-five to forty-eight gill rakers.

The goldfish and the gibel carp are difficult to distinguish. The colour of the goldfish always tends towards golden tints after its first year of life, and the internal lining of the body cavity is light (the peritoneum of the gibel carp is black).

The subspecies hybridize and the hybrids are fertile; both form fertile hybrids with the crucian carp. All three *Carassius* taxa hybridize with the carp, *Cyprinus*, but hybrid males are sterile. (The carp has 100 chromosomes compared with only 94 in *Carassius*.)

The distribution of all three *Carassius* taxa has been extended by human agency and the goldfish has been spread to many temperate regions of the world. It is now found in lakes and rivers of much of Europe and the United States, as well as Australia, New Zealand, South Africa and South America. Many of these free-living populations are composed of fish which have not developed their bright golden colour, and have reverted to the colouring of the wild form. However, many of the fully coloured, common goldfish are raised in semi-natural, outdoor conditions in northern Italy and the United States.

Domestication and early history

It is recorded in the ancient Chinese book *Shu Yi Ji* (*Records of the Unusual*) by Ren Fang (A.D. 460–508): 'When Huan Chong of the Jin dynasty toured Lushan mountain (Kiangsi) he found in its lake fish with red scales.' Thus the ancestor of goldfish was noted as early as A.D. 265–419. In the Sui and the Tang dynasties keeping goldfish for ornament came to be a common custom among the upper circles of the aristocracy. Domestication of goldfish, however, did not begin until the Sung dynasty (A.D. 960–1278). Two places of origin of goldfish have been recorded. One was the pond below Yuebo (Moonlit Waves) House in Jiaxing, Zhejiang province. According to local annals, 'there used to be a goldfish pond below

Yuebo House in Xiushan county. Governor Ding Yanzhan of the Tang dynasty acquired goldfish here and it was later turned into a Buddhist monks' pond for freed fish.' The other place was Hangzhou, where goldfish were found in the ravines below Luohe (Six Harmonies) Pagoda and also in the pond of Xingjiao Monastery opposite Jinci Monastery at the foot of Nanping Hill.

In the Southern Song dynasty (1127–79), Emperor Zhao Gou, very much given to keeping pet animals, opened a special pond for goldfish in his Deshou palace in Hangzhou. Following suit, officials and literary gentlemen went in for breeding goldfish in ponds and this became the fashion of the day. Consequently there came into being a group of people specializing in breeding goldfish. They learned to feed goldfish with tiny reddish 'insects' (perhaps daphnia). These fish breeders paid full attention to the cultivation of new breeds, thus exercising artificial selection. New breeds were thus preserved and so marked the turning-point from semi-domestication to full domestication.

Keeping goldfish in special ponds within house compounds as was practised by the aristocrats of the Southern Song dynasty differed from keeping goldfish in Buddhist monks' ponds for freed fish in the following ways. 1. In the special ponds only goldfish were bred whereas in monks' ponds crucian and other wild fish were kept as well as goldfish. In the special ponds crossbreeding between goldfish and other fish was therefore impossible and interspecific competition was prevented. As a result, better conditions were provided for the production and preservation of new breeds. 2. In the special ponds goldfish were not only protected and given cake titbits to eat, as were fish in monks' ponds, but were also fed with daphnia, which goldfish found especially to their liking. 3. With the extensive breeding of goldfish by goldfish lovers and professional breeders, artificial selection was carried out (Chen Zhen 1959).

Breeding techniques made much headway in the later years of the Ming dynasty. Goldfish were transferred from ponds to vats and basins. As this was labour-saving and economical, more and more people could afford to keep goldfish, placing fish-bowls in their homes as ornaments. The breeding of goldfish in vats and basins, moreover, made it possible for breeders to watch them more closely and for new breeds to be preserved from crossing. Unwitting selection was giving way to conscious artificial selection. Thus arose the large number of breeds in modern times, with their extremely varied forms and colours.

The limited space in vats or basins alleviated the struggle for survival. As a result the wild fish's laterally compressed body form, with pointed or thin head and narrow stomach, a form the facilitated swimming, gradually became short and fat, with swollen stomach. Most of the fins, especially the caudal fins, were enlarged and branched into several laminae. Their movements became slow and unhurried because, under artificial breeding, it was no longer necessary to hunt for food, to escape from predators, or to defend themselves. The pigment cells in the scales underwent changes and broke up into various colours. Even though goldfish have a history of artificial breeding extending over hundreds of years they have only three kinds of pigment cells in their scales: black, orange and light bluish (as do other fish). (Matsui 1972, says black, red and yellow.) The kaleidoscopic colours of domesticated goldfish are merely a result of varied composition and distribution of the three pigments, their varied intensities and concentrations, or the high intensity of one pigment and the weak intensity, or total disappearance, of another (Wang Yuchen and Qi Jingfen 1964).

Recent history

In China, after the reign of Emperor Daoguang (A.D. 1821–50) of the Qin dynasty, there were more and more goldfish lovers. The last years of the Qin dynasty were the heyday of goldfish breeding, which became quite a fashion in Beijing. When the Lunar New Year was approaching, people would buy goldfish and place fish-bowls on tables as New Year's Day ornaments. After the 1911 revolution goldfish yards were started and breeding and selling of goldfish came to be a specialized trade. Each yard boasted several big fish-ponds which lay close to one another. With green water and red goldfish the ponds looked very attractive. With the founding of the People's Republic in 1949, the Communist Party and the government have paid close attention to preserving such cultural heritages. Government departments in charge of

horticulture devote great efforts to goldfish breeding so that not only many lost varieties have been recovered but a number of new breeds have emerged. Before 1949, there were no more than 70 varieties in China but by 1958 they numbered 154. Goldfish breeding is no longer a pastime for the rich and the aristocratic only, but a recreational activity enjoyed by the broad masses of the people.

It is recorded that goldfish were first introduced into Japan about the year 1500 but it was nearly 200 years later before active breeding is described (Matsui 1972).

Goldfish were first introduced into England about the year 1705 (although they may have arrived in Portugal in 1611). Large numbers were imported in the 1730s and by the middle of the century they were widespread in England. They first appeared in Scandinavia before 1740, France about 1750, Holland in 1753–54, Italy in 1755, Germany about 1780, Russia before 1791 and the United States at the beginning of the 19th century (Hervey and Hems 1968, Lever 1977).

Fancy goldfish

Fancy goldfish in China were first mentioned in 1089 and by the 16th century there were many colour varieties as well as varieties with elongated and divided fins.

The main variations in external characters seen in Britain up to 1970 were as follows:

Body: long or short; slim or deep.
Fins: long or short; pointed or rounded.
Anal fin: single or divided.
Caudal fin: single or divided; forked or unforked.
Dorsal fin: present or absent.
Head: normal or hooded.
Eyes: normal or protruding sideways or upwards.
Scales: normal or domed.
Nasal septa: normal or enlarged.

In addition, scales can be of three types: metallic, nacreous or matt. The metallic scales are the normal, fully reflective type. The matt group have no reflective substance (guanine) in the scales or in the dermis below. The nacreous type, commonly called 'calico', is a mosaic of normal and transparent scales. The matt type is homozygous for a dominant gene; it is rare and delicate. The nacreous type is heterozygous.

Genetic studies on the other abnormalities

suggest that protuberant eye is due to a recessive gene with incomplete penetrance. Fin variations have a quantitative type of inheritance. There appears to be no information on the genetics of hood or of colour except that solid black appears only in telescope-eyed fish.

The recognized varieties in Great Britain, as well as a few other oriental ones, are described below (Hervey and Hems 1968, Matsui 1972).

Globe-eye (also called Telescope-eye or Pop-eyed). It has a divided caudal fin in addition to the protuberant eyes. It has been known in China (where it is called Dragon Eyes) since at least the middle of the 18th century and in Europe from the same date.

Fantail. This variety has a divided caudal fin of medium length and a high dorsal fin. It may or may not have telescope eyes, but they are not acceptable in the United Kingdom. It has been known in China since the 16th century and was one of the first goldfish to reach Europe in the mid-18th century.

Veiltail. The dorsal fin is high and the caudal and anal fins are usually paired. The pectoral and ventral fins are elongated and the caudal fin hangs down in long folds. It has been known in China since the 16th century. The European veiltail is of American origin and was taken to Germany between the wars. The name dates only from 1947; it is a translation of the German *Schleierschwanz* and replaces a number of less descriptive names. It may or may not have telescope eyes.

Broadtail Moor. This is a variant of the Veiltail which is black in colour and has protuberant eyes. It has always metallic scales.

Celestial. The eyes point upwards and there is no dorsal fin. The caudal fin is divided and usually deeply forked. The colour is usually pink, gold or white. In China it is called the Skygazer and also the Stargazer or Heavenward Dragon. It was developed in the mid-18th century.

Bubble-eye. The eyeballs are near the top of the head and below them there is a sac of gelatinous substance. The dorsal fin is lacking. It is also called Waterbubble-eye.

Pearl-scale. This is the same shape as the Fantail but the scales are domed which gives the fish an armoured appearance.

Shubunkin. This is a calico variety of variegated colour, hence the alternative names Speckled or Harlequin. It was launched in Japan in 1900 (the name means 'Vermilion Variegated') but there was a similar variety in China some 200 years earlier. Some types have normal body shape and others have larger fins, especially a large deeply-forked tail with rounded lobes. In England the Bristol Shubunkin has been produced. Its main colour is blue with black and orange speckling.

Bramblehead. Wart-like excrescences develop first on the top of the head and spread over the cheeks and opercula. The best specimens are deep orange or brassy yellow with a deep orange hood but they may also be red, pink or white. There is no dorsal fin and the caudal fin is divided and held stiffly. Alternative names are Lionhead, Hooded and Buffalohead; the German name is *Tomatenkopf* (= tomato head). In Japanese it is called Ranchu and in Shanghai, Tigerhead. It was developed in Japan, from the Korean variety, in the 19th century.

Oranda. This is a Japanese variety developed from a cross of Bramblehead and Veiltail about 1840. The Chinese name is Goosehead, or Frog-head, or Bouquet. The dermal growth is restricted to the top of the head. Red and white are the commonest colours. The dorsal fin is high and the caudal fin is divided.

Comet. This has a high dorsal fin, slender body and a long, deeply forked, tail fin. The commonest colour is deep red but it may also be yellow or other colours. It was developed in the United States in the early 1880s on the basis of Japanese stock.

Pompon (called Velvet Ball in China). This variety has the nasal septa abnormally developed into bunches of fleshy lobes. It is also called Narial Bouquet.

The following Chinese and Japanese breeds are little-known or unknown in Europe.

Egg-fish. So-called because of its short, rounded body and the absence of dorsal fin, it normally has a narrow, pointed head and a stiff, divided tail fin. It is often golden but may also be white or pied.

Watonai (or Long-bodied Ribbontail). A Japanese variety developed about 1883 from a cross between Veiltail and common goldfish.

Blue Fish. Completely dark blue – a colour long sought in Europe.

Brown Fish. Similar to the Moor except in colour. Also called Chocolate Fish or Purple Fish according to the shade and intensity of colour.

Tumbler. Dates from about 1757 and owes its name to the perpetual somersaults it turns due to a bend in its backbone.

Out-turned Operculum (or Curled Gill). A sufficiently descriptive name.

Future prospects

Goldfish breeding requires only simple equipment and simple techniques; goldfish have beautiful and graceful body forms that charm everyone. Following the improvement of the people's material and cultural life, more and more people in China are going in for keeping goldfish. It promises a bright future.

Besides being ornamental, goldfish serve as subjects of experiment in the study of genetics and variation of higher animals. As early as 1925, Professor Chen Zhen, a noted Chinese biologist, made a systematic study of goldfish. He confirmed through crossbreeding that goldfish and crucian are of the same genus, and gave scientific explanations of the varieties and colours of goldfish. On the basis of his studies, Professor Chen Zhen crossed the blue variety with the purple one and succeeded in creating a new variety: purplish blue.

Acknowledgements

The editor would like to thank Mr A. Wheeler of the British Museum (Natural History) for his assistance in rewriting the section on taxonomy and Mr W. L. Wilson, Honorary President of the Goldfish Society of Great Britain, for reviewing the section on fancy goldfish.

References

Chen Zhen (1959) [*Domestication and Variation of Goldfish*]. Kexue (Science) Press. (In Chinese)

Fu Yiyuan (1980) [*Goldfish*]. Zhejiang People's Publishing House. (In Chinese)

Hervey, G. F. and **Hems, J.** (1968) *The Goldfish.* Faber: London

Lever, C. (1977) *The Naturalized Animals of the British Isles.* Hutchinson: London

Matsui, Y. (1972) *Pet Library Goldfish Guide.* Pet Library: Harrison, New Jersey

Syvorov, E. K. (1948) [*Basic Ichthyology*]. (In Russian)

Wang Yuchen and **Qi Jingfeng** (1964) [Culture and regeneration of goldfish]. *Beijing Zoo Year Book,* 1964. (In Chinese)

Wheeler, A. (1969) *The Fishes of the British Isles and North West Europe.* Macmillan: London

61

Salmonids

S. Drummond Sedgwick

Formerly H.M. Inspector of Salmon and Freshwater Fisheries for Scotland

Wild species

Nomenclature

The family Salmonidae is made up of three main groups of species, the salmons, the trouts and the charrs. The salmons include six Pacific species belonging to the genus *Oncorhynchus*. These species in order of commercial importance are pink, *O. gorbuscha*; sockeye, *O. nerka*; Coho or silver, *O. kisutch*; Chum, *O. keta*; and Chinook or spring, *O. tschawytscha*. The sixth species, known as *O. masu*, which is smaller in size, is native to rivers in the northern Japanese island of Hokkaido and on Sakhalin.

The Atlantic salmon, *S. salar*, is placed together with the trout in the genus *Salmo*. The true trouts comprise three main species, the brown or German trout, *S. trutta*; the rainbow trout, *S. gairdneri*; and the cutthroat trout, *S. clarki*. There are also a number of other species whose status remains controversial. These include the golden trout, *S. aureolus* or *aquabonita*, the Gila trout, *S. gilae*, and the Far Eastern trout, *S. mykiss*, which are close relatives of the rainbow trout; and the marbled trout of Slovenija, *S. marmoratus*, which is a racially distinct variety of brown trout.

The charrs belong to the genus *Salvelinus* which is made up of the Arctic charr, *S. alpinus*; the brook charr, *S. salvelinus*; the Great Lake charr, *S. namaycush*; and the Dolly Varden charr, *S. malma*. The three latter species are commonly called 'trout' in North America.

There are two additional members of the family Salmonidae belonging to the separate genus *Hucho*. These are the hucho, *H. hucho*, and the taimen, *H. taimen*.

General biology

The Salmonidae are carnivorous fish dependent for their food supply on the animal food chain in fresh water and in the sea. They are slimly built and can swim rapidly either to escape from danger or in active pursuit of their prey. All members of the salmon family are distinguished by the possession of an adipose fin which is a fold of tissue without fin-rays situated on the back between the dorsal and caudal fins. The body is covered by comparatively large cycloid scales partially embedded in pockets in the skin. The embedded portion shows scale growth in the form of concentric ridges, formed closer together during winter periods when the lowered water temperature slows down the rate of metabolism, from which it is possible to deduce the age of the fish. Most species have well-developed teeth, not only on the jawbones but also on other bony structures in their mouths. The young fish or parr of all species are distinguished by dark markings like fingerprints along their sides. The adult fish develop distinctive secondary sex characteristics which can change the body colour and shape.

The principal species have races adapted to spending part of their life-cycle in a marine environment. All members of the salmon family must spawn in fresh water, generally in fast-flowing streams. Eggs are large and comparatively few, varying in number from about 1400 to 3000 per kg of body weight. In most species these are deposited in a nest or redd excavated in gravel by the female fish. Two of the Pacific salmons must migrate to sea within a short time of hatching in order to survive. The sea-going races of all species deliberately migrate to sea after a juvenile period of freshwater growth which may be one or more years. Atlantic salmon and some species of Pacific salmon, have races which spend their entire lives in freshwater. A few species or subspecies, including those of the genus *Hucho*, do not migrate to sea.

The anadromous salmonids on spawning migration usually return to their parental river. The homing ability varies between different species and is most strongly developed in the Pacific salmons. The Atlantic and Pacific species of salmon both migrate for long distances to their oceanic feeding grounds but remain capable of locating and recognizing their parent rivers. The means by which the water of the home river is recognized is thought to be through the olfactory sensory system but the method of navigation through the open seas remains unknown.

Geographical distribution and economic importance

Members of the salmon family are among the world's most valuable species of fish and they have always been a focus of human interest, both as a basic source of food and politically for their economic worth to the fishing nations. Competition for Pacific salmon on their oceanic feeding grounds and during homeward migration has forced the U.S.A., Canada, the Soviet Union and Japan to agree to stringent international controls for the conservation of stocks.

Atlantic salmon are not only valuable as food but are of particular interest as sporting fish for angling. Trout and charr in both anadromous and freshwater forms are fished commercially throughout their range as well as being of interest to anglers.

The natural distribution of all species of salmonids is in the northern hemisphere, northwards from the temperate zone. There are no native members of the salmon family in the southern hemisphere, although wild stocks have been successfully introduced to climatically suitable waters in South America, South Africa and Australasia. The Pacific salmons are distributed in rivers flowing into the Pacific and eastern Arctic Oceans from the eastern Siberian and Laptev Seas to Kamchatka and from Alaska to California.

Wild brown trout are naturally distributed in rivers or lakes in Europe, North Africa, the Middle East and Asia. They have been widely introduced into North America as well as into parts of the southern hemisphere. Sea-going forms of the species occur in rivers round the British Isles, and from northern Spain to Arctic Scandinavia including rivers flowing into the Baltic Sea. Stocks also exist in rivers on the Dalmatian coast of the Mediterranean and in the rivers draining into the Caspian Sea. Wild stocks in some areas have been drastically reduced or eliminated as a result of man-made environmental changes.

Rainbow trout were originally distributed in lakes and rivers on both sides of the western mountain chain along the Pacific seaboard of North America from Alaska to Lower California. The seagoing race called 'Steelhead' occurs in the rivers flowing into the Pacific Ocean. There is a lacustrine form which grows to great size and a non-migratory race in rivers, often in the same waters as the anadromous type. The cutthroat trout, which is also native to rivers in the American northwest, does not grow to large size. It can adapt to life in salt water and fish will periodically migrate into coastal waters, but it is not truly anadromous.

Arctic charr are circumpolar in their distribution. Sea-going races are present in rivers across the north coast of the American continent from Alaska to Labrador, in Greenland rivers and in Iceland. They are also found in northern Europe and Asia from north Norway to Siberia. Some rivers have non-migratory stocks and in many lakes, particularly in the southern part of its range in Europe, the species has developed racially distinct, relict populations.

The other species of charr are all native to North America. Brook charr in both migratory and non-migratory forms are found in cooler lakes and rivers in the eastern half of the continent. The Great Lake charr was originally, as its name suggests, a fish of the Great Lakes, but populations there have been decimated through the colonization of the system by the marine lamprey, *Petromyzon marinus*, and by pollution. It is still present in many of the large lakes in Canada. Races of this species have adapted to life in rivers in northern Canada but there is no anadromous form. Dolly Varden charr occur only in the rivers of the northwest of Canada and in Alaska.

Domestication and early history

The degree of domestication that has so far been achieved with any member of the salmon family is not comparable to that obtaining with mammals and birds, or even certain other fish such as carp. In most species there is still little difference between individuals in the wild and fish bred and grown to maturity in captivity on fish farms. The methods employed in intensive salmonid farming correspond in many ways to those developed in the production of broiler chickens or turkeys, but as far as the availability of domestic stock is concerned, fish farmers are still at the stage of looking for wild species with characteristics likely to lend themselves to intensive culture in an artificial environment, which may lead on towards the eventual development of domestic varieties. The only salmonid that has so far been truly domesticated is the rainbow trout, *S. gairdneri*.

The credit for having first successfully stripped spawn from trout, fertilized it artificially with sperm from a male fish and incubated and hatched out the eggs, is given to a German named Jacobi. His work is believed to have been carried out some time in the latter half of the 18th century. During the early years of the 19th century, experiments were made in Europe and North America in hatching eggs taken from wild parents belonging to several species of both non-migratory and anadromous salmonids.

The initial development of trout and salmon culture on a commercial scale took place mainly in the period 1850–70. A detailed description of the artificial fertilization of brown trout eggs, and illustrated instructions on how to build and operate a hatchery, by a Dr Henrik Kroner, were published in Denmark in 1852. In the same year, work commenced on the construction of a large fish farm in France at Huningue near Basle. This consisted of earth ponds covering a surface area of 35 ha, supplied with spring water at a temperature of 10 °C. The fish produced were mainly used for restocking angling waters and commercial net fisheries. In the 1860s and 1870s trout farms were constructed in Scotland at Howietoun near Stirling and at New Abbey in Dumfriess and Galloway, and in Surrey, England. Commercial brown trout culture also started experimentally at this time in a number of other countries in Continental Europe. Federal and state fisheries' organizations, as well as private individuals, were

also developing methods of culturing rainbow trout in North America and artificially hatching species of Pacific salmon.

In the latter half of the 19th century, brown trout in Europe and rainbow trout in North America were genuinely 'farmed' in the sense that, although originally from wild stocks, brood fish were kept and the fish spent their whole life-cycle in an artificial environment. The salmon and trout, either hatched and planted out or reared in captivity, were normally released in order to increase wild stocks. Trout were sometimes put into natural ponds and small lakes from which they could easily be recaptured with nets to be used directly as human food. This might be done on a commercial scale but the fish were seldom given any food and generally had to fend for themselves.

Although the techniques developed for fertilizing, incubating and hatching salmonids were successful, brown trout did not prove to be a species that could be reared intensively. The rainbow trout, however, was introduced into Europe in the latter years of the 19th century, and at once appeared to be a much more satisfactory candidate for farming. It was with rainbow trout that the first successful efforts were made to domesticate a fish of the salmon family which could be intensively stocked until ready to be slaughtered for the table market.

Rainbow trout farming gradually gained commercial importance during the years up to 1939. The methods employed in Europe involved the use of rectangular earth ponds, while in the United States narrow channels known as race-ways were generally used for growing-on the fish for sale. Brown trout had orginally been fed on a variety of animal proteins mainly derived from mammals. Minced liver had become the standard starter food for small fry. Hard-boiled hen eggs had first been tried but had proved too expensive to use commercially. The same fish foods were used initially in rainbow trout culture.

The most rapid development of rainbow trout farming took place in Denmark. None of the fish farms was far from the sea and it was found that raw, minced sea-fish or fish-offal was a better source of protein in trout feed than mammalian or avian tissue. More became known about the dietary needs at each stage of the growth of farm trout and how these might be met most economically. Food mixtures were made up using fresh, moist or wet constituents, but by the early 1940s experimental dry pelleted foods were being manufactured, based on fishmeal to which some other source of protein, fats and small amounts of digestible carbohydrate were added, together with essential vitamins.

Industrial production of rainbow trout and other salmonids for the table market expanded rapidly in the late 1940s and 1950s. Interest in domesticated fish production remained centred on rainbow trout until the 1960s when other species, notably Atlantic salmon, proved to be suitable for intensive culture.

Recent history and present status

The rainbow trout at the present time remains the most important species of salmon cultivated world-wide in a recognizably domestic form. The strains used by trout farmers have generally been developed from stocks bred in captivity for many generations. The original brood stock raised on trout farms in North America and the consignments of eggs exported to Europe, to Japan and to countries in the southern hemisphere came partly from sea-going or 'Steelhead' races and partly from the non-migratory populations of the so-called 'Shasta' variety.

Most farm stocks are today of inextricably mixed origins and have combined to form a recognizable fish-farm type. Stocks are generally the product of some generations of selective breeding in which individual fish among the progeny of a particular brood pair have been selected for desirable characteristics such as rapid growth, good conversion, resistance to disease and early or late sexual maturity. It has proved possible to stabilize some characteristics in particular brood stocks but it is most likely that these were racially inherent and that the fish are merely reverting to type. Attempts to breed for consistent characteristics on genetic principles have so far proved largely inconclusive.

The rainbow trout being farmed at the present time are a truly domesticated form of the species in that they are adapted to survive under conditions of intensive culture and are not stressed by life in close proximity to human beings whom they recognize as the source of their food supply. Two basic types are at present being produced, fish which grow quickly and are marketed at weights of 170–220 g and fish which are grown-on to large size and can weight 3–5 kg at slaughter. These fish are generally farmed in cages or enclosures in the sea or in tanks on the shore in a pumped supply of salt water.

Interest has turned recently to the prospect of domesticating other species of salmonids. At present commercial development is concentrated on Atlantic salmon, *S. salar*, and Coho salmon, *O. kisutch*. Both species are proving more difficult to domesticate than rainbow trout and at present farm stocks, particularly when juvenile, tend to react adversely to the presence of people nearby and can be unduly stressed by any form of handling. The density at which they can be safely kept is also less than for rainbow trout.

The selection of Atlantic salmon for domestication tended at first to be arbitrary and was made at random. So far attempts to create domestic strains with desirable characteristics using genetic techniques have not proved successful. Selection based on apparent racial differences appears, however, to be more successful and peculiarities can be retained through successive generations. The probable reason for the racial differences is that Atlantic salmon have been returning to the same parental rivers on spawning migration since the last Ice Age which has induced the development of separate and distinct characteristics in the salmon inhabiting different rivers.

Future prospects

The selection of wild species of salmonids as potentially useful for domestication as farm animals has mainly depended on market value in relation to the technical difficulty and cost of maintaining them in captivity. The propensities of a particular species as a source of permanently differentiated domestic types are now assuming greater importance.

Farm rainbow trout are now accepted as a domestic animal but much work remains to be done towards controlling the genetic make-up of stocks. The species tends to mature early and fish which have become sexually mature are of little use as table-market fish. It has proved possible to direct sexual differentiation towards the development of either male or female characteristics by the addition of appropriate hormones to the diet of the young fish during their first few weeks of feeding. More recently, experimental chromosome control has demonstrated that it is possible to produce polyploid, sterile fish.

The search for potentially domesticable salmonids has extended to species other than those at present farmed commercially. These include Arctic charr, *S. alpinus*, brook charr, *S. salvelinus*, and pink salmon, *O. gorbuscha*. The last species has the quickest growth of any member of the salmon family. It reaches weights of 2.5–4.5 kg in 2 years after hatching, but domestication is complicated because this species must migrate to sea immediately after it reaches the stage of development at which it is ready to feed. This means that culture involves the provision of salt water pumped to tanks on shore until the fish are sufficiently large to be transferred to floating cages or enclosures in the sea.

It is possible to hybridize many species of salmonids. Some hybrids grow more quickly and are easier to domesticate than either parent. Eventually the most useful domesticated salmon may prove to be a hybrid. An aspect of natural distinction which is of growing interest to fish-culturists is the possible domestication of a race of Atlantic salmon which does not migrate to sea, but nevertheless grows to a size comparable to the sea-feeding type, while remaining in fresh water until it reaches maturity. Such races of Atlantic salmon occur in Europe and in North America. Growing fish in fresh water has many advantages over mariculture such as the avoidance of osmotic problems associated with modern dry-pellet diets, the risk of storm damage and the growth of marine organisms on fish-holding structures.

A recent development in salmon production has been in the direction of ranching. This involves the production of juvenile fish of an anadromous race which are reared artificially up to the stage at which they are capable of migrating to the sea. They are then released to salt water at a specific place to which they can be expected to return when they reach maturity. In the meantime they feed wild in the sea at no cost to the producer. Research is in progress to produce a semi-domestic variety which can be used to give the highest potential survival in the sea and the greatest possible percentage return as adult fish.

The domestication of members of the salmon family has involved relatively complicated techniques which remained unknown until the 19th century. It is therefore a new development for human society. No other species has so far proved capable of the degree of domestication achieved with rainbow trout. The search still goes on to find other species or to produce hybrids that are equally amenable to life in direct association with human beings, as domestic animals in fresh water or in the sea, but there is still a long way to go.

References

Buckland, F. T. (1863) *Fish Hatching*. Tinsley: London

Buss, K. and **Wright, J. E.** (1956) Results of species hybridisation with the family Salmonidae. *Progressive Fish-Culturist*, **18** (4): 210–15

Day, F. (1887) *British and Irish Salmonidae*. Williams and Norgate: London

Gjedrem, T. (1976) Genetic improvements in salmonids. *Proceedings of the Royal Society of Edinburgh* (B), **75** (4): 253–61

Huet, M. (1975) *Textbook of Fish Culture*. Fishing News Books: Farnham, Surrey, England

Jensen, K. W. (1966) Saltwater rearing of rainbow trout in Norway. E.I.F.A.C. 66/sc 11–4: 43–8. European Inland Fisheries Advisory Commission: c/o FAO, Rome

Johnstone, R., Simpson, T. H. and **Youngson, A. F.** (1978) Sex reversal in salmonid culture. *Aquaculture*, **13**: 115–34

Møller, D. (1970) *Artsstukturen i Atlantisk Laks – Betydning for Kulturarbeidet*. [*Racial Characteristics in Atlantic Salmon – Significance for Fish Culture*.] Swedish Salmon Research Institute Report LFI. MED. 5

Purdom, C. E. (1972) *Genetics and Fish Farming*. Bulletin No. 25, Ministry of Agriculture, Fisheries and Food, Fisheries Laboratory: Lowestoft, England

Refstie, T. and **Gjedrem, T.** (1975) Hybrids between Salmonidae species: hatchability and growth rate in the freshwater period. *Aquaculture*, **6**: 333–42

Sedgwick, S. D. (1973) *Trout Farming Handbook*. Warne: London

Sedgwick, S. D. (1979) Salmon farming and smolt rearing in Norway. *Fish Farmer*, **2** (6): 7–8

Sedgwick, S. D. (1982) *The Salmon Handbook*. Deutsch: London

62

Tilapia

J. D. Balarin

Formerly Institute of Aquaculture, Stirling University, Scotland

Introduction

Taxonomy

The group tilapia (approximately 110 species) belongs to the family Cichlidae (over 700 species) and is characterized by all of its members being bilaterally compressed and only having one opening to the nares. Originally considered as one genus the tilapia have now been separated into two distinct genera (Trewavas 1973). Although there is still some adherence to the older names, the new classification is generally gaining acceptance and is therefore adopted here – *Tilapia species*: herbivorous substrate spawners that guard their eggs (about 10 species), and *Sarotherodon species*: generally planktophagous and mouth brooders of eggs and fry (about 100 species).

That these two genera do not interbreed readily in the wild is further evidence of a behavioural and genetic difference. It is hoped that the taxonomy of this group will finally be resolved when Trewavas's monograph is published (in 1983). (See note added in proof).

General biology

Illustrated life-cycles comparing the biology of the two genera are given in Figs 62.1 and 62.2.

Essentially tropical freshwater fish, tilapia are lowland dwelling and have become well adapted to the successful exploitation of almost all available niches in lacustrine and riverine environments (Fryer and Iles 1972). Such is the group's versatility that certain species such as *Sarotherodon mossambicus*, *S. spilurus* and *Tilapia zillii* are able to thrive in seawater and some (e.g. *S. alcalica grahamii*) can tolerate temperatures of up to 42 °C in the soda lakes of Kenya.

The resistance to poor water-quality and disease and a tolerance of a wide range of environmental conditions, including partial dessication, mean that tilapia can survive under conditions lethal to most other species. This and an ability to convert organic domestic agricultural wastes efficiently into fish protein as well as to exploit all available food sources, a rapid growth rate, ease of reproduction and an amenability to intensification, are some of the basic reasons why tilapia have been selected as ideal for warm-water fish-farming (Balarin and Hatton 1979). In particular this hardiness and ease of breeding mean that the species can be grown with minimal technical expertise and therefore can be made readily available to the subsistence farmer. However, the ability of tilapia to breed in captivity is not always a boon. Prolific precocious breeding can be triggered by typical fishpond conditions and the resultant overpopulation by fry leads to stunting due to competition for food. Conditions analogous to the drying up of a water body would appear to stimulate an innate survival mechanism. Of particular significance are high temperatures, prolonged photoperiod, reduced space availability, high light intensity, reduced food supply and a slight rise in salinity. Fish in such conditions mature earlier and produce numerous smaller individuals, which are often better adapted to surviving under the prevailing conditions. In captivity in the tropics, therefore, breeding can take place continuously up to six or eight times a year. Intensive fry production systems, making use of constant high temperatures, are therefore able to produce fingerlings all the year round (Balarin and Haller 1982).

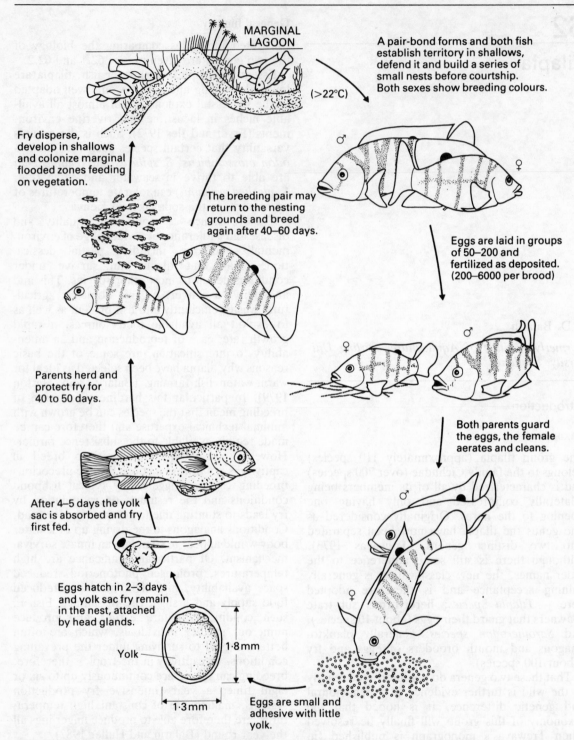

MARGINAL LAGOON

A pair-bond forms and both fish establish territory in shallows, defend it and build a series of small nests before courtship. Both sexes show breeding colours.

(>22°C)

Fry disperse, develop in shallows and colonize marginal flooded zones feeding on vegetation.

The breeding pair may return to the nesting grounds and breed again after 40–60 days.

Eggs are laid in groups of 50–200 and fertilized as deposited. (200–6000 per brood)

Parents herd and protect fry for 40 to 50 days.

Both parents guard the eggs, the female aerates and cleans.

After 4–5 days the yolk sac is absorbed and fry first fed.

Eggs hatch in 2–3 days and yolk sac fry remain in the nest, attached by head glands.

Yolk

1·8mm

1·3mm

Eggs are small and adhesive with little yolk.

Fig. 62.1 Diagrammatic sequence of the life-cycle of a typical substrate spawner, e.g. *T. zillii*. (After Fryer and Iles 1972, and Caulton 1979)

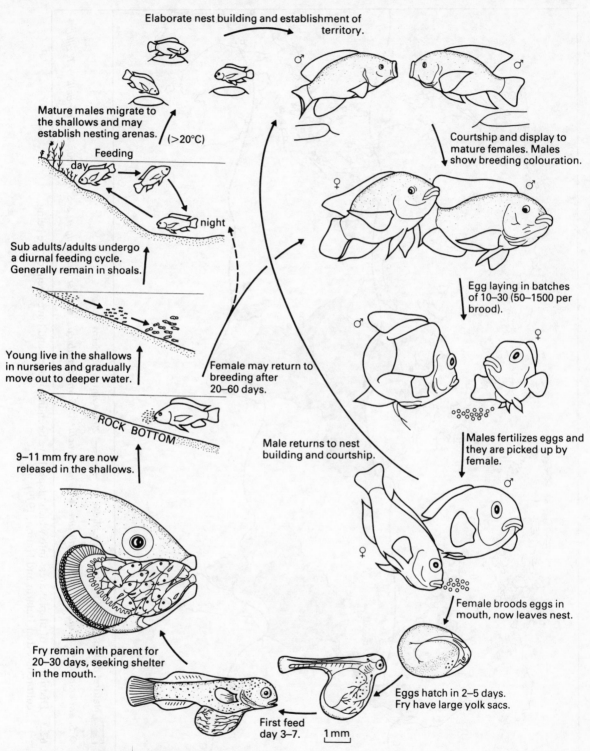

Elaborate nest building and establishment of territory.

Mature males migrate to the shallows and may establish nesting arenas. (>20°C)

Feeding
day
night

Sub adults/adults undergo a diurnal feeding cycle. Generally remain in shoals.

Young live in the shallows in nurseries and gradually move out to deeper water.

ROCK BOTTOM

9–11 mm fry are now released in the shallows.

Female may return to breeding after 20–60 days.

Male returns to nest building and courtship.

Courtship and display to mature females. Males show breeding colouration.

Egg laying in batches of 10–30 (50–1500 per brood).

Males fertilizes eggs and they are picked up by female.

Female broods eggs in mouth, now leaves nest.

Fry remain with parent for 20–30 days, seeking shelter in the mouth.

First feed day 3–7.

1 mm

Eggs hatch in 2–5 days. Fry have large yolk sacs.

Fig. 62.2 Diagrammatic sequence showing the life-cycle of a typical mouth brooder, e.g. *S. mossambicus*. (After Fryer and Iles 1972, and Caulton 1979)

393

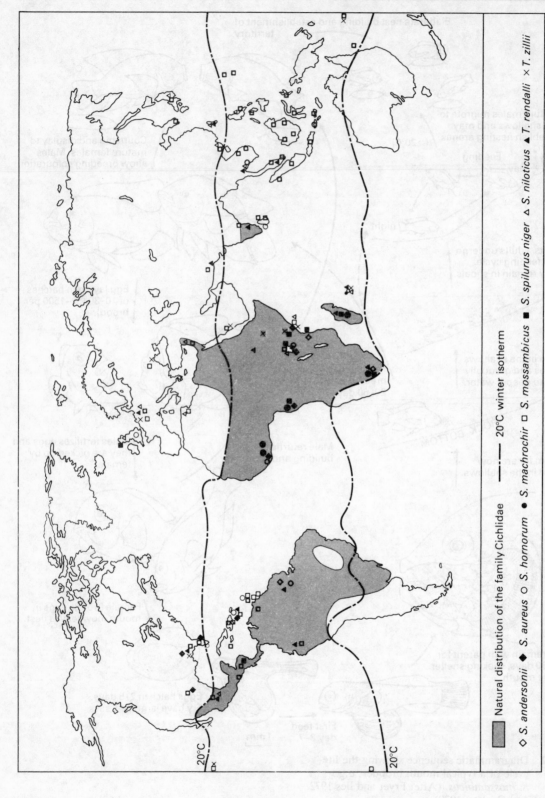

Fig. 62.3 Distribution of tilapia: areas into which tilapia have been introduced and have been, or are being cultured. (After Balarin and Hatton, 1979)

Natural distribution of the family Cichlidae — ─·─·─ 20°C winter isotherm

◇ *S. andersonii* ◆ *S. aureus* ○ *S. hornorum* ● *S. machrochir* ● *S. mossambicus* □ *S. spilurus niger* △ *S. niloticus* ▲ *T. rendalli* × *T. zillii*

Geographical distribution

Cichlids are widely distributed throughout Africa and Latin America as well as India and Ceylon, and tilapia, originally endemic to Africa, are now equally widespread. Their natural and introduced range (Fig. 62.3) falls almost entirely within the area defined by the 20°C winter isotherms. Outside this range tilapia can only exist in heated systems or at least require to be overwintered in heated units. The keen interest in tilapia as a food fish has meant that it has been transplanted almost all over the world and fifteen of the more economically important species can now be found in over a hundred countries (Balarin 1980).

Wild species

Wild species have evolved considerably from their early form and because of their plastic nature have adapted to exploiting numerous different habitat types. However, the interference by man, with his penchant for mixing stocks both in the wild and under culture conditions, has resulted in various hybrid strains. Often these are not morphologically distinguishable from either parent. Therefore, although there has been no major species change arising from the domestication of tilapia, electrophoretically strains can be shown to be genetically different. It would, however, seem to be only a matter of time before natural selection favours one of these strains to give rise to a new species.

History of domestication

In ancient Egypt, around 3000 B.C., tilapia were considered sacred, symbolizing the hope of rebirth after death and they were often depicted in bas reliefs. Scenes show *Sarotherodon niloticus* to be an important fishery species and Chimitz (1955) indicates that there is good evidence to believe that 2000 years B.C. tilapia culture was practised in well-built, drainable ponds. This would therefore be almost about the same time, if not earlier, that Ling (1977) indicated carp culture was first initiated in China. Other than biblical references (Isaiah, 19: 10) little of this early domestication has been documented.

It was not until Boulenger between 1909 and 1916, first catalogued the group, that scientists became interested. Early studies were mainly of a biological nature and in particular of the peculiar mouth-breeding habits. The trials conducted in Kenya in 1924 on *S. s. niger* mark the beginning of recorded, scientifically orientated tilapia culture. The practice soon spread throughout Africa (Balarin and Hatton 1979). Tilapia were transplanted into the Far East in the late 1940s, introduced into North America in the early 1950s and by the late 1950s became established in Latin America. Such has been the success of tilapia as a farm animal that in the past 35 years it has risen from obscurity to being one of the most important of cultured food-fish, ranking in importance alongside carp and trout.

The group's efficiency in aquatic weed and mosquito control, and its importance as fishmeal for stock feeds, as bait fish particularly in the tuna industry, as sport fish in angling waters and even as an occasional addition to a tropical aquarium, have further led to its world-wide transplantation. The interest in this fish is indicated by the 3300 works on aspects of its biology or culture listed in Balarin (1980), half of which had been published in the preceding decade.

Present status

Tilapia as a domesticated intensively-grown fish, is important in many tropical areas, contributing substantially to the protein diet of many people. In 1975 production in Africa amounted to 30 000 tonnes, but it is planned to increase it to 50 000 tonnes by 1985; in Asia, Taiwan alone currently produces 80 000 tons of tilapia per annum.

The adaptability of the species has, however, also led to it being termed a 'nuisance fish'. Introduction for a particular purpose and escapes from pond systems have resulted in the species becoming established in local waters, often displacing indigenous stocks. Tilapia are now almost endemic in every area into which they have been introduced, provided that winter temperatures remain above the lower lethal limit of 10–12°C.

In certain areas, the establishment of tilapia in the wild has benefited the fishery but, as was feared by Prowse in 1963, uncontrolled introduction has lead to a mixing of species through the hybridization of previously geographically isolated populations. Ultimately such mixing of stocks should produce better adapted strains; however, brood stock contamination may also arise and has resulted in generally poorer results in certain hybrid crosses which are known to produce all-male progeny. The benefit of stocking all-male fish is understandable from a pond management point of view. The prevention of unwanted reproduction would maintain a relatively constant and manageable stocking density. All-male hybrid crosses have therefore become commonplace in intensive tilapia culture since the technique was first discovered by Hickling in 1960. Uncontrolled translocation of tilapia species is, however, threatening the success of this method of reproduction control.

One of the earlier methods of controlling reproduction was through the stocking of a predator in the pond to crop the tilapia fry. The complexities of a balanced predator: prey relationship placed this technique beyond the capabilities of the peasant farmer. Hand-sexing as an alternative means of selection for monosex stocks was never completely accurate. Hormone treatment of sexually undifferentiated fry to produce all-male stocks, was then attempted. Though not entirely successful this latter technique is now finding commercial application in Israel and efforts are under way to make it more widely applicable.

Future prospects

Tilapia domestication can therefore be said to be still very much in its infancy despite the fact that it is an extremely important food-fish. The knowledge of the fish's biology is by no means complete. Considerable advances have been made in recent years and reported in such works as Balarin and Hatton (1979), Caulton (1979), Balarin and Haller (1982) and Pullin and Lowe McConnell (1982). The use of electrophoretic techniques to identify genetically pure strains and an understanding of the genetic mechanisms of sex determination will undoubtedly see a general trend toward the use of hybrids as farm fish. Once pure strains have been isolated it would certainly become possible to select for a particular size, shape, growth rate or colour to suit a particular market requirement.

A certain degree of selection has already taken place. Coloured mutant strains of *S. mossambicus* have been bred as aquarium fish and have further been selected for hybridization with *S. niloticus* to produce a red to orange coloured hybrid. Referred to as the 'cherry snapper', this hybrid is favoured over locally caught strains in both Japan and the United States. The appearance of dead tilapia has often led to a low consumer acceptance and a preference for the more coloured varieties has led an American company to produce six 'golden' to 'red' hybrids, which are now available for sale on the international market. As selection for a coloured hybrid can be costly it is probable that this practice will remain as the future trend in tilapia culture only for a select market. However, tilapia's future as a more widely accepted, cheap food-fish is likely to lie in the success of all-male production from hybrid crosses or hormone treatment or a combination of both. Domestication, or rather interference by man, has resulted in a genetic mixing of previously pure strains. Hybridization and selection for particular strains are recent innovations, and therefore it is only a matter of time before new species changes become evident.

References

Balarin, J. D. (1980) *A Bibliography of Tilapia. With notes on taxonomy, distribution and world status*. Institute of Aquaculture, Stirling University: Scotland

Balarin, J. D. and **Haller, R. D.** (1982) The intensive culture of tilapia in tanks, raceways and cages. In: *Recent Advances in Aquaculture*, ed. R. J. Roberts and J. M. Muir, pp. 267–355. Croom Helm: London

Balarin, J. D. and **Hatton, J. P.** (1979) *Tilapia. A guide to their biology and culture in Africa*. Institute of Aquaculture, Stirling University: Scotland

Caulton, M. S. (1979) *The Biology and Farming of Tilapia in Southern Africa*. Fisheries Development Corporation: Gingindlovu, South Africa

Chimitz, P. (1955) Tilapia and its culture. A preliminary bibliography. *FAO Fisheries Bulletin*, **8** (1): 1–33

Fryer, G. and **Iles, T. D.** (1972) *The Cichlid Fishes of the Great Lakes of Africa: their biology and evolution*. Oliver and Boyd: Edinburgh

Ling, S. W. (1977) *Aquaculture in Southeast Asia: a historical overview*. University of Washington Press: Seattle

Pullin, R. and **Lowe McConnell, R.** (1982) *Tilapia: its biology and culture*. International Center for Living Aquatic Resources Management: Manila, Philippines

Trewavas, E. (1973) On the cichlid fishes of the genus *Pelmatochromis* with a proposal of a new genus for *P. congicus*, on the relationship between *Pelmatochromis* and *Tilapia* and the recognition of *Sarotherodon* as a distinct genus. *Bulletin of the British Museum (Natural History) Zoology*: **25** (1): 3–26

Trewavas, E. (1983) *A Review of the Tilapian Fishes of the Genus* Sarotherodon. British Museum (Natural History): London

---

Note added in proof
Recently Trewavas has suggested a further separation of the *Sarotherodon* group into two genera:
Sarotherodon – biparentol mouth brooders
Oreochromis – maternal mouth brooders.

63
Potential domesticants: freshwater fish

R. Yamada

Department of Genetics, University of Hawaii, Honolulu, U.S.A.

In the following fish, with the exception of eels, the custom of collecting wild seed stock for pond cultivation has been rendered somewhat obsolete by the ability to control reproduction through hormone injections. This practice of artificial breeding is the incipient stage in domestication. However, unlike the common carp and several species of trout, rigid rules of quantitative genetic principles have yet to be optimally applied for genetic manipulation of desirable traits.

Eels

The catadromous eel, *Anguilla*, the only genus in the family Anguillidae, is composed of sixteen species found throughout the world (Usui 1974). However, only two commercially cultured species have established any social and economic significance: *Anguilla anguilla* in Europe and *A. japonica* in the Far East.

Eels are unable to breed in captivity; present stocks for culturing come from netting wild elvers during their migrations upstream (Usui 1974, Lane 1978, Moriarty 1978). This inability to breed eels prevents any genetic selection for domestication. Although many different eel investigators have administered hormonal treatments to induce sexual maturity, successful artificial fertilization and hatching of eggs has occurred only with *A. japonica* (Yamamoto and Yamauchi 1974). The additional knowledge required to close the life-cycle by rearing leptocephalus larvae under conditions simulating ocean migration and then providing suitable environmental stimuli for metamorphosis to the elver stage still remains years away. Hence, eel culture will continue to rely upon unimproved breeding stocks with prospects for domestication remaining quite remote.

Channel catfish

The channel catfish (*Ictalurus punctatus*) belongs to the North American freshwater catfish family, Ictaluridae. Its value as a game and food fish, especially in the southeastern United States, prompted experimental attempts at culture over 50 years ago (Bardach *et al.* 1972). Instrumental in the present expansion of the commercial catfish industry was the control of reproduction by hormonal techniques (Clemens and Sneed 1957, Sneed and Clemens 1959).

Past improvement of channel catfish traits (e.g. growth rate, survival, food conversion) has been attributed to desultory hybridization with other ictalurids as well as through mass selection methods (Giudice 1966, Burnside *et al.* 1975, Green *et al.* 1979). However, in recent years, a clearer picture towards improvement of desirable traits has been provided through estimates of heritability and genetic correlations (Reagan *et al.* 1976, El-Ibiary and Joyce 1978).

El-Ibiary and Joyce (1978) have suggested that domestication is a relatively recent event since certain traits in currently available 'domesticated' stocks differ little from their wild ancestors. This situation is expected to change rapidly in the near future with the advent of hatchery-reared commercial broodstock, controlled reproduction, and increased emphasis on proper genetic exploitation. The channel catfish is thus considered an excellent candidate for becoming an established domesticated species.

Chinese carps

The Chinese carps (order Cypriniformes) are

economically important food fishes extensively polycultured in Asia (Bardach *et al.* 1972). Three indigenous Chinese lowland species have gained international popularity and distribution: the silver carp (*Hypophthalmichthys molitrix*), the bighead carp (*Aristichthys nobilis*), and the grass carp (*Ctenopharyngodon idella*).

Although the ability to control reproduction was achieved in the late 1950s to early 1960s (Chaudhuri 1966), the scope of relevant genetic research has been noticeably limited. It appears that research trends are focused on the development of an intergeneric hybrid that combines the superior qualities of both parents (Andriasheva 1966, Berry and Low 1970, Beck *et al.* 1980). At present, no emphasis has been placed upon defining genetic variation of desirable traits for selective improvement within the species.

In the past, subsistence levels through polyculture were adequate and there was no urgency to improve the species genetically. As a consequence, the Chinese carps have not been properly domesticated. However, given their important ability to feed low on the food chain, it is realized that domestication can potentially provide the world with a cheaper and more efficient source of protein.

Indian carps

The Indian carps (order Cypriniformes) are an important food source in south Asia with the most desirable cultured species being the rohu (*Labeo rohita*), catla (*Catla catla*), mrigal (*Cirrhinus mrigala*), and calbasu (*Labeo calbasu*).

After the control of reproduction was achieved (Sinha *et al.* 1974), attempts at genetic improvement have been centred on interspecific and intergeneric hybridization within the Indian carp complex as well as Indian carp hybridization with the common and Chinese carps (Chaudhuri 1966, Hickling 1966, Berry and Low 1970). Studies on genetic variation of desirable traits and the consequences of selection within the species have yet to be made.

Although the Indian carps have been cultured for thousands of years, like the Chinese carps they have not been properly domesticated. However, given the impetus of India's urgent need for protein, it is only a matter of time before efforts at domestication are made.

Acknowledgement

The author is grateful to Dr John Bardach, East West Center, Hawaii, for his helpful suggestions.

References

Andriasheva, M. A. (1966) Some results obtained by the hybridization of Cyprinids. *FAO Fisheries Report*, **44** (4): 205–14

Bardach, J. E., Ryther, J. H. and McLarney, W. O. (1972) *Aquaculture: the farming and husbandry of freshwater and marine organisms*. Wiley-Interscience: New York

Beck, M. L., Biggers, C. J. and Dupree, H. K. (1980) Karyological analysis of *Ctenopharyngodon idella*, *Aristichthys nobilis*, and their F_1 hybrid. *Transactions of the American Fisheries Society*, **109** (4): 433–8

Berry, P. Y. and Low, M. P. (1970) Comparative studies on some aspects of the morphology and histology of *Ctenopharyngodon idellus, Aristichthys nobilis*, and their hybrid (Cyprinidae). *Copeia*, **4**: 706–26

Burnside, M. C., Avault, J. W. and Perry, W. G. (1975) Comparison of a wild and a domestic strain of channel catfish grown in brackish water. *Progressive Fish-Culturalist*, **37** (1): 52–4

Chaudhuri, H. (1966) Breeding and selection of cultivated warm-water fishes in Asia and the Far East – a review. *FAO Fisheries Report*, **44** (4): 30–66

Clemens, H. P. and Sneed, K.E. (1957) The spawning behaviour of the channel catfish, *Ictalurus punctatus*. *Special Scientific Report – Fisheries* No. 219. United States Fish and Wildlife Service: Washington, D.C.

El-Ibiary, H. M. and Joyce, J. A. (1978) Heritability of body size traits, dressing weight and lipid content in channel catfish. *Journal of Animal Science*, **47** (1): 82–8

Giudice, J. J. (1966) Growth of a blue × channel catfish hybrid as compared to its parent species. *Progressive Fish-Culturalist*. **28** (3): 142–5

Green, O. L., Smitherman, R. O. and Pardue, G. B. (1979) Comparisons of growth and survival of channel catfish, *Ictalurus punctatus*, from distinct populations. In: *Advances in Aquaculture*: papers presented at the FAO Technical Conference on Aquaculture, Kyoto, Japan, 26 May–2 June 1976, ed. R. V. R. Pillay and W. A. Dill, pp. 123–7. Fishing News Books: Farnham, Surrey, England

Hickling, C. F. (1966) Fish-hybridization. *FAO Fisheries Report*, **44** (4): 1–11.

Lane, J. P. (1978) Eels and their utilization. *Marine Fisheries Review*, **40** (4): 1–20

Moriarty, C. (1978) *Eels – a natural and unnatural history*. Universe Books: New York

Reagan, R. E., Pardue, G. B. and Eisen, E. J. (1976) Predicting selection response for growth of channel catfish. *Journal of Heredity*, **67**: 49–53

Sinha, V. R. P., Jhingran, V. G. and Ganapati, S. V. (1974) A review on spawning of the Indian major carps. *Archiv für Hydrobiologie*, **73** (4): 518–36

Sneed, K. E. and Clemens, H. P. (1959) The use of human chorionic gonadotropin to spawn warm water fishes. *Progressive Fish-Culturalist*, **21** (3): 117–20

Usui, A. (1974) *Eel Culture*. Fishing News Books: Farnham, Surrey, England

Yamamoto, K. and Yamauchi, K. (1974) Sexual maturation of Japanese eel and production of eel larvae in the aquarium. *Nature*, **251** (5472): 220–1

64

Potential domesticants: sea fish

C. E. Purdom

Ministry of Agriculture, Fisheries and Food Fisheries Laboratory, Lowestoft, Suffolk, U.K.

Introduction

There are no truly domesticated marine fish but several species are now cultivated for food and many more are being investigated for use in fish farming. All of these species may be regarded as potential domesticants and the purpose of this chapter is to review the evaluation of their use by man and the likelihood that one or another may become truly domesticated in the sense that some characteristics are moulded away from the wild type and towards a set of artificial criteria relevant to man's use of the animal.

Sea fishing produces the bulk of world fish supplies. During the past decade, about 70 million tonnes of fish were caught each year (FAO 1978) compared to the production of about 4 million tonnes by fish-farming (Pillay 1979), mostly in fresh water. The sea remains the last major hunting-ground although the extension of national coastal zones is beginning to make the 'hunt' more exclusive and less of a common property, and modern approaches to stock management, employing such concepts as sustainable yield and stock harvesting, seem more akin to the gentler pursuit of farming – or at least ranching.

Sea or ocean ranching is a term used in another context, namely the release of hatchery-reared fish into the natural environment for subsequent capture at a marketable size. This evergreen concept is highly fashionable now (Joyner 1980); a century ago it was equally popular (Earrl 1880) and considered by some to be the means by which the effects of over-fishing could be reversed (Herdman 1893). In actual fact, the bulk of sea-fish farming employs the opposite practice, for which no comparable term exists, of taking juvenile fish from the sea for fattening in tanks or other enclosures.

Yellowtail, mullet and milkfish

The main species employed are yellowtail (*Seriola quinqueradiata*), mullet (*Mugil cephalus*) and milkfish (*Chanos chanos*). These three species are primarily farmed in southeast Asia and account for all but a few per cent of the world production of marine fish. Production rates for yellowtail and milkfish are around 100 000 tonnes each per year and mullet farming probably contributes a similar amount but firm statistics are not available since national outputs are small and spread over a much wider geographic range than for the other two species.

Mullets and milkfish have been farmed for centuries by the simple method of the retention of natural fry in shallow sea- or brackish-water lagoons in which the principal management practice is water control and fertilization to maximize the production of algae (Chen 1976).

Milkfish are herbivorous throughout life and their basic diet is benthic algae but this can be supplemented by the addition of waste vegetable materials to the lagoons and production levels of up to 3 tonnes of fish per hectare can be achieved. Fry of about 15 mm in length are netted from the sea throughout the summer and on-grown in nursery lagoons prior to release in production lagoons which may measure several hectares in area.

Natural production of fry is, of course, highly variable and is a principal limiting factor in milkfish culture. Attempts are being made in Taiwan and elsewhere to rear the fish artificially and the animal's herbivorous status may well be a distinct advantage here. As we shall see later, provision of appropriate feeds is a major obstacle in the farming of other marine fishes.

Several species of mullets are farmed, comprising two genera, *Mugil* and *Liza*. The most important by far is the grey mullet, *Mugil cephalus*. This is farmed in much the same way as are milkfish, i.e. in shallow brackish or saltwater ponds in which the water is fertilized and to which is added supplementary feed in the form of vegetable waste. Mullets can also be grown in freshwater and are frequently used with carp in polyculture, i.e. the rearing of two or more species of fish together where each exploits some particular aspect of the environment. Mullets are not obligate herbivores but readily take small invertebrates if available. In their early phases, they are, in fact, carnivorous and this does present some difficulty in artificial production since raising live foods to feed to juvenile fish is very expensive. Nevertheless, and although the bulk of mullet seed seems likely to come from the sea for the foreseeable future, methods have been devised for raising tens of thousands of mullet artificially and this species is more likely than either milkfish or yellowtail to achieve some level of domestication in the future.

The third important fish, the yellowtail, is a top carnivore and a member of the mackerel family. It is farmed only in the Inland Sea of Japan and the juvenile fish are taken by net in early summer when they measure about 15 mm in length (Fujiya 1979). Farming methods are quite different from those used in mullet and milkfish farming. Yellowtail are farmed intensively, i.e. at high density and with supplied food in the form of trash fish; no reliance is placed on natural productivity. Floating cages or shore-based enclosures are used with net mesh sizes appropriate to the size of the fish. Subsistence-level yellowtail culture is an old-established practice but the present commercial approach dates back only about 25 years. There is some need for an artificial rearing method for fry but as yet no progress has been made.

So none of the three principal marine farm fish can be regarded as domesticated since each generation of production depends upon supplies of wild juvenile fish and there is no feedback in genetic terms from the cultivation processes. Hatchery techniques have been developed for mullets (Yashouv 1969, Chen 1976) and are being developed for milkfish (Anonymous 1978) and yellowtail (Fujiya 1979) but there is little evidence that they will significantly contribute to farm

stock for decades to come. Of the three fish species, the most likely to attain domesticated status is the mullet.

Flatfishes

The other large area of interest in marine fish farming concerns the flatfishes, but here the emphasis has been on research rather than on production and current levels of output are still only around 100 tonnes per year. For the species involved, however, the complete life-cycles are under man's control and the potential for domestication is therefore high and, in some short-term respects, already developing.

The idea of farming flatfish sprang from the 19th-century concept of fish conservation by artificial production of fry for release into the sea – the principle of sea ranching mentioned earlier. Hatcheries for the artificial propagation of marine fish for this purpose were set up in the United States, in the United Kingdom and in mainland Europe. Thousands of millions of newly hatched fry were liberated into the Atlantic Ocean and adjacent waters each year until, starting shortly after the turn of the century, realization dawned that the practice was unproductive and the various hatcheries were progressively closed down. In Europe, the species commonly hatched for liberation were cod (*Gadus morhua*) and plaice (*Pleuronectes platessa*) and it was the second of these which was subsequently chosen as of potential value for marine fish farming. In the early phases of research, natural zooplankton was used as food but the major breakthrough in flatfish cultivation came with the demonstration that newly hatched brine shrimp (*Artemia salina*) was an adequate live food for the larval phases (Rollefsen 1939). Brine shrimp eggs are a more-or-less readily available commodity and provide nauplii which are the basic initial foods now used in a variety of aquacultural practices. Starting in the mid-1950s progress in the development of mass rearing techniques was highly successful (Shelbourne 1970) and by the mid-1960s methods were available for the production of commercial quantities of juvenile plaice. Work by the White Fish Authority* in Great Britain demonstrated

---

* Now incorporated in the Sea Fish Industry Authority.

the feasibility of growing plaice to marketable size but, unfortunately, the relatively slow growth rate and low value made the farming of plaice uneconomic. Effort was therefore switched to a more valuable flatfish, the sole (*Solea solea*), and the basic hatchery techniques as employed with plaice were satisfactory but once again the processes of on-growing were not economically viable although research continues on the husbandry of this species.

One of the problems with sole was that it is a cautious, nocturnal browser and farmed fish are easier to manage if they are voracious visual feeders. A further prime species of flatfish, the turbot (*Scophthalmus maximus*), seemed to fit the latter criterion and an assessment of its potential began in 1970 (Purdom *et al.* 1972). Results were encouraging and initial commercial interest began with the use of wild juvenile stock captured by push-net from the beaches. Stocks of young turbot were not numerous as in the case of yellowtail, mullet and milkfish and the development of a rearing technique was essential to the proper development of turbot farming. The hatchery problems were different from those of plaice and sole. The small size of the eggs and subsequent larvae precluded the use of brine shrimp as a starter diet but other live organisms proved acceptable. Techniques just about adequate for routine production of commercial quantities now exist based on the sequential use of rotifers, brine shrimp nauplii and metanauplii and inert weaning diets (Howell 1979).

Turbot is the only species of flatfish currently being farmed commercially and production, in Europe, is about 100 tonnes per year. Domestication is in its early stages with the development of year-round spawning by photoperiod control (Bye and Htun-Han 1978) – more an environmental than a genetic adaptation – and control of sex-ratio by chromosome engineering – a genetic manipulation now commercially applicable to salmonids but still in the research phase with turbot. Likely trends in domestication are selection for increased egg size and for tolerance of cooler rearing conditions: in addition, hybridization with brill (*Scophthalmus rhombus*) seems of commercial value.

Other flatfish species examined in the past for farming include the lemon sole (*Microstomus kitt*) and the halibut (*Hippoglossus hippoglossus*).

Both failed largely because of early rearing problems.

Other species

Other roundfish species which are beginning to be used in fish farming include sea bass (*Dicentrarchus labrax*) (Girin 1979), red seabream (*Pagrus major*) (Fujiya 1979) and pompano (*Trachinotus carolinus*) (Bardach *et al.* 1972). All require warm-water environments and are valuable food fish in low supply. Cod is farmed at a low level in the colder waters around Great Britain but it is a relatively cheap fish and not likely to be developed further unless market forces change.

Marine fish farming thus exhibits a range of activities from the commercially successful but primitive rearing of naturally caught fry in southeast Asia to the pilot scale projects with flatfish in northern Europe employing more or less sophisticated hatchery techniques for the supply of seed. Domestication is just beginning with the turbot and is, perhaps, feasible with mullets in the not-too-distant future. Sea fish, however, seem unlikely ever to become highly domesticated in comparison with conventional farm mammals and birds.

References

Anonymous (1978) *Aquaculture Development for Hawaii*. Department of Planning and Economic Development: State of Hawaii, Honolulu

Bardach, J. E., Ryther, J. H. and **McClarney, W. O.** (1972) *Aquaculture: the farming and husbandry of freshwater and marine organisms*. Wiley-Interscience: New York

Bye, V. J. and **Htun-Han, M.** (1978) Daylength control – the key to flexible production. *Fish Farmer*, **1**: 10–14

Chen, T. P. (1976) *Aquaculture Practices in Taiwan*. Fishing News Books: Farnham, Surrey, England

Earrl, R. E. (1880) A report on the history and present condition of the shore cod fisheries of Cape Ann, Massachusetts, together with notes on the natural history and artificial propagation of the species. *Report of the U.S. Commissioner of Fish and Fisheries*. (1878) Part 6, pp. 685–740

FAO (1978) *Yearbook of Fishery Statistics*: *catches and landings*, Vol. 46. FAO: Rome

Fujiya, M. (1979) Coastal culture of yellowtail (*Seriola*

quinqueradiata and red seabream (*Pagrus major*) in Japan. In: *Advances in Aquaculture*, ed. T. V. R. Pillay and W. A. Dill, pp. 453–8. Fishing News Books: Farnham, Surrey, England

Girin, M. (1979) Feeding problems and the technology of rearing marine fish larvae. In: *Finfish Nutrition and Fishfeed Technology, Proceedings of a World Symposium*, Vol. 1, ed. J. E. Halver and K. Tiews, pp. 359–66. Heenemann Verlagsgesellschaft: Berlin

Herdman, W. A. (1893) Need of a sea fish hatchery. *Report for 1892 on the Lancashire Sea-Fisheries Laboratory*, pp. 29–32

Howell, B. R. (1979) Experiments on the rearing of larval turbot *Scophthalmus maximus* L. *Aquaculture*, **18**: 215–25

Joyner, T. (1980) Salmon ranching in South America. In: *Salmon Ranching*, ed. J. E. Thorpe, pp. 261–76. Academic Press: London and New York

Pillay, T. V. R. (1979) The state of aquaculture, 1976. In: *Advances in Aquaculture*, ed. T. V. R. Pillay and W. A. Dill, pp. 1–10. Fishing News Books, Farnham, Surrey, England

Purdom, C. E., Jones, A. and Lincoln, R. F. (1972) Cultivation trials with turbot *Scophthalmus maximus*. *Aquaculture*, **1**: 213–30

Rollefsen, G. (1939) Artificial rearing of fry of sea water fish. Preliminary communication. *Rapport et Procès-verbaux des Réunions, Conseil permanent international pour l'Exploration de la Mer*, **109** (3): 133

Shelbourne, J. E. (1970) Marine fish cultivation: priorities and progress in Britain. In: *Marine Aquaculture*, ed. W. J. McNeil, pp: 15–36. Oregon State University Press: Corvallis, Oregon

Yashouv, A. (1969) Preliminary report on induced spawning of *M. cephalus* (L.) reared in captivity in freshwater ponds. *Bamidgeh*, **21** (1): 19–24

65

Honeybees

Eva Crane

Director, International Bee Research Association, Gerrards Cross, Bucks, England.

Introduction

Definitions and nomenclature

The domestication of bees is in a somewhat different category from that of animals whose mating can be arranged for and witnessed by the farmer. In 1610, John Guillim expressed it thus, in *A Display of Heraldry*: 'The Bee I may well reckon a domestick Insect, being so pliable to the Benefit of the Keeper.'

In this chapter domesticated colonies of bees are defined as those living in man-made hives, and wild colonies as those living in natural sites (or adventitiously in man-made structures not intended to house bees, such as the roof spaces in buildings). Many wild (feral) colonies start as a swarm from a domesticated colony.

The main species involved is the European honeybee, *Apis mellifera*, which is the source of most of the honey produced in the world, some 800 000 tonnes a year. In Asia a similar but somewhat smaller species, *A. cerana*, is kept similarly in hives. Both these species build a nest of parallel wax combs in a dark cavity. The other two *Apis*

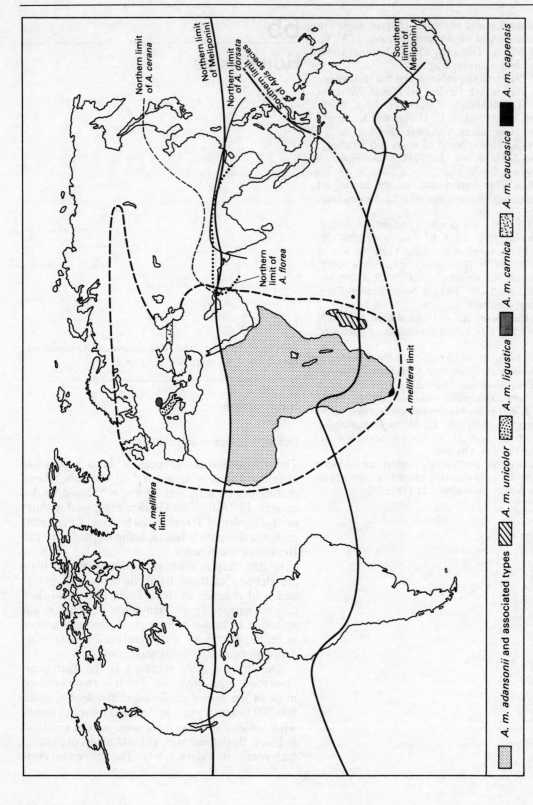

Fig. 65.1 The natural distribution of the four honeybee species (*Apis*) and of the stingless bees (Meliponini)

species, *A. dorsata* and *A. florea*, build a single comb in the open, and therefore cannot be hived in the same way, but honey and wax are nevertheless harvested from them.

In the tropics various species of Meliponini have been domesticated; like the Apini, whose only genus is *Apis*, these form permanent colonies and therefore store honey that they consume in dearth periods. Bumble bees (*Bombus*) have been 'kept' occasionally for their honey, and some of the larger species for pollination purposes. Other bees that nest gregariously, although not socially, can be domesticated, and species effective as pollinators of specific crops have actively been sought in the past few decades.

Natural geographical distribution of *Apis* species

Fig. 65.1 summarizes the known natural distribution of the honeybees. They are an Old World genus, originally tropical (Crane 1978). *Apis florea*, the smallest and the most primitive, is distributed throughout most of tropical Asia and as far west as southern Iran, and also Oman – the only place where its domestication has been recorded. It does not live higher than 500 m. A colony contains up to 30 000 bees, which produce a few hundred grams of honey a year, but – perhaps because the yield is low – the honey has commonly been prized as a medicine, and commands a high price.

The distribution of *A. dorsata* is closely restricted to the full tropics, and does not extend north of the Himalayas. This bee is, however, found up to 1200 m, occasionally up to 2000 m. Its workers have the largest body-size of all the *Apis*, about 18–19 mm long. The single comb may be up to 2 m × 1 m, and the honey yield up to 30 kg. In India more honey is harvested each year from the wild nests of *A. dorsata* than from the 650 000 colonies of *A. cerana* in hives.

Apis cerana is indigenous throughout tropical Asia and also farther north: in Himalayan valleys with a temperate climate such as Kashmir, and as far west as Afghanistan; in the east throughout mainland China and as far north as Korea, Japan and the Far Eastern province of the Soviet Union. It can live up to at least 2500 m. Different races or ecotypes have evolved in the regions outside the tropics.

Apis mellifera, the most highly advanced of the four species, presumably evolved from stock that originated in the tropics. Its natural distribution extends from Scandinavia to the southern tip of Africa, and from Portugal to the Ural and Caucasus mountains, and to Turkey and Iran. It has three main divisions, each consisting of many subspecies/races/ecotypes, in tropical Africa, in North Africa and the Middle East, and in Europe. Until after A.D. 1600 honeybees were confined to the Old World. Their spread throughout the New World is described later.

Several races of *A. mellifera* evolved in areas that remained isolated after the end of the last Ice Age, and some of these have been especially important in bee breeding (see Ruttner 1973). Cut off from the north by the Alps, Italian bees (*A. m. ligustica*) developed in a region with a long brood-rearing season and rich food resources. They have a great capacity to rear brood, which is advantageous in other areas with similar conditions, but not where food sources are uncertain. Carniolan bees (*A. m. carnica*), from a much harsher region in the eastern Alps, are more frugal in brood rearing and consumption of stores, and in recent decades they have been gaining in popularity in northern climates. Caucasian bees (*A. m. caucasica*), from the mountain range that stretches between the Black Sea and Caspian Sea, are rather gentle, conservative bees that have a longer tongue than most, and so are good pollinators of red clover with its deep corolla. But some Caucasian bees use so much propolis in the hive that it hampers beekeeping operations.

We do not yet have a very clear picture of the races of *A. mellifera* in tropical Africa, and the same is true of *A. cerana* in Asia. We know still less about races of *A. dorsata* and *A. florea*, but different ecotypes certainly exist.

General biology

All the honeybees live in permanent colonies, normally consisting of one female reproductive (queen) and many sterile females (workers), with a smaller number of males (drones) during the reproductive season. All species have the same haploid chromosome number, 16. A nest consists of one or more vertical combs of beeswax, which is secreted from abdominal wax glands. The cells of the comb are hexagonal, and in them brood is reared and food stored. All species reproduce by swarming, and all defend their nests by stinging an attacker. The colonies are permanent, and the food stored provides a

Table 65.1 Summary of age-linked stages in the life of a worker bee in summer (The ages entered for the adult bee are examples only. All are flexible in normal colony conditions, and highly flexible in abnormal conditions; an individual bee may show several different behaviour patterns on the same day.)

Age (days)	Stage	Food required	Other conditions	Behaviour
Brood Stage: Day 0 = Day Egg is Laid				
0–3	Egg	None	Temp. c. 34 °C	None
3–8	Larva	Bee milk, then pollen + honey	Temp. c. 34 °C	Eats, moves in open cell
8–9	Larva	None	Temp. c. 34 °C	Spins cocoon in sealed cell
9–21	Prepupa, pupa	None	Near 34 °C	None
Adult Stage:				
Day 0 = Day 21 of Brood Stage = Day of Emergence from Cell				
0–20 'House bee'* subdivided as follows:				
0–5	'Young bee'	Pollen + honey		Cleans cells
5–10	'Nurse bee'	Honey/nectar	Hypopharyngeal glands secrete bee milk	Feeds larvae
10–15	'Building bee'	Honey/nectar	Wax glands developed	Builds comb, caps cells
15–20	'Guard bee'	Honey/nectar	Venom glands developed	Guards hive (a few days only, or not at all)
20–30	'Honey-making bee'	Honey/nectar	Hypopharyngeal glands secrete invertase	Elaborates nectar, etc. into honey
20–35+ to death	'Field bee'	Honey/nectar	Flight muscles developed; attracted to light, not darkness	After short orientation flights, forages for pollen, nectar, etc., also (some bees) for water, or for propolis (and works with propolis in hive)

* remaining in the hive; preferring darkness to light. From Crane (1980a)

reserve for consumption during the next dearth period. Food is collected by foraging on plant resources within a few kilometres of the nest (less for the smaller species than for the larger ones) (see Table 65.1)

Carbohydrates are provided mainly by flower nectars, but also by extrafloral nectars, and honeydew excreted by Hemiptera. Protein is provided by pollen, which is also the bees' main source of minerals, vitamins and other trace substances. In *A. mellifera*, and probably in the other species, collection of pollen by a colony is intimately connected with its brood rearing: after a dearth period substantial brood rearing is usually initiated by the availability of new pollen, and pollen collection is stimulated by the pres-

ence in the nest of unsealed brood, i.e. larvae that need feeding. Pollen is not normally stored in large amounts. On the other hand, nectar and other energy foods may be collected greatly in excess of requirements, and collection can be stimulated by a large amount of empty storage space in the nest. The world's honey industry is based on this behaviour characteristic.

The dearth period, for which the food store is needed, is caused by drought or excessive rainfall in the tropics, and by cold in the temperate zones. The single-comb builders, *A. florea* and *A. dorsata*, cannot survive a cold winter, nor can the southern races of *A. cerana* or the tropical races of *A. mellifera*. These bees can, however, survive the relatively short warm dearth periods experienced in their native regions. In some regions *A. dorsata* colonies regularly migrate between two areas, each of which provides food resources for part of the year. The same is true of *A. mellifera* in parts of tropical Africa, for instance between the top and bottom of the Rift valley. *Apis florea* may also migrate through short distances, and *A. cerana* shows a similar tendency, but to a less pronounced degree.

As well as this migration, although not always easily distinguishable from it, the tropical honeybees may abandon their nests if subjected to various types of disturbance; this is often referred to as absconding. An *A. florea* colony may return to the nest a short time afterwards.

The ecotypes of *A. cerana* that colonized regions north of the eastern end of the Himalayas, and in the Himalayan valleys, are able to survive a cold winter by forming a cluster within the nest and regulating the temperature inside it. This characteristic is more highly developed in European ecotypes of *A. mellifera*, although not in its tropical ecotypes in Africa.

Colony reproduction in all honeybees is by swarming, in which the colony divides into two viable units. Unlike migrating and absconding swarms, reproductive swarms consist of only part of the adult population, the other part remaining in the parent nest with a substantial amount of brood. This brood includes several queens, each in a specially constructed elongated cell. One of the queens will remain to head the colony in the original nest, the others being either killed by her or leaving (unmated) with small 'after-swarms' or casts. A few days after emergence from her cell, the surviving queen flies out and mates with a number of drones and soon starts egg-laying.

The 'prime' swarm that leaves the nest includes the mated queen that headed the colony; she has been slimmed down by reduced feeding and is thus able to fly – for the first time since her own mating flight. The swarm clusters on some support not far from the nest, often quite high above the ground, and finally (usually on the same day) flies off to a new nesting cavity that has been found by scout bees from the swarm. The modern beekeeper aims to manipulate his colonies to prevent swarming or, if swarming does occur, to 'take' the clustered swarm and hive it, and prevent after-swarms by destroying queen cells.

The issue of a swarm represents an economic loss, the season's honey yield for the parent colony being approximately halved. Some earlier systems of management, for instance the skep beekeeping of northern Europe where heather (*Calluna*, *Erica*) provided a late honey harvest, encouraged swarming. One spring colony might produce a progeny of four or even more swarms in the course of the summer, each of which might produce a small honey harvest. One or two of these were left with some honey for winter consumption, and the rest killed off when the honey was taken.

The above refers to *A. mellifera*; there are some differences with the other species which there is not space to describe here. All *Apis* species mate in flight. The sex pheromone that attracts drones to a nubile queen (9-oxodecenoic acid) is the same in all species, and this presents problems when one species is introduced into the territory of another.

An account has been given elsewhere (Crane 1980a) of the foraging behaviour of honeybees, and of their production of honey from the materials collected. Table 65.1, from that source, summarizes the chronology of the individual worker bee's life and activities.

Wild bees and their exploitation

Exploiting wild nests of honeybees (*Apis*)

The following rough time-scale shows the antiquity of bees and honey compared with that of man (Crane 1975):

For 150–100 million years	Flowering plants have existed and produced nectar and pollen
For 50–25 million years	Solitary bees have existed, also early primates (monkeys)
For 20–10 million years	Social bees have produced and stored honey
For a few million years	Man has existed and has eaten honey
For ten thousand years	Records have survived of man's exploitation of honey

The nests of the social bees were certainly

raided by animals throughout the 10–20 million years of their existence. This would have been especially true of honeybee nests, which contain the most honey, and we can get an idea of the methods used through observation of similar animal marauders today. In general they seem willing to suffer many stings to get at the honey, but certain primates, including baboons, try to get the honey free from bees. Chimpanzees have been observed poking a long twig into a hole leading to a nest and withdrawing it coated with honey. This is one method used by primitive man, whose honey hunting was thus a continuation of much earlier animal behaviour: robbing bees' nests using a tool to extract the honey from them, and trying to avoid at least some stings.

The earliest known records of honey hunting by man are rock paintings in eastern Spain, made soon after the end of the Ice Age, probably around 7000 B.C. Ladders are a feature of most such paintings, so many of the nests must have been under a rock overhang on a cliff face or in a similar position; I do not know of any painting that shows a nest in a tree. Bees and their nests are shown in about eighty Bushman rock paintings in southern Africa, and in one a man is using a smoker when raiding a nest. The dates of these paintings are not known.

In the Middle Ages there was intense exploitation of honeybees in the forest areas of northern Europe, where the bees nested in hollow trees. This developed into a sort of beekeeping (Galton 1971), in which each bee tree would be owned by a beekeeper and marked with his sign. Felling of bee trees was prohibited, whether they had bees in them or only empty holes that might be used later by a swarm. Honey and beeswax were harvested through a door cut in the side of the tree, and marketed through a vast trade network; Novgorod was a great centre, and wax was sent down the Danube to Bohemia, down the Dnieper to Byzantium, and down the Volga to the Orient.

This tree beekeeping led to the development of beekeeping proper, when logs containing bees' nests were cut from the tree and removed to some convenient place (an apiary) where they could be tended and guarded. Beekeeping had also developed separately, several thousand years earlier, in the Middle East, whence it spread throughout Africa, Europe south of the Alps, and into Asia. Before following the course of these two developments of beekeeping with honeybees, the exploitation of other bees must be mentioned. The honeys from the four *Apis* species, and from the bees mentioned below, have been discussed elsewhere (Crane 1975).

Exploited nests of non-*Apis* bees

Stingless bees. The most important group of domesticated bees after the honeybees is the tropical Meliponini or stingless bees; their distribution is shown in Fig. 65.1 (p. 404). The two main genera are *Trigona* and *Melipona*, the latter being the larger bees. Most known sources of information on their biology and exploitation have been recorded (Crane 1978). The nests usually contain an irregular collection of central brood cells and peripheral honey pots, somewhat like nests of bumble bees (*Bombus*). Stingless bees have been most extensively domesticated in Central America (Schwarz 1948), being kept in logs, gourds, pots and other containers, and also in wooden hives designed appropriately for the species concerned (Nogueira-Neto 1970). Meliponiculture has also been promoted in parts of Africa, e.g. Angola, but not to any extent in Asia or Australia, where the bees also occur.

Honey yields vary from species to species, being often 1 kg or less each year, more rarely 10–20 kg. The honey is usually pressed out from the comb.

These stingless bees were the only source of honey and beeswax in Central and South America until honeybees were introduced in the 1800s. They must therefore have produced the large amounts of beeswax used by American Indians – notably in Colombia – in making the golden ornaments which were cast by the 'cire perdue' (lost wax) method.

Honey hunting must certainly have preceded beekeeping with stingless bees, and an account of honey hunting by the Guayaki Indians in Paraguay published in 1939 (see Crane 1975) shows the important part played by honey in the life of these people.

Bumble bees These social bees (*Bombus* species) are widely distributed in the northern hemisphere, where the queens overwinter alone and found colonies in spring that last through the next summer. (In the tropics there are species that form permanent colonies.) Honey from wild colonies has been collected and eaten in many areas, although the yield is small, since no stores are needed for overwintering. I know of only one record of the domestication of bumble bees for their honey, in Transylvania (Romania/Hungary), where villagers set up clay pots as hives (Crane 1975).

In the last few decades the domestication of bumble bees has been intensively studied in some areas, especially in northern Europe, because of their value as pollinators. The large species fly at lower temperatures than honeybees, and are thus useful pollinators of fruit blossom in spring. Also, some have much longer tongues than honeybees; these bees can collect nectar from flowers with a long corolla, and pollinate flowers such as tetraploid red clover (*Trifolium pratense*). Apart from direct domestication, populations of wild colonies can be usefully increased by locating red clover seed plots near land with many suitable nesting sites, and by leaving land round the plots uncultivated. This method has proved very effective in Finland, for example.

Apart from social bees, two species that are highly gregarious have been domesticated, both for pollination of alfalfa (lucerne, *Medicago sativa*). Pre-emergence conditions are adjusted so that adults are flying when the alfalfa blooms. *Megachile rotundata* (Hobbs 1972) nests in hollow stems, and will accept drinking straws of the right size, or similar hollows drilled in wood. Banks of the (horizontal) nest sites are set up under a shelter on alfalfa plots, together with trays containing overwintered immature bees on the point of emergence. The females use the nest sites provided, forage on the alfalfa, and pollinate it efficiently. *Nomia melanderi* (Bohart 1972) nests in alkaline soils, so artificial 'bee beds' are set up close to the alfalfa plots, and stocked with the prepupae; if necessary the beds are warmed. The most notable developments with both species have been in the northwest of North America, on both sides of the U.S.A.–Canada border.

Other candidates for similar domestication include species of *Osmia*, *Xylocopa*, *Anthophora*, and other *Megachile*.

The development of beekeeping

Traditional hives

The earliest known evidence of the use of hives is in Egypt. It consists of paintings or engravings that ornament a temple dating from 2400 B.C., two tombs from about 1450 B.C. and one from 600 B.C. These scenes are sufficiently similar to show that beekeeping did not change very much in the 1800 years spanned. It appears in 2400 B.C. as a well-developed craft, and we do not know its earlier history or chronology. (See Crane 1983).

Hives of much the same shape made of mud, cow-dung and straw are still used in the high mountains of Simen, Ethiopia. At the appropriate time of year (after the flowering season) the beekeeper takes the honey, and this may well be the only occasion on which he opens the hives. These are on the ground, not stacked up, and the beekeeper lies so that he faces the back of the hive, which he breaks off, and puffs smoke in from a primitive open smoker. The smoke drives the bees to the front end of the hive where their flight hole is, and induces many of them to leave it. The beekeeper can then, without too many stings, cut out the combs near the back of the hive that contain honey. The combs near the flight entrance are likely to contain the brood nest, and if the beekeeper is sufficiently enlightened he will leave them intact; with luck the queen will be unharmed, and the colony will survive. The beekeeper may strain the honey, or may eat it as it is – honey, pollen, wax, and any brood that is present. The pollen and brood provide protein and other nutrients; the wax is not digested by mammalian species.

This type of beekeeping, in long horizontal hives, was the norm in the ancient world, in Egypt, Crete, Greece and Rome. It remains so today where traditional beekeeping is practised in these regions, throughout Africa and the Middle East, in parts of southwestern Asia (Iran, Afghanistan, Turkey, Kashmir) and in some areas farther east. (In much of the rest of Asia, honey hunting was not replaced by beekeeping until development programmes were set up after the Second World War.)

All the ancient Greek hives found in excavations have one end open, with a circular closure, but no flight hole at the other end, so the beekeeper could not drive the bees through and out of the hive. The Greek hives are made of baked clay, as are most traditional hives today in the Middle East (except in Egypt where Nile mud is still used); they are usually stacked, and most have an opening at each end. In Egypt nowadays a stack contains 400 hives or more – mud cylinders about 120 cm long and only 8 cm in diameter. In tropical Africa, to which cultural practices were transmitted along the Nile valley, most of the cylindrical hives are made of log or bark, and are hung in trees to protect them from the many enemies that are present. I have measured such hives from widely scattered areas, and found that many are about 90 cm long, 15–20 cm longer than an arm's reach, so that a beekeeper taking honey from one end would be likely to leave the last four combs behind. I do not think this happened by chance. There are many variants in hive shape, in material (woven wicker and coiled straw are common), and in flight entrance position.

The other basic type of early traditional hive, the upright log of the forests in northern Europe, spread as far south as Spain and Portugal, but there the much lighter bark of the cork oak (*Quercus suber*) is used. If an upright hive is open at the bottom, the honey can be taken by turning it over and cutting out accessible honey comb from the bottom. In many forest areas, instead, a door was cut in the side of the log (as in the living tree) to give the beekeeper access to the combs, and the log was closed at both top and bottom.

A log large enough to house a colony of bees is heavy, and at some time – probably in late centuries B.C. – a tall wicker basket came to be used as a hive. It stood upright, its open mouth on a stand of wood or stone, and the bees flying out from a gap at the bottom edge. The beekeeper could gain access to the combs only from the open mouth, which presented him with maximal exposure to the bees, and in this aspect the wicker skep would seem to be a retrograde step. There is no evidence of clothing that protected the face against stings until about A.D. 1400.

These wicker hives cost less effort to make (and to move) than log hives. They spread as far south as the Pyrenees, as far east as Romania and Novgorod, and as far west as Britain, where their last recorded use was in Herefordshire in the 1920s. The use of upright hives hardly crossed the Alps – the horizontal hives deriving ultimately from ancient Egypt occupied all the territory beyond.

In due course, as agriculture developed, coiled-straw work was used for baskets, and these provided a more weather-tight container for bees. The wicker skeps had to be coated (cloomed), usually with a mixture of mud and cow-dung. Straw skeps were probably first used in north Germany, somewhere west of the Elbe, and they gradually replaced the wicker skep over most of its range.

The spread of honeybees into unoccupied territories

In the story of the domestication of honeybees so far, their distribution has been confined to the Old World, and their management has consisted of little more than cutting out combs of honey once a year. The 1600s were to change both these situations, and to initiate an explosive development that in the 1800s revolutionized the exploitation and domestication of bees.

The earliest likely record that has come to light of hives of bees making the Atlantic crossing is in 1621. The hives taken to North America were almost certainly straw skeps, but most records of early hives there are of upright logs (known as bee gums) or tall hives made of wooden boards; wood was generally plentiful.

In many of the places to which the bees were taken, they prospered sufficiently to yield honey for their owners, and to throw swarms, which spread across the country in advance of the settlers. The European honeybee had probably become fairly common throughout the eastern part of North America by 1800. Pellett (1938) gives details. Hives were taken to the west both overland and by sea; they were in California by 1853, and from there the first hives reached British Columbia in 1858. Honeybees were already in Florida by 1763, when they were taken thence to Cuba. They reached South America surprisingly late; e.g. the earliest known record for Brazil is 1839, and for Chile and Peru probably 1857.

Honeybees were first landed in Australia in 1810, in New Zealand in 1839, in Hawaii in 1857, and in other Pacific islands much more recently, e.g. Papua New Guinea in the 1940s.

All this expansion was into territory with no indigenous *Apis* species, and therefore no competition from them. But except in some Pacific islands, there were other indigenous bees, and also indigenous plants that produced pollen and nectar and were pollinated by bees. Many parts of the New World offered very rich forage to the adaptable honeybee, and even today the average honey yield per hive there is three times as high as in the Old World.

All the importations of honeybees into new territories (except the Pacific) were made before the start of modern beekeeping, which is the next phase in the history of the domestication of bees.

Modern hives

We shall shortly consider the complications and limitations to breeding and selection of honeybees. These are on quite a different scale from those encountered with domesticated mammals and birds; on the other hand a breakthrough was made in the last century in the effective *management* of bees for honey production, which increased the potential yield by a factor of 10 or 20 per hive and by a factor of several hundreds per beekeeper. This, rather than breeding, is what differentiates modern from traditional beekeeping.

The crucial advance was made by the Rev. L. L. Langstroth in Philadelphia, U.S.A., in 1851: he used vertical-sided boxes, with appropriately separated wooden frames (for combs) supported like files in a suspension system. Provided that the ends of the frames were similarly distanced from the hive walls, the bees would 'respect' the space and not build comb across it. The whole frame was removable.

In 1853 Langstroth set out his ideas and achievements in *The Hive and the Honeybee*, and the book played a seminal role in spreading the use of the movable-frame hives throughout the world. This in turn led to the invention of embossed beeswax comb 'foundation' to fit into the frame; the centrifugal extractor that spun the honey out of the strengthened framed combs, leaving them ready for re-use; and the queen excluder, a grid penetrable by workers but not by the queen, used to separate the brood chamber from the honey chamber.

What Langstroth's development makes possible is the manipulation of a colony of bees so that it achieves a high population – 50 000 adults or more, as well as brood – which can produce honey greatly in excess of the colony's requirements; it has stores for the winter, and the beekeeper gets his harvest as well. Each spring, honey chambers (supers) are superimposed on the brood chamber(s) in which the colony wintered, more being added as the colony expands, so that the colony continues to grow but is never crowded enough to swarm. More brood space is provided as necessary, kept separate from the honey store by a queen excluder. Reproduction of the colony by swarming is prevented, and excess honey is produced instead.

Bee breeding

Honeybee genetics and sex determination

The genetic mechanism of inheritance in social insects is very different from that in most animals considered in this book. Some background information is therefore needed before the practicalities of breeding and selection of honeybees can be considered. Both silkworms and honeybees have been discussed briefly by Hoy (1976).

Honeybees (*Apis* species) reproduce by a haplo–diploid system. The queen and worker castes are female; they develop from fertilized eggs, and have the diploid number of chromosomes (32). Differentiation between queen and worker – which includes both body characteristics and behaviour – is not genetic; it results from differences in the food of the larva during the first 3 days after hatching from the egg. Larvae become workers (non-reproductive) except for a few which are given richer food, and more of it; these become queens (reproductives). The critical factor that triggers the rate of food intake by the larva appears to be the sugar content (Beetsma 1979). The drone is male, produced by arrhenotoky from an unfertilized egg, and has the haploid number of chromosomes (16).

In the queen's reproductive system meiosis of the oocytes proceeds in the usual way, resulting in gametes with 16 chromosomes. In the drone's reproductive system, the first meiotic division in spermatogenesis is abortive, producing a cytoplasmic bud; the second meiotic division is modified so that only one spermatozoon is formed, and it contains all 16 chromosomes from each spermatocyte. According to one commonly accepted theory of sex determination (Whiting 1945) there is a single sex locus (X), with multiple alleles, and sex is determined thus: individuals that are heterozygous for the sex alleles (diploid) become females, and those that are hemizygous (haploid) become males. An individual developing from a fertilized egg that is homozgous at the sex locus is a diploid male. No diploid males are reared in nature, because the workers eat them soon after they hatch from the egg. In a long series of experiments Woyke (1978) has succeeded in rearing diploid drones to maturity, and triploid queens and workers were produced by 1980, but this work has not (yet) impinged on practical bee breeding. Another theory, the genic balance theory, has been proposed by Kerr and his co-workers (see Kerr 1974).

Mutations that are useful as markers for genetic studies have been produced, and where possible maintained; thirty-five are listed by Rothenbuhler (1975), but none of these is directly useful in practical bee breeding.

Every haploid drone inherits all his genes from one individual, the queen that is his mother. Female progeny of a single queen inherit genes from their mother, and from one of a number of drones that mated with her in flight – the donor of the spermatozoon fertilizing the egg in question. The only exception occurs if the queen is instrumentally inseminated with semen from a single drone (see below).

Natural mating

The young queen honeybee mates in flight, with a number of drones in quick succession. The drones may have flown more than 5 km from their colonies, and queens nearly as far. In tests with genetically marked bees (the cordovan mutation, in which black colouration is replaced by brown), 25 per cent of queens mated with drones from 16 km away. So, however carefully the genetic make-up of a queen is controlled, that of her worker offspring cannot be assured so long as there are any adventitious colonies containing drones within, say, 20 km of the queen's colony. Sites used for mating stations in different countries are: desert oases (excellent); small islands (often windy, but satisfactory if well off shore); deep valleys with high mountains on either side to give protection (but very few such bee-free valleys are available).

The difficulties in ensuring 'pure' mating were appreciated long before details of the mating itself had been elucidated. But it was not understood until 1954 that queens normally mate with a number of drones (commonly 6–10), on a single flight. In 1963 it was shown that drones are not attracted to a virgin queen unless she is flying higher than a critical distance above the ground, which can vary from about 5 to 40 m, for reasons still not fully understood. The workers' flight space is below that of the reproductives, and this helps drones to locate a queen in the air.

Attempts have been made in Zimbabwe to predetermine the drones mating with a queen by using a flight cage up to 11 m high, but only occasional successes have been reported.

Instrumental insemination

An inviting approach to breeding is insemination of the queen in the laboratory. A technique was first devised in the United States by L. R. Watson in 1927, and this has been improved by many others since, in the United States and elsewhere. During the past decade the procedure has been a routine one, and many thousands of queens are instrumentally inseminated each year in bee-breeding programmes.

The queen is anaesthetized with carbon dioxide and immobilized in a tube, from which the tip of her abdomen protrudes, in the field of a binocular microscope. A specially designed syringe is mounted appropriately, and is charged with semen from several drones, which is then injected into the queen's vagina, beyond the valvefold which is held out of way by a hook (Laidlaw 1977).

Changes in honeybees under selection

Social bees were producing honey for 10–20 million years before man existed. In the course of natural selection, the honeybees that survived had genetic characters enabling colonies to store enough honey to last them through the dearth periods they encountered, to defend their nest against enemies (many mammals and birds will eat the contents of a bees' nest), and to resist hazards of climate, disease and predation.

When and where man persistently took honey, and disrupted the bees' nests in doing so, there was also selection in favour of bees that were best able to defend their nest (were more 'aggressive'). In traditional African beekeeping, empty hives were populated by swarms. This was also true in early European beekeeping; the beekeeper 'took' the clustered swarm and put it into a hive. There was thus selection towards swarming tendency, and also towards swarming early in the active season, because early swarms had the best chance of building up population and stores before the dearth period. ('A swarm of bees in May/Is worth a load of hay ... A swarm in July /Isn't worth a fly.')

In modern beekeeping, queens are reared from eggs that are the progeny of a selected (breeder) queen. They do not originate in a colony preparing to swarm, but in a colony chosen by the beekeeper for the purpose. By judicious choice of breeder queens for colony characteristics such as those listed below, progress in selection through the female line can be made (Rothenbuhler 1980). Selection can include the male line if newly reared queens are put into (small) colonies in a completely isolated mating apiary, with other chosen colonies organized to rear many drones. But such progress in selection is not *permanent* unless this procedure is repeated for every queen reared. Normally drones of unknown ancestry are present (in other colonies) within flying reach of the young queen's flight range, and will contribute their genes to the queen's female progeny, in a way and to an extent that the bee breeder cannot know.

Selection became much more effective when queens could be inseminated instrumentally with semen from drones of specified ancestry, and selection for characteristics 8 and 10 below was not possible before this could be done. Different bee breeders have selected *Apis mellifera* for one or other of the following characteristics, according to their requirements. Queen-rearers, in general, will continuously aim for characteristics 1–4; 9 and 10 are specializations, and so is 8. at present, although its importance will increase.

1. Improved honey production.
2. Populous colonies with a low tendency to swarm (in earlier management based on use of swarms, selection had the opposite aim).
3. Gentleness, bees remaining on the combs when colonies are inspected – as they must be with modern management – and not flying off and stinging the operator.
4. Other colony characteristics such as overwintering, compact brood nest, early spring development.
5. Light (golden) colour of queens, which has found inordinate favour especially in North America, and whose only merit is that a queen is easy to spot on the comb.

6. Resistance to brood diseases, e.g. American foul brood (*Bacillus larvae*), European foul brood (a complex of bacterial pathogens), chalk brood (*Ascosphaera apis*). For American foul brood the mechanism(s) could be: resistance of honeybee larvae to *Bacillus larvae*; antibiotic content of the food fed to them by workers; or hygienic behaviour of the workers in cleaning-out infected brood cells.
7. Resistance to adult bee diseases, e.g. acarine disease (*Acarapis woodi*), 'hairless-black syndrome' (viral).
8. Low variability in colony performance, which allows all colonies in an apiary to be given the same treatment at the same time.
9. Heavy early brood rearing, needed for package bee production.
10. Effectiveness in pollinating a difficult crop, notably alfalfa.

Selected characteristics cannot be maintained in subsequent generations through natural matings, unless these occur in isolated areas. Moreover, bee breeding can impose a heavy burden in the maintenance of colonies with the required genetic characteristics. Some advances have, however, been made with storage of frozen drone semen.

The idea of crossing different strains of honeybees to exploit hybrid vigour is attractive, and heterotic effects have been well studied (e.g. Cale and Gowen 1956). Experiments have been done with inter-racial crosses, e.g. in France (Fresnaye and Lavie 1977) and in the Soviet Union (e.g. Mel'nichenko and Trishina 1977). In the United States, hybrids have been produced as a commercial operation since 1949 (Witherell 1976). In this system four inbred lines A, B, C, D, are produced from colonies selected after several years of evaluation and breeding. These lines are crossed in pairs whose characteristics complement each other (e.g. AB, CD), and the resultant hybrids are themselves crossed to give a 'double hybrid' (AB × CD). Two such double hybrids have been developed by Dadant and Sons, and are now sold by Genetic Systems, Inc. They are known as Starline (from all Italian stock) and Midnite (from Carniolan and Caucasian stocks).

The beekeeper who buys these hybrid queens in spring, and replaces ordinary Italian and Caucasian with them, can get up to twice as much honey from the colonies concerned. But only F_1

hybrids benefit from heterosis, and in subsequent generations deterioration is very rapid, so the beekeeper must buy new queens as necessary, usually each year, if he is to maintain the high yields. There are no improved, fixed hybrid races or strains in honeybees, because of the mating behaviour of these insects.

Present situation and future prospects

Today the domestication of bees results in the annual production of around 800 000 tonnes of honey, of which 75 per cent is consumed in the country of origin and 25 per cent sold on the world market. The largest producers are all large countries; the United States Department of Agriculture estimates for 1979 are: United States 98 000 tonnes, Chinese People's Republic 85 000, Soviet Union 80 000. Of these, only China is a major net exporter, and indeed it is becoming the world's largest honey exporter. Mexico and Argentina export most of their production, which is 56 000 and 28 000 tonnes respectively. The same is true of various other countries of Latin America that together produce some 35 000 tonnes. Australia and New Zealand between them produce 24 000 tonnes and export 6000. Other countries producing 10 000 tonnes or more are as follows. In Europe: France, West Germany, Greece, Spain; in Asia: India, Turkey; in Africa: Angola, Ethiopia, Madagascar. Most consume all or most of their production, and France and West Germany also import honey. West Germany is the largest importer, followed by Japan and the United Kingdom, for which 1978 figures were 57 000, 24 000 and 17 000 tonnes respectively. Most of the honey on the world market is imported into countries of Western Europe. A high per caput consumption of honey tends to be linked with a strong beekeeping tradition in past centuries, either directly as in northern Europe, or indirectly as in New World countries populated from northern Europe: United States, Canada, Australia, New Zealand. Japan's relatively high honey consumption started only after the Second World War, and was part of a change in eating habits brought about by American influence.

Prospects for increasing world honey produc-

tion in the future depend mostly on events in those regions of the tropics and subtropics where beekeeping has a large undeveloped potential. Few dramatic gains can be expected in temperate zones, except perhaps in the intensive short-summer growth regions in northern Canada and eastern Siberia, where it will depend on crops being grown that are good honey plants. In the tropics there is a wealth of honey plants, but the honey can be harvested only if there are more hives, better management systems, better roads, and suitable vehicles to carry the hives from one honey flow to another.

Beeswax is most easily produced in the tropics where temperatures are high, and the beeswax harvest could be increased greatly by paying proper attention to its production and collection. Whether or not the harvesting of other hive products – pollen, royal jelly, propolis and bee venom – will increase appreciably depends on future demand for them at attractive prices. Methods have been developed for harvesting all of them commercially (Crane 1980b).

The breeding of honeybees for honey production, and for other more specialized purposes, will continue. It is likely to increase yields in areas with good food resources for bees and a suitable climate, and where sophisticated systems of bee management are used. But it does not provide the same cost-effective pathway to increasing yields as with mammals, because of the instability of any lines or strains bred.

One type of beekeeping that will surely increase greatly is the provision of colonies of honeybees – and of other bees – for pollinating crops (Free 1970, McGregor 1976). In small-scale agriculture, scattered colonies often provide enough pollination to ensure good seed and fruit set in entomophilous plants. In large-scale agriculture, which will become more and more the norm, it is essential to move pollinators to certain crops in order to obtain maximum yields. The most versatile pollinators are the honeybees, and hives of honeybees can easily be transported to provide a large number of pollinators where required. Where an insect other than a honeybee is especially effective, it is almost always another species of bee, and methods will be devised for domesticating it, when and where the demand arises.

References

Beetsma, J. (1979) The process of queen-worker differentiation in the honeybee. *Bee World*, **60** (1): 24–39

Bohart, G. E. (1972) Management of wild bees for the pollination of crops. *Annual Review of Entomology*, **17**: 287–312

Cale, G. H., Jr and **Gowen, J. W.** (1956) Heterosis in the honey bee (*Apis mellifera* L.). *Genetics*, **41** (2): 292–303

Crane, E. (ed.) (1975) *Honey: a comprehensive survey*. Heinemann and International Bee Research Association: London

Crane, E. (1978) *Bibliography of Tropical Apiculture*; also *Satellite Bibliographies*. International Bee Research Association: London

Crane, E. (1980a) *A Book of Honey*. Oxford University Press: Oxford

Crane, E. (1980b) Apiculture. In: *Perspective in World Agriculture*, pp. 260–94. Commonwealth Agricultural Bureaux: Farnham Royal, Bucks, England

Crane, E. (1983) *The Archaeology of Beekeeping*. Duckworth: London

Free, J. B. (1970) *Insect Pollination of Crops*. Academic Press: London and New York

Fresnaye, J. and **Lavie, P.** (1977) Selection and crossbreeding of bees in France (*Apis mellifica* L.) In: *Genetics, Selection and Reproduction of the Honey Bee*, pp. 212–18. Apimondia Publishing House: Bucharest

Galton, D. (1971) *Survey of a Thousand Years of Beekeeping in Russia*. Bee Research Association: London

Hobbs, G. A. (1972) Beekeeping with alfalfa leafcutter bees in Canada. *Bee World*, **53**: 167–73

Hoy, M. A. (1976) Genetic improvement of insects: fact or fantasy? *Environmental Entomology*, **5**(5): 833–9

Kerr, W. E. (1974) Advances in cytology and genetics of bees. *Annual Review of Entomology*, **19**: 253–68

Laidlaw, H. H. (1977) *Instrumental Insemination of Honey Bee Queens*. Dadant: Hamilton, Illinois, U.S.A.

McGregor, S. E. (1976) Insect pollination of cultivated crop plants. *Agricultural Handbook, United States Department of Agriculture*, No. 496. Washington, D.C.

Mel'nichenko, A. N. and **Trishina, A. S.** (1977) Ecological and genetical bases of the heterosis in the honey bee (*Apis mellifera* L.) In: *Genetics, Selection and Reproduction of the Honey Bee*, pp. 203–9. Apimondia Publishing House: Bucharest

Nogueira-Neto, P. (1970) *A Criacão de Abelhas Indigenas sem Ferrão (Meliponinae)* (2nd ed). Chacaras e Quintais: São Paulo

Pellett, F. C. (1938) *History of American Beekeeping*. Collegiate Press: Ames, Iowa

Rothenbuhler, W. C. (1975) The honey bee, *Apis mellifera*. In: *Handbook of Genetics*, ed. R. C. King, Vol. 3: *Invertebrates of genetic interest*. Plenum Publishing: New York

Rothenbuhler, W. C. (1980) Necessary links in the chain of honey bee stock improvement. *American Bee Journal*, **120**: 223–5, 304–5

Ruttner, F. (1973) [*Techniques for the Rearing and Selection of Honeybees*]. Ehrenwirth Verlag: Munich. (In German)

Schwarz, H. F. (1948) Stingless bees (Meliponidae) of the western hemisphere. *Bulletin of the American Museum of Natural History* No. 90: 1–546

Whiting, P. W. (1945) The evolution of male haploidy. *Quarterly Review of Biology*, **20**: 231–60

Witherell, P. C. (1976) A story of success – the Starline and Midnite hybrid bee breeding programs. *American Bee Journal*, **116** (2): 63–4, 82

Woyke, J. (1978) Biology of reproduction and genetics of the honeybee (E21-ENT-28). Final technical report (1971–78). Agricultural University of Warsaw

66

Silkworm moths

Yataro Tazima

National Institute of Geneties,
Mishima, Japan

Introduction

Nomenclature

Among silk-producing insects, those used commercially are divided into two groups, mulberry silkworms and non-mulberry silkworms. This division coincides with the taxonomic classification. The former is represented by *Bombyx mori*, a unique member of the Bombycidae, that feeds only on mulberry leaves. The latter group comprises several species of *Antheraea* and *Philosamia* of the family Saturniidae that feed on plants other than mulberry, *i.e.* *Antheraea mylitta* (Indian tussore), *A. pernyi* (Chinese tussore), *A. yamamai* (Japanese tussore), *A. assama* (muga silkworm) and *Philosamia cynthia ricini* (eri silkworm). Tussore is also spelt 'tussah', 'tusser' or 'tussor'; it comes from Hindi *tasar* = shuttle. The range of food plants is a little wider in the latter group but each species has its favourite food plants as shown in Table 66.1. Non-mulberry silkworm species are not domesticated but they are raised outdoors on the appropriate trees or bushes, except the eri silkworm, which can be raised indoors if circumstances permit.

Biology

The mulberry silkworm *B. mori* and its wild relatives are adapted to a temperate climate, while some of the non-mulberry silkworms inhabit tropical regions.

As members of the Lepidoptera they all undergo a complete metamorphosis during their life-cycle – egg, larva, pupa, and moth. At the end of the larval stage, they spin cocoons with fine thread excreted from silk glands so that they are protected from predators during their pupal stage. Some species or races complete only one' generation a year (univoltine), but the majority have two or more generations, depending on the environmental conditions such as temperature, day length and nutrition. Whether they hibernate as eggs or in another developmental form depends on species, race, and environmental conditions.

Geographical distribution

According to statistics of the International Silk Association, annual world production of mulberry silk amounted to 55 315 tonnes in 1980, of which China produced 42.4 per cent, Japan 29.2 per cent, U.S.S.R. 8.0 per cent, India 7.2 per cent and Korea 5.9 per cent. The remaining 7.3 per cent was produced by countries in the Mediterranean area and Latin America.

The production of non-mulberry silk is believed to be only one-tenth of that of mulberry silk. In India the total production of non-mulberry silk in 1975 was reported to be 543 tonnes, of which tussore produced 72.7 per cent, eri 22.3 per cent and muga 5.0 per cent (Jolly *et al.* 1979).

Since silkworm rearing (sericulture) produces the material for valuable fabrics, it brings a fairly good income to the farmers. Indeed, Japan owes her present economic prosperity to the sericultural industry practised in the early years of the Meiji era. She earned the funds for industrialization by exporting the silk to the United States. The exports of silk amounted to more than 35 per cent of the value of total exports during the first 33 years and 25 per cent during the next 25 years after the government permitted the trade in 1859. Recently the situation has changed drastically, because of the acute scarcity of labour and the increased cost of production. Japan is now importing silk from China and Korea.

Table 66.1 Useful silkworms

Family	Scientific name	Chromosome number	Food plants*
Bombycidae	*Bombyx mori*	$n = 28$	*Morus*
	Bombyx mandarina	$n = 27, 28$	
	Theophila religiosae	$n = 31$	
	Rondotia menciana		
Lasiocampidae	*Pachypasa otus*		
Saturniidae	*Antheraea mylitta*	$n = 31$	*Terminalia, Shorea, Zizyphus*
	Antheraea pernyi	$n = 49$	*Quercus*
	Antheraea yamamai	$n = 31$	*Quercus*
	Antheraea assama	$n = 15$	*Machilus, Litsaea et al.*
	Antheraea roylei	$n = 30$	
	Attacus atlas		
	Philosamia cynthia ricini	$n = 14$	*Ricinus, Heteropanax, Ailanthus*
	Philosamia pryeri	$n = 14$	
	Philosamia walkeri	$n = 13$	

* Shown only for silkworm species that are currently used commercially

Wild species

The most widely-reared species is the mulberry silkworm, *B. mori*, and its wild relatives are few; there are only three species, all poor cocoon spinners. In contrast, the wild species of the *Antheraea* family are numerous and many of them spin beautiful cocoons. The systematic relationship of these species is shown in Table 66.1.

All species mentioned in this table were once used for production of silk but some were abandoned when *B. mori* was introduced. Such species include *Rondotia* used in China (Nunome 1979) and *Pachypasa* of Cos Island (Greece) (Richter 1929, Bock and Pigorini 1938). The ancestor of the present-day *B. mori* is considered to be *B. mandarina* inhabiting China, Korea and Japan. Later the domesticated varieties of this species spread all over the world.

In the past, six *Bombyx* species have been described, i.e. *B. mori, B. textor, B. croesi, B. fortunatus, B. arracanensis*, and *B. sinensis* (= *B. meridionalis*). However, the author of this article considers that all these should be considered a single species under the name *B. mori*. This is because several existing tropical races, assumed to be descendants of these species, can mate with *B. mori* and their F_1 are perfectly fertile. Furthermore, no differences were observed in the venation and wing pattern among these species (Toyama 1909).

Domestication and early history

Wild ancestor

Present day *B. mori* and its assumed ancestor *B. mandarina* have different chromosome numbers. After studying *B. mandarina* collected from several localities in Japan, Kawaguchi (1928) reported the haploid chromosome $n = 27$ in this species in contrast to $n = 28$ already known for *B. mori*. In F_1 males of the cross of this species and *B. mori* he observed that one of the 27 chromosomes paired with two chromosomes of *B. mori* at the first maturation division. Based on this observation he assumed that one of the *B. mandarina* chromosomes might have split into two during the long history of domestication, and that *B. mandarina* could have been the ancestor of the present-day domesticated silkworm.

In 1959, however, Astaurov *et al.* reported that *B. mandarina* from the Ussuri district of Manchuria and the Shanghai district of central China possessed 28 chromosomes in haploid. This

Table 66.2 Differences in the distribution of some isozyme genes between *Bombyx mori* and *Bombyx mandarina*. Figures indicate percentage of total numbers of larvae examined. From Yoshitake *et al.*, unpublished

		Bombyx mori	Bombyx mandarina
Number of larvae examined		768	942
Blood	O	7.3	2.2
acid phosphatase	A	4.2	0.0
(*Bph*)	B	9.4	2.2
	C	51.0	45.0
	D	28.1	50.6
Blood	O	28.1	17.1
esterase	A	62.5	19.2
(*Bes*)	B	3.1	1.9
	C	6.3	59.8
	D	0.0	2.0
Integument	A	1.5	0.0
esterase	B	20.7	0.0
(*Ies*)	C	48.2	7.8
	AB	7.5	0.0
	AC	22.1	0.0
	D	0.0	79.0
	E	0.0	13.2

finding led them to assume that the ancestral form of the domesticated silkworm was more likely to have been a 28-chromosome variety of *B. mandarina* than a 27-chromosome variety. Cytological observation by Li revealed that the Taiwan variety has 28 chromosomes, whereas Imai observed that the South Korean variety belongs to the 27-chromosome group (Hirobe 1968). Recent information brought by Chiang (personal communication) indicates that those inhabiting Sinchuan district, interior China, also have 28 chromosomes.

Although in an indirect way, evidence supporting Astaurov's view was obtained from a comparative study of isozyme gene frequencies between *B. mori* and the Japanese variety of *B. mandarina*. The isozyme genes investigated were those for blood phosphatase, blood esterase and integument esterase. Using agar-gel electrophoresis, Yoshitake *et al.* (unpublished) compared the frequency distribution of these isozyme genes among several silkworm races. The data for *B. mori* were obtained using 277 different races

including Japanese races and these can be regarded as good representatives of domesticated *B. mori*, whereas *B. mandarina* data are based on 942 individuals collected from 48 localities all over Japan. The results are given in Table 66.2. The distribution pattern of isozyme genes for blood acid phosphatase was not very different between the two species but it was markedly different for blood esterase. Further the integument esterase pattern observed was completely different. It is, therefore, hardly conceivable that *B. mori* descended from the Japanese variety of *B. mandarina*. The finding seems to support Astaurov's view. We may safely conclude on these findings that the domestication of the ancestor of present-day mulberry silkworm must have taken place in China from a *B. mandarina* variety with 28 chromosomes.

Time and place of domestication

In a Chinese classical history *Shi-ji*, written by Shi-Ma-Qian (145–86 B.C.), it is recorded that 'in

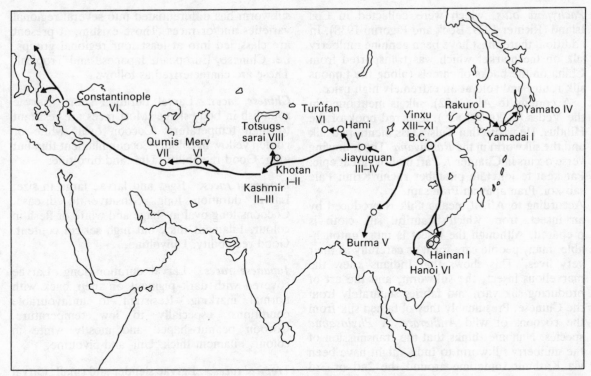

Fig. 66.1 Presumed routes and dates of transmission of sericulture from China. (From Nunome 1979)

the dynasty of Huang-di, the first Emperor of ancient China, the Empress loved sericulture'. Later this was quoted by Wang-Zhen in his technical book *Nong-Shu* stating that 'Empress Xi-Ling-Shi established and developed sericulture for the first time' (Nunome 1979).

Whatever reliability this story may have, other evidence supports the view that the domestication of the mulberry silkworm took place in China during the very early stages of history. The most substantial evidence was obtained from the discovery of Chinese characters engraved on oracle bones which were excavated from Yin ruins (1200–1050 B.C.). Among them several characters were identified as indicating the silkworm, mulberry, (silk) thread and cloth. Some characters appear to indicate even the different markings of the larva, suggesting that several varieties of silkworm already existed. Among many excavated articles, an axe and a cup were also discovered with a piece of cloth adhering to them. The cloth was identified as silk by Sylwan (1937) as well as by Nunome. From these findings

Nunome deduced that the domestication of the silkworm took place in China in the Yin period at the latest.

Spread of sericulture east and west from China

It has been traditionally said that sericulture was monopolized by China until the 2nd or 3rd century A.D. when mulberry seeds and silkworm eggs were said to have been smuggled out. Thereafter, sericulture spread to various regions of the world. However, based on an archival survey and study of articles excavated from ancient ruins, Nunome (1979) and Hirobe (1968) estimated the routes and dates of transmission of sericulture as follows: It reached Khotan, near the Tarim basin of Xinjiang province, around the 2nd–3rd century; Korea and Japan in the 1st century; Constantinople in the 6th century, Burma in the 5th century, as illustrated in Fig. 66.1.

It is almost certain that luxurious silk garments attracted the great interest of Roman aristocrats as early as several centuries B.C. when the mulberry silkworm had not yet been introduced to western countries. Several documents tell us that silk garments were produced with silk prepared from cocoons of the Lasiocampid moth,

Pachypasa otus, which were collected in Cos Island (Richter 1929, Bock and Pigorini 1938). In addition, there must have been genuine mulberry silk on the market which was transported from China on the backs of camels (along the famous silk route) and sold at an extremely high price.

According to Ali (1952), silk is mentioned in the Vedas (1500–600 B.C.), a sacred book of the Hindus. He wrote that Tulsi Dass mentioned silk and the silkworm in the *Ramayana*. The following verse occurs in Chapter 7, Part 6 of the great epic: Pát keet te joi, táte pitámber ruchir/Krimi Pále sab koi, Pran apáwan Prán sam.
According to Ali, it means 'Silk is produced by an insect, from which beautiful silk cloth is prepared. Although the insect is most untouchable, many people care for it as carefully as their very lives.' This shows that Indians knew the marvellous insect, the silkworm, and the art of producing silk yarn and fabrics separately from the Chinese. Presumably they obtained silk from the cocoons of wild *Antheraea* or *Phylosamia* species. Nunome thinks that the transmission of the mulberry silkworm to India might have been via Kashmir sometime around the 2nd or 3rd centuries A.D.

Although there is no historical evidence, the present author holds the view that there must have been another route of transmission of sericulture to India, i.e. by sea. The date of this transmission may not have been so far back as indicated in Fig. 65.1; possibly it was in the Middle Ages. As will be mentioned later, tropical races of *B. mori* differ from other races in several respects, but among races collected from Vietnam, Cambodia, Thailand, Burma and India several common characteristics are found: small and slender body shape, spindle-shaped cocoon covered with plenty of floss, adapted to tropical climate. It is conceivable from these characteristics that these races may have descended from a common ancestor, presumably a multivoltine race, reared in the Canton district, and spread along the coast.

Recent history and present status

Differentiation of regional races

Being reared under various climatic and social conditions over thousands of generations, the silkworm has differentiated into several regional varieties and/or races. Those existing at present are classified into at least four regional groups, i.e. Chinese, European, Japanese and Tropical. These are characterized as follows.

Chinese races. Larval duration short. Larvae roundish in body shape. Most races are resistant to high temperature. Cocoon oval, white or golden yellow in colour. Cocoon filament thin but long. Good reelability. Uni- and bivoltine.

European races. Eggs and larvae large in size. Larval duration long. Sensitive to disease. Cocoons long oval in shape and white or flesh in colour. Filament thick but high sericin content. Good reelability. Univoltine.

Japanese races. Larval duration long. Larvae covered with dark pigment on their back with normal marking. Resistant to unfavourable conditions, especially to low temperature. Cocoon peanut-shaped and mostly white in colour. Filament thick. Uni- and bivoltine.

Tropical races. Larvae slender and small. Larval duration short. Resistant to high temperature. Cocoon spindle-shaped and flossy, and white, yellow or green. Multivoltine.

Genetic analyses have been carried out by many investigators to study the genetic differences among regional races with regard to various characteristics; e.g. number of degenerating crochets on the tip of the abdominal legs, haemocyte typing, cocoon colour pigments and isozyme patterns (Hirobe 1968).

The most extensive work has been on isozymes. As mentioned in the previous section, Yoshitake and his collaborators (unpublished) investigated, using agar-gel electrophoresis, the distribution of several isozyme genes among 227 silkworm races. Those races were extracted from the world-wide collection maintained at the National Sericultural Experiment Station, Japan. Eight isozymes were investigated: blood acid phosphatase, midgut alkaline phosphatase, blood esterase, integument esterase, silk-gland esterase, brain esterase, digestive juice proteinase and blood phenoloxidase. The results revealed that the racial group possessing the highest diversity in genic constitution was the Chinese univoltine and that of the lowest diversity was the Tropical multivoltine.

Table 66.3 Classification of races according to pigment content and locality

Proportion of carotenoids	Races indigenous to:						
	Europe	China	Japan	Korea	Okinawa	Indo-China	India
>0.8	22	11	2				
0.8–0.6	3	3					
0.6–0.4	1	1	2				
0.4–0.2		1	1	1	1	1	1
<0.2							1
0		5*	1	1	1		2
Total	26	21	6	2	2	1	4

*Indigenous to Canton district
From Fujimoto and Hayashiyu 1961

Similar results were obtained from an investigation on the correlation between the pigment content of cocoons and geographical distribution of races. In the silkworm, several cocoon colours are known, such as white, yellow, golden yellow, straw, flesh, pink and green of various shades. Two main kinds of pigment produce these colours; namely ether-soluble yellowish carotenoids and water-soluble green flavonoids. These pigments come from the food, mulberry leaf, and selectively penetrate into silk substances in the gland, under the control of relevant genes, first through the intestinal mucosal cells and then the silk-gland cells. Genetic analysis of the genes involved is described elsewhere in detail (Tazima 1964).

By applying a very simple method of double extraction of pigments, first with 1 per cent aluminium chloride aqueous solution and then with a mixture of methanol and hexane, Fujimoto and Hayashiya (1961) estimated the amount of carotenoids and flavonoids with a photometer, and represented the values in terms of lutin and carotene contents respectively. The proportions of carotenoids to the total amount of pigments for sixty-two races are given in Table 66.3.

The results indicate that the coloured cocoons of most European races contain chiefly carotenoids, while those of south and east Asia have mostly flavonoid pigments. The proportion of carotenoids ranged widely among races indigenous to China and Japan.

According to Vavilov, the highest diversity is observed in the area where a cultivated plant originated. Following this idea Yoshitake (1968) sketched the genealogy of silkworm races as illustrated in Fig. 66.2.

This genealogy does not necessarily agree with that assumed from the historical survey. For instance, no relation is seen in this genealogy between Korean and Japanese races. The contradiction might have been caused either by random drift in genetic constitution that occurred afterwards or by direct transmission of silkworm races from China to Japan.

Silkworm breeding in Japan

The breeding of silkworms has long been directed towards the production of superior breeds either by means of selection alone or by combining outcrossing with selection in the subsequent generations. Several superior races have been produced by this method. The most important aim was to evolve a race that assured stabilized crops and the second was to improve both quantity and quality of the silk produced.

In the early 20th century Toyama (1909) observed that the F_1 hybrid between Japanese and Siamese varieties showed 30 per cent increase in cocoon yield. The results were so remarkable that he strongly advocated the use of the F_1 hybrid for commercial purposes. The National Sericultural Experiment Station of Japan carried out extensive studies on hybridization. The results were amazingly rewarding except for the incidence of double cocoons. It was revealed that in F_1 hybrids, compared with the parental races: 1. duration of the feeding period became shorter; 2. mortality was lower; 3. incidence of double

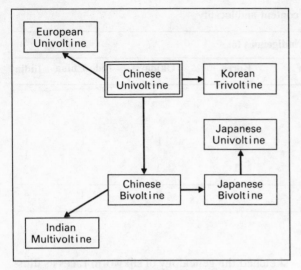

Fig. 66.2 Genealogical relationship assumed between regional races of the domesticated mulberry silkworm. (From Yoshitake 1968)

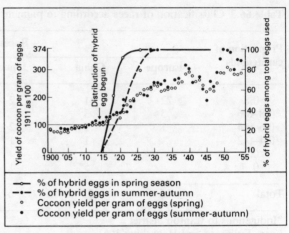

— o— % of hybrid eggs in spring season
— • — % of hybrid eggs in summer-autumn
 o   Cocoon yield per gram of eggs (spring)
 •   Cocoon yield per gram of eggs (summer-autumn)

Fig. 66.3 Relationship between propagation of hybrid eggs and cocoon yield. (From Yokoyama 1957)

cocoons was higher; 4. cocoon fibre length was longer; 5. cocoon shell weight was heavier; and 6. the size of the cocoon fibre larger. Finally, the reports concluded that among several combinations, the hybrids between Chinese and European showed the best results.

After this, the use of F_1 hybrids soon became popular and the productivity of the sericulture industry was raised markedly, as shown in Fig. 66.3 (Yokoyama 1957).

Thereafter the major effort for silkworm breeding was directed toward improvement of stock races that gave best combining ability for hybrid production. For this purpose large numbers of hybrid combinations had to be tested under uniform rearing as well as reeling conditions. Breeders of private enterprises competed in evolving superior races and contributed greatly to the increase in farmers' income. This, however, soon caused the next problem concerning the production of hybrid eggs. As the stock race of a hybrid was improved, the number of eggs produced by the mother moths decreased. This was solved by adopting three-way crossing and/or double-cross hybrids, as was done in the case of hybrid corn in the United States. The adoption of multiple-cross hybrids. however, increased the work of breeders, because it required large-scale testing for several economic

characters. Furthermore, double-cross hybrids were in general inferior to F_1 hybrids with regard to their performance. With great effort and ingenuity, silkworm-breeders achieved success in overcoming the defects. Silkworm races now commercially reared in Japan are almost all double-crosses (Gamo 1976).

Establishment of autosexing breeds

For the production of double-crossing eggs, it is necessary to separate females from males so as to prevent the moths from breeding within the race. This work has long been done by examining the imaginal discs of the genital organs, which show up as tiny spots on the ventral side of the larval abdomen. But the work needs a lot of labour by well-trained specialists with sharp eyes. If an autosexing breed were established, the labour could be reduced enormously.

The breeding of autosexing silkworm races was achieved by utilizing a translocation between the W-chromosome, which carried the female gene(s), and a piece of the second chromosome which carried larval marking genes. A translocation which could be used for this purpose was first discovered by the present author (1941). In this strain sexes are discriminated easily by the presence or absence of a marker gene for larval pattern. However, the females of this strain showed a defect in their growth, in comparison to males segregated in the same batch. The defect was assumed to be caused by the presence of the extra chromosome piece involved in the translocation. Every possible effort was made to remove the excess part of chromosome by applying X-irradiation. This was followed by the improvement of cocoon quality by repeated backcrossing to currently used strains (Tazima 1964). Thus, improved strains have been obtained and these are now widely used for industrial purposes.

With the same idea, several other translocations have been induced in the silkworm, all aiming to tag the W-chromosome with a visible marker, e.g. egg colour gene, zebra pattern gene of larva, yellow cocoon colour gene. Efforts to evolve commercially useful races with these translocations are now in progress. Among those one step ahead are the Soviet workers whose auto-sexing breed, with an egg-colour marker, is said to have already been distributed to the farmers.

Future prospects

Development of more resistant races

Experience in France, Italy and Japan has shown that sericulture cannot thrive in countries where industrialization is advanced. Sericulture is now shifting to less industrialized regions where plenty of labour is available at moderate cost. Such conditions can be found in India and Africa, where climatic conditions are usually harsher than where sericulture once thrived.

In this environment, breeding of silkworm races should take into consideration such factors as resistance to high temperature, resistance to wet or dry weather and resistance to diseases.

Development of races adapted to an artificial diet

In Japan, where some people still have a keen interest in sericulture, a new technique to rear silkworms with less labour is rapidly developing, namely, to rear them, at least during the young larval stage, on an artifical diet. Experience gained in recent years indicates that most Japanese races can feed on such a diet, while many Chinese races can barely do so. However, selection was found to be effective even in the Chinese races for increasing the frequencies of individuals that survive on an artifical diet (Gamo 1976).

Genetic analysis revealed further that the potential for feeding on an artificial diet was inherited as a dominant character and that it is controlled by at least one pair of major genes and several pairs of modifiers. Therefore, most F_1 hybrids between Japanese and Chinese races can feed on such a diet. The adaptability to the diet was, however, different depending on the hybrid combination. In order to ensure a perfect crop, development of more adapted races is urgently needed.

Application of genetic engineering techniques

Recent progress in molecular genetics has made possible the synthesis of new organisms which have never before existed. Among domestic animals, the silkworm seems to present the best possibility for applying these techniques. Ohshima and Suzuki (1979) have already achieved success in cloning silkworm fibroin genes in bacterial cells. Another subject of research for which these techniques may be extraordinarily useful is the introduction of genes into the silkworm genome from other insect species which do not cross with silkworms.

Acknowledgement

I acknowledge with many thanks the capable assistance of Mrs Sumiko Yamamoto in the preparation of the manuscript. My sincere thanks are also due to Dr Nancy Wanek who carefully read it through.

References

Ali, Hakim(1952) *The Story of Sericulture*. Department of Sericulture: Jammu and Kashmir, Srinagar

Astaurov, B. L., Garishba, M. D. and Radinskaya, I. S. (1959) [Chromosome complex of Ussuri geographical race of *Bombyx mandarina* M. with special reference to the problem of the origin of the domesticated silkworm, *Bombyx mori*, L.] *Cytology*, 1: 327–32. (In Russian)

Bock, Fr. and Pigorini, L. (1938) *Die Seidenspinner, ihre Zoologie, Biologie und Zucht*. Springer: Berlin

Fujimoto, N. and Hayashiya, K. (1961) Studies on the pigments of cocoon. VIII. On the classification of the silkworm races and their geographical distribution from the view-point of the proportion of cocoon pigment contents. *Journal of Sericultural Sciences of Japan*, 30: 83–8.

Gamo, T. (1976) Recent concepts and trend in silk-worm breeding. *Farming Japan*, **10**: 11–12

Hirobe, T. (1968) Evolution, differentiation and breeding of the silkworm – the silkroad, past and present. *Proceedings of the XIIth International Congress of Genetics*, Tokyo, August 19–28, 1968, Vol. 2, Supplement, Genetics in Asian Countries, pp. 25 –36

Jolly, M. S., Sen, S. K. and **Sonwalkar, T. N.** (1979) Non-mulberry silks. *FAO Agricultural Services Bulletin*, No. 29. FAO: Rome

Kawaguchi, E. (1928) Zytologische Untersuchungen am Seidenspinner und seinen Verwandten. I. Gametogenese von *Bombyx mori* L. und *Bombyx mandarina* M. und ihre Bastarde. *Zeitschrift für Zellforschung*, **7**: 519–52

Nunome, J. (1979) [*Origin of Sericulture and Silk Fabrics*]. Yuzankaku: Tokyo. (In Japanese)

Ohshima, Y, and **Suzuki, Y.** (1979) Cloning of silk fibroin gene and its flanking sequences. *Proceedings of the National Academy of Sciences, U.S.A.*, **74**: 5363–7

Richter, Gisela M. A. (1929) Silk in Greece. *American Journal of Archaeology* (new ser.), **33**: 27–33

Sylwan, V. (1937) Silk from Yin dynasty. *Bulletin of the Museum of Far Eastern Antiquities, Stockholm*, **9**: 119–26

Tazima, Y. (1941) [A simple method of sex discrimination by means of larval markings in *Bombyx mori*]. *Journal of Sericultural Sciences of Japan*, **12**: 184–8. (In Japanese)

Tazima, Y. (1964) *The Genetics of the Silkworm.* Logos Press: London

Toyama, K. (1909) [*Treatise on Silkworm Eggs*] Maru-yama-sha: Tokyo. (In Japanese)

Yokoyama, T. (1957) On the application of heterosis in Japanese sericulture. *Cytologia* (Supplement): 527–31

Yoshitake, N. (1966) [Differences in the multiple forms of several enzymes between wild and domesticated silkworms]. *Japanese Journal of Genetics*, **41**: 259–67. (In Japanese)

Yoshitake, N. (1968) [Phylogenetic aspects on the origin of Japanese races of the silkworm, *Bombyx mori*, L.] *Journal of Sericultural Sciences of Japan*, **37**: 83–7. (In Japanese)

67

Crustacea

J. F. Wickins

MAFF Fisheries Experiment Station, Conwy, Wales

Introduction

Many Crustacea are of economic importance to man; few, however, have been bred in captivity and none has yet been domesticated. The species most likely to be domesticated are within the order Decapoda and are directly useful to man as food. Several important groups – krill, crawfish, crabs – are excluded from this account because either they are difficult to breed or are not widely cultured.

Systematics

The Decapoda contains the suborders Natantia – shrimps and prawns (these common names being without taxonomic significance) and the Reptantia – crabs, marine crawfish or spiny lobsters, freshwater crayfish, lobsters and scampi. Most of the major crustacean groups are represented in Palaeozoic strata but early penaeid shrimp and lobster-like astacid forms are not known until the Permian and Triassic. The first major expansion occurred in the early Jurassic and gave rise to the caridean prawns, while the oldest freshwater crayfish occurred during a further expansion in the late Jurassic. The

Table 67.1 Chromosome numbers in selected Crustacea

Species	Chromosome number	Reference
Penaeus aztecus	$2n = 88$	Milligan (1976)
Penaeus duorarum	$2n = 88$	Milligan (1976)
Penaeus setiferus	$2n = 90$	Milligan (1976)
Penaeus japonicus	$2n = 92$	Niiyama (1959)
Procambarus clarki	$2n = 192$	Niiyama (1959)
Homarus americanus	$n = 69$	Roberts (1969)
Artemia	Sexual forms: diploid $2n = 42{-}44$	Abreu-Grobois and Beardmore
	Parthenogenetic forms: various states of ploidy	(1980)

homarid lobsters are also represented in the early Triassic but the genus *Homarus* dates only to the Cretaceous with two extant species. Freshwater prawns appeared during the more recent and rapid diversification which occurred in many decapod groups in the mid-Tertiary.

Biology

Several cultivated decapods demonstrate features favouring domestication studies – short life-cycle, high fecundity, the existence of distinct ecotypes – but in the majority of species growth rate is particularly sensitive to environmental conditions, which makes heritability estimates extremely difficult.

The sexes are separate in most valuable decapods although the Pandalidae contains several protandrous hermaphrodite prawns (e.g. *Pandalus platyceros*). Sex determination in Crustacea is frequently of the XO or XY type with the majority showing male heterogamety (Niiyama 1959). In contrast, during a preliminary study of thirteen individual freshwater prawns (*Macrobrachium rosenbergii*), no evidence was found of sex chromosomes, intraspecific chromosomal polymorphism or polyploidy (Hanson and Goodwin 1977). Similarly in the prawns *Penaeus aztecus, P. duorarum, P. setiferus, P. japonicus* and the crayfish *Procambarus clarki* (Niiyama 1959, Milligan 1976) there was no evidence of sex chromosomes. Chromosome numbers recorded for selected crustaceans are given in Table 67.1.

Age at first maturity ranges from 5–10 months in tropical prawns to 2–7 years in crayfish and lobsters. Copulation occurs generally, but not exclusively (see Dunham and Skinner-Jacobs 1978), between hard-shelled males and newly-moulted 'soft' females. Penaeids release up to 1.5×10^6 eggs directly into the sea while female carideans and reptantians incubate fewer eggs ($10{-}10^5$) carried beneath the abdomen for up to several months. In many cultivated decapods the number of larval stages that precede metamorphosis increases in proportion to fecundity (Wickins 1982).

Marine Crustacea with planktonic dispersal phases do not appear to be extensively subdivided into identifiable stocks as, for example, anadromous salmonids. In contrast some populations of the 'freshwater' prawn *Macrobrachium rosenbergii* may show distinct ecotypic differences between river basins only 600 miles apart (Hanson and Goodwin 1977).

Wide salinity tolerance is also a feature of many cultivated natantians and a few crayfish but some intraspecific differences have been recorded. The larvae of *M. rosenbergii* require brackish water (10–15 ‰ salinity) although post-larvae and juveniles of some ecotypes differ in their response to fresh and low-salinity water – 0–10‰ – (Sarver *et al.* 1979). Reports that *Penaeus monodon* from Philippine stocks grow as well in 'fresh' water (0 –4‰ salinity) as in brackish to saline water (10– 40‰) in pens have not been confirmed in laboratory studies with *P. monodon* from Thailand and Fiji (Cawthorne et al. 1983). It is possible that differences in salinity tolerance also exist between geographically separated stocks of *P. monodon*.

The scope for variability in growth is doubtless enchanced by periodic ecdysis. The crustacean exoskeleton is capable of only limited expansion and increase in linear dimensions therefore only occurs at moulting. The intermoult period varies in nature from a few hours to several months between species, and within species with size and age. Moult frequency shows greater variation than size increment at moult but both are sensitive to changes in environment, diet and crowding (see Wickins 1976). The large chelae typically found in territorial or predatory decapods often have important commercial value, in lobsters and crayfish, for example, but fighting and the development of large hierarchical size variations occur when such animals are crowded together. Considerable plasticity of growth may thus be exhibited by captive crustaceans.

Distribution

The lobsters *Homarus americanus* and *H. gammarus* are naturally distributed along the North American and European North Atlantic seaboards respectively where they are extensively fished. Mean U.S. annual landings are about 1.7×10^5 tonnes. Since 1965 *H. americanus* has been transplanted to the Pacific coast of Canada to form breeding populations and to Europe and Japan for commercial resale and for ranching trials.

Crayfish are widely distributed throughout the world in fresh water but are not indigenous to Africa. Areas where interest in their cultivation is greatest are Europe (*Astacus*), southern United States (*Procambarus*), and parts of Australia (*Cherax*). Introductions to Europe, Japan, central Africa and Hawaii of one or more of the North American species *Orconectes limosus*, *Procambarus clarki* and *Pacifastacus leniusculus* have been made, too often with undesirable consequences (Huner 1977). World natural and farm catches of crayfish probably do not exceed 10^5 tonnes.

Shrimps and prawns are found in fresh and salt water from the tropics to the subpolar regions. The bulk of the world's catch (more than 10^6 tonnes per year) consists of penaeids and comes from the South China Sea, the Arabian Sea and the Gulfs of Mexico and Carpentaria.

Recent history

Early attempts to culture Crustacea are summarized below from Wickins (1982).

The warm-water fish-ponds of the Far East traditionally produced an incidental by-catch of prawns. The advent of refrigeration and improved transport to high-priced city markets led to the development of prawn monoculture and the demand for reliable supplies of post-larvae for stocking. The pioneering studies of Motosaku Fujinaga (= Hudinaga) who first reared the larvae and juveniles of *Penaeus japonicus* began in 1933 at Oyanoshima, Amakusa, Japan. It was 6 years before the first post-larvae were reared and, although work was suspended during the Second World War, a further 30 years passed before the techniques were sufficiently established for commercial use. The choice of cultivable Natantia expanded in the mid-1960s when Shao-wen Ling working in Malaysia first cultured the larvae of *Macrobrachium rosenbergii*. Breeding was not a problem with caridean prawns but penaeids were not reared to maturity and bred in captivity until the mid-1970s. Interest in prawn culture spread to the United States in the 1950s and to Great Britain and France by 1964.

Hatching and rearing lobster larvae to augment natural fisheries were widespread practices in Europe and North America in the late 19th century until about 1917. Interest in culturing lobsters to a marketable size arose when studies in Maine and California, U.S.A., showed that salable lobsters could be raised in 2 years, compared to 6–8 years in nature, by growing them in thermal effluent or recirculation systems at about 20°C.

The ability to produce hybrids distinguishable from natural lobsters by crossbreeding between *Homarus americanus* and *H. gammarus* led to new efforts to assess the prospects for culture-based fisheries in France, the United States and recently in Great Britain.

Probably the only traditional crustacean 'cultivation' in Europe was the attempt made in the 15th and 16th centuries to manage natural freshwater crayfish fisheries or to 'fatten' crayfish prior to consumption. Interest was greatly stimulated in the 1880s by the decline of the fisheries following industrialization and the spread of the crayfish 'plague'. Subsequently many transplantations of North American and European species were made throughout Europe to revive the fisheries or create new ones. Since the 1940s a rapidly expanding culture industry has developed in the southern United States and annual production is estimated at 5000 tonnes.

Present status

Currently the main candidates for domestication

are the giant Malaysian prawn *Macrobrachium rosenbergii* and the North Atlantic lobsters *Homarus americanus* and *H. gammarus*. Longer-term propects could include species of *Penaeus* and possibly some freshwater crayfish. It should be noted that the brine shrimp *Artemia* (Branchiopoda, Anostraca) is widely cultured because of its importance as live food for fish and crustacean larvae. The genetics of *Artemia* have been extensively studied (see review by Barigozzi 1974) and the binomen *Artemia salina* is now taxonomically no longer valid as at least .five sibling species have been classified (Bowen *et al.* 1978).

The profit motive has so far provided the impetus for the majority of work on crustacean cultivation and breeding, the selection of species being largely dominated by the influence of market forces rather than any inherent suitability for cultivation.

Heritability studies

Estimates of heritability and assessment of genetic variation, within and between natural stocks, have only recently begun with *Macrobrachium rosenbergii, Homarus americanus* and *H. gammarus* (Hedgecock *et al.* 1977, 1979). Considerable plasticity of characteristics has been demonstrated in these species under different culture conditions. Examples include the 'bull-runt' phenomenon of communally reared *Macrobrachium rosenbergii* (Hanson and Goodwin 1977) and the possible existence of a short-lived, water-borne substance that can inhibit growth in juvenile lobsters sharing the same water supply (Nelson *et al.* 1980). Present estimates of heritability therefore can only be approximate. Examples include heritability of growth rate in *Homarus americanus* $h^2 = 0.30$ (mean upper limit), which means roughly one-month reduction in time to reach marketable size in one generation of selection of the top 25 per cent as brood stock (Hedgecock *et al.* 1976). Similar studies with *Macrobrachium rosenbergii* are in progress at the time of writing and a preliminary estimate for the sires is $h^2 = 0.20$, the dam component being 0.10 (S. Malecha, personal communication).

Genetic variation

It is expected that the potential for rapid genetic change or heritability is higher in groups with a recent evolutionary history of expansion which implies a wide natural genetic variation. A high number of species per genus gives a good indication of such groups and, according to Nelson (1977), the ratio in twenty-seven decapod genera was significantly correlated with average heterozygosity. Indeed, the penaeid and caridean prawns contain many close genera and species but *Homarus*, although dating from the Cretaceous, has only two extant species. Lobsters thus may have less potential for domestication than, for example, the genera *Penaeus* (more than 25 species) and *Macrobrachium* (more than 100 species). It has been suggested (Nelson 1977) that, due to protandry, obligate outcrossing is likely in *Pandalus platyceros*. This large, culturable species should therefore possess a rich store of genetic variation, highly desirable in a candidate for domestication.

In *Homarus* a low level of electrophoretically detectable protein variation was found between *H. americanus* and *H. gammarus* (4–5.6% of loci were heterozygous per individual genome). This contrasted with a somewhat greater variation in morphological characters between the species at all stages from eggs to adult. Interspecific hybridization may be an important means of introducing variability into lobster broodstock.

In a study of *Macrobrachium rosenbergii* from eleven localities, within-population variability was also low (2.8% heterozygous loci per individual Hedgecock *et al* 1979). Populations from Australia, New Guinea, Philippines and Palau were, however, well differentiated from those from Sri Lanka, India, Thailand, Java and Sarawak, with genetically fixed enzyme differences at over 20 per cent of the gene-enzyme systems. The major discontinuity in the gene pool corresponded to the well-recognized zoogeographic boundary – Wallace's line. This divergence represents a valuable and diverse genetic resource for the culturist.

Future prospects

Access to extant natural populations with large stores of genetic variation, high fecundity, some protandry and the possibility of labile sexual differentiation, are features of decapods shared by few other domesticated animals. On the other hand, husbandry and broodstock management techniques are only just beginning to become adequate for domestication programmes (D. Hedgecock, personal communication). Moulting, combined with territorial instincts and plasticity of growth rate, makes decapods one of the least suitable groups for cultivation and eventual domestication. For some time to come, significant improvements in growth rate and survival will be made by improvements in husbandry, diet and water-quality control, rather than by selection or crossbreeding. The rate of advance towards domestication will depend very much upon the quality of entrepreneural interest and the level of financial support this generates for continued research into crustacean breeding.

Acknowledgements

It is a pleasure to thank Dr D. Hedgecock of the Bodega Marine Laboratory, California, Dr S. Malecha of the Anuenue Fisheries Research Center, Hawaii, and Mr J. Hughes of the State Lobster Hatchery, Massachusetts, who generously provided information and comments on their latest findings.

References

Abreu-Grobois, F. A. and Beardmore, J. A. (1980) International study on *Artemia*. II. Genetic characterisation of *Artemia* populations – an electrophoretic approach. In: *The Brine Shrimp Artemia*, Vol. 1: Morphology, Genetics,Radiobiology, Toxicology, ed. G. Persoone, P. Sorgeloos, O. A. Roels, E. Jaspers. Universal Press: Wetteren, Belgium

Barigozzi, C. (1974) *Artemia*: a survey of its significance in genetic problems. *Evolutionary Biology*, 7: 221–52

Bowen, S. T., Durkin, J. P., Sterling, G. and Clark, L. S. (1978) *Artemia* haemoglobins: genetic variation in parthenogenetic and zygogenetic populations. *Biological Bulletin*, 155: 273–87

Cawthorne, D. F., Beard, T., Davenport, J. and Wickins, J. F. (1983) Responses of juvenile *Penaeus monodon* Fabricius to natural and artificial sea waters of low salinity. *Aquaculture*, 32: 165–74

Dunham, P. J. and Skinner-Jacobs, D. (1978) Intermoult mating in the lobster *Homarus americanus*. *Marine Behaviour and Physiology*, 5: 209–14

Hanson, J. A. and Goodwin, H. L. (1977) *Shrimp and Prawn Farming in the Western Hemisphere*. Dowden, Hutchinson and Ross: Stroudsburg, Pennsylvania

Hedgecock, D., Nelson, K. and Shleser, R. A. (1976) Growth differences among families of the lobster *Homarus americanus*. *Proceedings of the seventh*

annual meeting World Mariculture Society, San Diego, California, U.S.A., pp. 347–61

Hedgecock, D., Nelson, K., Símons, J. and Shleser, R. (1977) Genic similarity of American and European species of the lobster *Homarus*. *Biological Bulletin*, 152: 41–50

Hedgecock, D., Stelmach, D. J., Nelson, K., Lindenfelser, M. E. and Malecha, S. R. (1979) Genetic divergence and biogeography of natural populations of *Macrobrachium rosenbergii*. *Proceedings of the tenth annual meeting World Mariculture Society*, Honolulu, Hawaii, U.S.A., pp. 873–9

Huner, J. V. (1977) Introduction of the Louisiana red swamp crayfish *Procambarus clarkii* (Girard): an update. In: *Freshwater Crayfish*, ed. O. V. Lindquist, Vol. 3. University of Kuopio: Finland

Milligan, D. J. (1976) A method for obtaining metaphase chromosome spreads from marine shrimp with notes on the karyotypes of *Penaeus aztecus*, *Penaeus setiferus* and *Penaeus duorarum*. *Proceedings of the seventh annual meeting World Mariculture Society*, San Diego, California, U.S.A., pp. 327–32

Nelson, K. (1977) Genetic considerations in selecting crustacean species for aquaculture. *Proceedings of the eighth annual meeting World Mariculture Society*, San José, Costa Rica, pp. 543–55

Nelson, K., Hedgecock, D., Borgeson, W., Johnson, E., Dagett, R. and Aronstein, D. (1980) Density-dependent growth inhibition in lobsters, *Homarus* (Decapoda, Nephropidae). *Biological Bulletin*, 159: 162–76

Niiyama, H. (1959) Comparative study of chromosomes in decapods, isopods and amphipods. *Memoirs Faculty of Fisheries Hokkaido University*, 7: 1–60

Roberts, F. L. (1969) Possible supernumerary chromosomes in the lobster, *Homarus americanus*. *Crustaceana*, 16: 194–6

Sarver, D., Malecha, S. and Onizuka, D. (1979) Development and characterization of genetic stocks and their hybrids in *Macrobrachium rosenbergii*: Physiological responses and larval development rates. *Proceedings of the tenth annual meeting World Mariculture Society*, Honolulu, Hawaii, U.S.A., pp. 880–92

Wickins, J. F. (1976) Prawn biology and culture. In: *Oceanography and Marine Biology: an annual review*, ed. H. Barnes, Vol. 14, pp. 435–507. Aberdeen University Press

Wickins, J. F. (1982) Opportunities for farming crustaceans in western temperate regions, In: *Recent Advances in Aquaculture*, ed. J. F. Muir and R. J. Roberts, pp. 87–177. Croom Helm: London

68

Bivalve molluscs

C. M. Yonge

Department of Zoology,
University of Edinburgh, Scotland

Oysters

Edible oysters are bivalve molluscs of the family Ostreidae. They are attached by the left valve and comprise flat species of the genus *Ostrea* which incubate their eggs, and cupped oysters of the genus *Crassostrea* which liberate their more numerous eggs into the sea where fertilization and development occur. Owing to the temperatures controlling spawning, the former are restricted to temperate waters in both hemispheres, the latter occupying warmer waters, but there is considerable overlap. All species inhabit intertidal or shallow inshore areas with cupped oysters in particular often extending into estuarine waters of reduced salinity. Owing to structural differences these oysters can withstand higher turbidity.

Often present in enormous numbers with the products of successive spawnings forming massive reefs, oysters and other bivalves were among the richest sources of food for primitive man in coastal regions as indicated by accumulations of shells in extensive kitchen middens. The taste for oysters persisted as civilization developed and populations increased. Populations of natural

beds became depleted, to be made good – independently in European and Oriental seas – by methods of cultivation which represent the domestication of species of both flat and cupped oysters. *Ostrea edulis*, the British 'native' oyster, was initially the dominant European species extending from Norway to the Black Sea and forming large reefs along the Biscay coast and around the British Isles. Along the coast of Portugal it was displaced by the cupped oyster *Crassostrea angulata*. According to Pliny, *Ostrea edulis* was first cultivated by Sergius Orata in the 1st century A.D. Oysters were brought from near the modern Brindisi to be 'fattened' in Lago Lucrino which communicates by a narrow channel with the sea a little north of Naples. Knowledge of the methods employed in these 'ostriaria' is confined to their portrayal on the surface of glass vases from that period. Oysters appear to have been suspended on ropes. No attempt can have been made to breed oysters; centuries passed before the mode of reproduction became known. The waters of Lago Lucrino, now used for cultivation of mussels, are rich in inorganic nutrients and so in the microscopic plant plankton on which oysters feed.

These simple methods of cultivation persisted in southern Italy, certainly in Lago del Fusaro near Naples and around Taranto, oysters being attached to pyramids of stones surrounded by stakes in the former and suspended from ropes in the latter area. Young 'spat' oysters settled on the stakes and on suspended bundles of twigs. These methods were observed by the French embryologist, Coste, instructed by Napoleon III to devise methods for re-establishment of the disastrously depleted French oyster beds. Coste perceived the dual importance of providing settling surface for the hosts of swimming larvae during the early summer and for maintaining these where they could be protected alike from enemies and from the smothering effects of silt. On this basis highly successful methods for cultivation of *O. edulis* were developed.

'Collectors' in the form of half-cylinder roofing tiles covered with a thin coating of lime were arranged, concave side down, in alternate double rows within open wooden containers. These were placed in the sea during the early summer when the oysters released the incubated larvae. Spat settlement occurred and the young left to grow until the following spring when they were scraped off to be kept in sheltered *ambulances* before being either spread on the surface of intertidal *parcs* or grown to maturity in the enriched waters of artificial basins or *claires*. The *parcs*, extensively developed in the shallow Bay of Arcachon, south of Bordeaux, and off the Ile d'Oléron near Rochefort, are surrounded by palisades of stakes to prevent entrance of rays which, with crabs, starfish and marine snails or oyster drills, are the major oyster predators. The *parcs* are exposed at low spring tides when cultivation proceeds, pests being removed and oysters raised clear of falling sediment. Of recent years methods have been simplified with rejection of the *ambulances* and adoption of new types of collectors, usually stones and plastic surfaces.

Initially only *O. edulis* was cultivated but, following the largely accidental introduction of the gastronomically inferior Portuguese cupped oyster in the 1870s; it was largely replaced by this species in regions south of Brittany where temperature is too low for regular spawning. During recent years this species (*Crassostrea angulata*) has been almost completely destroyed by a disease which affects the all-important gills, which are also the organs of feeding. In turn it is being replaced by the related Japanese *C. gigas*.

Other culture methods were developed in the Oosterschelde in The Netherlands but these were doomed to destruction when the Dutch decided completely to enclose the mouth of the Schelde with exclusion of the sea. In Norway oysters have been cultivated in pools behind terminal moraines at the head of fjords. Surface layers of fresh water act like the glass of a greenhouse retaining high temperatures with rich plankton in middle depths where oysters are suspended in trays well clear of the deoxygenated bottom layer. In Great Britain, apart from recent establishment of hatcheries as described below, cultivation has been largely confined to the spreading of additional settling surface or 'cultch' (usually empty shells).

Two species of oysters occur off North America, the cupped *Crassostrea virginica* from the Gulf of St Lawrence to the Gulf of Mexico and the small flat *Ostrea lurida* along the north Pacific coast. Originally enormously abundant, stocks of the former have been greatly reduced in many areas, surviving best off Louisiana where oyster beds are privately owned. Hatcheries have been established in Long Island. On the Pacific

coast the local species has been almost completely displaced by imported Japanese cupped oysters, *Crassostrea gigas*.

This is one of a number of oysters occurring in the western Pacific but is much the commonest and has been cultivated in Japan, initially on palisades of bamboos, since the 17th century. Today it forms an impressive industry, the oysters suspended from rafts in southern areas around Hiroshima, and attached to stakes in shallow water to be exposed at low water of spring tides in the north. Settling surfaces consist of large scallop shells strung together and separated by spacers. As well as providing for local consumption, young oysters are 'hardened' by exposure for export, originally to North America but now widely elsewhere. This is the hardiest of oysters and the easiest to produce in hatcheries and is largely displacing other species throughout Europe. Other oysters, including species of the very similar *Saccostrea*, are cultivated in the Indian Ocean and in Australasia.

To make good the combined effects of overexploitation and disease, hatcheries have been developed in Great Britain and the United States. By exposure to suitable temperatures, oysters can be induced to spawn whenever needed, the larvae being fed on specially cultured strains of diatoms and minute green flagellates. Plastic settling surfaces are provided, although the spat are almost immediately detached to grow free. Because so much easier to rear, *Crassostrea gigas* has almost completely replaced *Ostrea edulis* in British hatcheries, their products even supplying French oyster *parcs*.

There are many different varieties of the various species of commercial oysters. These differ in general form but also in physiological characters such as tolerance to salinity, the temperature at which they spawn and the rate at which they grow. There are three subspecies of the American oyster, *Crassostrea virginica*, which spawn at different temperatures thus explaining the exceptionally wide distribution of this oyster from north temperate into almost tropical waters. There is evidence that such differences are genetically controlled and this also applies to resistance to infection by the diseases to which oysters are subject and which can wipe out entire populations. Breeding experiments have been carried out both in the United States and Japan. While little can be done to change the nature of wild populations, where oyster stocks depend on the produce of hatcheries – as they are likely increasingly to do – there are obvious possibilities for controlling the genetic characters of the stock. In this way oysters may be produced which grow quickly and are resistant to disease with the ability to flourish under the local conditions of salinity, temperature and turbidity.

Mussels

Edible mussels are species of the bivalve family Mytilidae, the triangular shell attached ventrally by tough byssus threads produced in a gland at the base of the foot. The common European *Mytilus edulis* has been cultivated in the shallow muddy Anse de l'Aiguillon near La Rochelle on the Biscay coast since the 13th century when nets erected in the hope of snaring seabirds became covered with mussels. From this the modern *bouchot* system was gradually developed. Parallel rows of stakes interwoven with twigs to form a compact hedge extend for hundreds of kilometres. Mussel larvae attach particularly to the lowest rows, the resultant mussels being later thinned out and the excess transferred to higher rows. Cultivators use small flat-bottomed boats or *acons* pushed by one foot encased in a large boot.

An extensive mussel industry in The Netherlands is based on transplantation from beds where mussels naturally settle to areas of richer plankton where growth is greater. Beds in the Waddenzee are open for collection by dredging for limited periods, the young mussels being transferred to privately-owned growing plots in shallow water. Formerly in the Oosterschelde, these are now largely also in the northern Waddenzee. Mussels are marketable when 2–3 years old.

Mussels are reared suspended on ropes in shallow water in Lago del Fusaro and elsewhere in the Mediterranean but by far the greatest cultivation is that recently developed in the fjord-like rias of Galicia in northwest Spain. The sheltered surface waters are covered with fleets of what were originally pontoons, now largely catamarans, with surrounding staging from which hang hundreds of ropes formerly of esparto grass now of nylon. All end well clear of the bottom and so of most predators. The ropes are quickly covered with mussels which, twice thinned out, become marketable in 18 months. With between 1–2 million ropes each producing 50 kg of mussels, production is enormous.

The extremely palatable green mussel, *Mytilus smaragdinus*, is cultivated on ropes suspended from rafts or floating platforms of bound bamboos in Bacoor Bay near Manila and elsewhere in the Philippines. In these mid-tropical conditions they are mature in 4–5 months.

Scallops

Of recent years the Japanese have cultivated the large scallop, *Patinopecten yessoensis*, in the north with such success that production is around three-quarters that of the Japanese oyster industry and exceeds this in value (£68 million in 1976). Special types of mid-water collectors have been developed with the young scallops 'ongrown' on the bottom or in suspended lantern nets. Using similar methods it is hoped to culture the large Atlantic scallop, *Pecten maximus*, in north European seas.

References

Oysters
  Korringa, P. (1976) *Farming the Flat Oysters of the Genus Ostrea*. Elsevier: Amsterdam
  Korringa, P. (1976) *Farming the Cupped Oysters of the Genus Crassostrea*. Elsevier: Amsterdam
  Yonge, C. M. (1960) *Oysters*. Collins: London
Mussels
  Korringa; P. (1976) *Farming Marine Organisms Low in the Food Chain*. Elsevier: Amsterdam
Japanese methods – including scallops and oysters
  Imai, T. (1978) *Aquaculture in Shallow Seas*. Balkema: Rotterdam. (Trans. from Japanese)

69

Edible snails

L. J. Elmslie
FAO Consultant, Rome, Italy

Wild species

Two families of land snails are of importance as food for humans, the Helicidae, basically in Europe, and the Achatinidae in West Africa and the Far East.

The most famous food snail internationally is *Helix pomatia* (also called *H. edulis*). *Helix aspersa, H. lucorum, H. aperta* and *Eobania* (or *Helix*) *vermiculata* are also traded internationally and several other species are eaten locally.

Achatina achatina and *Archachatina* spp are important sources of animal protein in coastal West Africa. *Achatina fulica* meat is exported from the Far East.

There are no distinct cultivated races or varieties of the wild edible species. All species vary considerably in shell colour and size. Taylor (1914) discusses such variations in *H. pomatia* and *H. aspersa*.

The edible Helicidae, and especially the very adaptable *H. aspersa*, have spread very widely from their original European or Mediterranean habitats, basically following the pattern of European colonization. In some cases *H. pomatia* and *H. aspersa* may have been deliberately introduced

for food. The requirement of *H. pomatia* for calcareous soil and warm dewy summer nights without excessive heat has restricted its spread compared with *H. aspersa*.

Achatina fulica, which is of East African origin, has spread widely in eastern Asia and the Pacific. In some cases it was introduced deliberately on account of its alleged therapeutic properties. It has become a pest in some areas and has recently reached the southern United States.

Edible snails are hermaphrodites. The Helicidae are not self-fertile and reciprocal fertilization normally takes place two to three times a year. Eggs are laid in the soil 5–30 days later in clutches of 50–80 eggs on average. *Helix pomatia* eggs hatch in 20–30 days, the hatch rate and hatch time depending on the environment. Growth after hatching depends on environment and species; probably to a lesser extent on food supply and quality. *Helix pomatia* may reach maturity in as little as 18 months or as long as 4 years. According to Cain and Sheppard (1957) some species may remain fertile for several years after mating.

Unlike the Helicidae, *Achatina fulica* is believed to be capable of self-fertilization. In the laboratory it has reached sexual maturity at 6 months of age. All snails suspend activity when conditions are unfavourable. *Helix pomatia* goes underground in late autumn and closes the shell opening with a thick white calcareous epiphragm or operculum. *Helix aperta* does the same thing in the heat of summer. Both species are highly prized as food, and very expensive, in the operculated condition. Other species, e.g. *H. aspersa*, do not form much of an operculum in normal circumstances.

Ford (1975) discusses variation of shell colour in relation to environment for *Cepaea* (or *Helix*) *nemoralis*. It is very probable that similar adaptations to local environment exist for characteristics less obvious than shell colour. The introduction of snails to a farm paddock is often followed by considerable mortality, presumably due to the animals' difficulty in adjusting to the new environment. Each species of snail has different requirements for both soil type and micro-climate.

Exploitation

That prehistoric man ate snails is known from the

presence of shells, of species which do not normally live together, in kitchen middens. Evans (1969) discusses the history of snail eating.

According to Pliny, the Roman Fulvius Lippinus was the first snail farmer. He fattened snails from Illyria, Africa and elsewhere on a mixture of new wine boiled down with spelt. Clearly he was supplying a luxury article.

In the valleys of northern Italy wild *H. pomatia* have for centuries been collected in spring and placed in paddocks. There they eat the natural vegetation, which may be supplemented if necessary. In late autumn the snails go underground and in December they are dug up and sold in the operculated condition. A similar system is followed with *H. aperta* in Sicily, except that the snails operculate in summer.

In parts of central Italy weedy cereal stubbles yielded considerable numbers of *H. aspersa*, which were collected and stored for several months. This system largely died out during the 1950s.

Changing patterns of agriculture have reduced the number of wild snails available in Western Europe and their importance as a food for the poor. At the same time increased interest in snails as a gourmet food has led to imports of live Helicidae from Eastern Europe and the Mediterranean and of *Achatina* meat from Taiwan. Chevallier (1978) reports that France imported about 7000 tons of snails in 1976.

This situation has led to attempts at complete life-cycle farming of snails in France, Germany and especially Italy. Avagnina (1979) makes recommendations for the rearing of *H. pomatia* and *H. aspersa*. The farm is first enclosed and cleared of predators. It is then divided into breeding paddocks and fattening paddocks. Both are sown with a mixture of plants which provide either a suitable environment or suitable food for the snails, or both. There were 4000 snail farmers in Italy using variants of this system in 1980. Only a minority of these were in production. Most were recent entrants to the business and were trying to rear *H. pomatia*. About 1000 of the larger farmers had plots averaging 0.5 ha each.

The emphasis so far has been on providing a suitable environment for dense snail populations and no selective breeding has been done up to the time of writing. This situation may change with the development of methods of producing young snails in a controlled environment for putting out

to pasture in the spring.

In France *H. aspersa* is the main species reared; results with *H. pomatia* have been disappointing.

References

Avagnina, G. (1979) *Principi di Elicicoltura.* Edagricole: Bologna

Cadart, J. (1975) *Les Escargots* (Helix pomatia *L. et* Helix aspersa *M.): biologie, élevage, parcage, gastronomie, commerce* (2nd edn). Lechavalier: Paris

Cain, A. J. and **Sheppard, P. M.** (1957) Some breeding experiments with *Cepaea nemoralis* (L). *Journal of Genetics*, **55**: 195–9

Chevallier, H. (1978) Il consumo di lumache in Francia e le prospettive dell'elicicoltura. *Quaderno* No. 7, I Centro di Elicicoltura, Borgo S. Dalmazzo, Cuneo, Italy

Chevallier, H. (1979) *Les Escargots – un élevage pour l'avenir.* Dargaud Editeur: Paris

Evans, J. G. (1969) The exploitation of molluscs. In: *The Domestication and Exploitation of Plants and Animals,* ed. P. J. Ucko and G. W. Dimbleby, pp. 479–84. Duckworth: London

Ford, E. B. (1975) *Ecological Genetics.* Chapman and Hall: London

Taylor, J. W. (1914) *Land and Freshwater Mollusca of the British Isles*, Vol. 3. Taylor: Leeds

Appendix: Taxonomy and nomenclature

G. B. Corbet and J. Clutton-Brock

British Museum (Natural History), London, England.

The system of giving latinized names to animal species seems complex and bewildering to many laymen and even to some professional zoologists. However, much of the difficulty arises from the enormous complexity of animal life itself and the system of nomenclature we now use has proved to be remarkably successful and adaptable, considering the great advances that have been made in our understanding of the diversity and variation of animals in the 200 years since Linnaeus first consistently used the present binomial system in 1758. The *International Code of Zoological Nomenclature* is indeed accepted and used internationally, and is carefully designed to provide a set of rules for the naming of species and other taxonomic groups or 'taxa' without interfering with the way in which the groups are identified, constructed or conceptualized.

However, attempts to extend the system to domesticated forms have created difficulties that scarcely arise in the case of wild species and some of these problems have also spilled back to affect the naming of those wild species that are ancestral to the domesticated forms. The success of the system for the naming of wild species rests upon the essentially discrete nature of species. The wolf species *Canis lupus*, for example, is a group of individuals that differ from each other in only minor ways, insufficient to prevent them from interbreeding and thereby sharing their characteristics within the group. The discreteness of species is not, of course, absolute. Wolves can and do hybridize with coyotes but not to the extent of making more than a very small proportion of individuals unidentifiable. On the other hand, when we consider fossils, it is clear that every intermediate condition between wolves and the common ancestors they share with other members of the dog family, and indeed with all other animals, must have existed, and it is only the fragmentary nature of the fossil record that alleviates the problem of dividing the evolutionary lineages into namable parts.

From the time of Linnaeus formal binomial latin names have been given to many domesticated animals. Linnaeus himself used *Canis familiaris* for the domesticated dog and *C. lupus* for the wolf. In the case of most of the other familiar domesticated animals he named the domesticated form but did not know, or did not separately name, the ancestral wild species – thus we have *Ovis aries* for the sheep, *Capra hircus* for the goat, *Bos taurus* for cattle, *Equus caballus* for the horse, *E. asinus* for the donkey and *Felis catus* for the cat. In other cases he clearly used the same name for both wild and domesticated forms, as in *Sus scrofa* for the wild boar and the pig and *Cervus tarandus* for both wild and domesticated reindeer.

For a long time names such as *Canis familiaris* and *Bos taurus* were used pretty consistently and unambiguously (if often unnecessarily) for the domesticated forms. However, during the past 30 years or so that situation has become more complicated. On the one hand the concept of a wild species has increasingly become that of an assemblage of individuals, often showing considerable variability but characterized especially by the ability to interbreed fully within the assemblage. At the same time knowledge of the ancestry of domesticated animals has improved, and with a few notable exceptions it can be argued with some confidence that the domesticated and wild forms constitute at least a potentially interbreeding assemblage.

This has led to a number of proposals that wild ancestor and domesticated derivative should be treated as members of the same species and named accordingly. Using this concept and applying the Law of Priority required by the International Code (whereby the earliest name satisfying the requirements of the Code is used for a species) leads to a number of confusing and ambiguous names: *Canis lupus* means 'wolf plus domestic dog' whereas previously it unambiguously meant 'wolf'; *Felis catus* means 'wild and domesticated cats' rather than just the domesticated form, and so on. To avoid these problems it has been suggested, e.g. by Bohlken (1961) and Groves (1971), that names based on domesticated forms should be excluded from application of the Code. This has much to commend it as a way of stabilizing the nomenclature of the wild species but it leads to further problems in relation to the Code (e.g. many names of domesticated forms are the types of generic names) and it leaves wide open the question of how to name the domesticated forms. Bohlken's solution, to add a suffix to the earliest name of the wild species leading, for example, to '*Canis lupus* forma *familiaris*' for the dog, is cumbersome and unlikely to supplant well-established names such as '*Canis familiaris*'.

One solution that has been used is to treat the domesticated form as a subspecies, using a trinomial, e.g. (in a case where the domesticated form was the first to be named) *Felis catus catus* for the domesticated cat and *Felis catus silvestris* for the wild cat (strictly speaking for one particular wild subspecies). However, this has many disadvantages, e.g.: 1. it leaves no way of designating wild cats as a category separate from domesticated cats; 2. it makes the species name *Felis catus* by itself ambiguous; and 3. it confuses the otherwise useful concept of the subspecies as a geographically discrete and morphologically distinctive segment of a species.

There is, in fact, no single system of nomenclature that has been proposed that can be applied to the whole gamut of domesticated forms, i.e. with all degrees of domestication, without introducing either ambiguity or excessive complexity. It therefore seems more helpful to adopt rather different procedures for different groups. Three categories are suggested here and these are dealt with separately below. For each category a list is given to show the recommended name or names for each species.

In these lists the original author and date are given following the name. Although this follows orthodox practice it should be stressed that, in accordance with the Code, this refers to the original use of the second name – the 'specific epithet' – and indicates only that the original concept (strictly speaking the original 'type' specimen) is included in the current concept of the group. It does not necessarily follow that the total content or concept of the species is unchanged. These original references will therefore include a description of animals *included* in the 'species' as now delimited but they will rarely provide descriptions of the current limits of the 'species'. Quotation of the author and date is therefore, in most contexts, an unnecessary piece of pedantry; except when the subject of nomenclature is under discussion they can be left out without any loss of clarity. When they are used it is a convention of the Code that author and date should be in parentheses if the species was originally attributed to a genus other than the one in which it is now placed.

Category 1. *Anciently domesticated forms that are distinctive, are rarely bred with their wild ancestors and have well-established scientific names that are normally understood as applying only to the domesticated forms.*

Domesticated forms in this category are best considered as *derivatives* of their wild ancestral species but not *parts* of them. The wild species can then be named by using the earliest available name based upon a wild form. As for naming the domesticated forms, in many contexts there is nothing to be gained by the use of latinized names. However, there are situations where it is helpful to use a more formal system, e.g. in tabulations or listings that include both wild species and domesticated forms especially when indexes or other kinds of data retrieval are involved. Since many Linnaean and later names have been used consistently for domesticated forms that are easily recognizable as such, their continued use is effective as a means of communication especially in an international context. Also, by not formally banishing such names from the scope of the Code they can continue to perform their role, as many of them do, as types of generic names.

It should be noted, however, that to comply with the Code and in the interest of stability, such names should follow the requirements of the Code, e.g. in relation to priority and to type specimens. Names such as '*domesticus*' and '*familiaris*', however apposite, have no validity unless they are attributable to an adequate original description that has date priority and conforms to the Code.

Although it is convenient to use binomial names for such domesticated forms as if they were orthodox species it is important to recognize that, in fact, they are not orthodox species. As groups these domesticated forms differ from wild species in a number of ways. In particular the members of a domesticated form are not necessarily monophyletic, i.e. they have not necessarily arisen from a single common ancestral stock within the parent species but may consist of a number of independently domesticated groups which nevertheless are interfertile and therefore not always discretely definable.

In the following list the fact that the domesticated forms are not orthodox biological species is emphasized by placing the 'specific epithet' in quotation marks. Those who prefer not to make this distinction and to treat these names more strictly within the provisions of the Code could dispense with the quotation marks but at the risk of confusion with the less frequent use of some of these names to represent the combined ancestral wild species and domesticated form.

Domesticated form	Probable wild ancestor
Bos '*taurus*' Linnaeus, 1758 (Including *B.indicus* Linnaeus, 1758) Common cattle (including zebu)	*Bos primigenius* Bojanus, 1827 Aurochs
Bos '*frontalis*' Lambert, 1804 Gayal, mithan	*Bos gaurus* H. Smith, 1827 Gaur
Bos '*grunniens*' Linnaeus, 1766 Domestic yak	*Bos mutus* (Przewalski, 1883) Wild yak
Bubalus '*bubalis*' (Linnaeus, 1758) Domestic water buffalo	*Bubalus arnee* (Kerr, 1792) Wild water buffalo, arni
Ovis '*aries*' Linnaeus, 1758 Domestic sheep	*Ovis orientalis* Gmelin, 1774 Asiatic mouflon
Capra '*hircus*' Linnaeus, 1758 Domestic goat	*Capra aegagrus* Erxleben, 1777 Bezoar goat
Camelus '*dromedarius*' Linnaeus, 1758 Arabian camel, dromedary	Unknown
Camelus '*bactrianus*' Linnaeus, 1758 Domestic Bactrian camel	*Camelus ferus* Przewalski, 1883 Wild Bactrian camel
Lama '*glama*' (Linnaeus, 1758) Llama	?*Lama guanicoe* Muller, 1776 Guanaco
Lama '*pacos*' (Linnaeus, 1758) Alpaca	Unknown
Sus '*domesticus*' Erxleben, 1777 Domestic pig	*Sus scrofa* Linnaeus, 1758 Wild boar
Equus '*caballus*' Linnaeus, 1758 Domestic horse	*Equus ferus* Boddaert, 1785 Wild horse, tarpan
Equus '*asinus*' Linnaeus, 1758 Donkey	*Equus africanus* Fitzinger, 1857 Wild ass
Canis '*familiaris*' Linnaeus, 1758 (Including *C.dingo* Meyer, 1793) Dog	*Canis lupus* Linnaeus, 1758 Wolf
Felis '*catus*' Linnaeus, 1758 Domestic cat	*Felis silvestris* Schreber, 1777 Wild cat
Mustela '*furo*' Linnaeus, 1758 Ferret	*Mustela putorius* Linnaeus, 1758 Western polecat
Cavia '*porcellus*' (Linnaeus, 1758) Domestic guinea-pig	*Cavia aperea* Erxleben, 1777 Cavy
Bombyx '*mori*' (Linnaeus, 1758) Domestic silk moth	*Bombyx mandarina* (Moore, 1872) Chinese wild silk moth

The policy inherent in the above list seems to offer the best chance of achieving clarity and it has the following particular advantages.

1. Groups that are generally recognized, and therefore frequently need to be referred to, can be given a relatively simple, long-established binomial name without the need for trinomials or suffixes, e.g. *Felis* '*catus*' for the domestic cat, *Capra aegagrus* for the wild goat.

2. Changes in the taxonomic concept of the wild species (some of which show very confusing variability open to a number of interpretations) do not interfere with the stability of the name of the domesticated form, and vice versa. For example, the domesticated sheep can be *Ovis 'aries'* whether the related wild sheep are considered to be one large species, *Ovis ammon*, or a series of smaller species, one of which, *Ovis orientalis*, is the most likely ancestor of the domesticated sheep.

3. In the case of subfossil finds the evidence for domestication may be very tenuous. Such specimens can be named, for example, *'Canis lupus,* possibly domesticated', resulting in greater clarity and conciseness than could be achieved by other systems that have been proposed since it is usually the degree of domestication rather than the identity of the wild species that is in most doubt.

It is, however, taxonomically wrong to attempt to qualify the degree of domestication either by ascribing subspecies names to prehistoric animal remains or by equating them with modern breeds. Even today, for example, it is not uncommon to read in archaeological reports of a fragmentary dog's skull that, because it has a basal length of less than 150 mm, is described as *'Canis familiaris palustris,* a small house-dog'. Subfossil remains of canids or any other animal should not be given trinomials on the basis of size when this falls within the normal range either for the wild species or its domesticated descendant. All that is necessary is to record the dimensions of the identified specimen in a prescribed manner.

Category 2. *Distinctive domesticated forms that are readily distinguishable from their wild ancestral species but have not generally been given separate specific names.*

These are partly species that were well-known and named as wild species before domesticated forms were developed, but the category also includes some anciently domesticated animals like the reindeer which are less sharply distinguished from the wild species than in the case of those in Category 1, and many birds for which there is no tradition of separate names for the domesticated and wild forms. Animals in this category resemble those in Category 1 (and differ from those in Category 3) in that the majority of individuals in captivity or in a state of domesti-

cation are readily distinguishable from members of the wild ancestral species and are rarely interbred with them. They mostly differ from Category 1 species in that the link with the ancestral wild species is more certain. The zoological justification and the practical need to treat them as separate domesticated 'species' derived from the wild ancestor are therefore less than in Category 1 species and the domesticated status of animals in this category is best indicated by adding a vernacular suffix to the name of the wild species, e.g. *'Rangifer tarandus* (domesticated)', *'Gallus gallus* (dom.)' or *'Anas platyrhynchos* (Aylesbury)'.

Wild species	Domesticated form
Bos javanicus d'Alton, 1823 Banteng	Bali cattle
Rangifer tarandus (Linnaeus, 1758) Reindeer, caribou	
Vulpes vulpes (Linnaeus, 1758) Red fox	Silver fox, etc.
Mustela vison Schreber, 1777 American mink	Ranch mink
Oryctolagus cuniculus (Linnaeus, 1758) Rabbit	
Mesocricetus auratus (Waterhouse, 1839) Golden hamster or Syrian hamster	
Mus domesticus Rutty, 1771 House mouse	Laboratory mouse
Rattus norvegicus (Berkenhout, 1769) Norway rat	Laboratory rat
Gallus gallus (Linnaeus, 1758) Red jungle fowl	Domestic fowl
Meleagris gallopavo Linnaeus, 1758 Turkey	
Anas platyrhynchos Linnaeus, 1758 Mallard	Domestic duck
Cairina moschata (Linnaeus, 1758) Muscovy duck	
Anser anser (Linnaeus, 1758) Greylag goose	Domestic goose
Anser cygnoides (Linnaeus, 1758) Swan goose	Chinese goose
Columba livia Gmelin, 1789 Rock dove	Domestic pigeon
Serinus canarius (Linnaeus, 1758) Canary	

Melopsittacus undulatus
(Shaw, 1805) Budgerigar
Carassius auratus (Linnaeus,
1758) Goldfish
Apis mellifera Linnaeus, 1758
Honeybee

For many of these species there are no doubt valid names in the synonymy that are based upon domesticated specimens and could be used for the domesticated 'species' in the same way as those in Category 1, leaving the above names solely for the wild species, e.g. *Mus albus* Bechstein, 1801 for the domesticated mouse in contrast to *M.domesticus* which, in spite of the name, is based upon a wild house mouse. But unless they are already well established it is unlikely that the use of such names would become sufficiently universal and stable to aid communication.

Category 3. *Wild species that are commonly bred or kept in captivity but in which the majority of domesticated individuals are not readily distinguishable as a group from the wild species and are not normally given separate names*.

For these species the name of the wild species is usually quite sufficient although a breed name or other vernacular suffix can be added to designate any animals that *are* distinguishable from the wild species. Only a few examples of commonly domesticated species are given below.

Elephas maximus (Linnaeus, 1758)	Asian elephant
Myocastor coypus (Molina, 1782)	Coypu, nutria
Chinchilla laniger (Molina, 1782)	Chinchilla
Numida meleagris (Linnaeus, 1758)	Guinea-fowl
Pavo cristatus Linnaeus (1758)	Peafowl
Struthio camelus (Linnaeus, 1758)	Ostrich

These various categories are not, of course, discrete and some of the allocations are somewhat arbitrary. *Columba domestica*, although undoubtedly a correct and potentially stable name for the domesticated pigeon, has not been used sufficiently widely to qualify for Category 1 although it is sufficiently useful and unambiguous to be a strong candidate. Likewise, the domestic fowl would fall very logically into Category 1 with the name *Gallus domesticus* Bechstein, 1805 if that

name were not so rarely used. Domesticated guinea-fowl are perhaps sufficiently distinctive as a group to qualify for Category 2 although that does not alter their nomenclature.

The ultimate test of any system of nomenclature is the proportion of people who, on reading or hearing a name, interpret it in the sense in which the writer or speaker intended. Such a system requires simplicity and a minimum of change to existing practice but at the same time a flexible and pragmatic approach to ensure that the names do not become fossilized and meaningless as our understanding of the biological concepts they stand for changes and advances.

References

Bohlken, H. (1961) Haustiere und zoologische Systematik. *Zeitschrift für Tierzüchtung und Züchtungsbiologie*, **76**: 107–13

Groves, C. P. (1971) Request for a declaration modifying Article 1 so as to exclude names proposed for domestic animals from zoological nomenclature. *Bulletin of Zoological Nomenclature*, **27**: 269–72.

Linnaeus, C. (1758) *Systema Naturae*, Vol. 1, (10th edn) Holmia. Facsimile (1956): British Museum (Natural History): London

Index of common names

Index of scientific names

no name ed 12/01
VNLS
exp